Lecture Notes in Electrical Engineering

Volume 221

For further volumes:
http://www.springer.com/series/7818

Mohan S · S. Suresh Kumar
Editors

Proceedings of the Fourth International Conference on Signal and Image Processing 2012 (ICSIP 2012)

Volume 1

Editors
Mohan S
Computer Science and Engineering
Dr. N.G.P. Institute of Technology
Coimbatore, Tamil Nadu
India

S. Suresh Kumar
Electronics and Communication Engineering
Dr. N.G.P. Institute of Technology
Coimbatore, Tamil Nadu
India

ISSN 1876-1100 ISSN 1876-1119 (electronic)
ISBN 978-81-322-0996-6 ISBN 978-81-322-0997-3 (eBook)
DOI 10.1007/978-81-322-0997-3
Springer New Delhi Heidelberg New York Dordrecht London

Library of Congress Control Number: 2012953282

Printed on acid-free paper

Springer is part of Springer Science+Business Media (www.springer.com)

Preface

Signal and Image Processing (SIP) is about the mathematical and computational methods that form the basis of the specialist modules covering the theory and application of SIP algorithms for the analysis, interpretation, and processing of data in diverse fields such as computer vision, robotics, acoustics, medical diagnosis, remote sensing, and telecommunications, and so on.

The SIP explores research and engineering issues related to the modeling of signals, developing and evaluating algorithms for extracting the necessary information from the signal, and the implementation of these algorithms. Specific research areas include filter design, fast transforms, adaptive filters, spectrum estimation and modeling, sensor array proccssing, image processing, motion estimation from images, speech processing, geophysics, computer-aided tomography, image restoration, robotic vision, and pattern recognition.

The 4th International Conference on Signal and Image Processing 2012 (ICSIP 2012), organized by Dr. N.G.P Institute of Technology, Coimbatore, India during Dec 13–15, 2012 provides academia and industry to discuss and present the latest technological advances and research results in the fields of theoretical, experimental, and application of signal, image, and video processing. ICSIP 2012 also aims to bring together engineers and scientists in signal, image, and video processing from around the world for better cohesion and collaboration in the fields of mutual interest.

The ICSIP 2012 called for a wide range of recent research papers in the field of SIP with applications, theories, and algorithms from all over the world from budding researchers. The submission was flooded with more than 250 papers from 12 different countries across the world. A total of 113 papers were selected based on blind review by experts from the respective fields for publication in the proceedings in two volumes by Springer India in Lecture Notes in Electrical Engineering.

The editors to express their gratitude to the founder of the ICSIP series Dr. P. Nagabhushan, Professor, University of Mysore, India for guidance right from the beginning. Also, the editors are thankful to Dr. S. Murali, President, Maharaja Institutions, Mysore, India for his valuable suggestions and timely input.

The editors are extremely thankful to all the keynote speakers (Prof. Kay Chen TAN-NUS, Singapore, Prof. Subhas, Massey University, New Zealand and Prof. P. Nagabhushan, UoM, India), tutorial speakers (Dr. Mathew Cherian-Kovai Medical Center and Hospital, India, Dr. Jharna Mazumdar, Former ADE DRDO, Ministry of Defence, India, Dr. Kumar Rajamani, GE, India, and Dr. Venkatesh Babu, IISc, Bangalore, India), panel members, reviewers, advisory, and organizing committee members. The editors convey their heartiest gratitude to Springer India team for accepting to publish the research findings in their Lecture Notes in Electrical Engineering, without whom this proceedings would not have been in existence today.

Above all, the editors are much thankful to the management of Dr. NGPIT, Coimbatore, India for hosting the prestigious conference ICSIP 2012. It was definite that the conference and the institution would complement each other in bringing high quality researchers in one place. The editors thank the Chairman-Dr. Nalla G. Palaniswami, Secretary- Dr. Thavamani D. Palaniswami, CEO-Dr. O. T. Buvaneswaran, Principal and all faculty members of Dr. NGP Institute of Technology, Coimbatore, who have been given great support by extending their seamless contribution to bring this proceeding in a big way.

Coimbatore, India

Mohan S
S. Suresh Kumar

About the Editors

Dr. Mohan S
B.E. (CSE), M.Tech. (CCT), Ph.D. (CS), MIEEE. With over 14 years of experience in Teaching, Research, and Industry (Infosys), Currently serving as Professor and Head, Department of CSE, Dr. NGP Institute of Technology, Coimbatore, Tamil Nadu, India. Part of organizing committees in ICCR 05 and 08, Organizing Chair in ICLDCM 2010.
Areas of interest: Computer Vision, Image/Video Processing
E-mail: s.mohan77@gmail.com

Dr. S. Suresh Kumar
Accomplished, Professor, Ph.D., C.Eng., MIEEE., MIEE., MIE., MAES., MISCA., MACCS., MSSI., MCSI., MISTE., with over 18 years of experience in education, research and development in a range of areas in Audio Signal Processing, Embedded Systems, and Power Electronics & Drives. Currently Director—Research at Dr. NGPIT, Coimbatore, Tamil Nadu, India.
E-mail: sskpsg@gmail.com

About ICSIP 2012

About the Proceedings

The proceedings include cutting-edge research articles from the Fourth International Conference on Signal and Image Processing (ICSIP), which is organized by Dr. NGP Institute of Technology, Kalapatti, Coimbatore. The Conference provides academia and industry to discuss and present the latest technological advances and research results in the fields of theoretical, experimental, and application of signal, image, and video processing.

The book provides latest and most informative content from engineers and scientists in signal, image, and video processing from around the world, which will benefit the future research community to work in a more cohesive and collaborative way.

Commitee Members

Patrons

Dr. Nalla G. Palaniswami, Chairman and Managing Director, Kovai Medical Center and Hospital, Coimbatore, India
Dr. Thavamani D. Palaniswami, Secretary, Dr. NGP Institute of Technology, Coimbatore, India
Dr. O. T. Buvaneswaran, CEO, KMCRET, Coimbatore, India
Dr. S. Dhandapani, Advisor, Dr. NGPIT, Coimbatore, India
Dr. K. Porkumaran, Principal, Dr. NGPIT, Coimbatore, India

Organizing Chairs

Dr. Mohan S, Professor and Head—CSE, Dr. NGPIT, Coimbatore
Dr. S. Suresh Kumar, Director-Research, Professor and Head—ECE, Dr. NGPIT, Coimbatore

Organizing Co-Chair

Prof. V. Vani, Professor and Head—IT, Dr. NGPIT, Coimbatore

Advisory Committee

Dr. Kay Chen Tan, National University of Singapore, Singapore
Dr. Subhas Chandra Mukhopadhyay, Massey University (Turitea), New Zealand
Dr. P. Nagabhushan, University of Mysore, Mysore, India
Dr. Murali S., Maharaja Institute of Technology, Mysore, India
Dr. Lakshminarayanan Samavedham, National University of Singapore, Singapore
Dr. R. M. Suresh, RMD college of Engineering, Chennai, India
Dr. Jharna Majumdar, Former Sc. G, DRDO, Bangalore, India
Dr. Mathew Cherian, KMCH, Coimbatore, India
Dr. R. Venkatesh Babu, SERC, IISc, Bangalore, India
Dr. Kumar Rajamani, GE Global Research, Bangalore, India

Organizing Committee Members

Faculty members of CSE, ECE and IT department, Dr. NGP Institute of Technology, Coimbatore, India

Contents

Performance Analysis of Morphological Operators Based Feature Extraction and SVD, Neural Networks as Post Classifier for the Classification of Epilepsy Risk Levels

R. Harikumar, T. Vijaykumar, C. Ganesh Babu and M. G. Sreejith

Abstract Most research to date using hybrid Models focused on the Multi-Layer Perceptron (MLP). Alternative neural network approaches such as the Radial Basis Function (RBF) and Elman network, and their representations appear to have received relatively little attention. Here we focus on Singular value Decomposition (SVD), RBF and Elman network as an optimizer for classification of epilepsy risk levels obtained from code converter using the EEG signals parameters which are extracted by morphological operators. The obtained risk level patterns from code converters are found to have low values of Performance Index (PI) and Quality Value (QV). These neural networks are trained and tested with 960 patterns extracted from three epochs of sixteen channel EEG signals of twenty known epilepsy patients. Different architectures of MLP, Elman and RBF networks are compared based on the minimum Mean Square Error (MSE), the better networks in MLP (16-16-1) and RBF (1-16-1) are selected. RBF out performs the MLP, Elman network, SVD Technique and code converter with the high Quality Value of 25 when compared to the Quality Values of 23.03, 22.2, 20.62 and 12.74 respectively.

Keywords EEG signals · Morphological operators · Code converter · SVD · RBF · MLP · Elman neural networks · Epilepsy risk levels

---

R. Harikumar (✉) · C. Ganesh Babu · M. G. Sreejith
ECE, Bannari Amman Institute of Technology, Sathyamangalam, India
e-mail: harikumarrajaguru@gmail.com

T. Vijaykumar
IT, Bannari Amman Institute of Technology, Sathyamangalam, India

Mohan S. and S. S. Kumar (eds.), *Proceedings of the Fourth International Conference on Signal and Image Processing 2012 (ICSIP 2012)*, Lecture Notes in Electrical Engineering 221, DOI: 10.1007/978-81-322-0997-3_1,

1 Introduction

Twenty–five percent of the world's 50 million people with epilepsy have seizures that cannot be controlled by any available treatment [1]. Epileptic seizures are a principal brain dysfunction with important public health implications, as they affect 0.8 % of humans. Many of these patients (20 %) are resistant to treatment with drugs. The ability to anticipate the onset of seizures in such cases would permit clinical interventions. Traditional signal analyses, such as the count of focal spike density, the frequency coherence or spectral analyses are not reliable predictors [2, 3]. This paper addresses the application of Morphological operators for feature extraction, SVD, and Neural networks Techniques towards optimization of Code converter outputs in the classification of epilepsy risk levels. We also present a comparison of these classifiers based on their Performance Indices and Quality Values.

2 Materials and Methods

2.1 Acquisition of EEG Data

Since the EEG records are over a continuous duration of about thirty seconds, they are divided into epochs of two second duration each by scanning into a bitmap image of size 400 × 100 pixels. A two second epoch is long enough to detect any significant changes in activity and presence of artifacts and also short enough to avoid any repetition or redundancy in the signal [4]. The EEG signal has a maximum frequency of 50 Hz and so, each epoch is sampled at a frequency of 200 Hz. Each sample corresponds to the instantaneous amplitude values of the signal, totaling 400 values for an epoch.

Figure 1 shows the overall epilepsy risk level classifier system. The principle of this research is to classify the epilepsy risk level of a patient from EEG signal parameters. This is accomplished as

1. Four parameters out of Seven features are extracted from EEG signals using morphological operators.
2. Code converter classification for epilepsy risk level at each channel from EEG signals and its parameters.

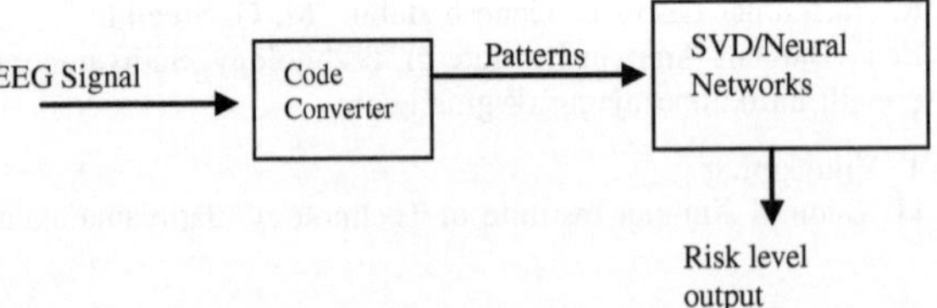

Fig. 1 SVD and neural networks post classifier for epilepsy risk level classification

3. Code converter results from each channel are optimized using SVD and Neural networks.
4. Performance of code converter classification, SVD and neural network optimization methods are analyzed.

1. The energy in each two-second epoch is given by [5]

$$E = \sum_{i=1}^{n} x_i^2 \qquad (1)$$

where x_i is signal sample value and n is number of samples. The scaled energy is taken by dividing the energy term by 1,000.
2. The variance is computed as σ given by

$$\sigma^2 = \frac{\sum_{i=1}^{n} (x_i - \mu)^2}{n} \qquad (2)$$

where $\mu = \frac{\sum_{i=1}^{n} x_i}{n}$ is the average amplitude of the epoch.
3. Covariance of Duration. The variation of the average duration is defined by

$$CD = \frac{\sum_{i=1}^{p} (D - t_i)^2}{pD^2} \qquad (3)$$

The following parameters are extracted using wavelet transforms.
4. The total number of positive and negative peaks exceeding a threshold is found.
5. Spikes are detected when the zero crossing duration of predominantly high amplitude peaks in the EEG waveform lies between 20 and 70 ms and sharp waves are detected when the duration lies between 70 and 200 ms.
6. The total numbers of spike and sharp waves in an epoch are recorded as events.
7. The average duration is given by [3]

$$D = \frac{\sum_{i=1}^{p} t_i}{p} \qquad (4)$$

where t_i is one peak to peak duration and p is the number of such durations.

2.2 Morphological Filtering of EEG Signals

Spike detection is significant for clinical diagnosis of epileptic disorders. In hospitals, a method of visual/manual detection of spikes in EEG recordings by an experienced neurologist is widely used. In the case of long EEG recording, this process is much time consuming. Thus automatic spike detection methods are needed. One of the methods of automatic spike detection in epileptic EEG is based on morphological filter. Automatic detection of spikes in the EEG is a complex task due to the variety of morphology of spikes, different amplitudes and their similarities to waves which are part of background noise and to artifacts such as eye blinks [6]. On the other hand there is a demand to process a big quantity of EEG data with their features expressed, such as frequencies, the parameters of spikes as time of appearance, amplitude, duration, shape. In the case of rolandic epilepsy it is important to establish the number of spikes during the sleep period and wakefulness period.

There are some groups of automatic methods for spike detection presented in relative studies: temporal approach- EEG is divided into waves and half waves and for each duration, amplitude and sharpness is calculated [7]. If all parameters exceed threshold, it is assumed to be spike; wavelet based method- spike is detected when a normalized wavelet power is above threshold; neural networks; morphological filtering; other [9]. Since the morphological filtering has quite high percentage of detection, this method was chosen to be used for spike detection [10].

2.3 Basic Operators of Mathematical Morphology

In this section, we give a brief description on mathematical morphology. Let f(t) and g(t) be the subsets of the Euclidean space E, where f(t) is the one dimensional time series data, i.e. the EEG signal needed to be analyzed, and g(t) is the predefined structuring element. Denote the domains of f(t) and g(t) as F and G respectively. $g^S(t) = g(-t)$ is the reflection of g rotated by 180° degree with respect to the origin. The *Minkowski addition* and *subtraction* with functional structuring element g(t) for one- dimensional time series data f(t) are defined as follows [9].

$$\textit{Addition}: (f \oplus g)(t) = \underbrace{\max}_{\substack{t-u \in F \\ u \in G}} \{f(t-u) + g(u)\}, \tag{5}$$

$$\textit{Subtraction}: (f \Theta g)(t) = \underbrace{\min}_{\substack{t=u \in F \\ u \in G}} \{f(t-u) - g(u)\}, \tag{6}$$

The morphological filter called *opening* and morphological filter called *closing* are defined as follows:

$$Opening\colon (f \circ g)(t) = [(f \Theta g^S) \oplus g](t), \tag{7}$$

$$Closing\colon (f \bullet g)(t) = [(f \oplus g^S) \Theta g](t), \tag{8}$$

Opening operation smoothes convex peaks of the original signal f(t) and closing operation smoothes concave peaks of f(t). Thus opening and closing can be applied to detect peaks and valleys in the signal.

2.4 Spike Detection Using Morphological Filter

Different operators smooth or extract different parts of the signal depending on the shape of a structuring element. Thus one of the tasks is the selection of the structuring element which separates the spiky areas of the signal. Combination of the morphological operators can produce a filter which separates an original signal into two signals: one signal is defined by the structuring element and the other is the residue of the signal. Thus the task is the selection of morphological filter. Spikes exist with positive and negative phase in an epileptic EEG. In order to detect bi-directional spikes, the morphological filters can be first applied with the opening operator followed by closing operator or vice versa:

$$\text{Open–closing operation: } OC(f(t)) = f(t) \circ g_1(t) \bullet g_2(t), \tag{9}$$

$$\text{Close–opening operation: } CO(f(t)) = f(t) \bullet g_1(t) \circ g_2(t), \tag{10}$$

where $g_1(t)$ and $g_2(t)$ are different structuring elements.

The open-closing filter provides an effect of the low pass filter. Thus, we can also construct a high-pass filter by taking the different between the input signal and it's open-closing. However the result of open-closing operation has lower amplitude than the original signal and the result of close-opening operation has larger amplitude [10]. If spikes are detected by selecting some threshold of the amplitude, the distortion of amplitude can cause pseudo positive or missing detection. Thus an average combination of open-closing and close-opening may be used for more precise extraction of bi-directional spikes

$$OCCO(f(t)) = [OC(f(t)) + CO(f(t))]/2.$$

Then the original EEG signal f(t) can be presented as follows:

$$f(t) = x(t) + OCCO(f(t)),$$

where ***OCCO*** represents the background activity, x(t) represents the transient activity or spiky part of the signal. According to [10] 'Spike is a transient, clearly distinguished from background activity, with pointed peak at conventional paper speeds and a duration from 20 to under 70 ms, i.e., 1/50 to 1/14 s, approximately. Main component is generally negative relative to other areas. Amplitude is variable. Generally, an EEG spike is different from the background activity; has a sharp peak and duration of 20–70 ms. Since different spike waves can have various amplitude and frequency, the structuring elements should be adjusted to the proper size where the spike component can be best extracted. In order to separate the spike component and background activity, the structuring element pair $g_1(t)$ and $g_2(t)$ is determined as two parabolas [11]:

$$g_i(t) = a_i t^2 + b_i, \quad \mathrm{i} = 1, 2$$

These elements can fit into the shape of background EEG waves but not into the spike waves. The result of separation of transient and background parts of the original signal f(t) depends on the selection of an amplitude and a width of $g_1(t)$ and $g_2(t)$, i.e. the selection of b_i and a_i, i = 1, 2. The amplitude and width of spike can vary on the same signal and on all signals of EEG record for the same person. These parameters are selected according to characteristics of spike waves. The criterion for structuring elements and optimization process is proposed in [11].

2.5 *Code Converter as a Pre Classifier*

The encoding method processes the sampled output values as individual code. Since working on definite alphabets is easier than processing numbers with large decimal accuracy, we encode the outputs as a string of alphabets [8]. The alphabetical representation of the five classifications of the outputs is shown in Table 1.

The ease of operation in using characteristic representation is obviously evident than in performing cumbersome operations of numbers. By encoding each risk level one of the five states, a string of seven characters is obtained for each of the sixteen channels of each epoch. A sample output with actual patient readings is shown in Fig. 2 for eight channels over three epochs. The Code converter's classification efficiency is evaluated from the following parameters. The Performance of the Code converter is defined as follows [5],

$$PI = \frac{PC - MC - FA}{PC} \times 100 \tag{11}$$

where PC—Perfect Classification, MC—Missed Classification, FA—False Alarm.

Table 1 Representation of risk level classifications

Risk level	Representation
Normal	U
Low	W
Medium	X
High	Y
Very high	Z

Epoch 1	Epoch 2	Epoch 3
XYYWYXX	WXXWYYY	WZYXWWW
YZZYXXX	YYYYXXX	YYYXYYY
YYZXYYY	YYYYYYY	YYYYYYY
YZZYXYY	XZZXYYY	YYYYYYY

Fig. 2 Code converters output with morphological operators based feature extration

The performance for code converter classifier is as low as 33.26 %. The code converter output emulates the occurrence of highly non periodic patterns in the classification of epilepsy risk levels. Therefore any closed form of solution will be failed in this purpose of optimization. SVD method for Optimization of Code converter outputs are discussed in the following section of the paper.

3 Singular Value Decomposition for Optimization of Code Converter Outputs

The singular value decomposition is a well known approach that may be used for such tasks as dimensionality reduction, and determining the modes of a complex linear dynamical system [12]. SVD of a matrix has one or more columns that arc identical, or that several groups of columns that are same which is useful in signal processing problems and applications. A SVD of an m × n matrix A = [a_1, a_2, a_3,…, a_n] is the composition of A into the product of three matrices as follows

$$A = U\sum V^T = \sum P\sigma_K u_{Kv_K^T} \tag{12}$$

where p = min(m,n), U = [u_1, u_2, u_3,…, u_m] is an m × n ortho normal matrix, V = [v_1, v_2, v_3,…, v_n] is an n × m ortho normal matrix, and Σis an m × n matrix with elements σ_k along the diagonal and zeros everywhere else. Matrix U is called left singular matrix, V is called right singular matrix, and Σis the singular value matrix [13]. If the singular values are ordered so that $\sigma_1 \geq \sigma_{2,\ldots.} \geq \sigma_{\text{p}}$, and if the matrix A has a rank r < p, then the last p-r singular values are equal to zero, and SVD becomes A = $\Sigma^{r}\, \sigma_k\, u_k\, v_k^{T}$.

4 Role of Neural Networks in the Optimization of Fuzzy Outputs

Artificial Neural Network (ANN's) is a powerful tool in pattern recognition problems. Specifically, they are useful for automating diagnostic tasks carried out by experts (supervised classification tasks). The ANN's capability of learning from examples eases this knowledge acquisition problem [14]. On the other hand, the ANN gives opaque knowledge representation.We have applied different architectures of Elman, RBF and MLP networks for optimization. Even though RBF is an unsupervised network, the cluster centers of the hidden layers are identified as the target codes (ZZYZZZ—epilepsy risk level) for a particular model. The weights of the linear connections between the hidden layer and output layer network are trained with error back propagation algorithm to minimize the square output error to zero. The simulations were realized by employing Neural Simulator 4.0 of Matlab v.7.0 [14]. Since our neural network model is patient specific in nature, we are applying 48 (3 $\times$ 16) patterns for each RBF model. There are twenty models for twenty patients. As the number of patterns in each database for training is limited, each model is trained with one set of patterns (16) for zero mean square error condition and tested with other two sets of patterns (2 $\times$ 16). After network is trained using these, the classification performance of test set is recorded. The testing process is monitored by the Mean Square Error (MSE) which is defined as [15].

$$MSE = \frac{1}{N}\sum_{i=1}^{N}(O_i - T_j)^2 \tag{13}$$

where Oi is the observed value at time i, Tj is the target value at model j; j = 1–10, and N is the total number of observations per epoch and in our case, it is 16. As the number of hidden units is gradually increased from its initial value, the minimum MSE on the testing set begins to decrease. The optimal number of hidden units is that number for which the lowest MSE is achieved. If the number of hidden units is increased beyond this performance does not improve and soon begins to deteriorate as the complexity of the neural network model is increased beyond that which is required for the problem.

5 Results and Discussion

The outputs are obtained for three epochs for every patient in classifying the epileptic risk level by the code converter, SVD, MLP, RBF, and. Elman Neural networks. To study the relative performance of these systems, we measure two parameters, the Performance Index and the Quality Value. These parameters are calculated for each set of the patient and compared. Table 2 depicts the performance index of all classifiers. From the Table 2 we identified that for code converter classifier have max detection of 62.6 % with false alarm of 19.13 %. Similarly for RBF Neural network optimization we obtained perfect detections of 100 % with no false alarms. As in the case of SVD, MLP and Elman Neural networks higher Performance Index of more than 90 % with less than 3 % false alarms are obtained. This indicates that the RBF classifier is a better one.

Table 2 Performance index for all classifiers

Classifiers	Morphological operators based feature extraction			
	Perfect classification	Missed classification	False alarm	Performance index
Code converter	62.6	18.25	19.13	33.26
With SVD optimization	91.22	7.31	1.42	89.48
With Elman optimization	95.75	1.86	2.38	95.48
With RBF optimization	100	0	0	100
With MLP optimization	96.68	2.31	1.02	96.53

5.1 Quality Value

The goal of this paper is to classify the epileptic risk level with as many perfect classifications and as few false alarms as possible. In Order to compare different classifiers we need a measure that reflects the overall quality of the classifier [13]. Their quality is determined by three factors.

1. Classification rate
2. Classification delay
3. False Alarm rate.

Thc quality value Q_V is defined as [3]

$$Q_V = \frac{C}{(R_{fa} + 0.2) * (T_{dly} * P_{dct} + 6 * P_{msd})} \tag{14}$$

where, C is the scaling constant, R_{fa} is the number of false alarm per set, T_{dly} is the average delay of the on set classification in seconds, P_{dct} is the percentage of perfect classification and P_{msd} is the percentage of perfect risk level missed.

A constant C is empirically set to 10 because this scale is the value of Q_V to an easy reading range. The classifier with the highest Q_V should be the best one. The quality value obtained by SVD and Neural Network optimization technique is shown in the Table 3. However, RBF Neural network settled at 100 % PI with full value of quality component. SVD method placed at high time delayed system with lesser false alarm when compare to Elman network. SVD quality value pegged at the lowest ebb of 20.62 only. Compare to Elman network MLP outperforms in terms of quality value slotted at 23.03.

Table 3 Quality value for average of twenty patients

Parameters	Code converter method	SVD optimization	Elman neural network optimization	RBF neural network optimization	MLP neural network optimization
Risk level classification rate (%)	62.6	91.22	95.75	100	96.68
Weighted delay (s)	2.34	2.26	2.02	2	2.07
False-alarm rate/set	19.13	1.42	2.38	0	1.02
Performance Index %	33.26	89.48	95.48	100	96.53
Quality value	12.74	20.62	22.27	25	23.03

6 Conclusion

This paper aims at classifying the epilepsy risk level of epileptic patients from EEG signals. The goal was to classify perfect risk levels with high rate of classification, a short delay from onset, and a low false alarm rate. Though it is impossible to obtain a perfect performance in all these conditions, some compromises have been made. As a high false alarm rate ruins the effectiveness of the system, a low false-alarm rate is most important. SVD, MLP, Elman, and RBF Neural network optimization technique is used to optimize the risk level by incorporating the above goals. The major limitation of this method is that if one channel has a high-risk level, then the entire group will be maximized to that risk level. This will affect the non-epilepsy spike region in the groups. The number of cases from the present twenty patients has to be increased for better testing. A comparison of EM and SVM will be taken for further studies.

Acknowledgments The authors express their sincere thanks to the Management and the Principal of Bannari Amman Institute of Technology, Sathyamangalam for providing the necessary facilities for the completion of this paper. **This research is also funded by AICTE RPS.:F No 8023/BOR/RID/RPS-41/2009-10, dated 10th Dec 2010.**

References

1. Lasemidis LD et al (2003) Adaptive epileptic seizure prediction system. IEEE Trans Biomed Eng 50(5): 616–627
2. Dingle AA et al (1993) A multistage system to detect epileptic form activity in the EEG. IEEE Trans Biomed Eng 40(12):1260–1268
3. Qu H, Gotman J (1997) A patient specific algorithm for detection onset in long-term EEG monitoring possible use as warning device. IEEE Trans Biomed Eng 44(2): 115–122
4. Celement C et al (2003) A comparison of algorithms for detection of spikes in the electroencephalogram. IEEE Trans Biomed Eng 50(4):521–526
5. Harikumar R, Dr, Sukanesh R, Bharathi PA (2005) Genetic algorithm optimization of fuzzy outputs for classification of epilepsy risk levels from EEG signal. J Interdisciplinary Panels 86(1): 1–10

6. Mohseni HR, Maghsoudi A, Shamsollahi MB (2006) Seizure detection in EEG signals: A comparison of different approaches. In: Proceedings of the 28th IEEE EMBS annual international conference New York City. 30 Aug– 3 Sep 2006, pp 6724–6727
7. Yuan Y (2010) Detection of epileptic seizure based on EEG signals. In: Proceeding of IEEE EMBS sponsored 3rd international congress on image and signal processing (CISP 2010), July 2010, pp 4209–4211
8. McCauley-Bell P, Badiru AB (1996) Fuzzy modeling and analytic hierarchy processing to quantify risk levels associated with occupational injuries-Part I: The development of fuzzy-linguistic risk levels. IEEE Trans Fuzzy Syst 4(2): 124–131
9. Tarassenko L, Khan YU, Holt MRG (1998) Identification of inter-ictal spikes in the EEG using neural network analysis. IEE Proce Sci Measure Technol 145(6): 270–278
10. Yamaguchi T, Fujio M, Inoue K (2008) Feature extraction of EEG waves based on morphological multiresolution analysis. In: Proceedings of 3rd international conference innovative computing information and control (ICICIC'08) IEEE, pp 22–28
11. Quian quiroga R, Nadasdy Z, Ben-Shaul Y (2004) Unsupervised spike detection and sorting with wavelets and super paramagnetic clustering. Neural Comput 16(8):1661–1687
12. Lee S, Hayes MH (2004) Properties of the singular value decomposition for efficient data clustering. IEEE Signal Process Lett 11(11): 862–866
13. Sadasivam PK, Dutt DN (1996) SVD based technique for noise reduction in electroencephalographic signals. Elsevier Signal Process 55: 179–189
14. Haykin S (1999) Neural networks a comprehensive foundation, 2nd edn. Prentice- Hall Inc, New Jersey
15. Hwang YS et al (1997) Recognition of unconstrained handwritten numerals by a radial basis function network classifier. Pattern Recogn Lett 18: 657–664

6. [illegible] Mageswari [illegible] Thangavelu MR (2004) [illegible] in EEG signals: A comparison of different wavelets. In: Proceedings of the 26th IEEE EMBS annual international conference, New York City, 30 Aug–3 Sep 2006, pp 67–70
7. Yang Y (2010) Detection of epileptic seizure based on EEG signals. In: Proceedings of IEEE 3rd international congress on image and signal processing (CISP 2010), [illegible]
8. McGoldrick E and P. [illegible] AB (1990) [illegible] design and analysis hierarchy processor for [illegible] levels associated with [illegible] Part 1. The development of the system [illegible]. IEEE Trans Fuzzy Syst [illegible]
9. Tanaka [illegible] T, Kim [illegible], Honda [illegible] (2005) [illegible] detection of epileptic spikes in the EEG [illegible] analysis. IEEE Trans Neural Syst Rehabil Eng [illegible]
10. Ahmad [illegible] (2005) Feature extraction of EEG signals based on wavelet [illegible] analysis. In: Proceedings of the international conference [illegible] communication, automation and control [illegible]
11. Ghandeharioun [illegible] Z, [illegible] (2014) Unsupervised [illegible] detection and [illegible] with wavelets and [illegible] Comput [illegible]
12. Lee [illegible] (2003) Properties of the singular value decomposition [illegible] dimensions of [illegible] [illegible]
13. Sabbagh [illegible] (2008) SVD based technique for [illegible] [illegible] signal [illegible]
14. [illegible]
15. Wang [illegible] (2009) [illegible] [illegible]

A Novel Approach for Different Morphological Characterization of ECG Signal

R. Harikumar and S. N. Shivappriya

Abstract The earlier detection of Cardiac arrhythmia of ECG waves is important to prevent cardiac disorders. A good system depends heavily upon the precise and consistent estimate of ECG signal morphology i.e. QRS complex, T and P wave. From the benchmark data bases: MIT-BIH Arrhythmia, QT and European ST-T database the ECG is fetched then the noise is removed from the digitized ECG signal. To analyze various power spectrum of ECG signal Stationary Wavelet Transform (SWT) is applied to the de-noised signal. Based upon the spectrum QRS complex T and P waves are detected and also delineated using different amplitude threshold values. This gives simple and reliable method for the detection and delineation of the constituent waves from a given ECG signal has been the fundamental goal of automatic cardiac arrhythmia detection. This algorithm allows delineation of different morphologies of QRS complex P and T wave.

Keywords ECG · Wavelet transform · Cardiac arrhythmia

1 Introduction

1.1 Electrical Cardio Gram

In recent years, ECG signal plays an important role in the primary diagnosis, prognosis and survival analysis of heart diseases. Electrocardiography has a

R. Harikumar (✉)
Bannari Amman Institute of Technology, Sathyamangalam, India

S. N. Shivappriya
Kumaraguru College of Technology, Coimbatore, India

Mohan S. and S. S. Kumar (eds.), *Proceedings of the Fourth International Conference on Signal and Image Processing 2012 (ICSIP 2012)*, Lecture Notes in Electrical Engineering 221, DOI: 10.1007/978-81-322-0997-3_2,

profound influence on the practice of medicine. This paper deals with the detection of QRS complexes of ECG signals using derivative based/Pan-Tompkins/wavelet transform based algorithms. The electrocardiogram signal contains an important amount of information that can be exploited in different manners. The Electrogram is composed of the wave and complexes. Waves and complexes in the normal sinus rhythm are the P-wave, PR-interval, PR segment, QRS complex, ST segment, QT interval and T wave.

The P-wave is caused by atrial depolarization. In normal sinus rhythm, the SA node acts as the pacemaker. The P-wave is usually smooth and positive. The P-wave duration is normally less than 0.12 s. The PR interval is the portion of the ECG wave from the beginning of the P wave (onset of the atrial depolarization) to the beginning of the QRS complex (onset of the ventricular depolarization).The PR segment is the portion on the ECG wave from the end of the P-wave to the beginning of the QRS complex. The PR segment is corresponds to the time between the end of atrial depolarization to the onset of the ventricular depolarization.

The QRS complex represents the time it takes for the ventricles in normal sinus rhythm, each P wave is followed by a QRS complex. The QRS complex proceeds ventricular contraction. ST segment represents the period of the ventricular muscle contraction before repolarization. The ST segment is normally iso-electric (no electrical activity is recorded). However, the ventricles are contracting. The QT interval begins at the onset of the QRS complex and to the end of the T wave. It represents the time of the ventricular depolarization until ventricular repolarization. T wave occurs due to ventricular repolarization.

In electrocardiograms (ECGs), most of the clinically useful information can be found in the wave intervals, amplitudes, or morphology. Therefore, efficient and robust methods for automated ECG delineation are of great importance. The QRS complex is relatively easy to detect and is thus generally used as a reference within the cardiac cycle. For P and T wave detection and delineation (i.e., determination of peaks and boundaries of the P and T waves), most existing methods perform QRS detection first. They then define temporal search windows before and after the detected QRS location to search for the P and T waves using filtering [1], basis expansions [2], or thresholding. Because of the low slope and magnitude of the P and T waves, as well as the presence of noise, interference, and baseline fluctuation, P and T wave delineation remains a difficult task.

Furthermore, in addition to delineation, accurate estimation of the waveform itself may be important, e.g. for T wave different pathological detection [4]. Wavelets are new promising approach to analyze and characterize the non-stationary signals such as ECG, EEG, and EMG etc. Wavelet transform (WT) decomposing signals into elementary building blocks that are well localized both in time and frequency, wavelet transform can characterize the local regularity of signals. This feature can be used to distinguish ECG waves from serious noise, artifacts and baseline drift.

1.2 Cardiac Arrhythmias

Any disturbances in the heart's normal rhythmic contraction are called an arrhythmia. Normal sinus rhythm is characterized by a regular rhythm and PR interval duration range of 0.12–0.20 s. Arrhythmias can be recognized by evaluating the systematic manner. Some arrhythmias are atrial fibrillation, atrial flutter, ventricular fibrillation and ventricular flutter. Among these we see the causes of the one of the arrhythmia: atrial fibrillation.

Atrial fibrillation (AF): This arrhythmia [3] will occur due to fast beating rate (300–500 beats/min) of the atrium. Here no P wave is observable. Here ventricle beats very slowly. This atrial fibrillation is very frequent arrhythmia which going to effect elderly people: 2–5 % of people over 60 years old and 10 % over 70 years old. It results in partial disorganization of atrial electrical activity, due to two electrophysiological properties. Due to AF upper heart chambers (atria) quiver instead of pumping blood effectively. Blood in the atria may pool and clot. If a clot breaks loose and advances to the brain, a stroke can result.

Atrial refers to the heart's two upper chambers, the left and right atria. The two lower chambers are called the ventricles. Fibrillating means quivering, or rapid beating. Irregular, rapid beating of the atrial chambers characterizes Atrial Fibrillation. This happens when the normal system that conducts electricity in the atria malfunctions. A storm of electrical activity across both atria causes them to fibrillate 300–600 Times/min.

The ventricles pick up only a small number of these impulses, but the ventricular rate can approach 180 or higher. Whether Atrial Fibrillation happens at high or low heart rates, its irregular rhythm means the ventricles can't pump blood efficiently to the rest of the body. Instead, blood pools in the heart and the body doesn't get enough. This can result in varying symptoms [5] from relatively mild ones, such as fatigue and cough, to serious ones, such as angina and stroke.

1.3 ECG Databases

Different ECG Databases are used for the present work such as Massachusetts Institute of Technology-Beth Israel Hospital (MIT-BIH) Arrhythmia databases, European Society of cardiology database and QT Database. The QT Database includes ECGs which were chosen to represent a wide variety of QRS and ST-T morphologies, in order to challenge QT detection algorithms with real-world variability. The records were chosen primarily from among existing ECG databases, including the MIT-BIH Arrhythmia Database, the European Society of Cardiology ST-T Database, and several other ECG databases collected at Boston's Beth Israel Deaconess Medical Center.

2 Materials and Methods

The Automatic Detection of ECG wave is important to cardiac disease diagnosis. The new pre-processing approach can remove the low-frequency components without introducing distortions in the ECG waveform.

2.1 Pan-Tompkins Algorithm

The QRS detection provides the fundamentals for almost all automated ECG analysis algorithms. Pan-Tompkins proposed a real-time QRS detection algorithm based on analysis of the slope, amplitude, and width of the QRS complexes [8] of typical cardiac signal as shown in Fig. 1. The algorithm includes a series of filters and operators that perform derivative, squaring, integration, adaptive thresholding operations and search procedures.

Band pass Filtering: The band pass filter for the QRS detection algorithm reduces noise in the ECG signal by matching the spectrum of the average QRS complex. This attenuates noise due to muscle noise, power line interference, baseline wander, T wave interference. The pass band that maximizes the QRS energy is in the 5–35 Hz range. The filter implemented in this algorithm is composed of cascaded high pass and low pass Butterworth IIR filters.

Derivative Operator: The next processing step is differentiation, standard technique for finding the high slopes that normally distinguish the QRS complexes from other ECG waves. The derivative procedure suppresses the low frequency components of P and T waves, and provides a large gain to the high-frequency components arising from the high slopes of the QRS Complex.

Squaring: The squaring operation makes the result positive and emphasizes large differences resulting from QRS complexes; the small differences arising from P and T waves are suppressed. The high frequency components in the signal related to the QRS complex are further enhanced. This is a nonlinear transformation that consists of point by point squaring of the signal samples.

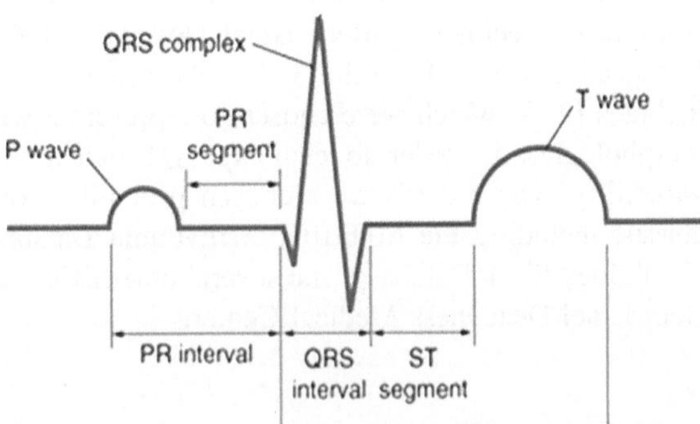

Fig. 1 Morphology of a mean PQRST-complex of an ECG recorded from a normal person

Integration: The squared waveform passes through a moving window integrator. This integrator sums the area under the squared waveform over a suitable interval, advances one sample interval, and integrates the new predefined interval window. The half-width of window has been chosen as 27 to include the time duration of extended abnormal QRS complexes, but short enough that it does not overlap both a QRS complex and a T-wave. MA (moving average) filter extracts features in addition to the slope of the R wave.

2.2 Wavelet Based Delineation

Time–frequency wavelet theory [7] is used for the detection of life threatening electrocardiography (ECG) arrhythmias. This is achieved through the use of the raised cosine wavelet transform (RCWT). The RCWT is found to be useful in differentiating between ventricular fibrillation, ventricular tachycardia and atrial fibrillation. Ventricular fibrillation is characterized by continuous bands in the range of 2–10 Hz; ventricular tachycardia is characterized by two distinct bands: the first band in the range of 2–5 Hz and the second in the range of 6–8 Hz; and atrial fibrillation is determined by a low frequency band in the range of 0–5 Hz.

A classification algorithm is developed to classify ECG records on the basis of the computation of three parameters defined in the time–frequency plane of the wavelet transform. Furthermore, the advantage of localizing and separating ECG signals from high as well as intermediate frequencies is demonstrated. The above capabilities of the wavelet technique [6, 9] are supported by results obtained from ECG signals obtained from normal and abnormal subjects.

Wavelet based morphological Detection and Delineation of ECG consists of following stages:

Pre-processing: This stage utilizes a filtering unit to remove the artifact signals from the ECG signal. These signals include baseline wandering, power line interferences, and a high frequency noise. After the preprocessing stage [10, 11], the Stationary Wavelet Transform is applied to the De-noised signal. Most of the energy of the ECG signal lies between 2^1 and 2^4 of the scales.

QRS complex Detection: First the noise is removed from the digitized ECG signal. Then SWT is applied. In the ECG signal R wave is having maximum amplitude, which is detected at larger scale (2^3). Fixing the R peak is simple in comparison to the location of the P and T wave boundaries. The level-3 SWT Low pass filter coefficients are given to the differentiator, squaring circuit and moving window average process from which the R peak is detected.

Setting the search window (100 ms) before and after the R-peak, two minima points are detected, they are respectively known as minima-1 and minima-2. Fixing the same search window before and after the minima-1 and minima-2, the QRS onset and offset (zero crossing) points are detected. Normally the onset and offset of the QRS complex have high frequency and low amplitude signal, which are detected at finer scale (2^1). The reason for detecting the onset and offset points at scale (2^1), rather than the original signal are to avoid the effect of baseline drift.

QRS Delineation (onset, offset and individual waves): One of the novelties with respect to [7] is the detection and identification of QRS individual waves. The algorithm departs from the position given by the detector, in this delineation process WT based algorithm departs from the position given by the detector, n_{qrs}, which must be flanked by a pair of maximum moduli with opposite signs at scale 2^2 namely at n_{pre} and n_{post}. The delineator looks before n_{pre} and after n_{post} for significant maxima of accounting for other adjacent slopes within the QRS complex. To consider a maximum modulus as significant, it must exceed the threshold, γ_{QRSpre} or $\gamma_{QRS_{post}}$ respectively for previous or subsequent waves. The morphology of QRS complex depends of the number of significant maxima moduli within the corresponding search window.

The onset (offset) of the QRS is before (after) the first (last) significant slope of the QRS, which is associated with a maximum of $|w_{2^2}\,x[n]|$. So first identify the samples of the first and last peaks associated within the QRS $w_{2^2}\,x[n]$ say n_{first} and n_{last}.

Onset and offset are determined by applying two criteria:

1. searching for the sample where is $|w_{2^2}\,x[n]|$ below a threshold (ξ_{QRSon} or ξ_{QRSend}) relative to the amplitude of the maximum modulus ($|w_{2^2}\,x[n_{first}]|$ or $|w_{2^2}\,x[n_{last}]|$);
2. searching for a local minimum of $|w_{2^2}\,x[n]|$ before n_{first} or after n_{last}.

T wave Detection: Fixing the search window (200 ms) after the offset of the QRS complex, the T wave is detected. Finding maxima of $|w_{2^4}\,x[n]|$ in that window; that maxima value which are exceeding this threshold value $\in_T$ are considered to be the T wave peak. Fixing the same search window before and after the T wave peak the onset and offset (zero crossing) points are detected. In order to avoid errors in detecting the onset and offset of these P and T waves due to baseline drift and motion artifacts, the scale 2^4 is selected.

T wave Delineation: Maximum of $|w_{2^4}\,x[n]|$ with amplitude greater than this threshold γ_T are considered as significant slope of the wave or T wave peak. Depending upon number and polarity of the found maximum, one out of six possible T wave morphologies: positive (+), negative (−), biphasic (+/− or −/+), only upwards, and only down wards. If the T wave is not found in scale 2^4 , repeat the above process over $|w_{2^4}\,x[n]|$. To identify the wave limits of T wave the same criteria as for QRS onset and offset with thresholds and applied to scale 2^k.

P wave Detection: Fixing the search window (200 ms) before the offset of the QRS complex, the T wave is detected. Finding maxima of $|w_{2^4}\,x[n]|$ in that window; that maxima value which are exceeding this threshold value $\in_p$ are considered to be the P wave peak. As like the T wave the onset and offset points of P wave also detected.

P wave Delineation: As like the T wave delineation process P wave Delineation also done.

2.3 Different Morphological Detection

Once the QRS complex has been detected, the T wave can be analyzed because ventricular repolarization always follows depolarization. Conversely, the P wave does not lend itself as easily to analysis because atrial and ventricular rhythms may be independent of each other. In the vast majority of cases, however, atrial and ventricular rhythms are associated so that P wave detection may be based on a backward search in time, beginning at the QRS complex and ending at the end of the preceding T wave. A method of wave delineation determines the boundaries of each wave within the PQRST complex so that, with the resulting time instants, different wave durations can be computed.

Once a wave has been delineated, [9] other measures characterizing the wave, such as amplitude and morphology, can be easily computed. Instead, many methods for wave delineation exploit the change in slope that occurs at a boundary to avoid the problems because of low-frequency noise. This type of delineation is to find the end of the S wave; the other wave boundaries of the PQRST complex can be found in a similar way. In this example, the search for the endpoint starts when the steepest upslope of the S wave occurs and continues until the derivative of the signal falls below a certain threshold value.

This wavelet based Delineation system gives more accurate detection peaks and fiducial points of QRS complex, P and T wave (positive (+), negative (−), biphasic (+/- or -/+) absence of wave, signal with low SNR only upwards, and only down wards) compare with the Pan-Tompkins Algorithm. With these different Morphological features mostly all of the arrhythmias are detected.

3 Validation Using Different Databases

As there is no golden rule to determine the peak, onset and end of the ECG waves, the validation of the delineator must be done using manually annotated databases. For these purposes, some easily-available or standard databases are used, namely the MIT-BIH Arrhythmia database (MITDB), the QT database (QTDB) and the European ST-T database (EDB). In this paper totally 151 different ECG signals (consists of normal, arrhythmic signals) are used for the detection and delineation of time and amplitude of QRS complex, P and T waves is done in more precise manner (Table 1).

The MITDB includes specially selected Holter recordings with anomalous but clinically important phenomena. The EDB files present ischemic episodes extracted from Holter recordings. These databases include annotations of QRS positions: R marks (MITDB) or QRS onsets (EDB). The QTDB includes some records from EDB and MITDB and also from several other MIT-BIH databases (ST Change, Supraventricular Arrhythmia, Normal Sinus Rhythm, Sudden Death, and Long Term).

Table 1 Characteristic of the validation databases

Databases	Files	Leads	*fs (Hz)*	Record duration
MITDB	25	2	360	30 min
QTDB	92	2	250	1 min
EDB	34	2	250	30 min

To assess the QRS detector [12, 13] calculate the sensitivity Se = TP/(TP + FN) and positive predictivity PP = TP/(TP + FP), where TP is the number of true positive (correct) detections, FN stands for the number of false negative(missed) detections, and FP stands for the number of false positive(wrong) detections. Error beats E = FN + FP.

4 Results and Discussion

The pre-processed ECG signal is given to SWT process after this detection and delineation of QRS complex T and P waves were done. The WT-based detector, in contrast to most QRS detectors found in the literature, allows to take advantage of the same wavelet analysis stage for ECG wave delineation, due to the particularly appropriate characteristics of time-scale analysis.Detection and Delineation ECG signal is done for 151 annotations of MITDB, QTDB and EDB of above mentioned record time. Detection and Delineation of P, QRS, T wave comparison table with their performance metrics and different Morphological features of individual ECG signal components are shown in Tables 2, 3 and 4.

Table 2 Validation of P wave

Database	P Wave					
	Detection			Delineation		
	Se (%)	PP (%)	E	Se (%)	PP (%)	E
MITDB	99.24	89.84	4,889	99.23	89.42	5095
QTDB	99.45	89.02	542	99.45	88.51	565
EDB	99.86	89.51	2,263	99.86	89.39	2,286

Table 3 Validation of QRS complex

Database	QRS complex					
	Detection			Delineation		
	Se (%)	PP (%)	E	Se (%)	PP (%)	E
MITDB	99.31	99.78	498	99.30	99.39	675
QTDB	99.45	99.77	24	99.45	99.76	25
EDB	99.89	99.81	65	99.89	99.32	176

Table 4 Validation of T wave

Database	T Wave					
	Detection			Delineation		
	Se(%)	PP(%)	E	Se(%)	PP(%)	E
MITDB	99.28	96.21	2080	99.28	96.02	2167
QTDB	99.45	98.85	63	99.45	98.64	73
EDB	99.89	98.44	416	99.89	98.16	489

Table 5 Various abnormalities and their characteristic

S. No.	Name of abnormality	Characteristic features
1	Atrial fibrillation	Absence of P wave
2	Dextrocardia	Inverted P-wave
3	Tachycardia	R–R interval <0.6 s
4	Bradycardia	R–R interval >1 s
5	Hyperkalemia	Tall T-wave and absence of P-wave
6	Myocardial Ischaemia	Inverted T-wave
7	Hypercalcaemia	QRS interval <0.1 s
8	Sinoatrial block	Complete drop out of a cardiac cycle
9	Sudden cardiac Death	Irregular ECG
10	Ventricular Fibrillation	Highly oscillated ECG

The wavelet analysis of ECG signal using different scales based on their power spectrum gives high performance on detection of different morphological features and also solves many technically computing problems. This performance improvement is, according to the multiscale approach, which permits to attenuate noise at rough scale, and then to refine the precision of the positions with the help of the finer scale. Table 5 shows different rhythmic changes of ECG signal component and their corresponding abnormalities.

5 Conclusion

The Wavelet Transform is a new promising technique in non invasive electro cardiology for providing improved methods for denoising, detection and delineation of cardiac arrhythmia signal. This work shows that the important features can be extracted both from the morphological and statistical characteristics of the ECG signal. Different morphological features were detected with Stationary Wavelet Transform. Mathematical approach fixes different amplitude threshold levels for the detection and delineation of QRS complex, P and T waves. A computer based ECG signal classifier can be developed by employing the extracted features—pathological features and statistical features for detection of all of the Cardiac arrhythmias and Cardiac Vascular Diseases.

Acknowledgments The authors thank the Management and the Principal of Bannari Amman Institute of Technology, Sathyamangalam and Kumaraguru college of Technology, Coimbatore for providing excellent computing facilities and encouragement.

A.1 6 APPENDIX

The amplitude thresholds in the presented algorithms can be grouped into three types. First, the thresholds used to decide if a pair of maximum moduli with opposite sign can account for a wave: $\in^1_{QRS}, \in^2_{QRS}, \in^3_{QRS}, \in^4_{QRS}$ (for QRS detection), $\in_T$ and $\in_P$. (in the T/P wave delineation). These thresholds are proportional to the RMS value of the WT at the corresponding scales. For QRS detection, in each expert of 2^{16} samples.

$\in^1_{QRS} = \text{RMS}\ (W_{2^i}x[n]),\ i = 1, 2, 3$

$\in^4_{QRS} = 0.5\text{RMS}\ (W_{2^4}x[n])$.

For T and P waves

$\in_T = 0.25\ \text{RMS}\ (W_{2^4}x[n])$ where the RMS is measured in each interval between

$\in_P = 0.02\ \text{RMS}\ (W_{2^4}x[n])$ two consecutive QRS. The morphology of QRS complexes and the type of T/P wave depend of number of significant maximum moduli. The thresholds to determine if they are significant, $\gamma_{QRS_{pre}}, \gamma_{QRS_{post}}, \gamma_T$ and γ_P are related to the amplitude of the global maximum modulus within the corresponding search window (sw).

$\gamma_{QRS_{pre}} = 0.06\ \max(|W_{2^2}x[n]|),\ n \in SW_{QRS}$ A third group of thresholds are used to $\gamma_{QRS_{post}} = 0.09\ \max(|W_{2^2}x[n]|),\ n \in SW_{QRS}$ determine the onset/offset of QRS complex. They are proportional to the amplitude of the WT at the first/last maximum modulus of the complex or wave.

$$\xi_{QRSon} = \begin{cases} 0.05W_{2^2}x[n_{first}], \text{if } W_{2^2}x[n_{first}] > 0 \\ 0.07W_{2^2}x[n_{first}], \text{if } W_{2^2}x[n_{first}] < 0 \end{cases}$$

$$\xi_{QRSend} = \begin{cases} 0.125W_{2^2}x[n_{last}], \text{if } W_{2^2}x[n_{last}] > 0 \\ 0.71W_{2^2}x[n_{last}], \text{if } W_{2^2}x[n_{last}] < 0 \end{cases}$$

References

1. Laguna P, Jañe R, Caminal P (1994) Automatic detection of wave boundaries in multi lead ECG signals: Validation with the CSE database. Comput Biomed Res 27(1):45–60
2. Chouhan VS, Mehta SS (2008) Threshold-based detection of P and T-wave in ECG using new feature signal. Int J Comput Sci Net Secur 8(2):144–152

3. Weng B, Wang JJ, Michaud F, Blanco-Velasco M (2008) Atrial fibrillation detection using stationary wavelet transform analysis. Conf Proc IEEE Eng Med Biol Soc. 2008:1128–1131
4. Clayton R, Murray A, Campell I (1994) Recognition of ventricular fibrillation using neural network. Med Biol Eng Comput 35: 611–626
5. Coast DA, Stern RM, Cano G, Briller SA (1990) An approach to cardiac arrhythmia analysis using hidden markov models. IEEE Trans Biomed Eng 37: 826–836
6. Dokur Z, Olmez T, Yazgan E (1999) Comparision of discrete wavelet and fourier transform for ECG beat classification. Electron Lett 35:1502–1504
7. Li C, Zheng C, Tai C (1995) Detection of ECG characteristic points using wavelet transforms. IEEE Trans Biomed Eng 42:21–28
8. Alfred L, Endt P, Oeff M, Trahms L (2001) Variability of the QRS Signal in High-Resolution ECG and MCG. IEEE Trans Biomed Eng 38:133–143
9. Martinez JP, Almeida R, Olmos S, Rocha AP, Laguna P A wavelet-based ECG delineator: Evaluation on standard databases. IEEE Trans Biomed Eng 51(4)
10. Pandit SV (1996) ECG Baseline drift removal through STFT. In: 18th annual international conference of the ieee engineering in medicine and biology society, Amsterdam
11. Meyer CR, Keiser HN (1977) Electrocardiogram baseline noise estimation and removal using cubic splines and state-space computational techniques. Comput Biomed Res 10:459–470
12. Minami K, Nakajima H, Toyoshima T (1999) Real-time discrimination of beats of ventricular tachyarrhythmia with Fourier transform neural network. IEEE Trans Biomed Eng 46:179–185
13. Osowski S, Linh TH (2001) ECG beat recognition using fuzzy hybrid neural. IEEE Trans Biomed Eng 44: 1265–1271

Performance Analysis of Wavelet Transforms and Principal Components as Post Classifier for the Classification of Epilepsy Risk Levels from EEG Signals

R. Harikumar, T. Vijaykumar, C. Ganesh Babu and M. G. Sreejith

Abstract The Objective of this paper is to analyze the performance of Principal components in optimization of code converter outputs in the classification of epilepsy risk levels from Electroencephalogram (EEG) signals. The Code converter is used to classify the risk levels of epilepsy based on extracted parameters like energy, variance, peaks, sharp and spike waves, duration, events and covariance from the EEG signals of the patient. Principal component method is applied on the classified data to identify the optimized risk level (singleton) which characterizes the patient's risk level. The efficacy of the above methods is compared based on the bench mark parameters such as Performance Index (PI), and Quality Valuc (QV). A group of twenty patients with known epilepsy findings are analyzed. High PI such as 97.35 % was obtained at QV's of 23.2, for Principal component optimization when compared to the value of 40 % and 6.25 through code converter classifier respectively. It was identified that Principal Component Method is a good post classifier in the optimization of epilepsy risk levels.

Keywords EEG signals · Wavelet transform · Code converter · Principal component analysis · Epilepsy risk levels

1 Introduction

Epilepsy, from which approximate 1 % of the people in the world suffer, is a group of brain disorders characterized by the recurrent paroxysmal electrical discharges of the cerebral cortex, that result in irregular disturbances of the brain functions, which are associated with significant changes of EEG signal. Electroencephalograms

R. Harikumar (✉) · T. Vijaykumar · C. Ganesh Babu · M. G. Sreejith
Bannari Amman Institute of Technology, Sathyamangalam, India

Mohan S. and S. S. Kumar (eds.), *Proceedings of the Fourth International Conference on Signal and Image Processing 2012 (ICSIP 2012)*, Lecture Notes in Electrical Engineering 221, DOI: 10.1007/978-81-322-0997-3_3,

(EEGs) are recordings of electrical potentials produced by the brain [1]. Analysis of EEG activity has been achieved principally in clinical settings to identify pathologies and epilepsies since Hans Berger's recording of rhythmic electrical activity from the human scalp. In the past, interpretation of the EEG was limited to visual inspection by neurophysiologist, an individual trained to qualitatively make a distinction between normal EEG activity and abnormalities contained within EEG records. A common form of recording used for this purpose is an ambulatory recording that contains EEG data for a very long duration of even up to one week [2]. It involves an expert's efforts in analyzing the entire length of EEG recordings to detect traces of epilepsy.

Because epileptic seizures, in general, occur frequently and unpredictably, automatic detection of seizures during long term EEG monitoring sessions is highly useful and needed. Over the past 20 years, numerous attempts to automate the detection of epileptic form activity have been made and comparatively good results have been obtained [3]. In this paper, we will propose another method for classification of epilepsy risk level based on extracted parameters using hard Thresholding of wavelet transforms and Principal Components from epileptic EEG signals.

2 Materials and Methods

The EEG data used in the study were acquired from twenty epileptic patients who had been under the evaluation and treatment in the Neurology department of Sri Ramakrishna Hospital, Coimbatore, India. A paper record of 16 channel EEG data is acquired from a clinical EEG monitoring system through 10–20 international electrode placing method. With an EEG signal free of artifacts, a reasonably accurate detection of epilepsy is possible; however, difficulties arise with artifacts. This problem increases the number of false detection that commonly plagues all classification systems. With the help of neurologist, we had selected artifact free EEG records with distinct features. These records were scanned by Umax 6,696 scanner with a resolution of 600 dpi.

2.1 Acquisition of EEG Data

Since the EEG records are over a continuous duration of about thirty seconds, they are divided into epochs of two second duration each by scanning into a bitmap image of size 400 × 100 pixels. A two second epoch is long enough to detect any significant changes in activity and presence of artifacts and also short enough to avoid any repetition or redundancy in the signal [3, 4]. The EEG signal has a maximum frequency of 50 Hz and so, each epoch is sampled at a frequency of

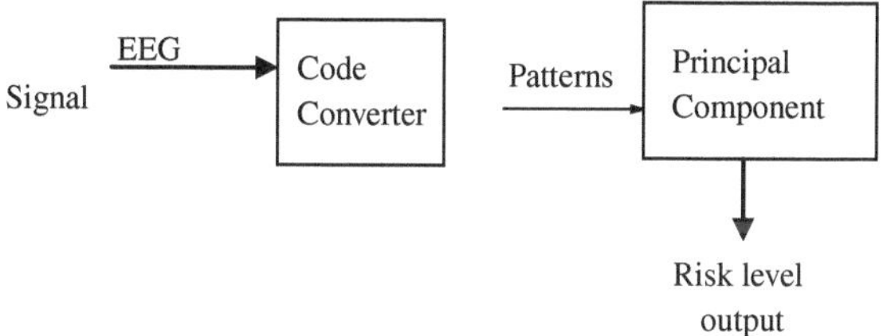

Fig. 1 Wavelet–principal component classifier

200 Hz. Each sample corresponds to the instantaneous amplitude values of the signal, totaling 400 values for an epoch.

Figure 1 shows the overall epilepsy risk level (Wavelet–Principal Component) classifier system. The motto of this research is to classify the epilepsy risk level of a patient from EEG signal parameters. This is accomplished as

1. Four parameters out of seven features are extracted from EEG signals using wavelet transforms, such as Haar, Db2, Db4, and Sym 8.
2. Code converter classification for epilepsy risk level at each channel from EEG signals and its parameters.
3. Code converter results from each channel are optimized using Principal components
4. Performance of code converter classification and Principal component optimization methods are analyzed.

1. The energy in each two-second epoch is given by [3]

$$E = \sum_{i=1}^{n} x_i^2 \tag{1}$$

where x_i is signal sample value and n is number of samples. The scaled energy is taken by dividing the energy term by 1,000.

2. The variance is computed as σ given by

$$\sigma^2 = \frac{\sum_{i=1}^{n} (x_i - \mu)^2}{n} \tag{2}$$

where $\frac{\mu=\sum_{i=1}^{n} x_i}{n}$ is the average amplitude of the epoch

3. Covariance of Duration. The variation of the average duration is defined by

$$CD = \frac{\sum_{i=1}^{p} (D - t_i)^2}{pD^2} \tag{3}$$

The following parameters are extracted using wavelet transforms

4. The total number of positive and negative peaks exceeding a threshold is found.

5. Spikes are detected when the zero crossing duration of predominantly high amplitude peaks in the EEG waveform lies between 20 and 70 ms and sharp waves are detected when the duration lies between 70 and 200 ms.
6. The total numbers of spike and sharp waves in an epoch are recorded as events.
7. The average duration is given by [3].

$$D = \frac{\sum_{i=1}^{p} t_i}{p} \tag{4}$$

where t_i is one peak to peak duration and p is the number of such durations.

Wavelet Transforms in Signal Decomposition

The wavelet transforms acts as a sort of mathematical microscope through which different parts of the signals are examined by adjusting the focus [5]. The wavelet transform (WT) of a function f(t) is an integral transform defined by [6],

$$wf(a,b) = \int_{-\infty}^{\infty} f(t)\psi_{a,b}^{*}(t)dt \tag{5}$$

where ψ^{*}(t) denotes the complex conjugate of the wavelet function ψ(t). The transform yields a time–scale representation similar to the time frequency representation of the Short-Time Fourier Transform (STFT). The set of the analyzing function the wavelet family is deduced from a mother wavelet ψ(t) by [7],

$$\psi_{a,b}^{*}(t) = \frac{1}{\sqrt{2}}\psi\left(\frac{t-b}{a}\right) \tag{6}$$

where a and b are the dilation (scale) and translation parameters respectively. The mother wavelet is a short oscillation with zero mean. The Discrete Wavelet Transforms (DWT) results from discredited scale and translation parameters e.g. $a = 2^j$ and $b = n.\ 2^j$ where j and n are integer numbers. There have been several investigations into additive noise suppression in signals using wavelet transforms. Johnstone and Donoho's 10 principal work is on Thresholding the DWT of a signal and then reconstructing it. The method relies on the fact noise commonly manifests itself as smaller values, and wavelet transforms provides a scale based decomposition. Thus, most of the noise tends to be represented by wavelet coefficients at the finer scales. Discarding these coefficients would result in a natural filtering out of noise on the basis of scale [8]. Because the coefficients at such scales also tend to be the primary carriers of edge information, by setting the wavelet coefficients to zero if their values are below a threshold. These coefficients are mostly those corresponding to noise. The edge related coefficients, on the other hand, are usually above the threshold. In this study, at first the effect of simple Haar wavelet is undertaken. Haar wavelet function is defined as [9].

$$\psi(t) = \begin{cases} 1; 0 \le t < 1/2 \\ -1; {}^1\!/_2 \le t < 1 \\ 0 : \text{otherwise} \end{cases} \tag{7}$$

2.2 Thresholding

Wavelet Thresholding is a signal estimation technique that exploits the capabilities of wavelet transform for signal denoising or smoothing. It depends on the choice of a threshold parameter which determines to great extent the efficacy of denoising. Typical threshold operators for denoising include hard threshold, soft threshold, and affine (firm) threshold. Hard threshold is defined as [7].

$$\rho_T(x) = \begin{cases} x, \mathit{if}|x| > T \\ 0, \mathit{if}|x| \le T \end{cases} \tag{8}$$

where T is the threshold level.

Selection of suitable wavelet and the number of decomposition levels is very important in analysis of EEG signals using DWT. The number of decomposition level is chosen based on the dominant frequency components of the signal. The levels are chosen such that those parts of the signal that correlates well with the frequencies necessary for classification of the signal are retained in the wavelet coefficients. In present study, since the EEG signals do not have any useful signal above 30 Hz, the number of decomposition levels was chosen to be 4. Thus, the EEG signals were decomposed into details D1–D4 and one final approximation A4. Usually, tests are performed with different types of wavelets and the one, which gives maximum efficiency, is selected for the particular application [9]. The smoothing feature of Daubechies wavelet of order 2 (Db2) made it more appropriate to detect changes of EEG signals. Hence, the wavelet coefficients were computed using the Db4 in the present study.

With the help of expert's knowledge and our experiences with the Ref. [3, 4], we have identified the following parametric ranges for five linguistic risk levels (very low, low, medium, high and very high) in the clinical description for the patients which is shown in Table 1.

2.3 Code Converter as a Pre Classifier

The encoding method processes the sampled output values as individual code. Since working on definite alphabets is easier than processing numbers with large decimal accuracy, we encode the outputs as a string of alphabets [10].The alphabetical representation of the five classifications of the outputs is shown in Table 1.

Table 1 Representation of risk level classifications

Risk Level	Representation
Normal	U
Low	W
Medium	X
High	Y
Very high	Z

Epoch 1	Epoch 2	Epoch 3
WYYWYYY	WYYWYYY	WZYYWWW
YZZYXXX	YYYYXXX	YYYXYYY
YYZXYYY	YYYYYYY	YYYYYYY
YZZYXYY	XZZXYYY	YYYYYYY
ZZZYYYY	WYYYXXX	YYYXYYY
YYZXXXX	WYZYYYY	YZZYYYY
ZZZYYYY	YYYYYYY	ZZZYYYY
YYYYXXX	YYYYXXX	YYYXZYY

Fig. 2 Code converters output

The ease of operation in using characteristic representation is obviously evident than in performing cumbersome operations of numbers. By encoding each risk level one of the five states, a string of seven characters is obtained for each of the sixteen channels of each epoch. A sample output with actual patient readings is shown in Fig. 2 for eight channels over three epochs. The Code converter's classification efficiency is evaluated from the following parameters. The Performance of the Code converter is defined as follows [3],

The Performance of the code converter method is defined as follows

$$PI = \frac{PC - MC - FA}{PC} \times 100 \tag{9}$$

Where PC—Perfect Classification, MC—Missed classification, FA—False alarm,

$$PI = [(0.5 - 0.2 - 0.1)/0.5] * 100 = 40\,\%$$

The perfect classification represents when the physicians and code converter classifier agrees with the epilepsy risk level. Missed classification represents a true negative of fuzzy classifier in reference to the physician and shows High level as Low level. False alarm represents a false positive of fuzzy classifier in reference to the physician and shows Low level as High level. The performance for code converter classifier is as low as 40 %. Table 2 shows the performance of wavelet transforms along with code converter outputs.

Table 2 Performance of code converter output based on wavelet transform along hard thresholding

Wavelet transforms with hard thresholding	Perfect classi fication	Missed classi fication	False alarm	Performance index
Haar	61.45	15.625	22.91	37.58
Db2	61.18	16.14	22.65	36.44
DB4	64.57	12.49	22.91	44.72
Sym8	63.52	11.44	23.95	44.81

Table 3 Rhythmicity of code converter

Wavelets	No of categories of patterns	Rhythmicity R = C/D
Haar	31	0.032292
Db2	41	0.042708
Db4	30	0.03125
Sym8	45	0.046875

2.4 Rhythmicity of Code Converter

Now we are about to identify the Rhythmicity of code converter techniques which is associated with nonlinearities of the epilepsy risk levels. Let the Rhythmicity is defined as [4]

$$R = {}^{C}/_{D} \tag{10}$$

where C = No of categories of patterns; and D = Total number of patterns which is 960 in our case. For an ideal classifier C is to be one and R = 0.001042. Table 3 shows the Rhythmicity of the Code converter classifier for hard Thresholding of each wavelet.

It is observed from the table 3 that the value of R is highly deviated from its ideal value therefore it is necessary to optimize the code converters outputs to endure a singleton risk level. In the following section we discuss about the Principal Component optimization of epilepsy risk levels.

3 Principal Components for Optimization of Code Converter Outputs

Principal Component Analysis (PCA) is a dimensionality reduction technique that has been applied to many kinds of data. In fact, PCA is the optimal such linear transform-that is, for any choice for the number of dimensions, PCA returns the subspace that retains the highest variance [11]. In this section, we describe how to use PCA to optimize the code converters outputs. PCA is a mathematical technique allows reducing the complex system of correlations in a smaller number of dimensions. χ being a table of P numeric variables (in columns) describing N individuals (in lines), w propose to seek a representation of N individuals (signals) e1, e2,…,en in a subspace of initial space. In other words, we have to define K new variables, combination of P of initial space, which could make loss less possible information. These K variables will be called principal axes [12].

For N observations, we will have a matrix of N × P size which is given by

$$e = [e_1 \quad e_2 \quad e_3 \quad \ldots \quad e_n] \tag{11}$$

The average signal is defined by:

$$\psi = \frac{1}{N}\sum_{m=1}^{N} e_m \tag{12}$$

For each element the difference:

$$\delta_i = e_i - \psi \tag{13}$$

The computation of covariance matrix is:

$$C = \frac{1}{N}\sum_{m=1}^{n} \delta_m \delta_m^T = \frac{1}{N} A \times A^T \tag{14}$$

$$With\ A = [\{\delta_1\, \delta_2 \ldots \delta_N\}] \tag{15}$$

However, the determination of the Eigen vectors of covariance matrix will require an excessive calculation [13]; the size of this matrix is (P × P). If υ_i is the Eigen vector of AXA^T its Eigen values are:

$$A^T A \upsilon_i = \mu\, \upsilon_i$$

Then the Eigen vectors of C are calculated by:

$$U_i = A\ \upsilon_i \tag{16}$$

Finally the principal component of each signal e_i is given by:

$$w_k = u_k^T \times (e_i - \delta) \tag{17}$$

The vectorw_k represents the new parameters completely de correlated and optimized for classification.

4 Results and Discussion

The outputs are obtained for three epochs for every patient in classifying the epileptic risk level by the code converter and Principal component approach. To study the relative performance of these two systems, we measure two parameters,

Table 4 Performance index for PCA classifier

Wavelets	Perfect classification	Missed classification	False alarms	Performance index
Haar	95.8	1.87	2.25	95.12
Db2	95.94	0.832	3.22	95.8
Db4	97.32	1.043	1.63	97.26
Sym8	95.56	0.833	3.6	95.49

the Performance Index and the Quality Value. These parameters are calculated for each set of the patient and compared.

Table 4 shows the Performance Index for PCA classifier. From the table IV we identified that for code converter classifier have max detection of 50 % with false alarm of 14 %.Similarly for PCA optimization we obtained perfect detections of 96 % with false alarms of 3.22 %. This indicates that the PCA classifier is a better one.

4.1 Quality Value

The goal of this paper is to classify the epileptic risk level with as many perfect classifications and as few false alarms as possible. In Order to compare different classifiers we need a measure that reflects the overall quality of the classifier [10]. Their quality is determined by three factors.

1. Classification rate
2. Classification delay
3. False Alarm rate

The quality value Q_V is defined as [3]

$$Q_V = \frac{C}{\left(R_{fa} + 0.2\right) * \left(T_{dly} * P_{dct} + 6 * P_{msd}\right)} \tag{18}$$

where, C is the scaling constant, R_{fa} is the number of false alarm per set T_{dly} is the average delay of the on set classification in seconds, P_{dct} is the percentage of perfect classification and P_{msd} is the percentage of perfect risk level missed. A constant C is empirically set to 10 because this scale is the value of Q_V to an easy reading range. The classifier with the highest Q_V should be the best one. The quality value obtained by PCA optimization technique is shown in the table 5.

However, estimation of the PCA projection from data has some limitations. First, its computational complexity makes it difficult to deal directly with high dimension data, e.g. Code converter outputs. Second, the number of examples available for the estimation of the PCA projection is typically smaller that the ambient dimension of the data and this can lead to over fitting of the projection. In order to alleviate these problems in classification applications, a simple variant of PCA known as Separable PCA can also be tried.

Table 5 Quality value for average of twenty patients

Wavelet transforms with hard thresholding	Quality values	
	Without PCA optimization	With principal component optimization
Haar	11.56	22.93
Db2	12.57	22.64
DB4	12.49	23.33
Sym8	12.84	22.55

5 Conclusion

This paper aims at classifying the epilepsy risk level of epileptic patients from EEG signals. The goal was to classify perfect risk levels with high rate of classification, a short delay from onset, and a low false alarm rate. Though it is impossible to obtain a perfect performance in all these conditions, some compromises have been made. As a high false alarm rate ruins the effectiveness of the system, a low false-alarm rate is most important. PCA optimization technique is used to optimize the risk level by incorporating the above goals. The major limitation of this method is that if one channel has a high-risk level, then the entire group will be maximized to that risk level. This will affect the non-epilepsy spike region in the groups. The number of cases from the present twenty patients has to be increased for better testing. A comparison of EM and SVM will be taken for further studies.

Acknowledgments The authors express their sincere thanks to the Management and the Principal of Bannari Amman Institute of Technology, Sathyamangalam for providing the necessary facilities for the completion of this paper. **This research is also funded by AICTE RPS.:F No 8023/BOR/RID/RPS-41/2009-10, dated 10th Dec 2010.**

References

1. Leon D, Iasemidis et al (2003) Adaptive epileptic seizure prediction system. IEEE Trans Biomed Eng 50(5):616–627
2. Alison Dingle A et al (1993) A multistage system to Detect epileptic form activity in the EEG. IEEE Trans Biomed Eng 40(12):1260–1268
3. Haoqu, Gotman J (1997) A patient specific algorithm for detection onset in long-term EEG monitoring possible use as warning device. IEEE Trans Biomed Eng 44(2):115–122
4. Celement C et al (2003) A Comparison of algorithms for detection of spikes in the electroencephalogram. IEEE Trans BioMed Eng 50(4):521–526
5. Prochazaka A, Kukal J, Vysata O (2008) Wavelet transform use for feature extraction and eeg signal classification.In: Proceedings of IEEE ISCCSP 2008, Malta, 12–14 March 2008, pp 719–722
6. Mirzaei A, Ayatollahi A, Gifani P, Salehi L (2010) EEG analysis based on wavelet- spectral entropy for epileptic seizures detection.In: Proceedings of 3rd international conference on biomedical engineering and informatics (BMEI 2010), Changai January 2010, pp 878–882

7. Xanthanopoulus P et.al (2010) A novel wavelet based algorithm for spike and wave detection in absence of epilepsy. In: Proceedings of IEEE International Conference on Bio informatics and Bio Engineering, Jan 2010, pp 14–19
8. Yuan Y (2010) Detection of epileptic seizure based on EEG signals. In: Proceedings of IEEE EMBS sponsored 3rd international congress on image and signal processing (CISP 2010), July 2010, pp 4209–4211
9. Tzallas AT, Tsipouras MG, Fotiadis DI A Time- Frequency Based Method for the Detection of Epileptic seizure in EEG Recording. 12th IEEE International Symposium on Computer based Medical Systems (CBMS'07) 23–27
10. Mohseni HR, Maghsoudi A, Shamsollahi MB (2006) Seizure detection in EEG signals: A comparison of different approaches. In: proceedings of the 28th IEEE EMBS annual international conference New York City, 30–3 Aug–Sep 2006, pp 6724–6727
11. Sophian A, Tian G, Taylor D, Rudin J (2003) A feature extraction technique based on principal component analysis for pulsed eddy current NDT.NDT&E. Int J 36: 37–41
12. Kheli M, Boundraa M, Kechida A, Drai R (2005) Classification of defects by the SVM method and the principal component analysis (PCA). World Acad Sci Eng Technol 9: 226–231
13. Bair E, Hastie T, Paul D, Tibshirani R (2006) Prediction by supervised principal components. J Am Stat Assoc 101473: 119–137

Edge Based LSBMR Scheme for Hiding Secret Messages with Genetic Algorithms

P. M. Siva Raja and E. Baburaj

Abstract Modern information hiding technology is an important branch of information security. Hiding capacity is very much important for efficient covert communication. The redundancies of digital media as well as the characteristic of human visual system make hiding technology a significant one. Steganography is the Art and Science of writing hidden messages in such a way that no one, apart from the sender and intended recipient suspects the existence of the message. Images are the mostly cover objects used for information hiding schemes. Image steganography is the most popular method for message concealment. Many different carrier file formats can be used, but digital images are the common, because of their frequency in the Internet. In this paper LSB Matching Revisited (LSBMR) image steganography using Genetic Algorithm (GA) is proposed, in which Genetic algorithm is used to select the embedding regions according to the size of the secret message and to optimize the threshold value of the selected image regions. Experimental analysis shows that the proposed algorithm outperforms the existing methods in terms of capacity and security.

Keywords Steganography · Message concealment · Information hiding · Region selection · Genetic algorithms

P. M. Siva Raja (✉)
Research Scholar, Sathyabama University, Chennai, Tamil Nadu, India
e-mail: selvamsivaraja@gmail.com

E. Baburaj
Department of Computer Science and Engineering, Sun College of Engineering and Technology, Nagercoil, India
e-mail: alanchybabu@gmail.com

Mohan S. and S. S. Kumar (eds.), *Proceedings of the Fourth International Conference on Signal and Image Processing 2012 (ICSIP 2012)*, Lecture Notes in Electrical Engineering 221, DOI: 10.1007/978-81-322-0997-3_4,

1 Introduction

As an important component of multimedia information security, information hiding has received wide attention in recent years. In fact, intellectual properties are becoming harder to protect and so are original contents, that's why we need techniques to be developed such as Image Steganography. Steganography is a technique for information hiding [1]. It aims to embed secret data into digital cover media, such as Images, Audio, and Video without being suspicious. When two parties exchange information over an insecure medium in such a way that even is an intruder is able to intercept, read and perform computation. The characteristic Invisibility, Storage capacity, Resilience against attack makes Steganography an important concept in hiding technology [2]. On the other hand the users of digital media are under threat due to the growing concern of copyright infringement, illegal distribution unauthorized tampering and security in communication [3]. Several data hiding methods have been developed for audio, image, video and graphics etc. and are also reported in literature (Special issue on copyright and privacy protection 1998; In data hiding problem, GA may be used for optimizing the fundamentally conflicting requirements of security and robustness [3–5]. Research of data hiding in digital media reports that there exists different tradeoff relations. Most of the data hiding algorithms developed so far focus on single requirement or provide suboptimal solutions to meet a group of requirements based on the applications [6, 7]. On one hand the digital information revolution and the thriving progress in network communications offer benefits of almost noise free communication, the ease of editing and the internet distribution of digital multimedia data. In this paper LSB Matching Revisited (LSBMR) image steganography using Genetic Algorithm (GA) is proposed, in which Genetic algorithm is used to select the embedding regions according to the size of the secret message and to optimize the threshold value of the selected image regions The rest of the paper is arranged as follows. Section 2 analyzes the limitations of the relevant steganographic schemes with the other techniques. Section 3 shows the details of data embedding and data extraction in the proposed scheme. Section 4 presents experimental results and discussions. Finally, conclusion and future enhancement are given in Sect. 5.

2 Related works

This section discusses data hiding scheme in order to highlight the effectiveness of this class of tools for performance improvement in data hiding. The objective of this review is to discuss merits and limitations of the relevant works.LSB based techniques pose a difficult challenge to a steganalyst in the passive warden model as it is difficult to differentiate cover images from stego images, given the small changes that have been made. Of course with an active warden, such techniques can be easily defeated by randomizing the LSB. In this LSB replacement and LSBM approaches [8], the embedding process is very similar. Given a secret bit stream to be embedded, a traveling order in the cover image is first generated by a PRNG, and then each pixel along the travelling order is dealt with separately. For LSB replacement, the, the secret bit simply overwrites the LSB of the pixel and for LSBM scheme, if the secret bit is not equal to the LSB of the given pixel, then plus or minus one is added randomly to the pixel while keeping the altered pixel in the range of [0, 255].According to the scheme of LSBMR, two secret bits can be embedded into each embedding unit and the threshold T for region selection can be determined [1, 9]. After data hiding, the resulting image is divided into non-overlapping blocks. The blocks are then rotated by a random number of degrees. In data extraction, the scheme first extracts the side information from the stego image. Based on the side information, it then does some parameter identification process and identifies the regions that have been used for data hiding [10, 11]. Finally, it obtains the secret message M according to the corresponding extraction algorithm. The main drawback of this scheme is the threshold for region selection can be varied. Embedding rate increases more regions can be released adaptively by decreasing the threshold T. The absolute difference is taken as the threshold value T [11–13].

3 Secret data hiding: issues

In the data embedding stage, the scheme first initializes some parameters, which are used for subsequent data preprocessing and region selection and optimize the threshold value of those selected regions. If the regions are enough for hiding the given secret message M, then data hiding is performed on the selected regions and then repeats the region selection process using Genetic Algorithms and optimize the threshold value of selected regions until the secret message M can be embedded completely. In data extraction, the scheme first extracts the side information from the stego image. Based on the side information, Algorithms does some post processing and identifies the regions that have been used for data hiding. Finally, it obtains the secret message M according to the specified extraction algorithm. In this paper, we propose an embedding scheme based on LSBMR with Genetic Algorithms Fig. 1.

3.1 Data Embedding

This section describes the process of data hiding method in digital images. The choice of cover images is important and influences the security in a major way. We consider gray scale image as cover and the similar type image like text information as message signal since it can preserve contextual information even after various signal processing operations. Algorithms for embedding process are as follows.

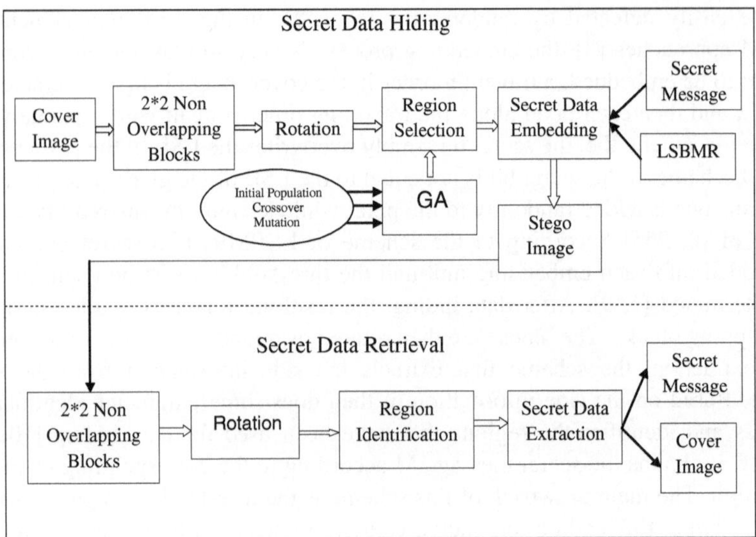

Fig. 1 Data hiding and data extraction scheme using LSBMR-ga cover image rotation region selection GA LSBMR stego image hiding

Step1: The cover image of size m x n is first divided into non overlapping blocks of Bz × Bz. For each small block rotate it by a random degree in the range of {0, 90,180,270} as determined by a secret key k1. Two benefits can be obtained by the random rotation, First, it can prevent the detector from getting the correct embedding regions without the secret key k1 and thus the security is increased much more.
Step2: The resulting image is rearranged as a row vector V by raster scanning and then the vector is divided non overlapping embedding units with every two consecutive pixels (x_i,x_{i+1}).
Step3: According to the scheme of LSMBR, two secret bits can be embedded into each embedding unit. Therefore, for a given secret message M, the threshold T for region selection can be determined by using genetic algorithm.

3.1.1 GA Operations

In our proposed GA, a chromosome is encoded as an array of 64 genes containing quantized DCT coefficients of each 8 * 8 pixel block of the image. Here we propose a new genetic algorithm approach to find the best position for data embedding and also optimize the quality of the steganographic image. In our optimization we need to handle four conflicting goals-longer hidden message, higher image quality, better robustness and larger data capacity. The first step to model this problem as a GA problem, is determining the chromosome, GA operators and fitness function. The procedure for the selection of embedding regions and optimization of threshold value of selected regions are as follows.

1. Initialization of Population: Chromosomal representation of the parameter values. The initial population is formed by taking Bz × Bz pixel blocks and Binary Encoding is used for forming the initial population. Binary encoding is the most common and simplest one. In binary encoding every chromosome is a string of bits, 0 or 1.

2. Selection: Best fitted pair of individuals is chosen by roulette-wheel selection process by adding up fitness values of the individuals to get fitness. Then randomly select individuals to cross 50 % of the fitness value in the cumulative way. The particular individual which crosses 50 % criteria in the cumulative process is chosen to be one of the matching pool pair. This process is again carried on to find another individual pool matching pair.

3. Crossover: Find the crossover site and to perform crossover between the selected pixel pair to get the new pair of more fitted individuals using arithmetic crossover. The selected matching pool is taken as input and finds the crossover site using arithmetic crossover. Exchange the portions lying on one side of crossover site of those matching pool pair, a new pair of individuals is an outcome.

4. Mutation: To mutate or change a particular bit in a pixel block with very small probability. Uniform mutation is used for the mutation process. A very small mutation probability is chosen, depending upon the probability value; change a bit from 1 to 0 or 0 to 1.

5. Objective Function: To estimate the fitness value of an individual, the initial population is taken as input; on each population 2D interpolation technique is applied to approximate the original matrix. The mean square error is evaluated by subtracting the interpolated matrix from the original matrix. The square of that MSE is considered to be the fitness value of that particular individual. The fitness function can be evaluated using the Eq. 1 as

$$f(x) = \left(\frac{1}{m\,x\,n}\right) \sum_{i}^{M} \sum_{j}^{N} (I_{ij} - I'_{ij})^{2}. \tag{1}$$

where mxn is the height and width of the cover image, I_{ij} is the pixel value of coordinate (x,y) in cover image, I'_{ij} is the corresponding pixel value in the rotated image.

Step4: According to LSMBR, two secret bits can be embedded into each embedding unit. Therefore for a given secret message M, the threshold T for region selection can be determined by using the optimization technique in Eq. (1). The embedding regions in a pseudorandom order determined by a secret key k_2. The secret key is the difference between the two pixel values. That value is greater than or equal to the threshold value. Performing data hiding on the selected embedding regions are as follows. In the LSB matching revisited, the choice of whether to add or subtract one from the cover image pixel is random. This method uses the choice to set a binary function of two cover image pixels to the desired value in Eq. (2). The embedding is performed using a pair of pixels as a unit, where the LSB of the first pixel carries one bit of information, and a function of the two pixel values carries another bit of information.

$$f(l,n) = LSB\left(\left[\frac{l}{2}\right] + n\right). \tag{2}$$

Data hiding is performed according to the following two properties:Property 1: $f(l-1,n) \neq f(l+1,n)$ Property 2:$f(l,n) \neq f(l,n+1)$where m_i and m_{i+1} denote two secret bits to be embedded. The function f (l, n) is a random value in {−1, +1} and (x, xi_{+1}) in the cover image pixel and $\left(x^{|}, x^{|}_{i+1}\right) = m_{i+1,}\left(x^{|}_{1}, x^{|}_{i+1}\right) =$ (x, xi_{+1}).

Step5: After data hiding, the resulting image is divided into non overlapping Bz × Bz blocks. The blocks are then rotated by a random number of degrees based on key $k_{1.}$

3.2 Data Extraction

The final stage of the algorithm is the retrieval process of the secret message M. To extract the secret message, first extract the side information as the block size Bz and the threshold T from the stego image. For that the stego image is divided into Bz × Bz non overlapping blocks and the blocks are then rotated by random degree based on the secret key k_1. The resulting image is rearranged as a row vector v_i. Finally to get the embedding units by dividing v_i into non overlapping blocks with two consecutive pixels. The travel through the embedding regions whose mean values are greater than or equal to the threshold T according to a pseudorandom order based on the secret key k_2 until all the hidden bits (secret message) are extracted completely based on Eq. 3. To extract the two secret bits m_i, m_{i+1} as follows.

$$M_i = LSB\left(x_i^{||}\right),\ mi_{+1} = LSB\left(\left[1^{|}/2\right] + n_{i+1}\right) \tag{3}$$

4 Results and Discussion

This section presents simulation results to demonstrate the effectiveness of the proposed data hiding method compared with existing relevant methods as mentioned in section II. The experiment is carried out using the Eight cover image of size (512 × 512), 8 bits/pixel to illustrate the effectiveness of proposed approach and two of them are shown in Fig. 2. The Peak Signal to Noise ratio (PSNR) is used to evaluate qualities of the stego images. Experimental results of two stego images are shown in figure. In addition, other noticeable image quality measures, such as Mean Square Error (MSE) and Receiver Optimization Curve (ROC) are also applied to our method indicating the significance to the contribution of this paper. Besides, from MSE values, it shows that the changed value of each pixel is almost same in stego images. To demonstrate the accomplished performance of our proposed approach in capacity and security for hiding secret data in the stegoimage. The analysis of selection technique is based on a comparison as the number of function evaluations. Population size, Crossover, Mutation, criteria for termination, Fitness function and number of generations are the parameters used for analysis.

4.1 Histogram Analysis

An image histogram is a type of histogram that acts as a graphical representation of the tonal distribution in a digital image. It plots the number of pixels for each tonal value. By looking at the histogram for a specific image a viewer will be able to judge the entire tonal distribution at a glance. In the field of computer vision, image histograms can be useful tools for thresholding. Because the information contained in the graph is a representation of pixel distribution as a function of tonal variation. Figure 3 shows the histograms of cover and stego images. It is clear that the histograms are almost identical. This is caused by the embedding the message bits in noisy regions of thc imagc.

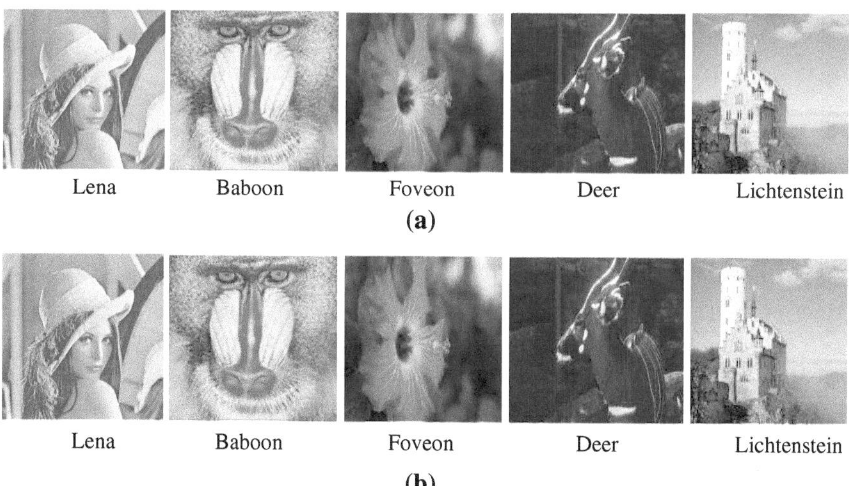

Fig. 2 **a** Cover images **b** Stego Images Lena baboon foveon deer lichtenstein

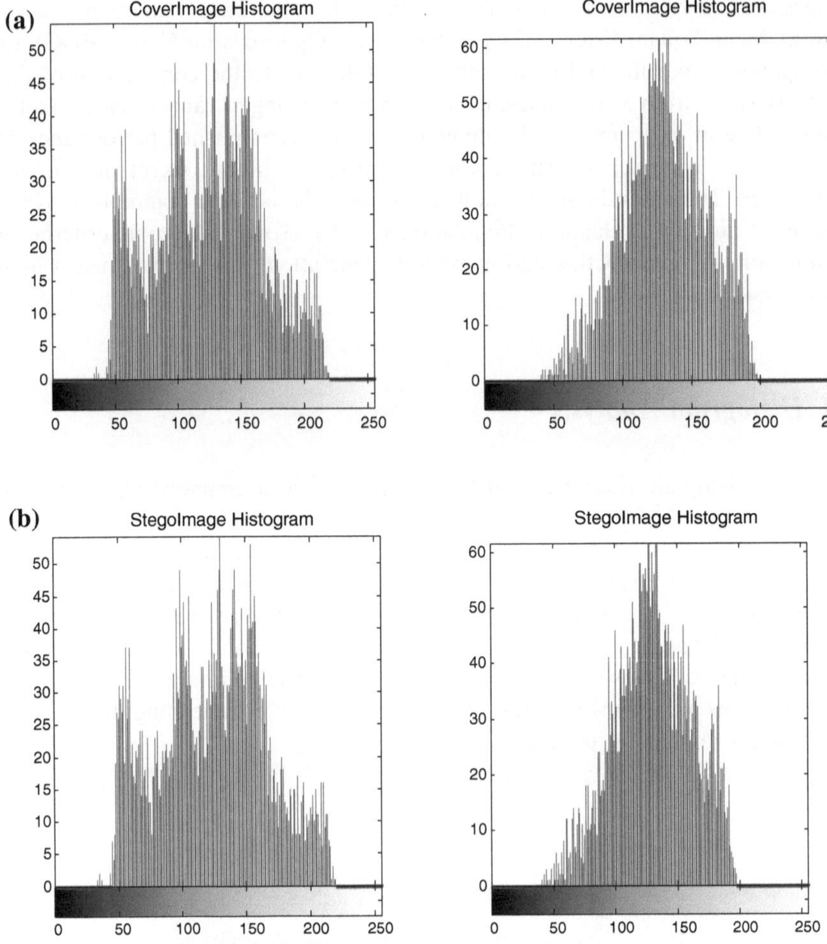

Fig. 3 **a** Histogram of Cover Images **b** Histogram of Stego Images

4.2 Optimization of Threshold Value

To evaluate the performance of the proposed selection methods, the experiment analysis of selection techniques based on a comparison of their respective performance estimated as the number of function evaluations. The raw data obtained from the optimization comprised the average, mean and best of 100 generations for each function of the set of functions under consecration. The parameters are used in experiments, The Population size is 100, Selection Methods is an Roulette Wheel selection, Simple Arithmetic Crossover with the probability of 0.08 per generations is used for crossover and the Mutation is the Multi Non Uniform with Probability 5 %, criteria for termination is defined as the executions stops on reaching the number of generations, then the Fitness function is an Objective Value.

4.3 ROC Curve Analysis

For comparing the embedding security of proposed method to that of other methods, Receive Operating Characteristic (ROC) curves have been used, to embed image databases with the minimum value of the maximum capacity of the LSBMR-GA method and the LSBMR method.

The Receiver Operating Characteristics (ROC) curves are shown in Fig. 4. It can be clearly observed that both specific steganalytic algorithms would fail in detecting the proposed method even when the embedding rate is as high as 75 %, while they obtain satisfactory results for detecting stego image using LSBMR and LSBMR-GA methods. As observed from figure the detection accuracy, shown as the area under the ROC curve, is lower for our proposed algorithm as compared to the LSBMR algorithms.

4.4 Comparing Capacity

Figure 5 shows the results of computing capacity for our proposed algorithm and the LSBMR method in 80 images. It is clear that the capacity of the proposed method is higher in most images. The mean capacity of the proposed method is about 1,024 bits higher than the mean capacity given by the LSBMR method.

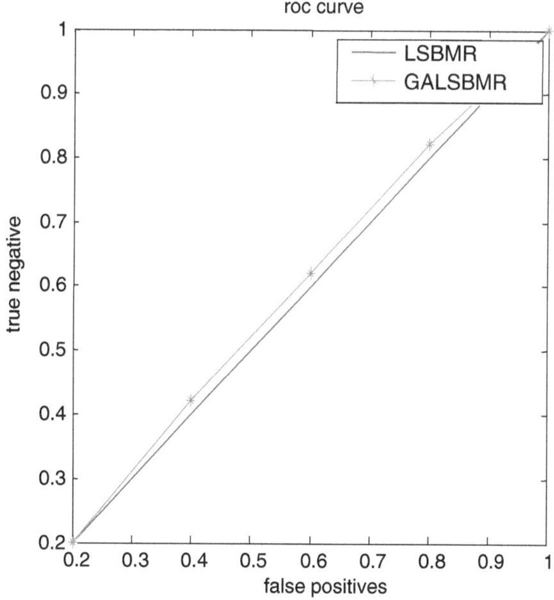

Fig. 4 ROC curve (embedding)

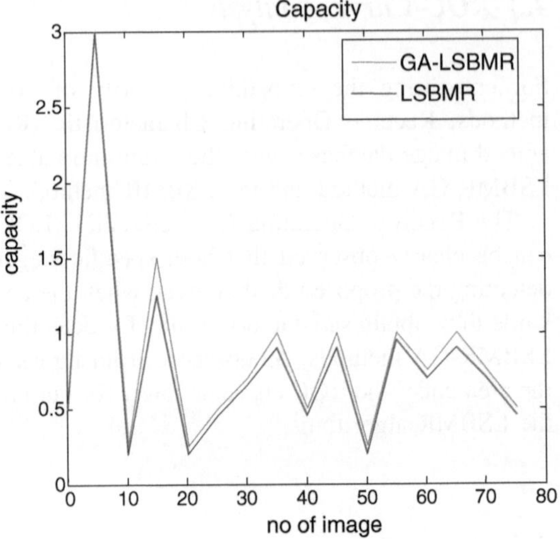

Fig. 5 Embedding Capacities of the proposed method (*Red*) and the LSBMR method (*Blue*)

The performances of the methods have been evaluated and compared on the basics of two measures are Mean Square Error (MSE) and Peak Signal to Noise Ratio (PSNR) (Table 1). The Eqs. 4 and 5 of these two measures are

$$MSE = \left(\frac{1}{m x n}\right) \sum_{i=1}^{m-1} \sum_{j=1}^{n-1} (I_{ij} - I'_{ij})^{2} \quad (4)$$

$$PSNR = 20 \log_{10} \left(\sum_{i=1}^{M} \sum_{j=1}^{N} (S_{ij})^2 \Big/ MSE \right). \quad (5)$$

Table 1 Results of PSNRS and MSE values for the proposed algorithm

Image	PSNR	MSE
Lena	48.55	0.00082
Baboon	48.63	0.00079
Foveon	47.89	0.00081
Deer	47.91	0.00089
Lichtenstein	48.92	0.00082

5 Conclusion

Image Steganography systems scale up and become increasingly complex, their optimization techniques face new challenges. Conventional image steganography methods become fragile and show poor performance. Lot of optimization techniques has long been proposed to improve the hiding capacity of the stego images. In this paper, we design on this idea of the optimizing the region using Genetic

Algorithm which is a conventional bio-inspired optimization technique used in engineering problems, by which regions in which the message is to be embedded on the cover image. This paper designs GA-LSBMR systems, with detailed design that meets all requirements. The results of our GA-LSBMR evaluation suggest that when the exact regions with correct threshold value are selected, the method offer high embedding capacity. Decoding reliability is improved with the increase in number of iterations, when set of parameter values are fixed. The algorithm is proven to be secured against stego test based on high order statistics. We will extend our work to consider the other steganography methods such as audio/video steganography in the spatial or frequency domains when the embedding rate is less than the maximal amount.

References

1. Yang CH, Weng CY, Wang SJ et al (2008) Adaptive data hiding in edge areas of images with spatial lsb domain systems. IEEE Trans Inf Forensics Secur 3(3):488–497
2. Dumitrscu S, Wu X, Wang Z (2003) Detection of LSB steganography via Sample pair analysis. IEEE Trans Sig Process. 51(7):1995–2007
3. Zhang T, Ping X (2003) A new approach to reliable detection of lsb steganography in natural image. Sig Process J 83: 2085–2093
4. Wang R-Z, Lin C-F, Lin J-C (2001) Image hiding by optimal LSB substitution and genetic algorithm. Pattern Recognition Society, Published by Elsevier Science Ltd. pp.671–683
5. Anderson RJ (1996) Stretching the limits of steganography. In: Proceeding first international workshop information hiding (IH'96), 39–48 1996
6. Ramezani M, Shahrokh (2010) Adaptive image steganoraphy with mod-4 embedding using image contrast. IEEE CCNC 2010 Proceedings
7. Luo W, Hunang F, Huang J (2010) edge adaptive image steganography based on LSB matching revisited. IEEE Trans Inf Forensics Secur 5(2)
8. Information hiding techniques for steganography and digital watermarking. S. Katzenbeizzer and F.A.Petticolas, eds. Artech House, 2000
9. Luo W, Huang F, Huang J (2010) Edge adaptive image steganography based on LSB matching revisited. IEEE Trans Inf Forensics Secur 5(2):201–214
10. Ramezani M, Ghaemmaghami S (2010) Towards Genetic Feature Selection in Image Steganalysis. In: 6th IEEE international workshop on digital rights management, Las Vegas, USA
11. Mielkainen (2006) LSB matching revisited. IEEE Sig process Lett 13: 285–287
12. Wang RZ, Lin CF (2001) Image hiding by optimal LSB substitution and genetic algorithm, pattern recognition, Vol. 34. ELSEVIER pp 671–883
13. Cachin C (1998) An information-theoretic model for steganography. In: Proceedings second in int'l workshop information hiding (IH'98), 306–318 1998

SMS Based HPCCDD Algorithm for the Identification of Leaf Spot Diseases

P. Revathi and M. Hemalatha

Abstract Advance computing techniques Capitalized as Mobile base identification of cotton diseases and control of pest, using Automatic system for rural area farmers. This technology base farm management system is Capitalized to identify and analyze the diseases, profitability, sustainability and safety for the land resource. The sharing of plant physical condition knowledge on a regional basis can support both crop manufacture and trade. In this model of work, a new computing technology has been proposed to help the farmer to take a superior decision concerning several aspects of the crop manufacturing process. Suitable evaluation and diagnosis of crop diseases in the field is very critical for the increased production. Foliar is the most important fungal disease of cotton and occurs in all growing Indian cotton regions. In this work we express technological strategies using mobile captured symptoms of cotton leaf spot images and classify the diseases using neural network. This system can identify diseases and provide pest recommendation to farmers.

Keywords HPCCDD algorithm · Cotton leaves spot diseases · Mobile camera · Capture · Image process

P. Revathi (✉) · M. Hemalatha
Department of Software System, KarpagamUniversity, Coimbator, India
e-mail: Kvrevathi1@gmail.com

M. Hemalatha
e-mail: hemabioinf@gmail.com

Mohan S. and S. S. Kumar (eds.), *Proceedings of the Fourth International Conference on Signal and Image Processing 2012 (ICSIP 2012)*, Lecture Notes in Electrical Engineering 221, DOI: 10.1007/978-81-322-0997-3_5,

1 Introduction

India is an agricultural country; wherein about seventy percentages of the population depend on agriculture. Farmers have wide range of diversity to select suitable Fruit and Vegetable crops. However, the cultivation of these crops for optimum yield and quality product is highly technical. It can be amended through the serve of technological sustain. The management of perennial fruit crops requires close monitoring especially for the management of diseases that can affect production significantly and subsequently the post-harvest life.

Cotton,"The White Gold" or the "Emperor of Fibers" enjoys a pre-eminent status among all cash crops in the country and is the principal raw material for flourishing textile industry. It provides livelihood to about sixty million people and is an important agricultural commodity providing remunerative income to millions of farmers both in developed and developing countries. In India, in spite of severe competition from synthetic fibers it is occupying the premier position with a seventy percentage contribution to the textile industry.

Cotton (Gossypium spp.) is a crop of warm climate and requires a regular contribution of water either naturally in the form of rainfall or assured through canals from the above exterior and/or from underground sources. Although cotton is not a water loving plant, it requires a regular supply of water for maintaining growth and balance between vegetative and reproductive phase.

1.1 The Image Analysis in Agricultural

The image processing techniques are extensively applied to agricultural science, and it has great perspective especially in the plant protection field, which ultimately leads to crop management.

Image analysis can be applied for the following purposes:

- To recognize diseased leaves, stem, fruit.
- To quantify affected area by disease.
- To find the boundaries of the affected area.
- To find out the color of affected areas.
- To find out the size and shape of fruits.

This work implements the automatic system to identify the diseases using impure images of various cotton leaf spots. Images are captured by digital camera mobile and processed using image emerges, image segmentation techniques to identify impure parts of the plants. Then the impure part of the leaf spot has been used for the categorization purpose we proposed system HPCCDD approach. The methods evolved in this system are both image processing and computing techniques.

1.2 Varieties of Cotton Leaf Spot Diseases

The diseases on the cotton leaves are classified as:

- Fusarium wilt
- Verticillium wilt
- Root rot
- Grey mildew
- Leaf blight
- Bacterial blight
- Leaf curl.

Symptoms of Cotton Diseases

Grey mildew (Areolate Mildew/Dahiya) - Ramularia areola

This disease [1] initially appears on older leaves as the plants reach maturity, in the form of irregularly angular, pale translucent spots, 1–10 mm (usually 3–4 mm) in diameter and with a definite and irregular margin formed by the veins of the leaf (called areolate). The lesions are light to yellowish green on the upper surface. As the spots grow older, the leaf tissues turn yellowish brown while a whitish frosty growth appears chiefly on the under surface but occasionally also on the upper surface. This is the conidial stage of the causal fungus. Lesions occur on the bracts subtending the bolls. As the leaf becomes chlorotic, the lesion turns reddish brown and defoliation takes place. Early and severe defoliation leads to premature boll opening and immature lint.

Bacterial blight (Xanthomonas axonopodis pv. Malvacearum)

Dark green, water soaked, angular lesions of 1–5 mm across the leaves and bracts, especially on the undersurface of leaves. Hence called angular leaf spot. Sometimes extensive dark green, water soaked lesions along the veins known as vein blight. Symptoms are usually more prevalent on lower leaves than on upper leaves. Lesions dry and darken with age and leaves may be shed prematurely resulting in extensive defoliation. Black lesions on the stem which girdle and spread along the stem or branch known as black arm. Dark green, water soaked, greasy, circular lesions of 2–10 mm across the bolls, especially at the base of the boll under the calyx crown. As the boll matures the lesions dry out and prevent normal boll opening. This phase of symptom is called as “Boll rot”.

Leaf Curl virus disease–Gemini virus

The initial symptom is characteristic Small Vein Thickening (SVT) on young upper leaves of plants. Later, upward curling of leaves occurs due to the uneven growth of veinal tissues on the abaxial side of the leaves. Subsequently, formation of the cup shaped or leaf laminar outgrowth called enations appear on the underside of the leaf. In severe cases and in plants affected at an early age, reduction of inter-nodal length leading to stunting and reduced flowering/fruiting is observed.

Alternaria leaf spot–Alternaria macrospora, A. alternate

Alternaria macrospora causes brown, gray brown or tan lesions 3–10 mm in diameter, especially on lower leaves. Sometimes with dark or purple margins and with concentric zones… Affected leaves develop an abscission layer, senses and drop to the ground. Circular dry brown lesions up to 10 mm across may also be seen on the bolls. A. alternata causes usually purple specks or small lesions with purple margins on leaves and bolls.

1.3 Objective of this Work

This Research work presents an effort of using intelligent computing technology (ICT) to overcome the information gap by:

- Creating a sophisticated agricultural environment to support the farmers to easily identify the diseases and get control of it.
- Attempting to automate disease identification process using Advance Color image processing.

This work contributes to the agriculture recommendation center has an information repository of the individual farms collected using a mobile phone for data and image capture. Continuous updating of data is carried out, and the interactions with recommendation center advisory are captured. Farmers can upload additional information and pictures of the crop using their mobile phones.The extension worker handles the calls from farmers and conferences in experts as and when necessary.

2 Literature Review

Existing papers are describing to diagnosis the cotton leaves using various approaches suggesting the various implementation ways as illustrated and discussed below [2]. Cotton Diseases Control has been developed in a BP neural network as a decision-making system [3]. Earlier paper the fuzzy feature selection approach fuzzy curves (FC) and surfaces (FS)–is proposed to select features of cotton disease leaves the image [4]. The images are subsequently classified using two indigenous techniques RPM and Dis Bin and compared with the classical PCA based technique [5]. Cotton foliar diseases presented a method for automatic classification of cotton diseases used Wavelet transform energy has been used for feature extraction while Support Vector Machine has been used for classification [6]. Existing the research work described in the features that could be extracted using self organizing feature map together with a back-propagation neural network is used to recognize the color of the image [7]. The cotton leaf disease segmentation is performed using modified self organizing feature map with genetic algorithms for optimization and support vector machines for classification. Finally, the resulting segmented image is filtered by Gabor wavelet which allows the system to analyze leaf disease color features more efficiently.

2.1 Analysis the Problem

Presently, in the recent agricultural system, various computation techniques have been developed to help farmers (or) agricultures to monitor the proper development of their crops. In our early agricultural system, during the harvesting process of the crops, the exposed eye observation of farmers or experts is the main approach adopted in practice for recognition and identification of crop diseases,under microscopic conditions in the laboratory. However, this requires continuous monitoring of experts which might be prohibitively expensive in large farms. Further, in some developing countries, farmers may have to go long distances to contact experts, this makes consulting experts too expensive and time consuming. The basic problems regarding with crop is on the field, a fast and accurate recognition and classification of the diseases is required by inspecting the impure leaf spot images also identify the severity of the diseases. There are two major characteristics of plant-disease detection machine-learning methods that must be achieved, they are performance and accuracy.

Proposed Research work will describe the process of automatic recognition of leaf spot diseases as this can give much assistance in monitoring huge fields of crops and detect the symptoms of diseases. In this work we have to find out the computer systems which analyze the input images using one of the (image processing Techniques) the homogenize techniques to detect the affected parts of the leaf edges and using RGB pixel counting feature extraction to recognize the diseases and then result (recognition of the diseases and pest recommended) is given as output to the farmers.

3 Proposed Method

The digital images are acquired from the situation using a digital mobile camera. Then image-processing techniques are applied to the acquired images to extract useful features that are necessary for further analysis. After that, several analytical discriminating techniques are used to classify the images according to the specific problem at hand. In this work farmers can take to the decision immediately at the time they want to get the best solution to diseases and pest recommendation is three languages Tamil, English, Hindi, Production can be improved, and the yield loss can be reduced, they minimum cost of ultimate system very useful to farmers and we can increase the economic status of the country.

Proposed Homogeneous Pixel Counting Algorithm for Cotton Diseases Detection (HPCCDD)

Step 1: RGB image acquires
Step 2: Create the Color transformation construction
Step 3: Convert the Color values in RGB to the space specified in the Color transformation model
Step 4: Apply Color Filter
Step 5: Masking green-pixels (damaged part of the leaf)
Step 6: Eliminate the masked cells inside the boundaries of the infected clusters
Step 7: Find Edge detection using (Canny with Sobel and homogeneous Techniques Proposed Algorithm HSBED)
Step 8: Calling the Pixel Ranging method to calculate the RGB features (every disease dissimilar)
Step 9: Texture Statistics Analysize (using the Proposed Algorithm SAPR)
Step 10: Configuring Disease Reorganization and Pest Recommendation.

Figure 1, Users acquire images of the leaves from the field using sensors and pass it to the computer system which analyze the input images using the homogenous edge detection algorithms and diseases wise the pixels Range function logic. It can also detect true the affected parts of the leaves to recognize the diseases the result (recognition of the diseases and pest recommended) is given as output to the farmers.

In this paper, Proposed HPCCDD a new algorithm has been proposed in the automatic detection of cotton leaf diseases. The neural network model has been chosen for the classification as it is a successful classifier for many real applications hence we propose a new model which reduces the number of hidden layers. The training and validation processes are among the important steps in developing an accurate process model. The data set for training and validation processes consists of two parts; the training feature set which are used to train the new model using an RGB range of values; whilst a testing feature set is used to verify the accuracy of the trained model. In the training phase, connection weights

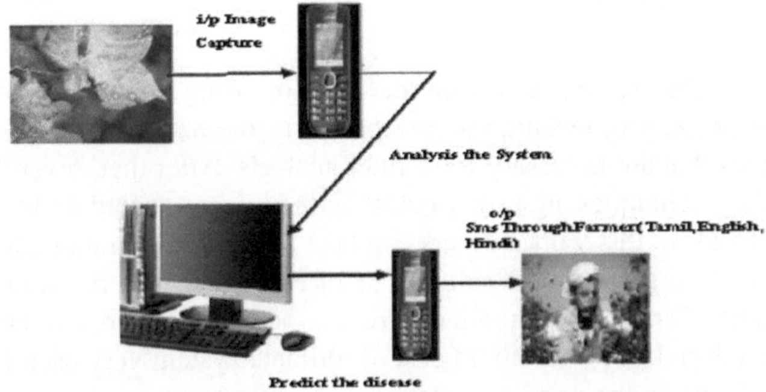

Fig. 1 Proposed system flow digram

which correspond to threshold values were always updated until they reached the defined iteration number (rows) or acceptable error. Before the data can be fed to the model, the proper network design has been set up. However, this phase was carried out simultaneously with the network training phase, in which the network was trained with RGB range values,disease name, symptoms and precautions. Finally the output of this model is given in terms of disease name, symptoms and precautions.

4 Experimental Results

Take the Image a leaf as an input image and convert into a grayscale image, Next using Canny and Sobel Edge detection techniques and with the use of homogenous operator the images are combined and through this we obtain the image clarity for the edges, and we get centered pixel of opposite neighboring one, two, three pixels of white lightning. This lightning identified is denoted as the diseases affected is represented in the edges of leaf boundary. The dissimilarity edge detector can take the dissimilarity of opposite pixels in instead of a 3×3 area.

The Homogeneity-based edge detector takes the result of any edge detector and divides it by the average range of the part. This division removes the effect of not level lighting in the image. The average range of an area is available by convolving the part with a mask containing all ones and separating by the size of the area.

First the farmer captures the cotton leaf spot disease symptom image is sent to the mobile camera, Then send to the server, Next the server already stored symptoms of diseases Image is there, then the system automatically analysis through the algorithms and train and test the using a neural network based approach classifier and recognition the image path and achieving an output result. Image and due to what diseases, what symptoms and recommendation to the pest details assist to the farmer's precaution the disease awareness. Finally send Sms to the farmers in three languages like Tamil, English, and Hindi.

Figure 2, show the first normal image, next RGB image converts into the grayscale image, then using edge detection techniques to identify the leaf spot image disease affected part. Then using color filter checking RGB pixel counting features used to identify the disease part. And recognition image part and achieving an output result. Image and due to predict the disease and recognition the symptom and recommended to the pest details help to the farmer's (Precaution the disease awareness).

5 Benefits of Society

- Diseases and pest recommendation to three language wise Tamil, English, and Hindi.
- Production can be improved and yield loss can be reduced.

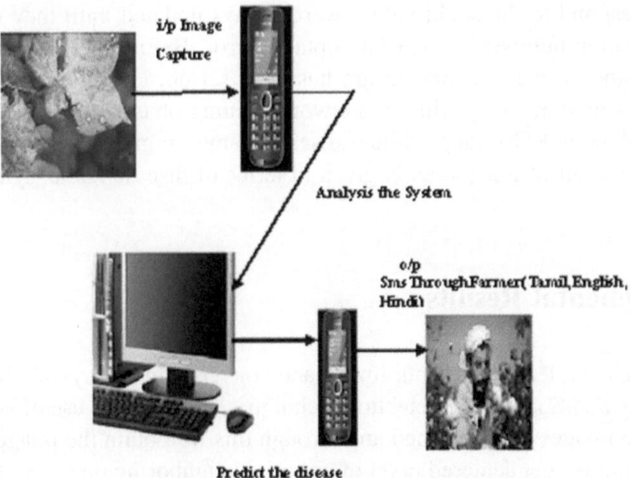

Fig. 2 Report of farmer and diseases description

- Software is minimum cost.
- Ultimate system very useful to farmers and we can increase the economic status of the country.
- Time consuming to agriculture Experts.
- They didn't go and visit long place.

6 Conclusion

The distributing of plant health knowledge on a regional basis can sustain both crop manufacture and trade. In this model of work, a new computing technology has been proposed to assist the farmer to take superior decision about many aspects of the crop manufacturing process. Manufacture can be amended and yield loss can be reduced. This Software is minimum cost an ultimate system, very useful to farmers and we can increase the economy of the country. And life security to the farmers and reduces their burden.

7 Acknowledgments

I would like to express my special thanks of gratitude to S.Nakkeeran Savage Perumal, Associate Professor (Pathology) Department of Plant Pathology of Tamil Nadu Agricultural University, and thankful to Karpagam University to do this Research work.

References

1. Chidambaram P (2007) Integrated disease management to reduce yield losses in quality cotton. pp 99–109
2. Li H, Ji R, Zhang J, Yuan X, Hu K, Qi L (2011) WEB-based intelligent diagnosis system for cotton diseases control. IFIP Adv Info Commun Technol 346:483–490
3. Zhang YC, Mao HP, Hu B, Xili M (2007) Features selection of cotton disease leaves image based on fuzzy feature selection techniques. In: Proceedings of IEEE, Beijing, China, pp 124–129
4. Hayat SA, Abdullah A, Muhammad, Chaudary A, Malik YJ, Gillani W "Automatic Cleansing and classification on Cotton leaves, bolls and flowers using CMYK color splitting"
5. Bernardes AA, Rogeri JG, Marranghello N, Pereira AS, Araujo AF, João Manuel RS (2011) Tavares. Identification of Foliar Diseases in Cotton Crop. SP, Brazil
6. Gulhane VA, Gurjar AA (2011) Detection of diseases on cotton leaves and its possible diagnosis. J Image Process(IJIP) 5(5):591–598
7. Meunkaewjinda A, Kumsawat P, Attakitmongcol K, Sirikaew A (2008) Grape leaf disease detection n from color imaginary using hybrid intelligent system. In: Procccdings of ECTI-CON

QPAC: A Novel Document Image Compression Technique Based on ANFIS for Calibrated Quality Preservation

Apurba Das and Arathi Issac

Abstract In the present paper, a novel compression technique is used to compress the document images without compromising abrupt quality degradation. An Adaptive neuro-fuzzy inference system (ANFIS) based classification scheme is used for segmenting the input document followed by some neighborhood smoothing is performed. A tuning based adaptive compression scheme is used for compressing the ANFIS classified data. A 3D trade-off between quality, compression ratio and relative occupancy of image region is addressed in a calibrated manner.

Keywords Adaptive neuro-fuzzy inference system · Segmentation · Classification · Decision tree · Compression · PSNR

1 Introduction

Document image compression is very much essential for effective communication, print, decompose and store the data. One document essentially contains text and image class of objects. Lossy compression degrades the quality of text blocks significantly. In the present paper a novel adaptive compression scheme utilizing a method of segmentation based on ANFIS system is proposed to achieve quality preserving compression. SompornChuai-aree et al. [1] introduced a text and image

A. Das (✉) · A. Issac
Imaging Tech Lab, HCL Technologies Ltd, Chennai, India
e-mail: apurba.das@hcl.com

A. Issac
e-mail: arathiissac@gmail.com

Mohan S. and S. S. Kumar (eds.), *Proceedings of the Fourth International Conference on Signal and Image Processing 2012 (ICSIP 2012)*, Lecture Notes in Electrical Engineering 221, DOI: 10.1007/978-81-322-0997-3_6,

classification scheme using Fuzzy c-means algorithm which uses 3 classes- text, image and background. Other researchers in the domain of document image processing [1–3] also proposed some the algorithms which classify the document image to 3 classes- text, image and background. Djvu compression [4] uses a wavelet based compression technique and JBIG2 to give a very high compression rate but do not preserve the text quality and hence not suitable for compressing document image, especially for printer related applications [5].

The current work proposes segmentation before compression of the document image into three classes- text, image and white space or background to achieve quality preserving compression. As the problem itself is trying to address a classification between merely overlapping classes, the strength of Fuzzy based classification is entrusted. The algorithm uses Adaptive Neuro-Fuzzy Inference System method of classification to segment the data. After segmentation, each of the class uses different compression schemes based on the properties of the classes. The text regions need to be compressed with lossless compression as lossy compression may change the shape of a letter by blurring the edges or corners which may be misleading for the next stage of image processing or optical character recognition. Image blocks are compressed using lossy compression techniques like jpeg. As jpeg itself uses 8×8 blocks to compute DCT, it is logical enough to segregate the entire document in several 8×8 blocks and perform segmentation between them. We have also attached a constrained of continuity by 8×8 neighborhood smoothening. The current work used a block based approach and hence less time consuming compared to the previous works described in [1, 2] which are essentially pixel based approaches.

In Sect. 2 we have discussed about the feature extraction in detail, next in Sect. 3,feature reduction by the Decision tree based feature hierarchy is explained briefly. Section 4 describes the ANFIS based segmentation of the document image. The segmented blocks are dealt adaptively in Sect. 5 to achieve quality preserving compression. Finally we have concluded our findings in Sect. 6.

2 Feature Extraction

Features are said to be appropriate when the ratio of inter-class to intra-class distance between the member blocks is very high after classification. Here we are interested to find suitable features which can play significant role segregating text and image blocks. SompornChuai-areeet al [2], Apurba et al. [6] and Lin et al. [3] used color statistics and texture based features for this purpose, respectively. The ratio of high frequency to low frequency ratio is also a distinct feature that can be used to achieve better segmentation results [6]. The primarily selected features for the current work are color variance, edge strength and high frequency (HF) to low frequency (LR) ratio and are explained in the coming sections.

The proposed algorithm extracts features from an image block rather than from an image pixel to minimize the computational cost. As printer uses subtractive color space (CMYK gamut), the segmentation is done on these 4-seperation image. It can be easily migrated to RGB based compression also, wherever needed. The CMYK image is then divided to blocks of size 8 × 8 which in quite in agreement with the block size used in JPEG compression. Eventually we are using DCT based compression (JPEG) for image region.

2.1 Color Variance

Color variance determines the measure of intensity spread with respect to mean intensity in each separation. The text region usually contains less number of colors in a document, and hence the variance is low. Image regions are composed of a number of colors and hence the color variance of these regions is high. In printing, texts are generally printed by contone (continuous tone) and images by halftones. The concept of half toning inspired us to choose the color variation as suitable feature. The white space regions (background) have a zero color variance. The color variance is calculated from all the 4 color channels separately and from each 8 × 8 non-overlapping image block. The obtained feature values are shown in Table 1.

2.2 Edge Strength

Here edge strength signifies the average intensity gradient at each block. The text region showed higher edge strength as compared to the image region. The edge strength is also calculated in the four separations. For computing edge strength, the gradient in the horizontal and vertical direction is calculated by applying a sobel mask over all the blocks in the document (Table 1).

Table 1 Extracted Features for ANFIS based classification of document blocks

Color variance				Edge strength				HF/LF	Class
C	M	Y	K	C	M	Y	K		
1	0.81	0.68	0.68	31894.40	43513.60	39278.81	95612.4	80.1	Img
1	1	1	1	34165.71	26448.83	70643.43	68674.1	1203	Img
0.93	1	0.78	0.90	40101.25	25139.14	79265.49	61729.8	225	Img
1	0.90	1	0.68	42496.28	68943.62	59810.62	122252.	576	Img
0.96	1	0.93	0.90	63787.06	23714.25	78959.19	68868.9	889	Img
0.62	0.71	0.21	0.53	106161.1	264441.0	502622.9	270948	7.29	Txt
0.59	0.46	0.59	0.65	158233.0	325491.6	482907.7	341368	16.1	Txt
0.71	0.84	0.18	0.75	178800.4	248388.4	505261.9	3253994	17.8	Txt
0.53	0.65	0.71	0.37	323508.7	332715.3	313657.7	424304	5.94	Txt
0.68	0.65	0.31	0.31	344368.0	311248.9	267760.6	4628990	15.7	Txt

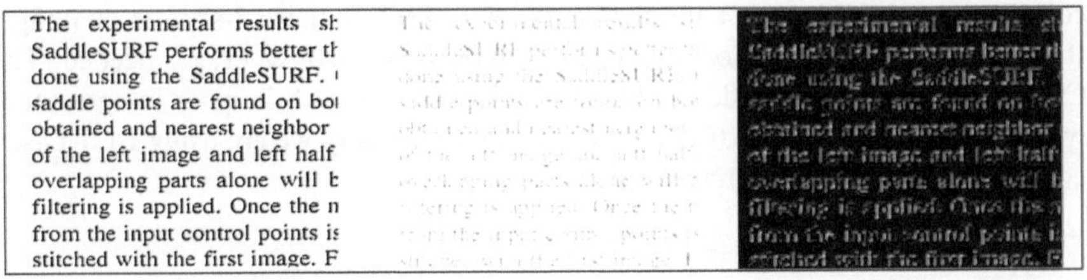

The experimental results sh
SaddleSURF performs better th
done using the SaddleSURF.
saddle points are found on bo
obtained and nearest neighbor
of the left image and left half
overlapping parts alone will b
filtering is applied. Once the n
from the input control points is
stitched with the first image. F

Fig. 1 Extraction of low frequency and high frequency components from a text document

2.3 High Frequency to Low Frequency Ratio

It has been found out that [6, 7]) the ratio of high frequency to low frequency components is giving promising results. The ratio of high frequency to low frequency is computed over the blocks of equivalent gray image (Table 1).

By applying suitable high pass and low pass filters, the high frequency and low frequency components are separated [6]. Figure 1 shows the low frequency and high frequency components extracted from an image.

Consequently, a nine dimensional feature vector is formed by the 4-seperation color variance and edge strength with 1-seperation high to low frequency ratio, i.e. $(4 \times 2 + 1 = 9)$.

3 Reduced Feature Selection

A decision tree is a tree like structure where each leaf node represents the classes, the internal nodes represent a test on the attribute or the feature value and the branches represent the attribute value. A decision tree tells us about the percent of usage of the particular feature in classification. Such features that have higher usage can be only used for classification and thereby reduce the feature size.

In the current problem, the initial feature space comprised of 9 features which are denoted by $A_0, A_1, \ldots\ldots, A_8$. The decision tree program C5, developed by Quinlan [8]was used for the classification of the feature points. The time required for training was very less. By using a decision tree it is found out that only 3 features contributes to the segmentation process. Even if the feature space is reduced from nine to three, it gives the same output and reduces the complexity in calculating the nine features. The features are high frequency to low frequency ratio,Edge Strength of Cyan and Color variance of Black. The attribute usage for high frequency to low frequencyratio, edge strength and color variance are 90, 72 and 48 % respectively. The decision tree is constructed for a three class problem for segmenting the data points into text, image and white block. The leaf nodes are class Image (C1), Text (C2) and White region (C3). The training data contains 175 training data points which are classified to three classes. The testing was done over

Table 2 Confusion matrix obtained before and after smoothening on ANFIS classified data

Before neighborhood smoothening				After neighborhood smoothening			
	C1	C2	C2		C1	C2	C3
C1	33	5	0	C1	35	3	0
C2	0	52	3	C2	0	57	0
C3	0	8	22	C3	0	1	29

125 test data points. The error rate was 6.4 %. The confusion matrix for a three class problem is shown in Table 2.

4 ANFIS Based Segmentation

Neuro-fuzzy systems harness the power of the two paradigms: Fuzzy logic and artificial neural network (ANN) by utilizing the mathematical properties of ANN in tuning rule-based fuzzy systems that approximate the way humans' process information. A specific approach in neuro-fuzzy development is the adaptive neuro-fuzzy inference system (ANFIS), which has shown significant results in modeling nonlinear functions [9, 10]. The ANFIS learns features in the data set and adjusts the system parameters according to a given error criterion [6]. In the present problem also, it hasbeen observed that the relation between the three selected features with the output class levels are nonlinear as shown Fig. 2.

4.1 Architecture of ANFIS

The ANFIS is a fuzzy Sugeno model put in the framework of adaptive systems to facilitate learning and adaptation. Such framework makes the ANFIS modeling more systematic and less reliant on expert knowledge. To present the ANFIS

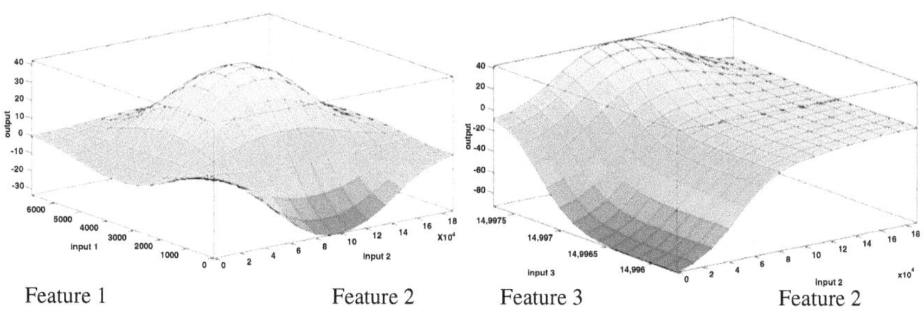

Fig. 2 Decision surface for three chosen features

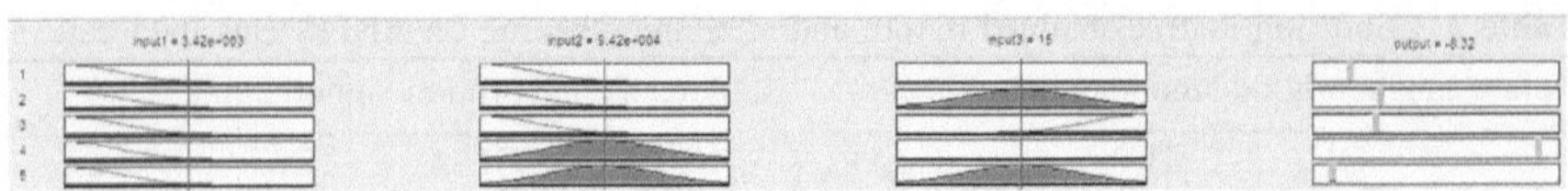

Fig. 3 List of fuzzy rules (5 samples from 27) for three chosen surfaces

architecture, 27 fuzzy if–then rules based on a first order Sugeno model are considered [6]:

Rule 1 If (in1 is in1mf1) and (in2 is in2mf1) and (in3 is in3mf1) then (output is out1mf1)

Rule 2 If (in1 is in1mf1) and (in2 is in2mf1) and (in3 is in3mf2) then (output is out1mf2)

Rule 27 If (in1 is in1mf3) and (in2 is in2mf3) and (in3 is in3mf3) then (output isout1mf27)

The number of rules is formed by taking all possible combinations of the number of features. So, the number of membership function here can be given by ${}^{3}C_{1} \times {}^{3}C_{1} \times {}^{3}C_{1} = 27$ which is shown in Fig 3. The algorithm used a hybrid training method which combines the least square and back propagation gradient descent method for training. The ANFIS structure for the current work which consists of the 27 aforesaid rules is shown in Fig 4.

The test data is assigned to some arbitrary class and was given to ANFIS for training. Some misclassifications are found in the classification which has been removed successfully by the neighborhood smoothening.

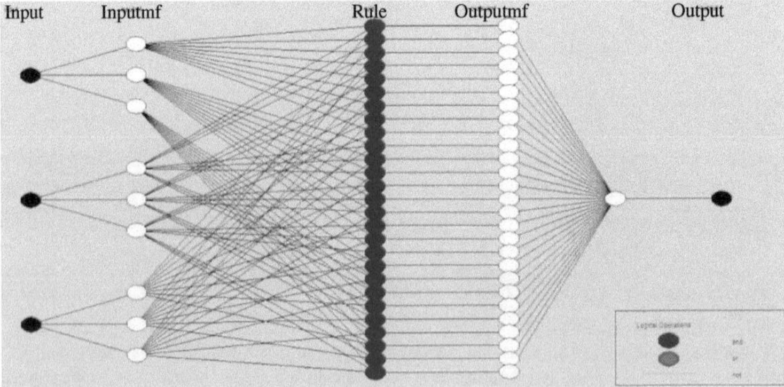

Fig. 4 ANFIS model structure

4.2 Neighborhood Smoothening

The ANFIS classification gave three classes named text, image and white regions as the output. The blocks under the classes may not be continuous blocks. So it is tedious to compress these blocks independently which will also reduce the compression rate. So, 8 neighborhood smoothening operation is performed over the blocks to refine the segmented output. For each block the 8 neighborhood is considered and the percentage of blocks in the neighborhood belonging to the same class is found. If this count is greater than 75 %, then the block remains in the same class. Otherwise the block is moved to the class where the higher percentage of its neighbors belongs to. After the neighborhood smoothening operation the blocks will be merged to get a continuous region. The confusion matrix for the ANFIS classification after employing neighborhood smoothening is shown in the Table 2.

5 Adaptive Compression

Most of the recently published algorithms [4, 11, 12], compromise quality to achieve high compression ratio. The present work uses adaptive compression schemes with respect to the class level of the segregated document to achieve a good compression ratio maintaining the quality of the document. A lossless compression method,Run length Encoding (RLE) is used for compressing text blocks and white blocks. The image region is compressed by lossy Jpeg compression algorithm.

5.1 Tuning in Compression

Adaptive compression addresses the text and image blocks in a different manner [5]. It is therefore obvious that, if the relative occupancy of image region in the document is higher, we can achieve better compression ratio without much quality degradation. The first run of ANFIS classifies the text and image regions. Hence, the engine decides the relative occupancy of image region and thereafter the compression ratio. In classical systems [4, 13], to maximize the compression ratio, some non-image blocks are re-grouped to image class, abruptly. In this proposed system the tuning in compression is performed depending on the fuzzy membership values of each blocks belonging to the non-image class. Non-image blocks having relatively lower membership values are chosen primarily as target blocks for regrouping. Next the continuity with existing image blocks are checked by neighborhood smoothening and hence the blocks to be regrouped are finalized. Since these identified blocks have the properties similar to that of the image

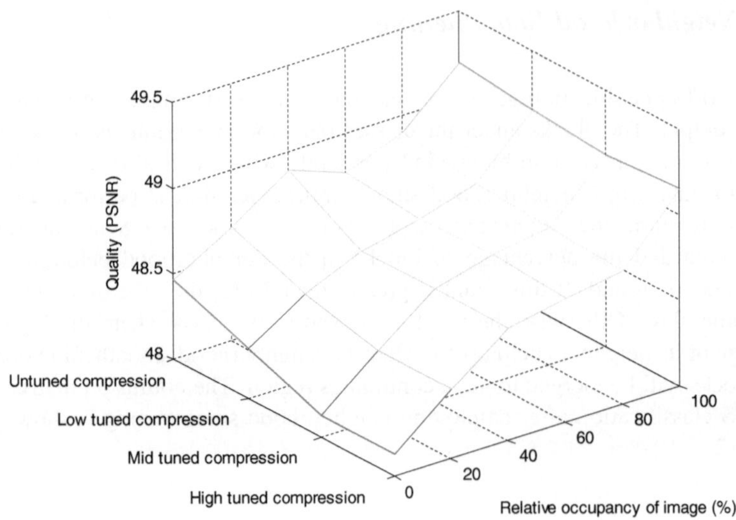

Fig. 5 3D dependency between compression parameters showing smooth slope

blocks, compressing these with jpeg will not cause much degradation in quality of the document.

The proposed scheme of adaptive compression clearly depicts a three dimensional (3D) relationship between the three entities namely quality, compression ratio and relative occupancy of images, in a calibrated manner. The first run compression by ANFIS is considered as un-tuned compression. The proposed engine ensures a moderately smooth relationship between these three significant parameters of compression, as shown in Fig. 5. We have chosen some documents with uniformly increased relative occupancy of image regions from 10 to 100 %, as shown in Table 3. We have tuned the compression ratio in three levels like low, moderate and high. Obviously, the absolute compression ratio of each level will not be unique, conversely, will be input document dependent.

A smooth dependency between the three parameters of adaptive compression is observed. If the entire document is an image, it is impossible to increase the compression ratio. The nature of the document image and its color distribution is responsible for the nominal variation in the quality, for 100 % occupancy of image region class. Generally the quality is decreasing with a smooth slope with respect to increase in compression. As the relative occupancy is increasing, the change in quality with respect to change in compression levels is decreasing, as expected. As the quality is preserved in a calibrated manner in the proposed adaptive compression scheme, it can be named as Quality Preserving Adaptive Compression (QPAC).

Table 3 Different images containing varying relative occupancy of image region

0 % image	20 % image	40 % image
60 % image	80 % image	100 % image

Table 4 Comparison with existing methods

Compression Alg.	Compression Ratio	Quality (PSNR)
Dj vu	22	23.47
LZW	12	27.29
Tiff	3.769	43.77
DCT based	4.93	42.96
QPAC	16	46.6

5.2 Comparison with the Existing Compression Schemes

The present scheme of adaptive compression is much concerned about the quality of the document image along with compression ratio.

The average compression ratio with respect to quality is tabulated in Table 4, to show the effectiveness of calibrated Quality Preserving Adaptive Compression (QPAC) compared to other algorithms. The parameters are calculated taking 20 document images of A4 size with different relative occupancy of image region.

6 Conclusion

The present work proposed a new calibrated Quality Preserving Adaptive Compression (QPAC), where the smoothness of quality preservation is ensured both in terms of compression ratio and relative occupancy of image region. An 'on-demand' tuning methodology for compression is also described utilizing the fuzzy membership values as scores. By applying all other different lossy and lossless compression technique in the image and non-image region we can further achieve higher compression ratio, obviously with an upper limit.

References

1. Chuai-aree S, Lursinsap C, Sophatsathit P, Siripant S (2001) Fuzzy c-mean: a statistical feature classification of text and image segmentation method. In: Proceedings of International Journal of Uncertainty, Fuzziness and Knowledge-Based Systems, ACM, vol 9, pp 661–671, Nov 2001
2. Shadkami P, Bonnier N (2010) Watershed Based Document Image Analysis. In: Proceedings of 12th International Conference, ACIVS, LNCS, vol 1, pp 114–124. Springer, Sydney, 13–16 Dec 2010
3. Lin M, Tapamo J, Ndovie B (2006) A texture-based method for document segmentationand classification. Jt Spec Issue Adv End User Data Min Tech 36:49–56
4. Bottou L, Haffner P, Howard P, Simard P, Bengio Y, LeCun Y (1998) high quality document image compression with DjVu. J Electron Imaging 7(3):410–425
5. Das A, Remya R (2012) A novel scheme of orientation and scale mapped RDC(OS-RDC) to improve compression in document images ensuring quality preservation. In: International Conference on Pattern Recognition, Tsukuba, Japan 2012
6. Das A, Parua S (2012) Psycho-visual evaluation of contrast enhancement algorithms by adaptive neuro-fuzzy inference system. Lect notes comput sci, springer 7143:75–83
7. Gonzalez RC, Woods RE (2002) Digital image processing, 2nd edn. Prentice Hall, USA
8. http://www.rulequest.com/index.html
9. Jang JSR (1992) Self learning fuzzy controllers basedon temporal back propagation. IEEE trans Neural Netw 3(5):714–723
10. Jang JSR (1993) ANFIS: Adaptive neuro-fuzzy inference system. IEEE Trans Syst Man Cybern 23(3):665–683
11. Koutchev R, Milanova M, Todorov V, Koutchev R (2006) Document image compression with IDP and adaptive RLE. In: Proceedings of 32nd Annual conference on IEEE Industrial Electronics, IEEE pp 2361–2366. Paris, 6–10 Nov 2006
12. Imura H, Tanaka Y (2009) Compression and string matching algorithm for printed document images. In: Proceedings of 10th International Conference on Document Analysis and recognition. pp 291–295, 26–29 July 2009
13. Wallace GK (1992) The JPEG still picture compression standard. IEEE Transa Consum Electron 38(1) (Feb)

A DCT Based Near Reversible Data Embedding Scheme for MPEG-4 Video

Sagar Gujjunoori and B. B. Amberker

Abstract The new category near reversible data embedding is emerging in the area of digital watermarking and steganography for providing security to the multimedia contents. Probably, the increased interest on this category of schemes is due to the limitations like more modifications, low embedding capacity of reversible schemes, apart from the applications of near reversible schemes. In this paper we propose a simple near reversible data embedding scheme which embeds the data during the process of MPEG-4 compression of digital video. We use the non-zero AC elements of quantized DCT coefficients in the middle frequency region of DCT blocks. We also give the formal definitions of the reversibility classifying the data embedding schemes. The experimental results not only shows the good performance in terms of capacity, visual quality, robustness but also the percentage of reversibility of host data is around 99.92 % for most of the test video sequences.

Keywords Near reversible · MPEG-4 · DCT · Embedding capacity · Visual quality · Non-zero AC coefficients.

The proliferation of high capacity digital recording devices and the rapid growth in using the Internet technology, though it is an indication of technological growth, is become a threat to the multimedia content owners. Since, this advancement made the illicit users to share and distribute the multimedia contents like image, audio, video, etc. easily and illegally. The Digital Watermarking and Steganography have become a backbone for the multimedia content owners for providing the security to their contents. In both the contests, the multimedia content is altered for providing the security. We refer the process of altering the multimedia content for providing the

S. Gujjunoori (✉) · B. B. Amberker
Department of Computer Science and Engineering, National Institute of Technology, Warangal, Andhra Pradesh, India
e-mail: gsr@nitw.ac.in

Mohan S. and S. S. Kumar (eds.), *Proceedings of the Fourth International Conference on Signal and Image Processing 2012 (ICSIP 2012)*, Lecture Notes in Electrical Engineering 221, DOI: 10.1007/978-81-322-0997-3_7,

security as the *data embedding* to capture both the Watermarking and Steganography contests. A data embedding scheme alters the cover contents such as image, audio, video, etc. for embedding the data. The data to be embedded can be secret information, identity of the content owner, information about the cover content, etc. depending up on the application for which the embedding scheme is designed.

In a generic Steganography (Watermarking) scheme, an *Embedder* $\mathbb{E}$ is used to embed the *data* $\mathbb{I}$ (e.g., either the secret message or the watermark) into a *Cover-object* $\mathbb{C}$. The *Detector* $\mathbb{D}$ is used to detect (extract) the data from the *stego-object* (watermarked object) $\mathbb{C}'$. Though, both the Steganography and Watermarking schemes alter the cover content for embedding the data $\mathbb{I}$, their primary goals are aimed at different purposes. The main purpose of a Steganography scheme is to embed the data $\mathbb{I}$ into the multimedia content $\mathbb{C}$ to obtain the new content $\mathbb{C}'$, practically indistinguishable from $\mathbb{C}$, by human beings, in such a way that the unauthorized person can not detect the existence of $\mathbb{I}$ in $\mathbb{C}'$. Whereas, the main purpose of a watermarking scheme is that the unauthorized person should not be able to remove or replace the $\mathbb{I}$ in $\mathbb{C}'$ [4].

The data embedding into the multimedia content introduces some modifications to the original cover content. Some schemes in the literature are not so specific to restore the original content after extracting the embedded data. Here the restoration of the cover content is not an important concern, rather their focus is merely on other requirements such as embedding capacity, visual quality, robustness, etc. These schemes are referred as *Irreversible* data embedding schemes. But there are some applications such as military communication, medical imaging, remote sensing, fine arts, multimedia archive management etc. which require the restoration of the original content after the extraction of the embedded data. The schemes which aim the above applications are referred as *Reversible* data embedding schemes [5, 9, 11].

Generally, achieving reversibility results in more modifications to the original content, and achieving the embedding capacity is a limiting factor. However, the reversibility for the above applications is a desired requirement. But, for some applications, it is not compulsory to achieve the reversibility. That is, for these applications certain level of modifications to the original content can be acceptable [2, 10]. Here the level of modifications allowed to original content is user defined. These schemes are called *near-reversible* or *near-lossless* or *semi-reversible* schemes. Recently, the interest in this category of data embedding schemes is increasing. Probably due to the limitations involved in reversible schemes such as more modifications, low embedding capacity. Moreover, the urgent need of near-reversible schemes is due to their applications, viz. copyright protection of remote sensing images [2].

Barni et al. [2]. have introduced the near-lossless paradigm for the first time. They showed that, by forcing a maximum absolute difference between the original and watermarked scene, the near-lossless paradigm makes it possible to decrease the effect of watermarking on remote sensing applications to be carried out on the images. Ahmed and Moskowitz [1] proposed a semi-reversible frequency domain digital watermarking technique that can be used to authenticate medical images in a distributed diagnosis and home health care environment. Tang and Huang [10] have proposed a near-reversible data embedding scheme based on projection histogram manipulations. The issues aimed in their scheme are embedding capacity, image fidelity, and robustness. Recently, Zhang et al. [12] have proposed a near reversible watermarking algorithm based on LSB replacement. Their scheme achieves a strong robustness and low calculating complexity.

Most of the literature on near reversible watermarking exist for the digital images. To the best of our knowledge, no near reversible schemes exist for video watermarking. Video is also used in the similar applications where images are used. For instance, in remote sensing application the video is captured while monitoring the damaged regions in the natural disasters such as tsunami, volcanic eruption, etc. where it may not require the exact reversibility after extracting the data, i.e. certain level of modifications are allowed. Compared to military applications which involve special zooming operations to analyze the video, here it may not require such special viewing conditions. Hence, the near reversible schemes for video need to be explored. Moreover, when the video is considered for communication and storage, the tight relationship of video with the compression techniques such as MPEG-2, MPEG-4, H.264, JVT, etc. can not be neglected [6]. The remote sensing applications involve low bit rate video encoding such as H.263, MPEG-4 etc.[8]. Therefore, designing the near reversible data embedding schemes for video in association with its compression technique is important. Further, coming up with a less complex near reversible data embedding scheme which achieves robustness, acceptable visual quality and better embedding capacity is a challenging task.

In this paper, we propose a near reversible data embedding scheme for video which embeds the data during the process of MPEG-4 compression. And we also give the formal definitions for the reversibility by classifying the schemes into irreversible, reversible, near-reversible. Our basic idea is to design a scheme which avoids altering the original content for representing (embedding) every bit of the data, rather few elements (pixels or transformed coefficients of original content) will be altered to represent the entire data. By observing most of the standard QCIF formatted videos, we identified that most of the AC coefficients are near around zero in the quantized DCT coefficients. By this observation, we designed the *log* functions which make use of the non-zero AC coefficients for embedding and extracting the data such that most of the original content can be restored after extracting the data. With this, we will be able to achieve better visual quality. As there are good amount of non-zero AC coefficients in the quantized DCT coefficients, we can achieve better embedding capacity. Further, the robustness can be achieved by embedding data into the compressed domain.

1 Proposed Scheme

We embed the data into the MPEG-4 video during the process of compressing the raw YUV video into MPEG-4 format. The MPEG-4 compression involves the formation of sequence of three kinds of frames: *I-*, *P-*, *B-frame*. The *I-frames* are called *reference frames* and *P-*, *B-frames* are called *predicted frames*. The *I-frames* are coded using *Intraframe* technique, i.e, they can be reconstructed without having the reference to any other frames. The *P-frames* are coded using *Interframe* technique called *forward prediction*. They are forward predicted from the recent *I-frame* or *P-frame*. The *B-frames* are also coded using *interframe* technique but they are both *forward predicted* from the recent and *backward predicted* from the future *I-frame* or *P-frame*, i.e, two other frames are necessary to reconstruct the *B-frames*. Hence, in the MPEG-4 compression the *I-frames* are the key frames without which the reconstruction of the compressed video is not possible. In this paper, as shown in Fig. 1, we choose the luminance component (Y) of the every *I-frame* for embedding the data. We take the 8×8 block of a luminance component (Y) of an *I-frame*, get the quantized *DCT* coefficients and embed the data into it. The dark line in the Fig. 1 shows the sequence of steps for embedding the data into the video during the process of MPEG-4 compression. Note that we present only the steps of interest in MPEG-4 compression and encoding is followed after embedding in Fig. 1.

1.1 Models, Notations and Definitions

The raw *YUV* video consist of sequence of frames. Let $F = \{\bar{f}_1, \bar{f}_2, ..., \bar{f}_n\}$ be the sequence of original frames of raw *YUV* video, where n is the total number of frames. Each frame $\bar{f}_i \in F$ consists of one *luminance*, two *chroma* components. Let $\bar{f}_i = \{Y, C_b, C_r\}$, where Y is the *luminance* component and C_b, C_r are the two *chroma* components of $\bar{f}_i$. All these components can be compressed using MPEG-4 encoder. All the frames in F are given as input to the MPEG-4 encoder. While the compression process is being carried out, the MPEG-4 encoder expresses the frames in F as the sequence of *I-*, *P-*, *B-frames*. Then $F = I \cup P \cup B$, where I is the set of *I-frames*, directly retrieved from F, treated as *reference frames* and P, B are the sets of *P-*, *B-frames*, which are the *predicted frames*.

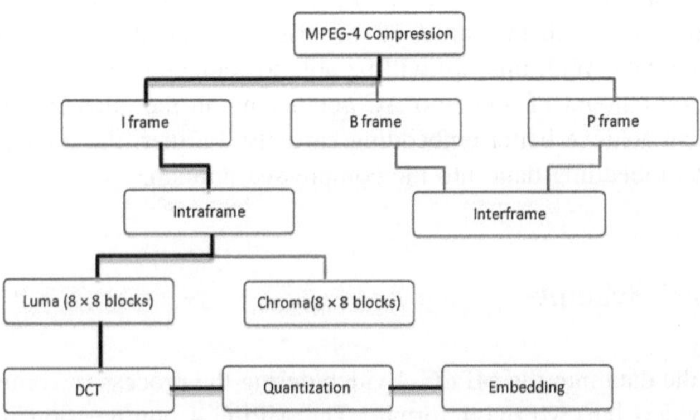

Fig. 1 Flowchart of embedding process in MPEG-4

Though, all the frames in F can be used for embedding the data, we use only *I-frames* for embedding. Let $I = \{I_1, I_2, ..., I_m\}$, where $m<n$. As we concern with I, let $I_i = \{Y^i, C_b^i, C_r^i\}$, where Y^i is the *luminance* component of I_i, C_b^i and C_r^i are the two *chroma* components of I_i. We consider Y^i for embedding the data. Here each Y^i, of size $n_1 \times n_2$, is partitioned into 8×8 blocks of intensity values. We assume that both n_1, n_2 are the multiples of 8. Let $Y^i = \{B_1^i, B_2^i, \cdots, B_l^i\}$, where B_j^i is the jth 8×8 block of Y^i and $l = \frac{n_1 \times n_2}{64}$. Here $\hat{m} = m \times l$ gives the total number of blocks in the set I. These 8×8 non-overlapping blocks are transformed into 2-dimensional *DCT* using (1).

$$F_{u,v} = \frac{\alpha(u)\alpha(v)}{4} \sum_{x=0}^{7} \sum_{y=0}^{7} B_j^i(x, y) \cos\left(\frac{(2x+1)u\pi}{16}\right) \cos\left(\frac{(2y+1)v\pi}{16}\right) \quad (1)$$

where

$$\alpha(e) = \begin{cases} \frac{1}{\sqrt{2}} & \text{if } e = 0, \\ 1 & \text{if } e \neq 0. \end{cases}$$

Here, $0 \leq u, v \leq 7$, and $B_j^i(x, y)$ represent the intensity value (pixel value) of block B_j^i at the coordinate (x, y) in the spacial domain and $F_{u,v}$ represent the coefficient at the coordinate (u, v) in the frequency domain. The inverse $DCT(IDCT)$ is obtained by (2) as follows, where $\alpha(e)$ are the same as in (1), and $0 \leq x, y \leq 7$.

$$B_j^i(x, y) = \sum_{u=0}^{7} \sum_{v=0}^{7} \frac{\alpha(u)\alpha(v)}{4} F_{u,v} \cos\left(\frac{(2x+1)u\pi}{16}\right) \cos\left(\frac{(2y+1)v\pi}{16}\right) \quad (2)$$

Let $\hat{B}^i = \{\hat{B}_1, \hat{B}_2, \cdots, \hat{B}_l\}$ be the set of 8×8 blocks of *DCT* coefficients of Y^i, and Q be a 8×8 block of the quantization table used in *Intraframe* coding. Let $C^i = \{C_1, C_2, \cdots, C_l\}$ be the set of 8×8 blocks of quantized *DCT* coefficients and $\hat{C}^i = \{C_1', C_2', \cdots, C_l'\}$ be the set of embedded blocks of Y^i. Let D_i $(1 \leq i \leq 9)$ be the set of quantized *DCT* coefficients from high frequency to low frequency of a 8×8 block as shown in Fig. 2 [3]. We use these sets for embedding the data. Let $(d_{i,1}, d_{i,2}, , d_{i,X(i)})$ be the sequence of quantized *DCT* coefficients in the set D_i, where $X(i)$ is given in Table 1.

Further, we use the following two metrics *NCC* and *BER* to define the reversibility of the data embedding scheme.

$$NCC = \frac{\sum_{x,y} Y_{x,y}^i \bar{Y}_{x,y}^i}{\sum_{x,y} (Y_{x,y}^i)^2} \quad (3)$$

where *NCC* gives the Normalized Cross-Correlation [7] between original luminance component Y^i of size $n_1 \times n_2$ and $\bar{Y}^i$ is its corresponding luminance component after extracting the data and here $Y_{x,y}^i$ is a pixel value of Y^i at (x, y), and its corresponding modified value is $\bar{Y}_{x,y}^i$. Here $NCC \in [0, 1]$ and $NCC = 0$ means that

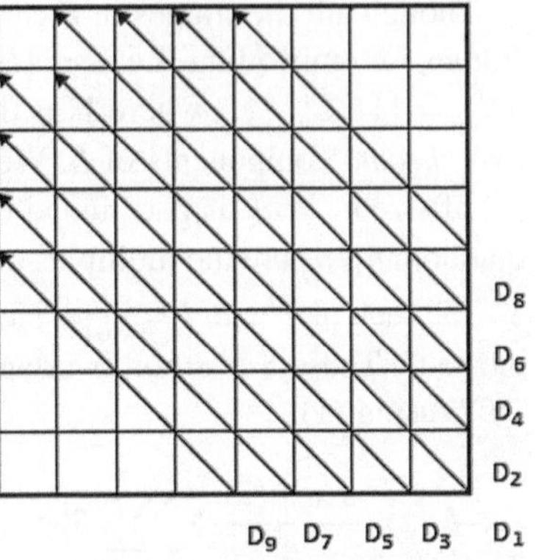

Fig. 2 Chosen sets for embedding

Table 1 The lengths of the chosen sets for embedding

i	1	2	3	4	5	6	7	8	9
X(i)	7	7	7	6	6	5	5	4	4

the Y^i and $\bar{Y}^i$ are not correlated, $NCC = 1$ means that the Y^i and $\bar{Y}^i$ are highly correlated.

$$BER = \frac{INCORRECT}{TOTAL} \tag{4}$$

where the BER gives the Bit Error Rate between the original data and extracted data, $INCORRECT$ = Number of incorrect bits in the extracted data, and $TOTAL$ = Total number of bits in the original data. Here $BER \in [0, 1]$.

We define the reversibility of a data embedding scheme with respect to the metrics *NCC* and *BER*. However, the reversibility of a scheme can be defined by considering other metrics such as *MSE*, *SC*, etc. In a broader view, the data embedding schemes can be classified into *Irreversible* and *Reversible.* To define the above classification, without loss of generality, we refer a steganography or a watermarking scheme as a *Data Embedding Scheme* ($\mathcal{E}$).

Definition 1 Irreversible: A data embedding scheme $\mathcal{E}$ is irreversible if it has $0 \leq NCC < 1$ and $0 < BER \leq 1$.

Definition 2 Reversible: A data embedding scheme $\mathcal{E}$ is reversible if it has $NCC = 1$ and $BER = 0$.

Another category of data embedding schemes, which recently emerging more, is the *Near-reversible* schemes which is a sub category of *irreversible* schemes.

Definition 3 Near-reversible: A data embedding scheme $\mathcal{E}$ is near-reversible if it has $NCC \approx 1$ and $BER \approx 0$.

In the **Definition 3**, the $NCC \approx 1$ we mean that *NCC* should be almost nearer to the maximum value 1 and $BER \approx 0$ means that *BER* should be almost nearer to the minimum value 0. However, the lower bounds for the *NCC* and *BER* of a near-reversible scheme can not be fixed and an application can fix the them as desired.

1.2 Data Embedding Procedure

We embed the data in the middle frequency components as in Chang et al. [3] scheme. The sets D_i $(1 \leq i \leq 9)$ are considered for embedding as shown in Fig. 2. Note that the DC coefficient at $d_{1,8}$ is not considered. For better visual quality, our design is proposed to reduce the modifications to the original coefficients in the selected sets. For that, we first identify the suitable blocks for embedding the data using (5).

$$T = \sum_{i=1}^{9} \sum_{j=1}^{X(i)} D_{i,j} = \sum_{i=1}^{9} (|d_{i,1}| + |d_{i,2}| + +|d_{i,X(i)}|) \tag{5}$$

where $D_{i,j}$ is the *j*th element in the *i*th set D_i, and $X(i)$ gives the number of elements in the set D_i as shown in Table 1. When $T > 0$, the block is said to be *suitable* for embedding, otherwise it is *not suitable*. Then, we embed the data into the suitable blocks using (6). In the (6), we modify all the non-zero AC coefficients in every set D_i of the suitable block for embedding the data bits.

Algorithm 1: Data Embedding Scheme

Input : $I = \{I_1, I_2, ..., I_m\}$ and the data, $\mathbb{I}$.
Output: The set of *I-frames* with embedded data.

1. Extract the Y^i from I_i
2. Partition $Y^i \rightarrow \{B_1^i, B_2^i, \cdots, B_l^i\}$
3. For each $B_j^i \in Y^i$, where $1 \leq j \leq l$
   (a) Find the *DCT* of B_j^i: $\hat{B}_j^i = DCT(B_j^i)$;
   (b) Quantize the *DCT* coefficients in $\hat{B}_j^i$ as below.
   for $i_1 \leftarrow 1$ **to** *8* **do**
   for $i_2 \leftarrow 1$ **to** *8* **do**
   $C_j(i_1, i_2) = \hat{B}_j^i(i_1, i_2)/Q(i_1, i_2)$;
   end
   end
   (c) Consider $D_k (1 \leq k \leq 9)$ sets of C_j as shown in Fig. 2.
   i. Compute T as shown in (5), If $T > 0$, then modify all the non-zero AC coefficients in every D_k to represent the data bits in $\mathbb{I}$ using (6).
   Let the resultant block be $C_j^{'}$.
4. Combine all the $C_j^{'}$ blocks into $\hat{C}^i$.
   $\hat{C}^i \leftarrow \{C_1^{'}, C_2^{'}, \cdots, C_l^{'}\}$
5. Restore the $\hat{C}^i$ back to $I_i = \{\hat{C}^i, C_b^i, C_r^i\}$
6. For all the $I_i \in I$, repeat step 1) to step 5).

$$e = \begin{cases} \frac{c}{|c|}\lfloor 2\log_2(2|c|)\rfloor & \text{if } I_i = 0, \\ \frac{c}{|c|}\lfloor 2\log_2(2|c|) - 1\rfloor & \text{if } \mathbb{I}_i = 1. \end{cases} \tag{6}$$

where c be a non-zero AC coefficient in the suitable set D_i and $\mathbb{I}_\daleth$ be a ith data bit, e be the modified version of c. Note that the index i of D_i and $\mathbb{I}_\daleth$ are independent. We present our proposed data embedding scheme in **Algorithm 1**.

1.3 Data Extraction Procedure

The data extraction is an inverse process of data embedding. We extract the data bits using (7). However, the extraction and restoration can be combined, here we gave them separately.

$$\mathbb{I}_i' = \begin{cases} 0 & \text{if} e\%2 = 0, \\ 1 & \text{Otherwise.} \end{cases} \tag{7}$$

where e is the modified coefficient during the embedding process and $\mathbb{I}_\daleth'$ is the ith extracted data bit.

Algorithm 2: Data Extraction Scheme

Input : I, the set of all the I-*frames* with embedded data.
Output: The set of restored I-*frames*, and the extracted data: $\mathbb{I}'$.

1. Extract the $\hat{C}^i$ from I_i
2. Partition $\hat{C}^i \rightarrow \{C_1', C_2', \cdots, C_l'\}$
3. For each $C_j' \in \hat{C}^i$
   (a) Consider $D_k (1 \leq k \leq 9)$ sets of C_j' as shown in Fig. 2 and when $T > 0$,
      i. Extract the data bits using (7).
      ii. Restore the modified coefficients using (8).
   Let the resultant block be E_j'.
   (b) Dequantize the elements of E_j' as follows:
   for $i_1 \leftarrow 1$ **to** *8* **do**
   for $i_2 \leftarrow 1$ **to** *8* **do**
   $R_j(i_1, i_2) = E_j'(i_1, i_2) \times Q(i_1, i_2)$;
   end
   end
   (c) $R_j' = IDCT(R_j)$
4. Combine all the R_j' blocks to get the $\hat{R}^i$.
   $\hat{R}^i \leftarrow \{R_1', R_2', \cdots, R_l'\}$
5. Restore the $\hat{R}^i$ back to $I_i = \{\hat{R}^i, C_b^i, C_r^i\}$
6. For all the embedded $I_i \in I$,
   repeat step 1) to step 5).

And after extracting the data bits, we restore coefficients which are modified during the embedding using (8).

$$c' = \begin{cases} \frac{e}{|e|} 2^{\frac{|e|}{2}-1} & \text{if } e\%2 = 0, \\ \frac{e}{|e|} 2^{\frac{|e|+1}{2}-1} & \text{Otherwise.} \end{cases} \quad (8)$$

where the c' is the restored version of the original c. The algorithm for data extraction and restoration is given in the **Algorithm 2**. Note that the data extraction and embedding is the near-reversible.

2 Results and Discussion

We use various QCIF formatted videos in our experiment, including MissAm, Akiyo, CarPhone, SalesMan, etc. Some of the test videos are shown in Fig. 3. The frame size of all these test videos is 176 × 144 pixels. We compress these test videos by the standard MPEG-4 encoder. The widely used measurement for evaluating the visual quality of a stego-video is PSNR (Peak Signal to Noise Ratio). The PSNR for each YUV channel of a frame is given by the Eq. (9).

$$PSNR_{dB} = 10 \log_{10} \frac{255^2}{MSE} \quad (9)$$

where

$MSE = \frac{1}{MN} \sum_{x=1}^{M} \sum_{y=1}^{N} (f_{x,y} - f'_{x,y})^2$ and $f_{x,y}, f'_{x,y}$ are the pixel values at the coordinate (x, y) of original and stego-video YUV channels respectively, each of size $M \times N$.

Another measurement used for evaluating the performance of a data embedding scheme is *embedding capacity*. We define the embedding capacity as the number of bits that can be embedded into a single Y^i. To assess the performance in terms of reversibility, we have calculated the *NCC* and *BER* as in (3) and (4) respectively. The tabulated results are shown in Table 2. As our scheme is the first proposed near-reversible for MPEG-4 video, it is inadequate to compare our proposed method with the methods available in the literature which are proposed for the digital images. However, Zhang et al. [12] have stated that their near-reversible scheme achieves about 93.6 % reversibility of host an image, which is an LSB based technique for didital image. Our proposed method not only out performs in

Table 2 Performance metrics of our proposed scheme for various test videos

	Capacity	*PSNR*	*NCC*	*BER*
MissAm	1157	34.3583	0.9996	0.1046
Akiyo	2329	30.0294	0.9995	0.1237
CarPhone	2981	27.4747	0.9992	0.1275
SalesMan	3787	28.8682	0.9993	0.0943

Fig. 3 The four original I frames of various test videos **a** MissAm **b** Akiyo **c** CarPhone **d** SalesMan

Fig. 4 The embedded I frames of test videos **a** MissAm **b** Akiyo **c** CarPhone **d** SalesMan

terms of embedding capacity by maintaining the visual quality, but also the percentile of reversibility of cover-object is around 99.92 % (*NCC*) and the $BER = 0.1$ for most of test video sequences. The embedded *I*-frames of various test video sequences are shown in Fig. 4.

3 Conclusion

With the invention of the new category near-reversible data embedding, there is a need for the formal definition of reversibility. In this paper, we have given the formal definition for the reversibility by classifying the embedding schemes as *Irreversible*, *Reversible* and *Near-reversible*. Our first proposed near-reversible scheme for MPEG-4 video out performs in terms of embedding capacity, visual quality, reversibility. We have shown that our less complex scheme achieves $NCC = 99.92\%$ and $BER = 0.1$, which indicates the scheme is almost reversible.

References

1. Ahmed F, Moskowitz IS (2006) A semi-reversible watermark for medical image authentication. In: 1st Transdisciplinary conference on distributed diagnosis and home healthcare. D2H2, pp 59–62.
2. Barni M, Bartolini F, Cappellini V, Magli E, Olmo G. (2002) Near-lossless digital watermarking for copyright protection of remote sensing images. In: IEEE international geoscience and remote sensing symposium, 2002. IGARSS '02., vol 3, pp 1447–1449.

3. Chang C-C, Lin C-C, Tseng C-S, Tai W-L (2007) Reversible hiding in DCT-based compressed images. Inf Sci 177:2768–2786
4. Cox I, Miller M, Bloom J, Fridrich J, Kalker T (2008) Digital watermarking and steganography. Morgan Kaufman, San Francisco
5. Fridrich J, Du R (2002) Lossless authentication of MPEG-2 video. In: ICIP'02, pp 893–896.
6. Furht B (1995) A survey of multimedia compression techniques and standards. Part I: JPEG standard. R'eal-Time. Imaging 1:49–67
7. Hsu C-T, Wu J-L (1999) Hidden digital watermarks in images. IEEE Trans Image Process 8(1):58–68
8. Pereira F, Koenen R (1996) Very low bit rate audio-visual applications. Signal Process: Image Commun 9(1):55–77
9. Sagar G, Amberker BB (2012) A DCT based reversible data hiding scheme for mpeg-4 video. In: Proceedings of international conference on signal, Image and Video Processing (ICSIVP) 2012, IIT Patna, pp 254–259, Jan 2012.
10. Tang Y-L, Huang H-T (2007) Robust near-reversible data embedding using histogram projection. In: In IIH-MSP 2007, vol 02, pp 453–456.
11. Zeng X, Chen Z, Zhang X (2011) Issues and solution on distortion drift in reversible video data hiding. Multimedia Tools Appl 52(2–3):465–484
12. Zhang B, Xin Y, Niu X-X, Yuan K-G, Jiang H-B (2010) A near reversible image watermarking algorithm. In: International conference on machine learning and cybernetics (ICMLC), vol 6, pp 2824–2828, july 2010.

3. Huang CH, Lin CC, [illegible] (20[illegible]) Reversible [illegible] DCT [illegible] compressed images. [illegible]
4. Cox I, Miller M, Bloom J, Fridrich J, Kalker T (2008) Digital watermarking and steganography. Morgan Kaufmann, San Francisco
5. Fridrich J, Du R (2002) Lossless authentication of MPEG-2 video. In: ICIP'02, pp 893–896
6. [illegible] (19[illegible]) [illegible] compressed [illegible] and standard. [illegible] IEEE [illegible]
7. [illegible] (19[illegible]) Hidden digital watermarks in images. IEEE Trans Image Process [illegible]
8. Petitcolas F, Anderson R (1998) [illegible] on [illegible] audio visual [illegible] Signal Process Image Commun [illegible]
9. [illegible] (2012) A [illegible] based [illegible] data hiding scheme for image [illegible] In: Proceedings of [illegible] signal, image and video [illegible] 2012 [illegible], pp 254–259, Jan 2012
10. [illegible] Huang [illegible] (20[illegible]) Robust [illegible] reversible data embedding using [illegible] [illegible] In: ICMSP [illegible] vol [illegible], pp [illegible]
11. [illegible] Zhang [illegible] (2011) [illegible] and [illegible] video. [illegible] Trans [illegible] vol 21, [illegible]
12. Zhang [illegible] (2010) A [illegible] reversible [illegible] watermarking [illegible] In: [illegible] conference on machine [illegible] (ICMLC [illegible]), vol [illegible], pp [illegible], July 2010

Shot Boundary Detection and Key Frame Extraction for Sports Video Summarization Based on Spectral Entropy and Mutual Information

Shanmukhappa Angadi and Vilas Naik

Abstract Video Summarization methods attempt to abstract the main occurrences, scenes, or objects in a clip in order to provide an easily interpreted synopsis of the video. This is an essential task in video analysis and indexing applications. New methods for detecting shot boundaries in video sequences and extracting key frames using metrics based on information theory are proposed in this work. The method for shot cut detection relies on the mutual information between the frames. The method for key frame extraction uses the difference of entropy value computed from eigen value matrix of consecutive frames to decide which frames to choose as key frame. The proposed method captures satisfactorily the visual content of the shot. The information theory measure provides the better results because it exploits the inter-frame information in a more compact way. It can also be successfully compared to other methods published in literature. The method for key frame extraction employs entropy measure computed on eigen values of frames to reduce complexity of computation. The proposed algorithm can capture the important yet salient content as the key frame. Its robustness and adaptability are validated by experiments with various kinds of video sequences.

Keywords Video summarization · Dynamic key frames extraction · Mutual information · Entropy difference measure · Information theory

S. Angadi (✉) · V. Naik
Department of CSE, Basaveshwar Engineering College, Bagalkot, India
e-mail: vinay_angadi@yahoo.com

V. Naik
e-mail: vilasnaik_h@rediffmail.com

Mohan S. and S. S. Kumar (eds.), *Proceedings of the Fourth International Conference on Signal and Image Processing 2012 (ICSIP 2012)*, Lecture Notes in Electrical Engineering 221, DOI: 10.1007/978-81-322-0997-3_8,

1 Introduction

The developments in software tools in the last few years' have made available myriad applications in areas such as multimedia databases quite feasible. The incredible rates at which these databases are being published have far exceeded the capabilities of current text-based cataloguing. New techniques and approaches and quick search algorithms have increased the potential of the media databases, which are now containing, not only text and image but video and audio as well. Extensive research efforts have been made with regard to the retrieval of video and image data based on their visual content such as color distribution, texture and shape. These approaches are mainly based on feature similarity measurement. The proposed work is an attempt to use information theory based parameters and similarity in terms of these parameters to represent visual content and changes in the content to achieve shot boundary detection and key frame extraction for summarization of sports videos.

The summarization, indexing and retrieval of digital video is an active research area. Shot boundary detection and key frame extraction are important tasks for analyzing the content of video sequences for indexing, browsing, searching, summarizing and performing other content based operations of large video databases.

The video shot is a basic structural building block of a video sequence and its boundaries need to be determined possibly automatically to allow for content based video abstraction. A video shot may be defined as a sequence of frames captured by *one camera in a single continuous action in time and space* [1]. It should be a group of frames that have consistent visual characteristics (including color, texture and motion). After shots are segmented, key frames can be extracted from each shot. *Key frame* is the frame which can represent the salient content of the shot. Depending on the content complexity of the shot, one or more frames can be extracted [2].

Key-frames are still images which best represent the content of the video sequence in an abstracted manner, and may be either extracted or reconstructed from original video data. Key-frames are frequently used to supplement the text of a video log, but there has been little progress in identifying them automatically. The challenge is that the extraction of key-frames needs to be automatic and content based so that they maintain the important content of the video while removing all redundancy.

Key-frames provide a suitable abstraction and framework for video indexing, browsing and retrieval. The use of key frames greatly reduces the amount of data required in video indexing and provides an organizational framework for dealing with video content. Much research has been done in key frame extraction. The simplest methods proposed choose only one frame for each shot usually the first one, regardless of the complexity of visual content. The more complicated approaches take into account visual content, motion analysis and shot activity [3].

These approaches either cannot effectively capture the major visual content or are computationally expensive.

Key-frame based video representation views video abstraction as a problem of mapping an entire Video segment to some small number of representative frames. The extraction of key frames must be automatic and content based so that they maintain the salient content of the video while avoiding all redundancy. Many of the systems described in the literature use a constant number of key frames from a fixed position in the shot or several frames separated by a fixed distance. Better key-frames could be chosen if shot content were considered in the selection.

In this work, a new approaches for shot boundary detection based on the mutual information and key-frame extraction based on the difference in entropy of eigen value matrix between consecutive frames is proposed. The mutual information is a measure of the information passed from one frame to another. Mutual information is used for detecting abrupt cuts, where the image intensity or color is abruptly changed. A large difference in content between two frames, that shows a weak inter-frame dependency leads to a small value of mutual information. The considerable change in the entropy beyond certain threshold at any pair of consecutive frames indicates movement from one significant segment of shot to another significant segment of shot and representative key frames will be extracted from both segments of shots to abstract video shot leading to summary of entire video.

The proposed work first detects shot boundaries and divides video into meaningful units by exploiting the capability of Mutual information to differentiate the frames in terms of their content. Then the algorithm detects the points of significant changes within a shot by looking at changes that are occurring in entropy values of each frame. The threshold is so defined that every significant change in the visual content within a shot are picked and from every such points the key frames are extracted. So as the overall algorithm extracts key frame from all parts of video a smooth summary of complete video can be constructed. The experimentation also reveal that performance parameters of this algorithm are comparable with many of recent algorithms which have been agreed to provide a good summary of the video.

The remaining part of the paper is organized into 6 sections. Section 2 presents the related work and background of the algorithms used. Section 3 describes the new proposed algorithm. In Sect. 4 the algorithm developed for shot boundary detection is presented. Section 5 discusses the scheme of key frame extraction. Section 6 discusses the results obtained, and Sect. 7 brings up the conclusion.

2 Related Work

Recent works in the story based video summarization area can be classified into three categories, based on the method of key frame extraction namely: sampling based, shot based, and segment based. In the sampling based approaches, key frames were extracted by randomly choosing from sampling of the original video.

It is the straightforward and easy way to extract key frames, yet such an arrangement may fail to capture the real video content, especially when it is highly dynamic. In the shot based approaches, the video is segmented into separate shots and one or more key frames is extracted from each shot. A sequence of frames captured by one camera in a single continuous action in time and space is referred to as a video shot [4]. Normally, it is a group of frames that have constant visual attributes, (such as color, texture, and motion). It is a more significant and straightforward method to extract key frames by adapting to dynamic video content. In shot based approaches a typical and easy manner is utilizing low level features such as color and motion to extract key frames. More complicated systems are based on color clustering, global motion, or gesture analysis which are found in [5, 6]. In segment based approaches, the video segments are extracted by clustering of frames and then the key frames are chosen as those frames of video that are closest to the centroids of each calculated clusters. Some systems based on this approach can be found in [7–9].

Even though shot boundary detection and key frame extraction are strongly related, the two problems have usually been addressed separately. Often a boundary detection algorithm is first used to detect the shots followed by a key frame extraction. In [10], the first frame of the shot is selected as the key frame. Work in [11] assumes that the shot boundaries have been detected and use an unsupervised clustering algorithm to find the key frame of the given shot. Another approach is to compare consecutive frames in video. Compare the difference in color histograms of consecutive frames with a threshold to obtain key frames [12].

Towards the more complicated end, techniques based on clustering have also been developed. The technique in [13] clustered visually similar frames and used constraints on key frame positions in the clustering process for selecting key frames. In the algorithm described in [14] the video is down sampled to decrease the number of frames and features are extracted from frames. The refined feature space obtained by applying SVD and feature points are clustered, the key frame is then extracted from each cluster.

Different methods can be used to select key frames. In general these methods assume that the video has already been segmented into shots (a continuous sequences of frames taken over a short period of time) by a shot detection algorithm, and extract the key-frames from within each shot. One of the possible approaches to key frame selection is to take the first frame in the shot as the key frame [15]. Few methods use the first and last frames of each shot [16, 17]. Other approaches time sample the shots at predefined intervals, as in [18] where the key frames are taken from a set location within the shot, or, in an alternative approach where the video is time sampled regardless of shot boundaries and frames at regular intervals are picked as key frames [19]. These approaches do not consider the dynamics in the visual content of the shot but rely on the information regarding the sampled sequences and boundaries. They often extract a fixed number of key frames per shot. In [20] only one key frame is extracted from each shot: the frames are segmented into objects and background, and the frame with the maximum ratio of objects to background is chosen as the key frame of the segment since it is assumed to convey the most information about the shot. In the work proposed in [21] the key frames are extracted based on compression policy. The algorithm proposed in [22] divides frames into clusters, and the key frames are selected from the largest clusters only. In [23] constraints on the position of the key frames in time are used in the clustering process; a hierarchical clustering reduction is performed, obtaining summaries at different levels of abstraction.

In order to take into account the visual dynamics of the frames within a sequence, some approaches compute the differences between pairs of frames (not necessarily consecutive) in terms of color histograms, motion, or other visual descriptions. The key frames are selected by analyzing the values obtained. A simple method for key frame extraction is a frame is selected as a key frame if its color histograms differs from that of the previous frame by a given threshold. In [24] frame differences are taken to build a "content development" curve that is approximated, using an error minimization algorithm, by a curve composed of a predefined number of rectangles. The method in [25] proposes a very simple approach, the key frames are selected by an adaptive temporal sampling algorithm that uniformly samples the y-axis of the curve of cumulative frame differences. The resulting non-uniform sampling on the curve's x-axis represents the set of key frames.

The compressed domain is often considered when developing key frame extraction algorithms since it easily allows to express the dynamics of a video sequence through motion analysis. A neural network based approach using motion intensities computed from MPEG compressed video is presented in [26]. A fuzzy system classifies the motion intensities in five categories, and those frames that exhibit high intensities are chosen as key frames [27]. In [28] perceived motion energy (PME) computed on the motion vectors is used to describe the video content. A triangle model is then employed to model motion patterns and extract key frames at the turning points of acceleration and deceleration.

The drawback to most of these approaches is that the number of representative frames must be prefixed in some manner depending on the length of the video shots for example [29]. This cannot guarantee that the frames selected will not be highly correlated. It is also difficult to set a suitable interval of time, or frames: large intervals mean a large number of frames will be chosen, while small intervals may not capture enough representative frames, those chosen may not be in the right places to capture significant content. Still other approaches that work only on compressed video, are threshold-dependent, or are computationally intensive (e.g. [29, 30]).

In this paper, a novel approach is proposed for selection of key frames that determines the complexity of the sequence in terms of changes in the pictorial content using entropy of eigen values of each frame. Similarity between frames is computed in terms of difference in entropy calculate using eigen value of frames within an identified shot. The frame differences are then used to dynamically and rapidly select a variable number of key frames within each shot. The method works fast on all kind of videos (compressed or not), and does not exhibit the complexity of some of the existing methods based on clustering. It can also output key frames while computing the frame differences without having to process the whole shot.

The proposed mechanism of mutual information based shot boundary detection and further key frame extraction from separated shots based on spectral entropy is presented in following section.

3 The Proposed Mechanism for Shot boundary detection and Key Frame Extraction

The overall key frame extraction algorithm may be summarized as follows

Step 1. Extract the individual frames from a video. Consider F_t as the tth frame and F_{t+1} as very next neighbor of F_t.

Step 2. Calculate the Mutual information. $I^{R,G,B}_{t,t+1}$ between pair of consecutive frames for full length of the video using the equations, viz.

$$I^R_{t,t+1} = \sum_{i=0}^{N-1}\sum_{i=0}^{N-1} \mathbf{C}^{\mathbf{R}}_{\mathbf{t,t+1}}(i,j) log \frac{\mathbf{C}^{\mathbf{R}}_{\mathbf{t,t+1}}(\mathbf{i},\mathbf{j})}{\mathbf{C}^{\mathbf{R}}_{\mathbf{t,t+1}}(i)\mathbf{C}^{\mathbf{R}}_{\mathbf{t,t+1}}(\mathbf{j})}$$

$$I_{t,t+1} = I^R_{t,t+1} + I^G_{t,t+1} + I^B_{t,t+1}$$

$C^R_{t,t+1}(\mathrm{i},\mathrm{j})$	Represents Co occurrence Matrix between Frames F_t and F_{t+1}
$I^R_{t,t+1}$	Mutual Information between F_t and F_{t+1} for the red component
$I^G_{t,t+1}$	Mutual Information between F_t and F_{t+1} for the green component
$I^B_{t,t+1}$	Mutual Information between F_t and F_{t+1} for the blue component
$\mathrm{II}_{t,t+1}$	Total Mutual Information between Frames F_t and F_{t+1}

Step 3. Consider a Window of 100 frames within the window Compare Mutual information values to a threshold. In this work threshold is 0.30 × Mean of all Mutual information values in the window.

Step 4. The shot boundary will be detected between any two frames $t, t+1$ if there Mutual information $I^{R,G,B}_{t,t+1}$ is less than threshold. (The detailed mechanism of shot boundary detection is presented in Sect. 4).

Step 5. Consider all the frames within a shot and compute the eigen values of each frame

$$[V_t D_t] = eig[f_t]$$

where Dt is diagonal matrix containing eigen values of frame image.

Step 6. Compute the entropy from diagonal matrix Dt as

$$E_t = \sum_{j=0}^{n} \mathrm{e_j} \log {}^{1}/_{\mathrm{e_j}}$$

where $E_t = Entropy\ of\ frame\ t$

$$\mathrm{e_j} = j^{th} eigen\ value\ of\ frame\ t$$

For all the frames within a shot.

Step 7. Start calculating difference in the entropy value of consecutive frame from first to last frame in the shot, If this difference $\mathrm{d_t} = \mathrm{E_t} - \mathrm{E_{t+1}}$ at any instant t satisfy the condition $\mathrm{d_t} > \mathrm{T_k}$

where $T_k = 2.0\times$ Mean of the entropy values in the Shot. At that instant of frame F_t select frame F_{t+1} as key frame.

Step 8. For every shot add first frame as key frame and the shots where the difference value d_t is always less than T_k select only first frame as key frame as change in visual content is negligible any frame can become representative key frame.

The proposed algorithm internally uses shot boundary detection mechanism based on mutual information between two successive frames is calculated separately for each of the RGB components. A small value of the mutual information $\mathrm{I_{t,t+1}}$ leads to a high probability of having a cut between frames $\mathrm{f_t}$ and $\mathrm{f_{t+1}}$. The details of the approach are presented in following section.

4 Shot Boundary Detection

The proposed approach for *Shot boundary detection* is based on the *Mutual Information* (MI) which is the measure of information transported from one frame to another one. It is used for detecting abrupt cuts, where the image intensity or color is abruptly changed. A large video content difference between two frames, showing weak inter-frame dependency leads to a low MI In this approach, the mutual information and the joint entropy between two successive frames is calculated separately for each of the RGB components. As the color intensity levels of the image sequence vary from 0 to $N-1$, at frame f_t three $N \times N$ matrices $\mathrm{C}^{R}_{t;t+1}$, $\mathrm{C}^{G}_{t;t+1}$ and $\mathrm{C}^{B}_{t;t+1}$ are created, that carry information on the grey level transitions between frames f_t and f_{t+1}.

In other words, considering only the R component, the matrix $C^R_{t;t+1(i,j)}$, with $0 < i < N-1$ and $0 < j < N-1$, corresponds to the joint probability: *a pixel with grey level i in frame* f_t *has color intensity level j in frame* f_{t+1}. $C^R_{t;t+1(i,j)}$ represent a co-occurrence matrix between frames f_t and f_{t+1}. The mutual information $I^R_{t;t+1(i,j)}$ of the transition from frame f_t to frame f_{t+1} for the R component is expressed by Eq. (1) as presented in [30]:

$$I^R_{t,t+1} = -\sum_{i=0}^{N-1}\sum_{j=0}^{N-1} \mathbf{C}^{\mathbf{R}}_{\mathbf{t,t+1}}(i,j) log \frac{\mathbf{C}^{\mathbf{R}}_{\mathbf{t,t+1}}(\mathbf{i},\mathbf{j})}{\mathbf{C}^{\mathbf{R}}_{\mathbf{t,t+1}}(i)\mathbf{C}^{\mathbf{R}}_{\mathbf{t,t+1}}(\mathbf{j})} \tag{1}$$

and the total mutual information is given by Eq. (2):

$$I_{t,t+1} = I^R_{t,t+1} + I^G_{t,t+1} + I^B_{t,t+1} \tag{2}$$

A small value of the mutual information $I_{t;t+1}$ as shown in the Fig. 1 leads to a high probability of having a boundary between frames f_t and f_{t+1}.

Basically, in this context boundary detection is an discontinuity detection in a one-dimensional signal. Several algorithms exist for discontinuity detection, based on threshold value. In order to detect possible shot boundary, an adaptive thresholding approach is employed in this work. The local mutual information mean values on a one-dimensional temporal window W of size N_W are obtained. In the proposed mechanism the value of N_W is considered as sequence of 100 frames from video. The boundary will be detected between any pair of frames f_t and f_{t+1} if the mutual information between them falls empirically below the 30 % of the mean of all the mutual information values in that window of 100 values.

5 Key Frame Extraction

Key-frames are still images which best represent the content of the video sequence in an abstracted manner, and may be either extracted or reconstructed from original video data. The proposed approach extracts key frames from each

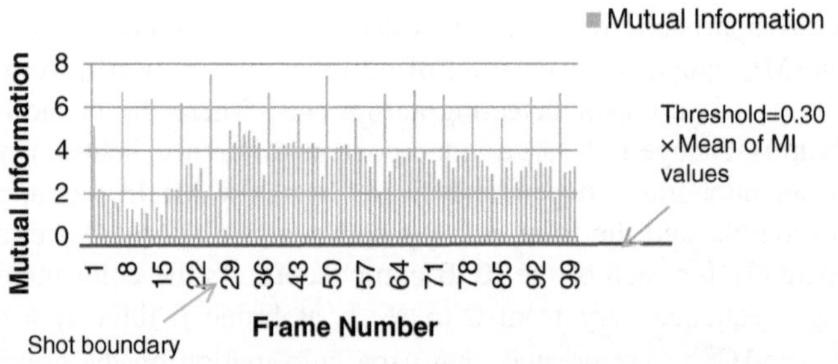

Fig. 1 Time series of the mutual information from a video sequence showing detection of shot boundary at low value of mutual information. *X-axis* frame number. *Y-axis*: mutual information

meaningful shot of the video segmented by identifying shot boundaries by Mutual information based algorithm. The method for key frame extraction employs entropy measure computed using eigen values of frames to reduce complexity of computation. The entropy measure exploits the inter-frame information flow in a more compact way. The proposed algorithm can capture the important yet salient content as the key frame. The entropy of each frame is computed by calculating entropy of Eigen value matrix of frame as in the Eq. (3)

$$[V_t D_t] = eig[f_t] \tag{3}$$

where

$$V_t = eigen\ vector\ matrix\ of\ frame\ t$$

$$D_t = diagonal\ matrix\ containing\ eigen\ values\ of\ frame\ t$$

The entropy is calculated for each frame as entropy of matrix D_t as shown in Eq. (4)

$$E_t = \sum_{j=0}^{n} \mathrm{e_j} \log {}^{1}\!/_{\mathrm{e_j}} \tag{4}$$

where

$$E_t = Entropy\ of\ frame\ t$$

$$\mathrm{e_j} = j^{th} eigen\ value\ of\ frame\ t$$

The proposed approach uses entropy values calculated as indicated by Eq. (4), the difference value of entropy between any two consecutive frames provide the information about content changes between consecutive frames in the shot [30]. If a video shot is S $= \{f1, f2, f3 \ldots\ldots\ldots ft\}$ where $f1, f2, f3 \ldots\ldots\ldots ft$ are individual frames within the shot S obtained by proposed method for shot boundary detection. Let the entropy values in this shot be E $= \{E1, E2, E3 \ldots\ldots\ldots Et\}$. In order to find if the content in the shot changes significantly, the difference of the values $d_t = E_t - E_{t+1}$ of entropies in this shot is calculated. The value d_t is compared to predefined threshold value T_k, in the proposed work the value of T_k is taken as $T_k = 2.0\times$ Mean of the entropy values in the Shot.

At any pair of frames f_t and f_{t+1} within shot if the condition $d_t > T_k$ is satisfied it indicates significant change in the content of shot so the frame after change is picked as Key frame. If $d_t < T_k$ it means the content during the shot changes are negligible [31]. In the proposed method for every shots the first frame is also selected as a key frame. For shots with no or small changes whatever frame can effectively express visual content so the first frame is considered as Key frame as in [32]. The results of proposed scheme have demonstrated that the key frames can effectively generate meaningful summary.

A video summary should not contain too many key frames since the aim of the summarization process is to allow users to quickly grasp the content of a video sequence. For this reason, algorithm is also evaluated with the compactness (compression ratio) of the summary that can be generated by extracted key frames. The compression ratio is computed by dividing the number of key frames in the summary by the length of video sequence. For a given video sequence St, the compression rate is thus defined as in Eq. (5).

$$CRatio(S_t) = 1 - \gamma_{NKF}/\gamma_{NF} \tag{5}$$

where

$$\gamma_{NKF} = Number\ of\ key\ frames$$

$$\gamma_{NF} = Number\ of\ frames\ in\ the\ Video\ Sequence$$

Ideally, a good summary produced by a key frame extraction algorithm will presents high compression ratio (i.e. small number of key frames) [33]. In the propose work, the compactness (compression ratio) is calculated for key frames extracted from 6 videos and results are presented in the ensuing section.

6 Experimentation and Results

The algorithm was implemented in the Matlab and experiments are performed on a Core2 Duo 2.40 GHz Windows machine with 2 GB of RAM. In all the experiments reported in this section, the video streams are in AVI format, with the digitization rate equal to 25 frames/sec. In order to validate the effectiveness of the proposed shot boundary detection and key frame extraction algorithm, experiments conducted are presented in the following. The performance of the proposed algorithm is evaluated on the YOU TUBE videos, Soccer clips and Open Videos data set. The performance of the proposed algorithm is evaluated using evaluation metrics.

Initially all the video clips have been split into simple frames in order to identify possible key-frames. These can be judged by a human who watches the entire video. Although key-frames are a semantic concept, relative agreement can be reached among different people. The work presented first performs shot boundary detection by identifying the points on time axis of video sequence where Mutual information between two frames falls to minimum value.

The effectiveness of proposed algorithm is presented by performance parameters of the proposed model on six different video samples. The performance of the proposed shot boundary detection model is evaluated using precision and recall as evaluation metrics. The precision measure is defined as the ratio of number of correctly detected cuts to the sum of correctly detected and falsely detected cuts of a video data and recall is defined as the ratio of number of detected cuts to the sum of detected and undetected cuts as indicated in Eq. (5)

$$\text{Recall} = \frac{\text{Number of correctly detected boundaries}}{\text{Number of True boundaries}}$$
$$\text{Precision} = \frac{\text{Number of correctly detected boundaries}}{\text{Number of Totally detected boundaries}} \qquad (5)$$

A video summary should not contain too many redundant frames since the aim of the summarization process is to allow users to quickly grasp the content of a video sequence. For this reason, key frame extraction algorithm is also evaluated with the compactness (compression ratio) of the summary that can be generated by extracted key frames. The compression ratio is computed by dividing the number of key frames in the summary by the length of video sequence. For a given video sequence **St**, the compression rate is thus defined as in Eq. (6)

$$CRatio(S_t) = 1 - \gamma_{NKF}/\gamma_{NF} \qquad (6)$$

where

$$\gamma_{NKF} = Number\ of\ key frames$$

$$\gamma_{NF} = Number\ of\ frames\ in\ the\ Video\ Sequence$$

The Fig. 2a shows the plot of mutual information values for window of 100 frames from a video. The threshold is plotted as per the proposed algorithm and mutual information value falling below the threshold is indicative of a shot boundary between those two frames for which the mutual information value is computed. The Fig. 2a indicates the detection of shot boundary between *frame* 28 and *frame* 29. The shot boundary detected from another video sequence between frame 49 and 50 is shown in Fig. 2b.

The Fig. 2c is the plot of entropy difference values plotted against frame index shows that the difference value d_t is more than threshold T_k at four instants of difference values that are between frames 6 and 7, frames 12 and 13, frame 18 and 19 and frames 24 and 25. So as per the proposed algorithm the frames 7, 13, 19 and 25 will be selected as key frames. The four frames extracted out of 28 frames (Fig. 3a) are given in the Fig. 3b.

The experimentation is conducted to test performance of the proposed mechanism on set of video sequence and results for 6 such video sequence from different sources is presented in Table 1.

The graph shown in the Fig. 4a and b reveal that number of key frames extracted depends upon number of frames in every shot separated using proposed method. The time taken by the mechanism is also proportional to number of frames in shot under consideration.

The Fig. 5 is the snap shot of a folder containing 134 key frames extracted from a video of 1,342 frames which can successfully construct the summary of video.

The proposed mechanism for key frame extraction is not a method which picks up the frames at some fixed regular intervals as many algorithms do, instead every content change within every shot is detected and key frames are extracted from instances of content change only. So the key frames extracted can construct a

(a)

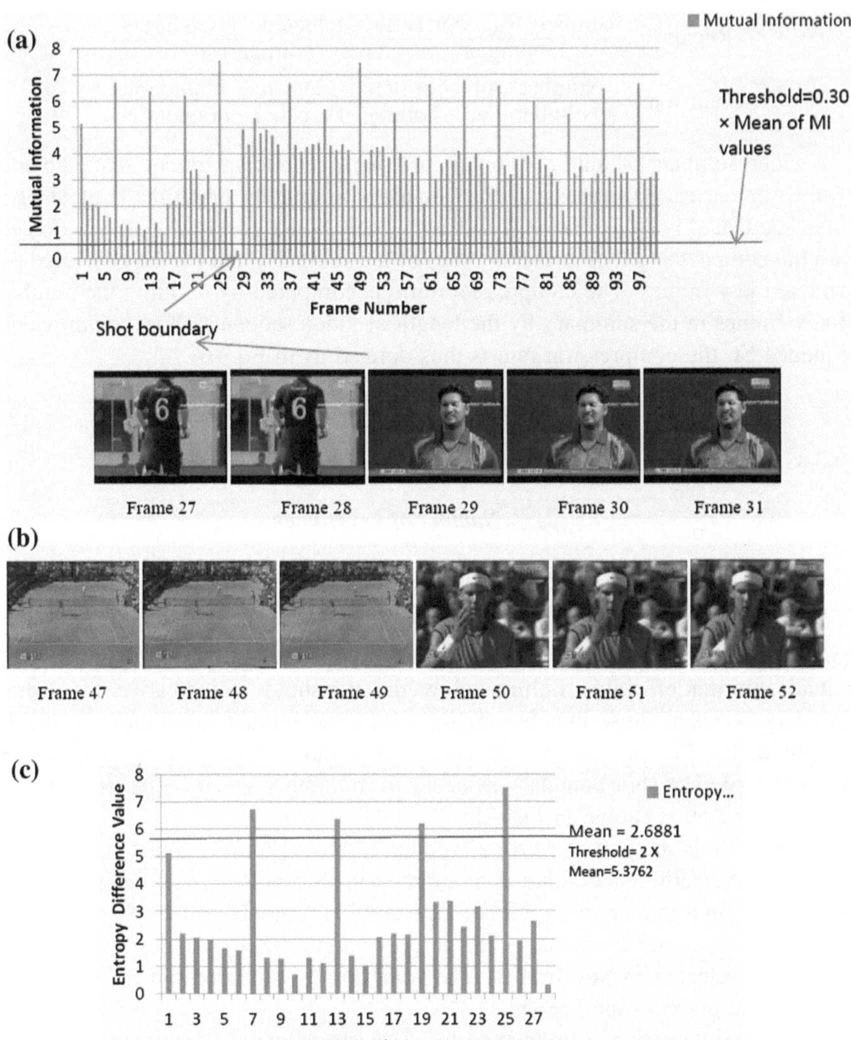

(b)

(c)

Fig. 2 **a** The plot of mutual information values against frame index and threshold and indication of shot boundary between frame 28 and frame 29. **b** A shot boundary detected between frame 49 and frame 50 using proposed algorithm. **c** The plot of spectral entropy difference values against frame index and threshold and indication of frames 7, 13, 19 and 25 as key frames

summary without redundant frame and covering all important information content of the video. The average values of performance parameters recall and precision up to 95 % for SBD algorithm and average compression ratio of 92 % for key frame extraction scheme indicates that the proposed mechanism is suitable for extraction of key frames to construct the summary of sports videos.

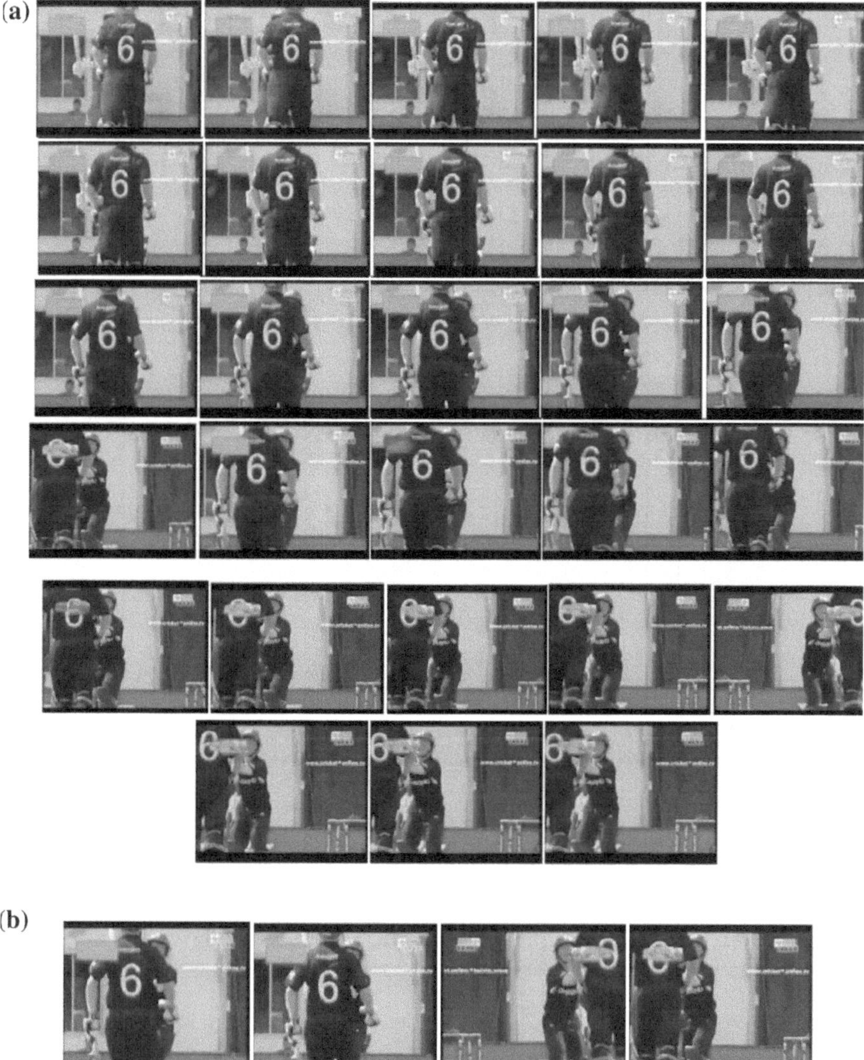

Fig. 3 **a** The 28 frames from representative shot separated from a video clip of 1,342 frames using mutual information based algorithm. **b** The 4 key frames extracted using proposed algorithm based on spectral entropy difference based algorithm

7 Conclusion

In this paper an algorithm for key frame extraction is presented, which selects the variable number of key frames from shots segmented using mutual information as difference and for similarity measure to differentiate two consecutive frames. The

Table 1 Performance parameters of proposed shot boundary detection algorithm and results of experimentation on keyframe extraction in terms of number of frames in a shot, number of key frames extracted and time taken by processor for extraction of key frames

Videos	Shot boundary detection		Key frame extraction							Compression ratio
	Recall	Precision	Shot no.	1	2	3	4	5	6	
1	1.000	1.000	No of frames	60	146	56	409	36	188	0.88
			Key frames	8	15	1	43	4	28	
			Time	23.89	53.02	22.05	148	13.49	67.01	
2	1.000	1.000	No of frames	52	170	188	395	174	131	0.91
			Key frames	8	19	12	42	18	14	
			Time	33.7	62.19	64.2	150.2	63.94	58.5	
3	0.857	0.950	No of frames	49	91	538	312	109	170	0.92
			Key frames	1	9	54	34	4	15	
			Time	16.1	31.53	176.19	99.2	56.97	72.46	
4	0.857	1.000	No of frames	60	146	58	409	36	188	0.90
			Key frames	5	16	6	42	3	28	
			Time	24.5	60.35	24.05	163.16	14.3	55.77	
5	0.903	0.975	No of frames	28	115	88	113	103	240	0.92
			Key frames	4	13	7	7	5	23	
			Time	10.84	39.92	32.62	57.42	43.15	86.59	
6	0.875	0.950	No of frames	640	660					0.90
			Key frames	68	70					
			Time	236	248					

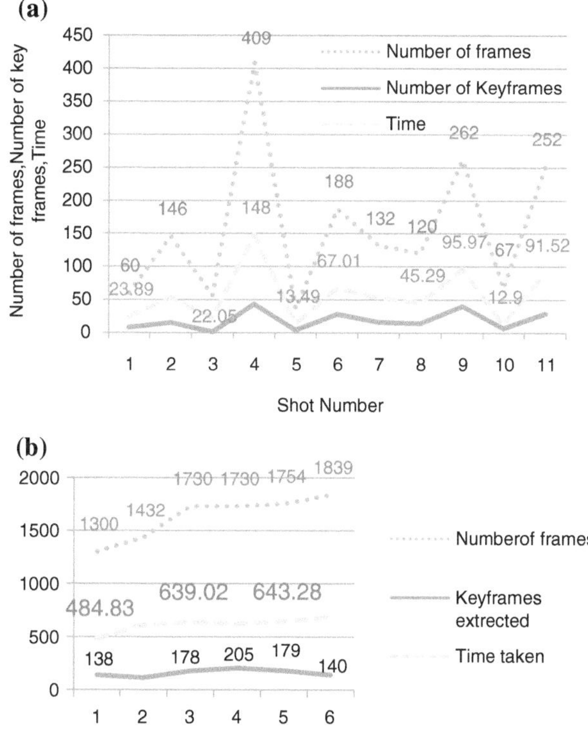

Fig. 4 **a** Graph of number of frames, number of key fames extracted plotted against shot number. **b** Plot of number of total number key frames extracted form different video sequences of different frame lengths and time taken by algorithm for extraction of key frames from different videos

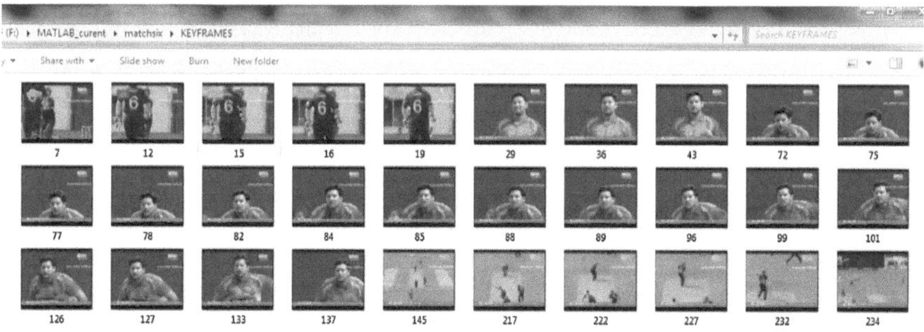

Fig. 5 The snapshot of folder containing total key frames extracted out of video of 1,342 frames

performance of the shot boundary detection algorithm used is presented in terms of *Precision* and *Recall*. The proposed key frame selection algorithm picks the key frames whenever there is change in the visual content of the shot, the change in the visual content is determined in terms of entropy change that occurs from frame to next frame in a shot. The entropy computation takes only eigen values of frames instead of complete image entropy, which will minimize computational complexity and time taken. The frames within a shot are extracted as key frame if their entropy difference value with previous frame crosses predefined threshold. The results have shown that the algorithm is able to summarize the video capturing all salient events in the video sequence. The compression ratio which is ratio of number of key frames utilized to build the summary to total number of frames in video for the algorithm is 92 %. The value of performance parameter reveal that the algorithm can preserve over all information while compressing video by over 92 %.

References

1. Yang S, Lin X (2005) Frame extraction using unsupervised clustering based on a statistical model. Tsinghai Sci Technol J 10(2):169–173
2. Zhuang Y, Rui Y, Huang TS, Metrotra S (2002) Adaptive key frame extraction using unsupervised clustering. In: Proceeding of IEEE international conference on image processing (ICIP'02), Chicago IL, pp 886–890
3. Lai J-L, Yi Y (2012) Key frame extraction based on visual attention model. J Vis Commun Image Process Elsevier 23(2012):114–125
4. Hanjalic A (2002) Shot-boundary detection: unraveled and resolved? IEEE Trans Circ Syst Video Technol 12(2):90–105
5. Girgensohn A, Boreczky J (2000) Time-constrained key frame selection technique. J Multimedia Tools Appl 11:347–358
6. Ju SX, Black MJ, Minneman S, Kimber D (1998) Summarization of videotaped presentations: automatic analysis of motion and gestures. IEEE Trans Circ Syst Video Technol 8(5):686–696
7. Turaga P, Veeraraghavan A (2009) Unsupervised view and rate invariant clustering of video sequences. Comput Vis Image Underst 3:353–371
8. Liu L, Fan G (2005) Combined key-frame extraction and object based video segmentation. IEEE Trans Circ Syst Video Technol 15(7):869–884
9. Setnes M, Babuska R (2001) Rule base reduction: some comments on the use of orthogonal transforms. IEEE Trans Syst Man Cybern Part C Appl Rev 31(2):199–206
10. Nagasaka A, Tanaka Y (1992) Automatic video indexing and full-video search for object appearances, in second working conference on visual database systems
11. Zhuangy Y, Rui Y, Huang TS, Mehrotra C (2007) An adaptive key frame extraction using unsupervised clustering. IEEE Trans Circ Syst Video Technol pp 168–186
12. Hanjalic A, Langendijk RL (2006) A new key-frame allocation method for representing stored video stream, 1st international workshop on image databases and multimedia search
13. Girgensohn A, Boreczky J (2001) Time-constrained key-frame selection technique. Multimedia Tools Appl
14. Chitra A, Dhawale and Sanjeev, Jain. SI (2008) Summarization a novel approach towards keyframe selection for video. Asian J Inf Technol
15. Tonomura Y, Akutsu A, Otsugi K, Sadakata T (1993) VideoMAP and videospaceicon: Tools for automatizing video content. In: Proceedings of ACM INTERCHI'93 conference, pp 131–141
16. Ueda H, Miyatake T, Yoshizawa S (1991) Impact: An interactive natural-motion-picture dedicated multimedia authoring system. In: Proceedings of ACM CHI'91 conference, pp 343–350

17. Rui Y, Huang TS, Mehrotra S (1998) Exploring video structure beyond the shots. In: Proceedings of IEEE international conference on multimedia computing and systems (ICMCS), Texas USA, pp 237–240
18. Pentland A, Picard R, Davenport G, Haase K (2003) Video and image semantics: advanced tools for telecommunications. IEEE Multimedia 1(2):73–75
19. Sun Z, Ping F (2004) Combination of color and object outline based method in video segmentation. In: Proceedings of SPIE storage and retrieval methods and applications for multimedia, 5307:61–69
20. Arman F, Hsu A, Chiu MY (1993) Image processing on compressed data for large video databases. In: Proceedings of ACM multimedia '93, Annaheim, CA, 1993:267–272
21. Savakis A, Rao RM (2003) Key frame extraction using in MPEG7 motion descriptors. In: Proceedings of 37th Asilomar conference on signals, systems and computers
22. Girgensohn A, Boreczky J (2000) Time-constrained keyframe selection technique. Multimedia Tools Appl 11:347–358
23. Gong Y, Liu X (2000) Generating optimal video summaries. In: Proceedings of IEEE international conference on multimedia and expo, 3:1559–1562
24. Zhao L, Qi W, Li SZ, Yang SQ, Zhang HJ (2000) Key-frame extraction and shot retrieval using nearest feature line (NFL). In: Proceedings of ACM international workshops on multimedia information retrieval, pp 217–220
25. Hanjalic A, Lagendijk RL, Biemond J (1998) A new method for key frame based video content representation. In: Image databases and multimedia search, World Scientific Singapore
26. Hoon SH, Yoon K, Kweon I (2000) A new technique for shot detection and key frames selection in histogram space. In: Proceedings of the 12th workshop on image processing and image understanding pp 475–479
27. Narasimha R, Savakis A, Rao RM, De Queiroz R (2004) A neural network approach to key frame extraction. In: Proceedings of SPIE-IS&T electronic imaging storage and retrieval methods and applications for multimedia, 5307:439–447
28. Calic J, Izquierdo E (2002) Efficient key–frame extraction and video analysis. In: Proceedings of IEEE ITCC'2002 multimedia web retrieval section, pp 28–33
29. Liu TM, Zhang HJ, Qi FH (2003) A novel video key-frame-extraction algorithm based on perceived motion energy model. IEEE Trans Circ Syst Video Technol 13(10):1006–1013
30. Cover TM, Thomas JA (2003) Elements of information theory, John Wiley & Sons Publications. ISBN 0-471-06259-6
31. Ciocca G, Schettini R (2004) An innovative algorithm for key frame extraction in video summarization. Int J Pattern Recogn Artif Intell 18(5):819–846
32. Truong BT, Venkatesh S (2007) Video abstraction: A systematic review and classification. ACM transactions on multimedia computing, communications and applications, vol 3, No. 1, Article 3
33. Ciocca G, Schettini R (2004) An innovative algorithm for key frame extraction in video summarization. J Real-Time Image Process 1(1):69–88

An Efficient Two Dimensional Multiple Real-Valued Sinusoidal Signal Frequency Estimation Algorithm

Prasad Kar Sambit and P. Palanisamy

Abstract In order to alleviate the effect of additive noise and to reduce the computational burden, we proposed a new computationally efficient cross-correlation based two-dimensional frequency estimation method for multiple real valued sinusoidal signals. Here the frequencies of both the dimensions are estimated independently with a one-dimensional (1-D) subspace-based estimation technique without eigendecomposition, where the null spaces are obtained through a linear operation of the matrices formed from the cross-correlation matrix between the received data. The estimated frequencies in both the dimensions are automatically paired. Simulation results show that the proposed method offers competitive performance when compared to existing approaches at a lower computational complexity. It has shown that proposed method perform well at low signal-to-noise ratio (SNR) and with a small number of snapshots.

Keywords 2-D frequency estimation · Sub-space based method: Real-valued sinusoidal model · Cross-correlation

P. K. Sambit (✉) · P. Palanisamy (✉)
Department of Electronics and Communication Engineering, National Institute of Technology, Trichy 620015, India
e-mail: sambitpk@gmail.com

P. Palanisamy
e-mail: palan@nitt.edu

Mohan S. and S. S. Kumar (eds.), *Proceedings of the Fourth International Conference on Signal and Image Processing 2012 (ICSIP 2012)*, Lecture Notes in Electrical Engineering 221, DOI: 10.1007/978-81-322-0997-3_9,

1 Introduction

In this paper, we consider the problem of estimating the frequencies of multiple two-dimensional (2-D) real-valued sinusoids in presence of additive white Gaussian noise. This problem is the more precise case of estimating the parameters of a 2-D regular and homogeneous random field from a single observed realization as of [1]. The real-valued 2-D sinusoidal signal models, also known as X-texture modes. These modes come into existence naturally in experimental, analytical, modal and vibrational analysis of circular shaped objects. X-texture modes are often used for modelling the displacements in the cross-sectional planes of isotropic, homogeneous, thick walled cylinders [2–4], laminated composite cylindrical shells [5], and circular plates [6]. These X-texture modes have also been used to describe the radial displacements of logs of spruce subjected to continuous sinusoidal excitation [7] and standing trunks of spruce subjected to impact excitation [8–10]. The proposed signal model offers cumbersome challenges for 2-D joint frequency estimation algorithms. Many algorithms for estimating complex-valued frequencies are well documented in the literatures [11, 12] and for 1-D real-valued frequencies in [13–15]. A vivid discussion on the problem of analyzing 2-D homogeneous random fields with discontinuous spectral distribution functions can be found in [16]. Parameter estimation techniques of sinusoidal signals in additive white noise include the periodogram-based approximation (applicable for widely spaced sinusoids) to the maximum-likelihood (ML) solution [17–19], the Pisarenko harmonic decomposition [20], or the singular value decomposition [21]. A matrix enhancement and matrix pencil method for estimating the parameters of 2-D superimposed, complex-valued exponential signals was suggested in [11]. In [22], the concept of partial forward–backward averaging is proposed as a means for en-hancing the frequency and damping estimation of 2-D multiple real-valued sinusoids (X-texturemodes) where each mode considered as a mechanism for forcing the two plane waves towards the mirrored direction-of-arrivals. In [23], 2-D parameter estimation of a single damped/undamped real/complex tone is proposed which is referred to as principal-singular-vector utilization for modal analysis (PUMA).

We present a new approach of solving the 2-D real valued sinusoidal signal frequencies estimation problem based on cross-correlation technique to resolve the identical frequencies. The proposed idea based upon the computationally efficient subspace based method without eigendecomposition (SUMWE) [24, 25]. The paper is organized as follows. The signal model, together with a definition of the addressed problem, is presented in Sect. 2. The basic definition and the proposed technique both are detailed in Sect. 3 followed by simulation results and conclusion in Sects. 4 and 5, respectively. Throughout this paper upper case, bold letters denote matrices where as lowercase bold letters are vectors. The superscript T denotes transposition of a matrix.

2 Data Model and Problem Definition

Consider the following set of noisy data:

$$\mathrm{r(m,n) = x(m,n) + e(m,n)} \tag{1}$$

where $0 \leq m \leq N_1-1$ and $0 \leq n \leq N_2-1$. The model of the noiseless data x(m,n) is described by,

$$x(m,n) = \sum_{k=1}^{D} a_k cos(\omega_k m + \varphi_{1k}) cos(v_k n + \varphi_{2k}) \tag{2}$$

The signal x(m,n) consists of D, two dimensional real-valued sinusoids described by normalized 2-D frequencies $\{\omega_k\ v_k\}$, (k = 1,2…,D), the real amplitude $\{a_k\}$ (k = 1,2…,D) and the phases ϕ_{1k} and ϕ_{2k} which are independent random variables uniformly distributed over $[0,2\pi]$. e(m,n) is a zero mean additive white Gaussian noise with variance σ^2. Further assumed that α_k and β_k are independent of e(m,n).

Let us define two $M \times 1$ snapshot vectors with assumption $M > D$, described as follows

$$\mathbf{y}_\omega(\mathrm{m,n}) \triangleq \frac{1}{2}[\mathbf{y}_1(\mathrm{m,n}) + \mathbf{y}_2(\mathrm{m,n})] \tag{3a}$$

$$\mathbf{y}_v(\mathrm{m,n}) \triangleq \frac{1}{2}[\mathbf{y}_3(\mathrm{m,n}) + \mathbf{y}_4(\mathrm{m,n})] \tag{3b}$$

where,

$$\mathbf{y}_1(\mathrm{m,n}) \triangleq [\mathrm{r(m,n)\ r(m+1,n)...r(m+M-1,n)}]^T \tag{4a}$$

$$\mathbf{y}_2(\mathrm{m,n}) \triangleq [\mathrm{r(m,n)\ r(m-1,n)...r(m-M+1,n)}]^T \tag{4b}$$

$$\mathbf{y}_3(\mathrm{m,n}) \triangleq [\mathrm{r(m,n)\ r(m,n+1)...r(m,n+M-1)}]^T \tag{4c}$$

$$\mathbf{y}_4(\mathrm{m,n}) \triangleq, left[\mathrm{r(m,n)\ r(m,n-1)...r(m,n-M+1)}]^T \tag{4d}$$

From the above set of equations we can obtain pair of expression for the two $M \times 1$ snapshot vectors by substituting equation (4a, b) in (3a) and (4c, d) in (3b) as follows,

$$y_\omega(m,n) = A(\omega)s(m,n) + g(m,n) \tag{5}$$

$$y_v(m,n) = A(v)s(m,n) + h(m,n) \tag{6}$$

where $A(\omega) = [\gamma(\omega_1)\ldots\gamma(\omega_D)]$ and $A(v) = [\rho(v_1)\ldots\rho(v_D)]$ are $M \times D$ matrices, and $s(m,n) = [a_1\cos(\omega_1 m + \alpha_1)\cos(v_1 n + \beta_1)\ldots a_D\cos(\omega_D m + \alpha_D)\cos(v_D n + \beta_D)]^T$ is the $D \times 1$ signal vector, $\gamma(\omega_i)$ and $\rho(v_i)$ are $M \times 1$ vectors defined respectively as

$\gamma(\omega_i)$=[1 $\cos(\omega_i)\ldots\cos((M-1)\omega_i)]^T$ and $\rho(v_i) = [1\ \cos(v_i)\ldots\cos((M-1)v_i)]^T$. The modified M × 1 error vectors g(m,n) and h(m,n) are defined respectively as $\mathbf{g}(m,n)\ \triangleq[g_1(m,n)\ g_2(m,n)\ldots\ g_M(m,n)]^T$ and $\mathbf{h}(m,n)\triangleq[h_1(m,n)\ h_2(m,n)\ldots h_M(m,n)]^T$, where $g_j(m,n) = 1/2[e(m+j-1,n) + e(m-j+1,n)]$ and $h_j(m,n) = 1/2[e(m,n+j-1) + e(m,n-j+1)]$. The matrices A(ω) and A(v) are full rank matrices because all the columns are linearly independent to each other.

2.1 Data Model Modification

We first obtained two new data models as follows,

$$\mathbf{z}_{\omega}(m,n) = \mathbf{A}(\omega)\boldsymbol{\Omega}_{\omega}\mathbf{s}(m,n) + \mathbf{q}_r(m.n) \tag{7}$$

$$\mathbf{z}_{v}(m,n) = \mathbf{A}(v)\boldsymbol{\Omega}_{v}\mathbf{s}(m,n) + \mathbf{q}_e(m.n) \tag{8}$$

by implementing following mathematical operations,

$$z_{\omega}(m,n) = \frac{1}{4}\sum_{j=1}^{4} z_j(m,n) \tag{9a}$$

$$\mathbf{z}_{v}(m,n) = \frac{1}{4}\sum_{j=5}^{8} \mathbf{z}_j(m,n) \tag{9b}$$

where

$$\mathbf{z}_1(m,n) \triangleq [r(m+1,n)\ r(m,n)\ r(m-1,n)\ldots r(m-M+2,n)]^T \tag{10a}$$

$$\mathbf{z}_2(m,n) \triangleq [r(m-1,n)\ r(m,n)\ r(m+1,n)\ldots r(m+M-2,n)]^T \tag{10b}$$

$$\mathbf{z}_3(m,n) \triangleq [r(m-1,n)\ r(m-2,n)\ r(m-3,n)\ldots r(m-M,\ n)]^T \tag{10c}$$

$$\mathbf{z}_4(m,n) \triangleq [r(m+1,n)\ r(m+2,n)\ r(m+3,n)\ldots r(m+M,n)]^T \tag{10d}$$

$$\mathbf{z}_5(m,n) \triangleq [r(m,n+1)\ r(m,n)\ r(m,n-1)\ldots r(m,n-M+2)]^T \tag{11a}$$

$$\mathbf{z}_6(m,n) \triangleq [r(m,n-1)\ r(m,n)\ r(m,n+1)\ldots r(m,n+M-2)]^T \tag{11b}$$

$$\mathbf{z}_7(m,n) \triangleq [r(m,n-1)\ r(m,n-2)\ r(m,n-3)\ldots r(m,n-M)]^T \tag{11c}$$

$$\mathbf{z}_8(m,n) \triangleq [r(m,n+1)\ r(m,n+2)\ r(m,n+3)\ldots r(m,n+M)]^T \tag{11d}$$

where $z_i(m,n)$ for i = 1,2…8 are M × 1 observation vectors. Ω_{ω} and Ω_v are two D × D diagonal matrices defined respectively as $\Omega_{\omega} = \mathrm{diag}\{\cos\omega_1\ldots\cos\omega_D\}$ and $\Omega_v = \mathrm{diag}\{\cos v_1\ldots\cos v_D\}$. The two M × 1 modified noise vectors $q_r(m,n)$ and

$q_e(m,n)$ are defined respectively as $q_r(m,n) = [q_{r1}(m,n)\ldots q_{rM}(m,n)]^T$, $q_e(m,n) = [q_{e1}(m,n) \quad q_{e2}(m,n)\ldots q_{eM}(m,n)]^T$ where $q_{ri}(m,n) = 1/4[e(m-i+2,n) + e(m+i-2,n) + e(m+i,n) + e(m-i,n)]$ and $q_{ei}(m,n) = 1/4[e(m,n-i+2) + e(m,n+i-2) + e(m,n+i) + e(m,n-i)]$ for $i = 1,2\ldots,M$.

2.2 Further Modification of Data Model

As like Sect. 2.1, we deduced another set of modified data models described by,

$$\mathbf{p}_{\omega}(m,n) = \mathbf{JA}(\omega)\Omega_{\omega}\mathbf{s}(m,n) + \mathbf{q}_{w}(m.n) \tag{12}$$

$$\mathbf{p}_{v}(m,n) = \mathbf{JA}(v)\Omega_{v}\mathbf{s}(m,n) + \mathbf{q}_{u}(m.n) \tag{13}$$

The above two data models were obtained by implementing similar kind of mathematical operations as that of equation (9a, b) that is,

$$\mathbf{p}_{\omega}(m,n) = \frac{1}{4}\sum_{j=1}^{4}\mathbf{p}_j(m,n) \tag{14}$$

$$\mathbf{p}_{v}(m,n) = \frac{1}{4}\sum_{j=5}^{8}\mathbf{p}_j(m,n) \tag{15}$$

where

$$\mathbf{p}_1(m,n) \triangleq [r(m-M+2,\ n)\ldots r(m-1,\ n)\ r(m,\ n)\ r(m+1,\ n)]^T \tag{16a}$$

$$\mathbf{p}_2(m,n) \triangleq [r(m+M-2,\ n)\ldots r(m+1,\ n)\ r(m,\ n)\ r(m-1,\ n)]^T \tag{16b}$$

$$\mathbf{p}_3(m,n) \triangleq [r(m-M,\ n)\ldots r(m-3,\ n)\ r(m-2,\ n)\ r(m-1,\ n)]^T \tag{16c}$$

$$\mathbf{p}_4(m,n) \triangleq [r(m+M,\ n)\ldots r(m+3,n)\ r(m+2,n)\ r(m+1,n)]^T \tag{16d}$$

$$\mathbf{p}_5(m,n) \triangleq [r(m,\ n-M+2)\ldots r(m,\ n-1)\ r(m,\ n)\ r(m,n+1)]^T \tag{17a}$$

$$\mathbf{p}_6(m,n) \triangleq [r(m,\ n+M-2)\ldots r(m,\ n+1)\ r(m,\ n)\ r(m\ ,n-1)]^T \tag{17b}$$

$$\mathbf{p}_7(m,n) \triangleq [r(m,\ n-M)\ldots r(m,\ n-3)\ r(m,\ n-2)\ r(m,\ n-1)]^T \tag{17c}$$

$$\mathbf{p}_8(m,n) \triangleq [r(m,\ n+M)\ldots r(m,\ n+3)\ r(m,\ n+2)\ r(m,\ n+1)]^T \tag{17d}$$

where $p_i(m,n)$ for $i = 1,2\ldots8$ are $M \times 1$ observation vectors and J is the $M \times M$ counter identity matrix, in which 1 s present in the principal anti-diagonal. The two $M \times 1$ modified noise vectors $q_w(m,n)$ and $q_u(m,n)$ are defined respectively as $q_w(m.n) = Jq_r(m.n)$ and $q_u(m.n) = Jq_e(m.n)$.

3 Proposed Algorithm

In this section, we present the algorithm for 2-D frequency estimation for multiple real-valued sinusoidal signals.

3.1 Estimation of First Dimension Frequencies

Under the assumption of data model, from (5) and (7) we easily obtain the cross correlation matrix R_{yz1} between the received data, $y_\omega(m,n)$ and $z_\omega(m,n)$ as,

$$\mathbf{R}_{yz1} = E\{\mathbf{y}_\omega(m,n)\mathbf{z}_\omega^T(m,n)\} = \mathbf{A}(\omega)\mathbf{R}_{ss}\Omega_\omega\mathbf{A}^T(\omega) \tag{18}$$

where R_{ss} is source signal correlation matrix defined by $\mathbf{R}_{ss} \triangleq E\{\mathbf{s}(m,n)\mathbf{s}^T(m,n)\}$. From (12), we have another data model that is $p_\omega(m,n)$ in backward way such that $p_\omega(m,n) = Jz_\omega(m,n)$, similarly from (5) and (12) we can obtain another cross-correlation matrix between the two received data

$$\mathbf{R}_{yp1} = E\{\mathbf{y}_\omega(m,n)\mathbf{p}_\omega^T(m,n)\} = \mathbf{A}(\omega)\mathbf{R}_{ss}\mathbf{J}\Omega_\omega\mathbf{A}^T(\omega) \tag{19}$$

In noise free case $R_{yp1} = JR_{yz1}$ but in practical case that is when signal is noise corrupted, then the relation holds true partially that is $\mathbf{R}_{yp1} \cong \mathbf{J}\mathbf{R}_{yz1}$. Considering the above assumptions we formulated an extended cross correlation of size $M \times 2M$ as,

$$\mathbf{R}_\omega = [\mathbf{R}_{yz1}\mathbf{R}_{yp1}] = [\mathbf{R}_{yz1}\mathbf{J}\mathbf{R}_{yz1}] = \mathbf{A}(\omega)\left[\mathbf{R}_{ss}\Omega_\omega\mathbf{A}^T(\omega)\mathbf{R}_{ss}\mathbf{J}\Omega_\omega\mathbf{A}^T(\omega)\right] \tag{20}$$

Since $A(\omega)$ is a full rank matrix, we can divide $A(\omega)$ into two sub matrices as $A(\omega) = [(A_1(\omega))^T\ (A_2(\omega))^T]^T$ where $A_1(\omega)$ and $A_2(\omega)$ are the $D \times D$ and $(M\text{-}D) \times D$ sub matrices consisting of the first D rows and last (M-D) rows of the matrix $A(\omega)$ respectively. There exists a $D \times M\text{-}D$ linear operator P_1 between $A_1(\omega)$ and $A_2(\omega)$ [26] such that $A_2(\omega) = P_1^T A_1(\omega)$, using the above assumptions we can segregate (20) into the following two matrices.

$$\begin{aligned}\mathbf{R}_\omega &= \left[(\mathbf{A}_1(\omega))^T(\mathbf{A}_2(\omega))^T\right]^T\left[\mathbf{R}_{ss}\Omega_\omega\mathbf{A}^T(\omega)\mathbf{R}_{ss}\mathbf{J}\Omega_\omega\mathbf{A}^T(\omega)\right] \\ &= \left[(\mathbf{A}_1(\omega))^T\mathbf{P}_1^T\mathbf{A}_1(\omega)\right]^T\left[\mathbf{R}_{ss}\Omega_\omega\mathbf{A}^T(\omega)\mathbf{R}_{ss}\mathbf{J}\Omega_\omega\mathbf{A}^T(\omega)\right] \\ &\triangleq \left[\mathbf{R}_{\omega1}^T\mathbf{R}_{\omega2}^T\right]^T\end{aligned} \tag{21}$$

where $R_{\omega1}$ and $R_{\omega2}$ consist of the first D rows and the last M-D rows of the matrix R_ω, and $\mathbf{R}_{\omega2} = P_1^T\mathbf{R}_{\omega1}$. Hence, the linear operator P_1 found from $R_{\omega1}$ and $R_{\omega2}$ as [26]. However, a least-squares solution [27] for the entries of the propagator matrix P_1 satisfying the relation, $\mathbf{R}_{\omega2} = P_1^T\mathbf{R}_{\omega1}$ obtained by minimizing the cost function described as follows,

$$\xi(\mathbf{P}_1) = ||\mathbf{R}_{\omega 2} - \mathbf{P}_1^{\mathrm{T}}\mathbf{R}_{\omega 1}||^2 \tag{22}$$

where $||.||_{\mathrm{F}}^2$ denotes the Frobenius norm. The cost function $\xi(\mathrm{P}_1)$ is a quadratic (convex) function of P_1, which can be minimized to give the unique least-square solution for P_1, that can be evidently shown as,

$$\mathbf{P}_1 = \left(\mathbf{R}_{\omega 1}\mathbf{R}_{\omega 1}^{\mathbf{T}}\right)^{-1}\mathbf{R}_{\omega 1}\mathbf{R}_{\omega 2}^{\mathbf{T}} \tag{23}$$

further by defining another matrix $Q_\omega = [P_1^T - I_{M-D}]^T$, such that $\mathrm{Q}_\omega^{\mathrm{T}}\mathrm{A}(\omega) = 0_{(\mathrm{M}\text{-}\mathrm{D})\times\mathrm{D}}$ which can be used to estimate the real valued harmonic frequencies of first dimension $\{\omega_{\mathrm{k}}\}$ for $\mathrm{k} = 1,2\ldots,\mathrm{D}$ as like [25]. Thus when the number of snapshots are finite the frequencies of first dimension can be estimated by minimizing following cost function, $\hat{f}(\omega) = a^T(\omega)\hat{E}a(\omega)$ where $a(\omega) = [1cos\omega\ldots cos(M-1)\omega]^T$ and $\hat{E}\triangleq\hat{Q}_\omega(\hat{Q}_\omega^T\hat{Q}_\omega)^{-1}\hat{Q}_\omega^T$. The orthonormality of matrix $\hat{Q}_\omega$ is used in order to improve the estimation performance while E is calculated implicitly using matrix inversion lemma as [24] and $\hat{E}\ and \hat{Q}_\omega$ are the estimates of E and Q_ω.

Steps for estimating ω_{k}:

- Calculate the estimate $\hat{R}_\omega$ of the cross-correlation matrix R_ω using (20).
- Partition $\hat{R}_\omega$ and determine $\hat{R}_{\omega 1}$ and $\hat{R}_{\omega 2}$.
- Determine the estimate of the propagator matrix P_1 using (23).
- Define $\hat{Q}_\omega = [\hat{P}_1^T - I_{M-D}]^T$ and from $\hat{Q}_\omega$ find out $\hat{E}\triangleq\hat{Q}_\omega(\hat{Q}_\omega^T\hat{Q}_\omega)^{-1}\hat{Q}_\omega^T$ using matrix inversion lemma.
- The first dimension frequencies that is, $\{\omega_{\mathrm{k}}\}$ for $\mathrm{k} = 1,2\ldots,\mathrm{D}$ estimated by minimizing the following cost function, $\hat{f}(\omega) = a^T(\omega)\hat{E}a(\omega)$.

3.2 Estimation of Second Dimension Frequencies

The method adopted for estimating the first dimension frequencies ω_{i} for $\mathrm{i} = 1, 2,\ldots,\mathrm{D}$, can be used for estimating the second dimension frequencies v_{i} for $\mathrm{i} = 1, 2,\ldots,\mathrm{D}$. That is the same procedure used in Sect. 3.1 of this Section applied to estimate second dimension frequencies. The second dimension frequencies obtained by doing similar kind of operation across the data models developed in (6), (8) and (13).

The proposed method has notable advantages over the conventional MUSIC algorithm [15], such as computational simplicity and less restrictive noise model. Though it required peak search but there is no eigenvalue decomposition (SVD or EVD) involved in the proposed algorithm unlikely MUSIC, where the EVD of the auto correlation matrix is needed. It also provide quite efficient estimate of the frequencies and the estimated frequencies in both dimensions are automatically paired.

4 Simulation Results

Computer simulation have been carried out to evaluate the frequency estimation performance of the proposed algorithm for 2-D multiple real-valued sinusoids in presence of additive white Gaussian noise. The average root-mean-square-error (RMSE) is employed as performance measure, apart from that some other simulations also conducted to show the detection capability and bias of estimation. Besides CRLB, the performance of the proposed algorithm is compared with those of 2D-MUSIC and 2D-ESPRIT [28] algorithms for real-valued sinusoids. Four type of analysis have been performed.

4.1 Analysis of Frequency Spectra

The signal parameters are $N_1 = N_2 = 50$ and the dimension of snapshot vector $M = 20$. Number of undamped 2-D real-valued sinusoids $D = 2$, amplitude $\{a_k = 1\}$ for $k = 1,2\ldots,D$. The first dimension frequencies, and the second dimension frequencies are $(\omega_1,\omega_2) = (0.1\pi,0.13\pi)$ and $(v_1,v_2) = (0.13\pi,0.16\pi)$ respectively. Note that the frequency separation is 0.03, which is smaller than the Fourier resolution capacity 1/M (=0.05). This means classic FFT-based method cannot resolve these two frequencies, and also this method can resolve identical frequencies present in different dimensions ($\omega_2 = v_1 = 0.13\pi$). Figure 1 displays spectra of the proposed algorithm at SNR = 10 dB. We can see from Fig. 1 that the frequency parameters in both the dimensions are accurately resolved. The estimated frequencies are shown in Table 1.

In second analysis that is shown in Fig. 2 where we considered the signal parameters are $N_1 = N_2 = 100$ and the dimension of snapshot vector $M = 50$, keeping all other parameters same as previous experiment. The detection of frequencies in both the dimensions are found to be more accurate. The estimated frequencies of this analysis are shown in Table 2.

4.2 Performance Analysis Considering RMSE

The same signal parameters as first analysis of Sect. 4.1 of this section is considered. We compared root-mean-square-error (RMSE) on the estimates for the proposed algorithm, MUSIC and 2-D ESPRIT algorithm as a function of SNR. Here a Monte-Carlo simulation of 500 runs was performed. Figure 3a shows the

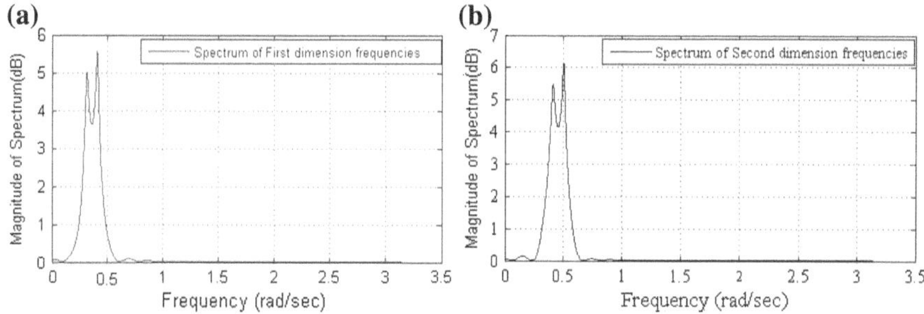

Fig. 1 Spectrum of frequencies in both dimensions (M = 20)

Table 1 Estimated frequencies considering M = 20

Frequency	Original frequency(rad/sec)	Estimated frequency(rad/sec)
ω_1	0.1π	0.1012π
ω_2	0.13π	0.1285π
v_1	0.13π	0.1289π
v_2	0.16π	0.1594π

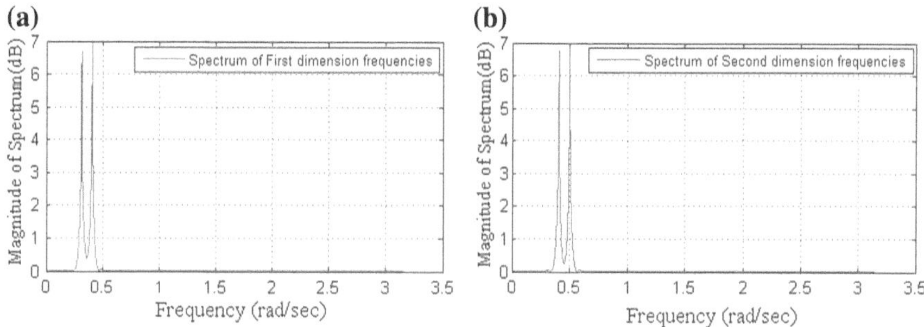

Fig. 2 Spectrum of frequencies in both dimensions (M = 50)

Table 2 Estimated frequencies considering M = 50

Frequency	Original frequency(rad/sec)	Estimated frequency(rad/sec)
ω_1	0.1π	0.1003π
ω_2	0.13π	0.1303π
v_1	0.13π	0.1303π
v_2	0.16π	0.1604π

RMSEs and the corresponding CRB of first 2-D frequencies $\{\omega_k\}$, while Fig. 3b shows the second 2-D frequencies $\{v_k\}$ (k = 1,2…,D). It is clearly seen that the proposed algorithm outperforms the ESPRIT algorithm and in lower SNR case the

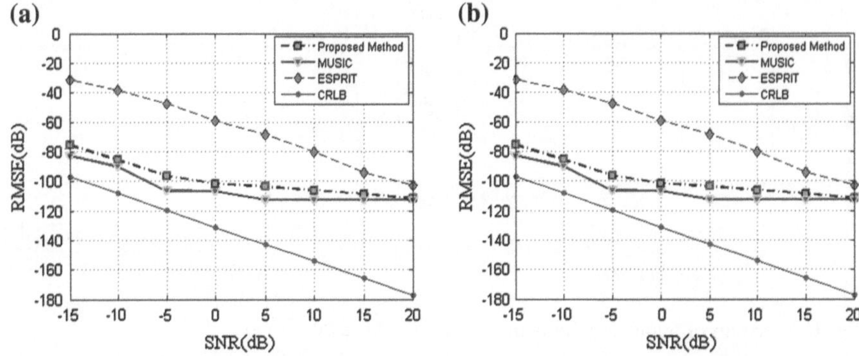

Fig. 3 **a** RMSE (dB) for first dimension frequencies vs SNRs (dB). **b** RMSE (dB) for second dimension frequencies vs SNRs (dB)

performance is similar to that of MUSIC algorithm. As SNR increases the proposed algorithm performs exactly same as that of MUSIC algorithm.

4.3 Performance Analysis Considering Probability of Correct Estimation and Bias of Estimation

In this analysis, we considered Probability of correct estimation of frequencies as performance measure. Taking the same signal parameters as of last two Sections, we determined the probability of correct estimation of 2-D real-valued sinusoidal signal frequencies for both dimensions by varying SNR. The obtained results are shown in Fig. 4a, b. From the above analysis, it is evident that proposed method performs far superior compared to 2-D ESPRIT and behaves in a same way as conventional MUSIC algorithm but without any eigendecomposition (EVD/SVD). Similarly we analyzed the bias of estimation for each dimensions and the results are plotted in Fig. 5a, b respectively. From bias analysis, it is clear that the proposed method performs much better than ESPRIT and almost similar to that conventional MUSIC algorithm in varied SNR ranges.

4.4 Performance Analysis Considering Computational Time

In this section we compared the performance of proposed method and conventional MUSIC algorithm based on their computational timing. Considering the same signal parameters at a fixed SNR of 10 dB we vary the snap shot vector dimension (M) and the results are plotted in Fig. 6. From Fig. 6 its clear that proposed method is less time consuming compared to conventional MUSIC algorithms.

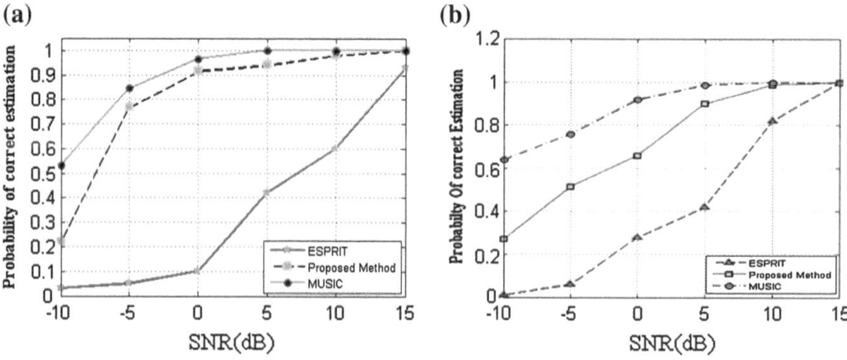

Fig. 4 **a** Probability of correct estimation for first dimension frequencies vs SNRs (dB). **b** Probability of correct estimation for second dimension frequencies vs SNRs (dB)

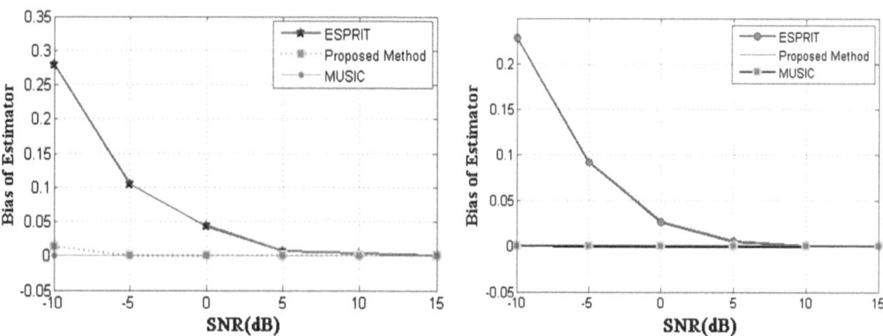

Fig. 5 **a** Bias of the estimator for first dimension frequencies vs SNR's (dB). **b** Bias of the estimator for second dimension frequencies vs SNR's (dB)

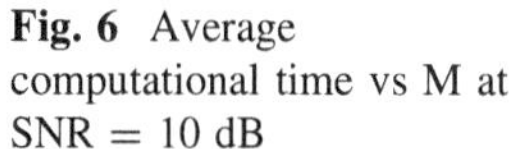

Fig. 6 Average computational time vs M at SNR = 10 dB

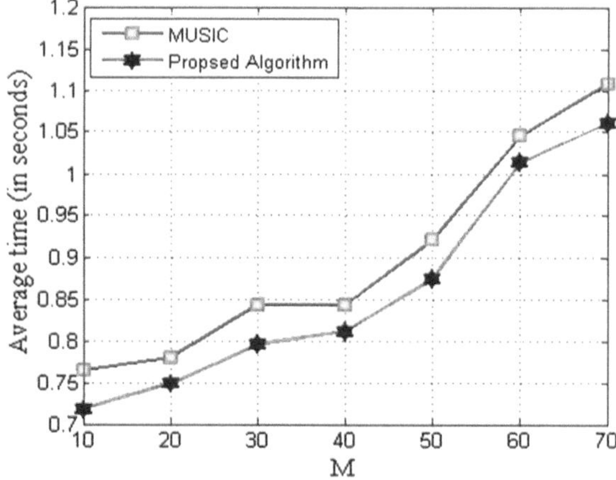

5 Conclusion

We have proposed a new approach based on subspace method without eigendecompostion using cross-correlation matrix for estimation of multiple real-valued 2-D sinusoidal signal frequencies embedded with additive white Gaussian noise. We have analytically quantified the performance of the proposed algorithm. It is shown that our algorithm remains operational when there exist identical frequencies in both the dimensions. Simulation results show that the proposed algorithm offers comparative performance when compared to MUSIC algorithm, but at a lower computational complexity and exhibit far superior performance when compared to ESPRIT algorithm. The frequency estimates thus obtained are automatically paired without an extra pairing algorithm.

References

1. Francos JM, Friedlander B (1998) Parameter estimation of two-dimensional moving average random fields. IEEE Trans Signal Process 46:2157–2165
2. Wang H, Williams K (1996) Vibrational modes of cylinders of finite length. J Sound Vib 191(5):955–971
3. Verma S, Singal R, Williams K (1987) Vibration behavior of stators of electrical machines, part I: theoretical study. J Sound Vib 115(1):1–12
4. Singal R, Williams K, Verma S (1987) Vibration behavior of stators of electrical machines, part II: experimental study. J Sound Vib 115(1):13–23
5. Zhang XM (2001) Vibration analysis of cross-ply laminated composite cylindrical shells using the wave propagation approach. Appl Acoust 62:1221–1228
6. So J, Leissa A (1998) Three-dimensional vibrations of thick circular and annular plates. J Sound Vib 209(1):15–41
7. Skatter S (1996) TV holography as a possible tool for measuring transversevibration of logs: a pilot study. Wood Fiber Sci 28(3):278–285
8. Axmon J, Hansson M (1999) Estimation of circumferential mode parameters of living trees. In: Proceedings of the IASTED international conference on signal image process (SIP'99), pp 188–192
9. Axmon J (2000) On detection of decay in growing norway spruce via natural frequencies. Licentiate of Engineering Thesis Lund University. Lund, Sweden, Oct 2000
10. Axmon J, Hansson M, Sörnmo L (2002) Modal analysis of living spruce using a combined Prony and DFT multichannel method for detection of internal decay. Mech Syst Signal Process 16(4):561–584
11. Hua Y (1992) Estimating two-dimensional frequencies by matrix enhancement and matrix pencil. IEEE Trans Signal Process 40(9):2267–2280
12. Haardt M, Roemer F, DelGaldo G (2008) Higher-order SVD-based subspace estimation to improve the parameter estimation accuracy in multidimensional harmonic retrieval problems. IEEE Trans Signal Process 56(7):3198–3213
13. Mahata K (2005) Subspace fitting approaches for frequency estimation using real-valued data. IEEE Trans Signal process 53(8):3099–3110
14. Palanisamy P, Sambit PK (2011) Estimation of real-valued sinusoidal signal frequencies based on ESPRIT and propagator method. In: Proceedings of the IEEE-international conference on recent trends in information technology, pp 69–73 June 2011

15. Stoica P, Eriksson A (1995) MUSIC estimation of real-valued sine-wave frequencies. Signal process 42(2):139–146
16. Priestley MB (1981) Spectral analysis and time series. Academic, New York
17. Walker M (1971) On the estimation of a harmonic component in a time series with stationary independent residuals. Biometrika 58:21–36
18. Rao CR, Zhao L, Zhou B (1994) Maximum likelihood estimation of 2-D superimposed exponential signals. IEEE Trans Signal Process 42:1795–1802
19. Kundu D, Mitra A (1996) Asymptotic properties of the least squares estimates of 2-D exponential signals. Multidim Syst Signal Process 7:135–150
20. Lang SW, McClellan JH (1982) The extension of Pisarenko's method to multiple dimensions. In: Proceedings of the international conference on acoustics, speech, and signal processing, pp 125–128
21. Kumaresan R, Tufts DW (1981) A two-dimensional technique for frequency wave number estimation. Proc IEEE 69:1515–1517
22. Axmon J, Hansson M, Sörnmo L (2005) Partial forward–backward averaging for enhanced frequency estimation of real X-texture modes. IEEE Trans Signal Process 53(7):2550–2562
23. So HC, Chan FKW, Chan CF (2010) Utilizing principal singular vectors for two-dimensional single frequency estimation. In: Proceedings of the IEEE international conference on acoustics, speech, signal process, pp 3882–3886, March 2010
24. Xin J, Sano A (2004) Computationally efficient subspace-based method for direction-of-arrival estimation without eigen decomposition. IEEE Trans Signal process 52(4):876–893
25. Sambit PK, Palanisamy P (2012) A propagator method like algorithm for estimation of multiple real-valued sinusoidal signal frequencies. Int J Electron Electr Eng 6:254–258
26. Marcos S, Marsal A, Benidir M (1995) The propagator method for source bearing estimation. Signal Process 42(2):121–138
27. Golub GH, Van Loan CF (1980) An analysis of the total least squares problem. SIAM J Numer Anal 17:883–893
28. Rouquette S, Najim M (2001) Estimation of frequencies and damping factors by two-dimensional ESPRIT type methods. IEEE Trans Signal Process 49(1):237–245

Speech Separation Based on Time Frequency Ratio of Mixtures and Track Identification

M. K. Prasanna Kumar

Abstract Analysis of non stationary signals like audio, speech and biomedical signals require good resolution both in time and frequency as their spectral components are not fixed. There are many applications of time frequency analysis in non stationary signals like source separation, signal denoising, automatic gain control, speaker recognition etc. This paper presents an application of time frequency analysis using STFT, Short Time Fourier Transform in speech and audio separation. This method is known as Blind Source Separation. The method is blind since the information about the sources and mixing type is not available. The method uses relative amplitude information and time frequency ratios of audio and speech mixtures in time frequency domain and ideal binary mask of source signals. A mixture of male speech, female speech and tones of musical instruments are considered for the separation first with a strong mixing matrix and next with a weak mixing matrix.

Keywords Short time Fourier transform · Binary masking · Automatic speech recognition · Time–frequency domain · Ideal mask · Ratio of mixtures

1 Introduction

When a signal mixture is represented as a linear combination of the original sources at every time instant, it is defined as an instantaneous mixing model. For recording in a real room the microphones in the environment picks up not only the

M. K. Prasanna Kumar (✉)
Department of Electronics and Communication Engineering,
Vivekananda College of Engineering and Technology,
Puttur, Karnataka, India
e-mail: prasanna251983@yahoo.com

Mohan S. and S. S. Kumar (eds.), *Proceedings of the Fourth International Conference on Signal and Image Processing 2012 (ICSIP 2012)*, Lecture Notes in Electrical Engineering 221, DOI: 10.1007/978-81-322-0997-3_10, © Springer India 2013

signals of the sources, but also the delayed and attenuated versions of the same signals. This type of mixtures are called convolved mixtures. There are many instances where it is necessary to separate signals from a mixture to obtain the original sources, by observing only the mixture i.e. without prior knowledge of the original sources. This is a typical requirement especially in speech enhancement applications. Blind Source Separation (BSS) [1] is one way of solving this kind of problems.

In this proposed model, an underdetermined speech and audio mixtures with strong and weak mixing matrices are considered at multiple observations. Speech separation is done in Time- Frequency domain by using relative amplitude information and time frequency ratios of Short Time Fourier Transform of individual observations. Performance evaluation can be done in three ways. (1) Visual Observation (2) Auditory Observation (3) Mathematical Observation. Visual observations can be done through spectrograms and histograms. Auditory observations can be done by hearing original and recovered speech signals. Finally mathematical analysis can be done by calculating and comparing Signal to Noise Ratio (SNR) at various observations.

2 Methodology

In the proposed model Blind separation is done in Time–Frequency domain using STFT (Short Time Fourier Transform). Since audio and speech signals are non stationary in nature it is not adequate to use Fourier Transform. BSS methodology is dependent on the type of input signals. If the input signal is stereo audio track then both left and right channel signals can be analyzed separately in Time Frequency domain for the elimination of unwanted signals. If the input signal is a mono track like most of the recorded speech signals then the mixture of multiple sources has to be analyzed at various sensors or observations in Time frequency domain to eliminate the unwanted signal. Figure 1 depicts the various stages involved in implementing the proposed work.

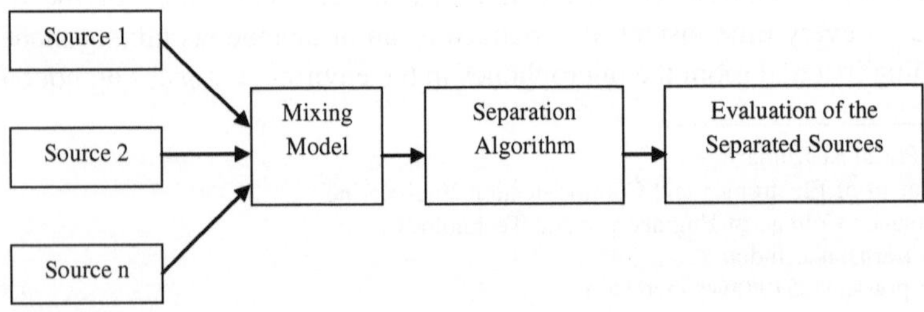

Fig. 1 Various stages involved in speech separation

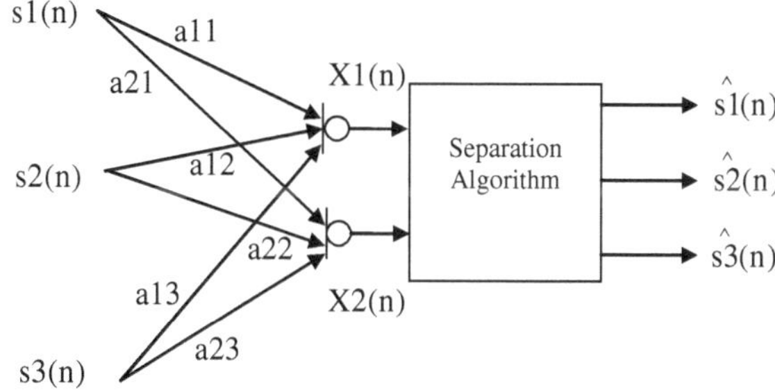

Fig. 2 Instantaneous underdetermined mixing model

3 Mixing Model

The speech mixture used is under determined mixture [2, 3] where number of source signals are greater than number of sensors or observations (Fig. 2).

$$S = 3 \text{ and } X = 2 \tag{1}$$

where S is number of sources and X is number of sensors or observations. The mixing model shown above can be modeled with following equations.

$$X = A.S \tag{2}$$

where X is the observation matrix, S is the source matrix, and A is the mixing matrix.

$$X = [X1(n), X2(n)]^T \tag{3}$$

where X1 (n) and X2 (n) are the observation vector at sensor1 and sensor 2 respectively.

$$s = [s1(n), s2(n), s3(n)] \tag{4}$$

where s1 (n), s2 (n) and s3 (n) are the source vector 1 source vector 2 and source vector 3 respectively.

$$A = \begin{bmatrix} a11\ a12\ a13 \\ a21\ a22\ a23 \end{bmatrix} \tag{5}$$

$$X1(n) = a11.s1(n) + a12.s2(n) + a13.s3(n) \tag{6}$$

$$X2(n) = a21.s1(n) + a22.s2(n) + a23.s3(n) \tag{7}$$

where a11, a12, a13, a21, a22, a23 are the coefficients of mixing matrix. Separation algorithm uses STFT and relative amplitudes of each time frequency cluster. Separation algorithm is discussed in next section. s1^, s2^ and s3^ are the recovered source 1, source 2 and source 3 respectively.

4 Separation Algorithm

Separation algorithm involves finding STFT (Short Time Fourier Transform) of speech mixtures individually. Once the individual STFT is obtained in the matrix format next step is to find the STFT ratio. STFT ratio is the ratio of STFT1 and STFT2 where STFT1 is the Short Time Fourier Transform of mix1 at observation 1 and STFT2 is the Short Time Fourier Transform of mix2 at observation 2. It is the element by element division which is indicated by the symbol './'. Since both STFT's are obtained in the matrix format STFT ratio is defined by STFT1./STFT2 (Fig. 3).

Figure 4 shows frequency of STFT ratios between observation 1 and observation 2. Both mixtures at observation 1 and observation 2 consist of all the three source signals with varying amplitude levels depending on the attenuation level. The first smaller peak represents the first source signal, middle larger peak represents second source which is assumed to be at equal distance from both observations. In the proposed model it is assumed that source 2 is at equal distance from both observations. Hence it is observed that larger peak corresponding to second source signal is obtained at a ratio of '1'. It indicates that source 2 is equally sensed by observation 1 and observation 2. The last smaller peak represents the third source signal. The first and last peaks are smaller compared to the middle peak since source 1 and source 3 are not sensed equally due to varying attenuation levels at observation 1 and observation 2. Once the tracks are identified [4, 5], proceed to substitute them by zeros. This method is usually known as binary

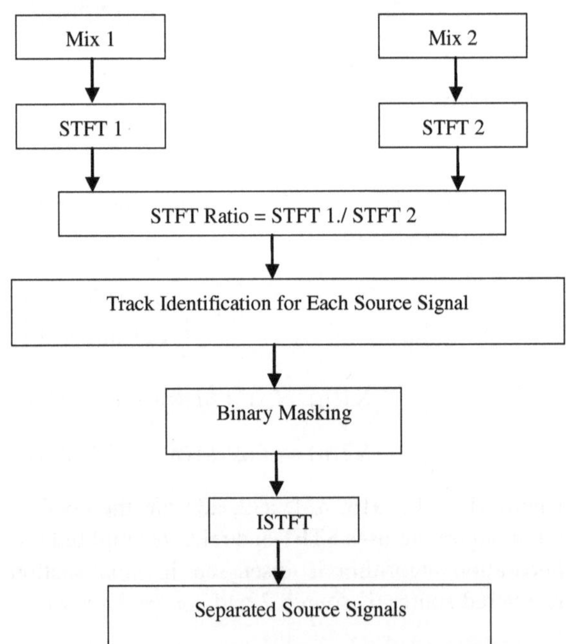

Fig. 3 Various stages in separation algorithm

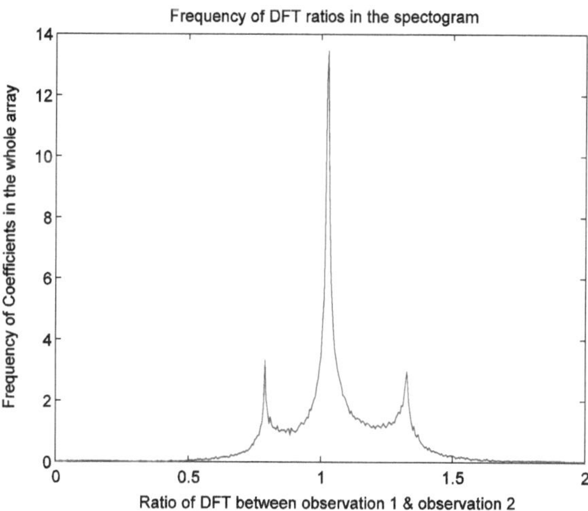

Fig. 4 Frequency STFT ratios

masking because the coefficients are multiplied by either one or by zero. This can be seen mathematically as follows

$$M = \begin{cases} 0 & \text{if} \quad \mathrm{a} < \mathrm{Ratio} < \mathrm{b} \\ 1, & \text{otherwise} \end{cases} \tag{8}$$

where M is the binary mask. From the Fig. 4 for first peak a = 0.7 and b = 0.8, for the second peak a = 0.9 b = 1.1 and for the last peak a = 1.3 and b = 1.4. After the binary mask has been applied, the signal has to be transformed back to the time domain. In order to do this, the ISTFT is used on each one of the observations.

5 Speech Separation

The separation algorithm discussed in Sect. 4 is successfully applied to a mixture with a male speech, a female speech and a musical tone. The entire analysis and synthesis results are shown in the following spectrograms. Since the mixture considered here is under determined it is required to consider lesser observations than source signals.

Figure 5 shows steps involved in separation of under determined speech mixtures with musical note for S = 3 and X = 2 in detail. Here S is the number of source signals and X is the number of observations or sensors. In Fig. 5 (a) represents the spectrogram of original male speech (b) spectrogram of original female speech (c) spectrogram of original saxophone tune (d) spectrogram of speech mixture (e) & (f) spectrogram of mixture after masking saxophone (g) spectrogram of mixture after masking female speech (h) spectrogram of mixture after masking both saxophone and female speech (i) spectrogram of speech

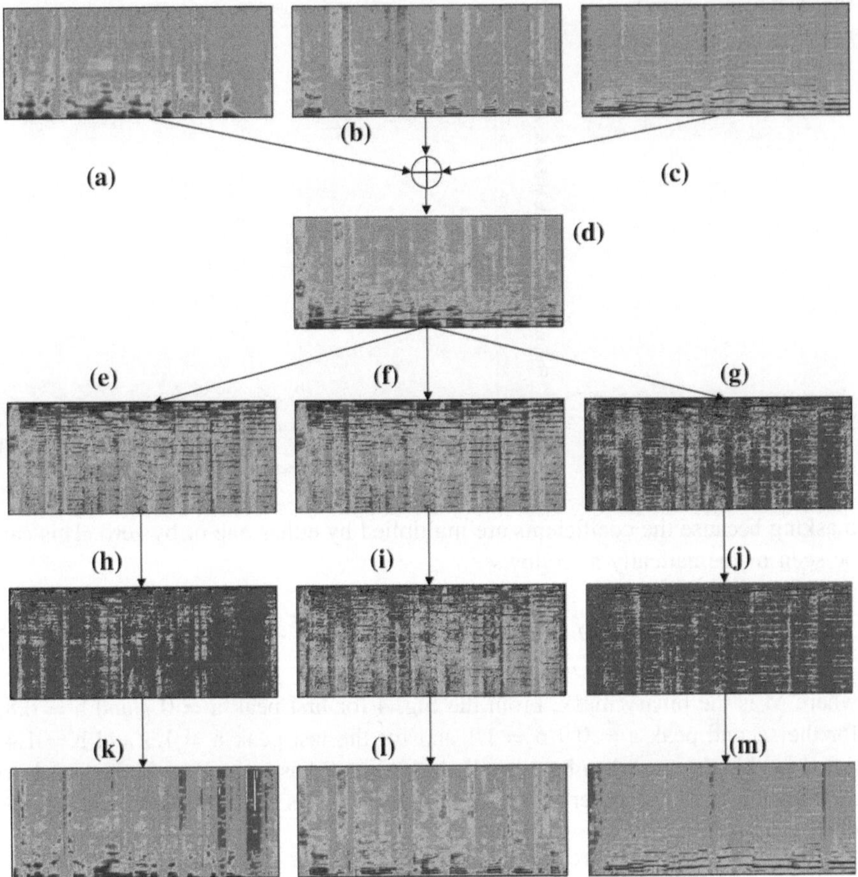

Fig. 5 Spectrograms at various stages of speech separation

mixture after masking both saxophone and male speech (j) spectrogram of speech mixture after masking both female and male speech (k) recovered male speech from the mixture (l) recovered female speech from the mixture (m) recovered saxophone tune from the mixture.

6 Performance Evaluation

In order to analyze the results of source separation, a method for performance evaluation should be used based on distortion measures [6]. These distortion measures take into account interference from other sources, signal to noise ratio (SNR) and artifacts introduced by the algorithm of source separation. Performance of separation algorithm was evaluated using SNR. Performance was evaluated for a strong mixing matrix and also for a weak mixing matrix. The results obtained are tabulated.

$$\text{SNR} = 20\,\log_{10}\frac{\|\text{output}\|}{\|\text{output - input}\|} \tag{9}$$

where output is the normalized array of output after source separation and input is the normalized array of original input speech signal. The above formula is applied for both male and female speech signal (Table 1).

The mixing matrix used for under determined mixture with S = 3 and X = 2 is given by

$$\text{Mixing matrix} = \begin{bmatrix} 0.9 & 0.8 & 0.7 \\ 0.7 & 0.8 & 0.9 \end{bmatrix} \tag{10}$$

Table 1 Signal to noise ratio at various observations

SNR in db		
	Observation 1	Observation 2
Male	26.7588	25.8489
Female	22.3014	38.5487
Saxophone	33.3538	22.0818

7 Conclusion

In the proposed model initially an instantaneous mixture of speech and audio signals were analyzed (with 20 kHz sampling rate), with different gains in relation to each source and sensor position. An ideal binary mask was obtained for the speech signals. Objective is to analyze the speech separation using an ideal mask and using only information of the relative amplitude, without the phase information (relative delay). Time–Frequency analysis using STFT gives greater flexibility in choosing the masking region. This is due to the Fact that large number of frequency components can be extracted for the better selectivity.

8 Future Work

There is no unique solution to Blind Source Separation Applications. The separation algorithm should be modified according to the application. Further research can be done on development of unique algorithms for various applications of speech separation. Since audio and speech signals are non stationary in nature FT (Fourier Transform) is inadequate for the separation of multiple signals. Hence transforms like STFT and Wavelets has to be used which are more complex. Further wavelet transforms can be applied in order to achieve good resolution both in time and frequency.

References

1. Abrard F, Deville Y (2005) Atime frequency blind signal separation method applicable to underdetermined mixtures of dependent sources. Signal Process 85(7):1389–1403
2. Yilmaz O, Rickard S (2004) Blind separation of speech mixtures via time-frequency masking. IEEE Trans Signal Process 52(7):1830–1847
3. Araki S, Makino S, Sawada H, Mukai R (2004) Underdetermined blind separation of convolutive mixtures of speech with directivity pattern based mask and ica. Fifth international conference on independent component analysis and blind signal separation, pp 898–905
4. Torkkola K (1996) Blind separation of convolved sources based on information maximization. IEEE Worshop on neural networks for signal processing, Kyoto, pp 423–432
5. Bell AJ, Sejnowski TJ (1995) An information-maximization approach to blind separation and blind deconvolution. Neural Comput 7(6):1129–1159
6. Vincent E, Gribonval R, F'evotte C (2006) Performance measurement in blind audio source separation. IEEE Trans Speech Audio Process 14(4):1462–1469

Designing a Perception Based Anti-Aliasing Filter for Enhancement of Down-Sampled Images

M. Sankaralingam, S. Arya and Apurba Das

Abstract In this paper, the problem of aliasing due to pixel based image down-sampling in CMYK color space is addressed. Such a problem exists when a high-resolution image or video is to be mapped to low-resolution. Signal processing theory tells us that optimal decimation requires low-pass filtering with a suitable cutoff frequency, followed by down-sampling which remove many useful image details blurring. Instead of operating in the entire image, the proposed method finds the edge maps and then applies anti-aliasing filters only on the edge map regions excluding the horizontal and vertical edges. The algorithm shows a significant reduction of the aliasing artifacts, commonly known as "jaggies". Perceptual relative color dominance which is calculated from psycho visual experiments is included in the anti-aliasing part to improve the performance of the algorithm. The number of color quantization levels is varied separation by separation for each color channel and psycho visual survey is conducted to find the perceptual color dominance.

Keywords Anti-aliasing · Image down-sampling · Color quantization · ANOVA · Psycho-visual perception

M. Sankaralingam (✉) · A. Das
Imaging Tech Lab, HCL Technologies Ltd, Chennai, India
e-mail: sankarm@hcl.com

A. Das
e-mail: apurba.das@hcl.com

S. Arya
Amrita Vishwa Vidyapeetham, Coimbatore, India
e-mail: aryasukumar88@gmail.com

Mohan S. and S. S. Kumar (eds.), *Proceedings of the Fourth International Conference on Signal and Image Processing 2012 (ICSIP 2012)*, Lecture Notes in Electrical Engineering 221, DOI: 10.1007/978-81-322-0997-3_11, © Springer India 2013

1 Introduction

Both displays and print devices have a limited number of pixels or dots per unit area, respectively and each pixel or dot can represent only a limited number of colors. A low sampling frequency results in a jagged or blocky outline in digital images, when the resolution is insufficient to show the detail. The effect is called aliasing. Techniques for reducing aliasing are often applied in computer graphics [1, 2]. This work concentrates on the aliasing due to image down-sampling where a high-resolution image or video is to be displayed or mapped onto a low-resolution display or print terminals. Anti-aliasing is used to give a smoothening effect on curved boundaries/edges of objects. When down-sampling is applied, the curved regions will be affected. A common approach of anti-aliasing is to divide the pixel grid into a finer grid of sub pixels, and with respect to the relative occupancy of target color, the color intensities are adjusted. To achieve smoother edge transitions, number of quantization levels are increased [3, 4], which degrades the performance in terms of execution speed, as it implies realizing a finer granularity sub pixel grid. Considering the cause of aliasing problem as low resolution, one easy solution is to increase the resolution, causing sample points to occur more frequently [5]. By the method of super-sampling, high frequencies components are eliminated [6], whereas in pre-filtering method, anti-aliasing is achieved by considering the image intensities over a particular region [7, 8]. That is why, this method works well for non-area primitives like points, lines etc., not for area primitives like triangles, especially in 3-D.

In the proposed work, the perceptual color dominance concept is combined with anti-aliasing for image down-sampling. The perceptual relative color dominance among the C, M, Y, K color planes is calculated by conducting psycho-visual experiments, as described in Sect. 2. In Sect. 3, the anti-aliasing approach is explained with experimental results. Finally in Sect. 4 we have concluded our findings with some directions to future work.

2 Relative Color Dominance by Perception

Perception based anti-aliasing uses the relative color dominance which is calculated from a set of psycho visual experiments. From the experiments the less dominant color plane is identified.

Four different color quantization patches, like the color calibration patches for printer, are used for this psycho-visual experiment as shown in Fig. 1.

Reduction in number of valid colors in images results perceptual degradation. We have reduced the number of quantization levels in each of the separations (C, M, Y, and K) and studied their effect on perceptual quality degradation with respect to the original image. Some sample quantization cases as shown in the Fig. 1 are as follows:

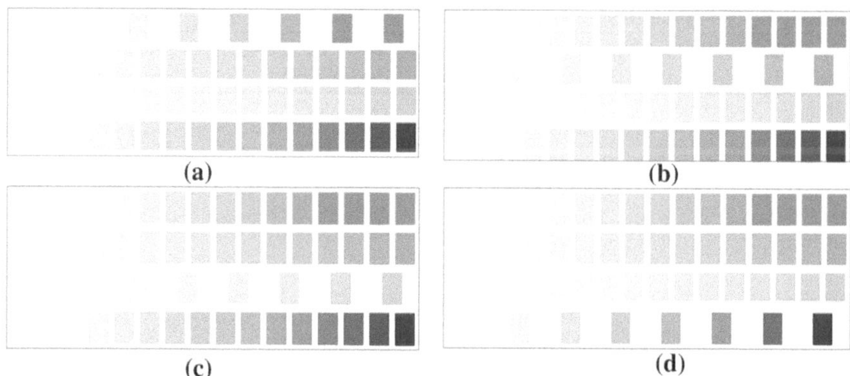

Fig. 1 Color quantization patch

Case 1: 16 level Quantization for Cyan and 32 for other planes.
Case 2: 16 level Quantization for Magenta and 32 for other planes.
Case 3: 16 level Quantization for Yellow and 32 for other planes.
Case 4: 16 level Quantization for Black and 32 for other planes.

A set of 500 images were taken from standard image databases like USC-SIPI [9], Oriental bird images [10], BSDS500 [11]. All the images were subjected to four cases of quantization as described and were used for the psycho visual survey [12]. For the survey, the subjects (observers) were seated 3 ft from the calibrated CRT in a darkened room in order to enhance the visualization quality. 50 trained persons for psycho-visual experiments and 50 untrained persons were taken as subjects. Each of the subjects is given same 30 sets of original image and four quantized images, with four separations (CMYK) attack. The subjects indicated the most qualitatively similar image to the original image by a mouse click. The result of the survey is presented in Table 1. In the next subsection, we have computed a statistical analysis on this table to understand the statistical significance of the outcome.

2.1 Anova

Here, a balanced one-way ANalysis Of VAriance (ANOVA) is performed for comparing the means of four columns of data in the 2-by-4 matrix X, as presented in Table 1 where each column represents an independent sample containing 2 mutually independent observations [12, 13]. The function returns the p value for the null hypothesis that all samples in X are drawn from the same population. If the p value is near zero, this casts doubt on the null hypothesis and suggests that at least one sample mean is significantly different from the other sample means. It is common to declare a result "statistically significant" if the p value is less than 0.05.

Table 1 Psycho-visual survay on color quantization

Quantization groups	Reduced levels in C plane	Reduced levels in M plane	Reduced levels in Y plane	Reduced levels in K plane
Untrained	212	372	650	266
Trained	90	467	705	238

Table 2 One way ANOVA on color quantization

ANOVA Table

Source	SS	df	MS	F	Prob>F
Columns	317583	3	105861	30.55	0.0032
Error	13859	4	3464.8		
Total	331442	7			

The ANOVA table (Table 2) has six columns: The first shows the source of the variability. The second shows the Sum of Squares (SS) due to each source. The third shows the degrees of freedom (df) associated with each source. The fourth shows the Mean Squares (MS) for each source, which is the ratio SS/df. The fifth shows the F statistic, which is the ratio of the MS's. The sixth shows the *p* value, which is derived from the cdf of F. As F increases, the *p* value decreases. Figure 2 displays box plots of each column of X. Large differences in the center lines of the box plots correspond to large values of F and correspondingly small *p* values. The very small p-value of 0.0032 indicates that differences between the column means are highly significant. The probability of this outcome under the null hypothesis (i.e., the probability that samples actually drawn from the same population would have means differing by the amounts seen in X) is around 32 in 1,000.

The test therefore strongly supports the alternate hypothesis, that one or more of the samples are drawn from populations with different means. The box plot also infers that, the yellow color plane is less relatively perceptual prominent with respect to other color planes, i.e., even if we make small color changes in yellow plane, it is not that much susceptible by human eye. This result indicates that the yellow plane may not be taken into consideration while applying the anti-aliasing, without any compromise in quality. It reduces the computational cost.

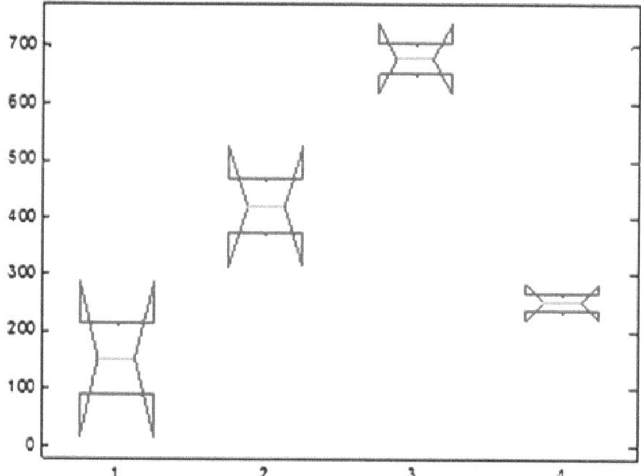

Fig. 2 Box plot of the ANOVA on color quantization

3 Proposed Methodology of Anti-Aliasing

3.1 Detailed Flow of the Algorithm

Anti-aliasing is one of the most important processes for generating high quality graphics and images. A low sampling frequency results in a jagged or blocky outline in digital images, when the resolution is insufficient to show the detail Anti-Aliasing works by utilizing the perceptual weakness of eye in identifying small color changes made. The effect of anti-aliasing can be realized here, by filter arrangement as shown in Fig. 3. Contour map of the high resolution image is calculated using Sobel mask [15–17] filtering on each color plane (C, M, Y, K) of the image separately, instead of computing the contour detection on the gray scale image of the input, as the object contours are different in each plane, with respect to its color description. Some contours identified in one plane may not be present in other plane, in that case no need to perform anti-aliasing in such areas. Since the jagged edge effects are seen on the curved contours only, anti-aliasing on horizontal and vertical contours can be excluded from this filtering process.

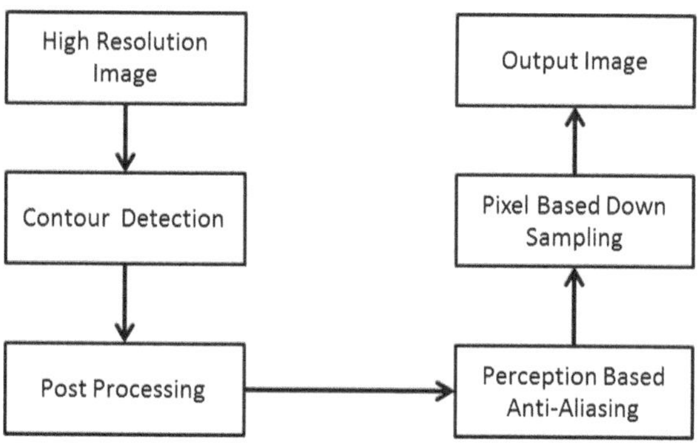

Fig. 3 Block diagram of proposed anti-aliasing filter

From the experimental results it has been observed that the anti-aliasing on the contour map alone will not give pleasing result [18]. To reduce the aliasing effect on the curved edges, morphological filtering based post processing operations are done on the contour map. In the proposed method, anti-aliasing is performed on the neighborhood of contours to reduce the jaggies.

A low pass filer is the basis for smoothing methods [19]. An image is smoothed by decreasing the disparity between pixel values by averaging nearby pixels. From the Table 3, it has been observed that Gaussian kernel based low pass filtering is better for anti-aliasing (AA) than weighted average filter based low pass filtering.

The visual effect of this blurring technique is a smooth blur resembling that of viewing the image through a translucent screen, distinctly different from the bokeh effect produced by an out-of-focus lens or the shadow of an object under usual illumination. The final stage of this algorithm is image down-sampling to the required output resolution. Here we are using pixel based down-sampling [14]. In this method the down-sampling is performed by selecting one out of every N pixel. The detailed data flow of perception based anti-aliasing filter is depicted in Fig. 4.

In the proposed anti-aliasing (AA), the Gaussian filter of size 5×5 is used. Left most column of Tables 3 and 4 show the image after pixel based down sampling where, jagged effects are clearly visible. From the lead of perception based experiment, we have finally removed the yellow plane from the purview of

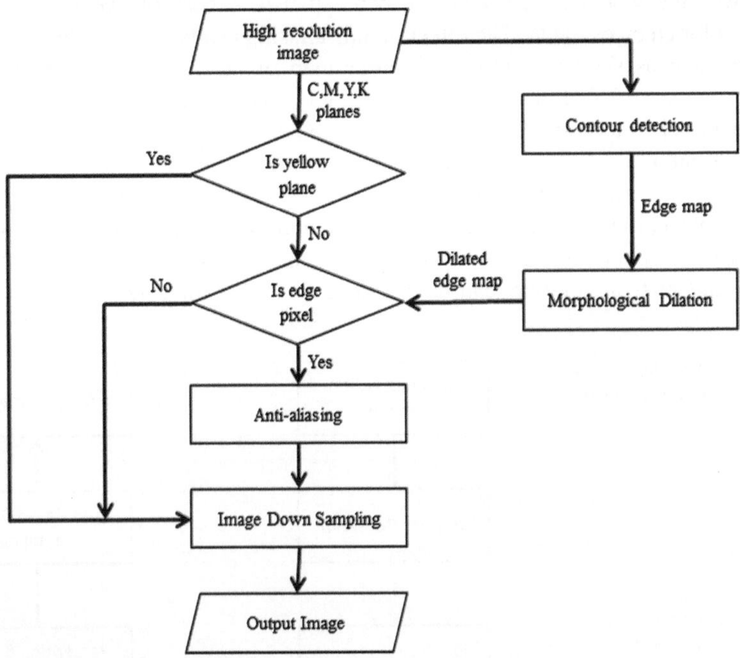

Fig. 4 Data flow of perception based anti-aliasing filter

Table 3 Performance comparison between weighted average and Gaussian AA filters

Downsampled image	Weighted average AA	Gaussian AA

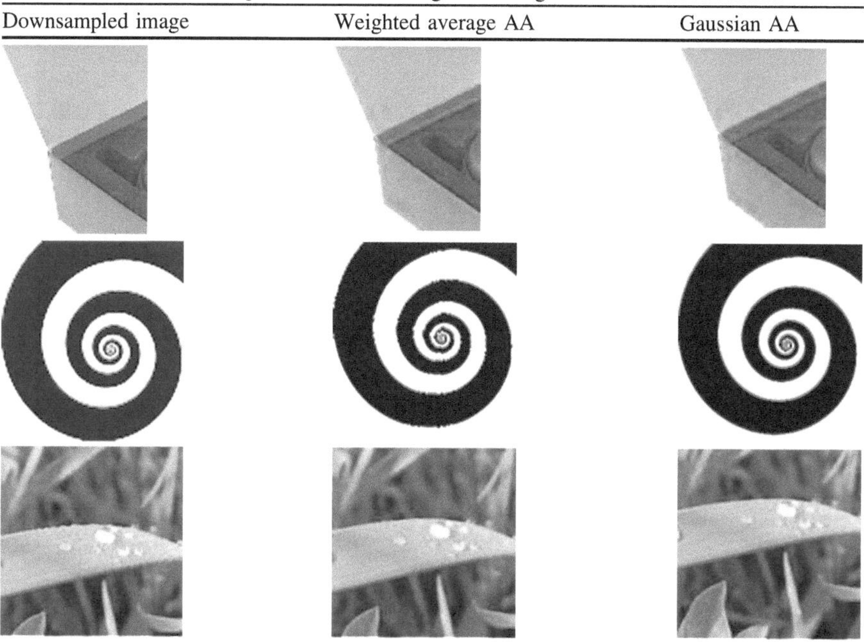

AA filter based enhancement to reduce the computational complexity, without compromising the visual quality of the down sampled image (Table 4).

3.2 Computational Complexity of the Proposed Method

Let's consider one image with X and Y number of row pixels and column pixels, respectively. Therefore, the computational complexity involved in conventional Anti-aliasing filter approach for the CMYK based color image is,

Complexity:

$$fx(4 \times X \times Y) \tag{1}$$

Computational complexity for the proposed method where CMK separations are only used, is as follows,

Complexity:

$$fy(3 \times Ex \times Ey) \tag{2}$$

$$Ex = X - X1 \tag{3}$$

$$Ey = Y - Y1 \tag{4}$$

Table 4 Quality preservation by removing yellow plane from AA filtering

Downsampled image	AA to CMYK planes	AA to CMK planes

where,

Ex is the number of edge pixels in a row of an image
Ey is the number of edge pixels in a column of an image
X1 is the number of non-edge pixels in a row of an image
Y1 is the number non-edge pixels in a column of an image

The complexity of the function (*f*) for the proposed method per pixel is defined as *fy*, since the method involves edge detection followed by filtering.

$$fy = f1 + f2 \tag{5}$$

where,

$$f1(filtering) = \{((hx * hy \rightarrow multiplications), (hx * hy - 1) \rightarrow additions)\} \tag{6}$$

$$f2(edge) = \{((4 \rightarrow logicalshift), (10) \rightarrow additions)\} \tag{7}$$

Here, *hx* and *hy* are the filter dimension in x and y directions and the complexity function can be written as follows,

$$Complexity(proposed) = 3 * (f1 + f2)\,\{Ex, Ey\} \tag{8}$$

and the complexity of the function (*fx*) in the conventional filtering approach is,

$$fx = 4f1 \tag{9}$$

$$Complexity(conventional) = 4 * f1\{\mathrm{X},\ \mathrm{Y}\} \tag{10}$$

Equations 8 and 10 are the evidences of realizing the anti-aliasing feature much efficiently with respect to the traditional one in terms of computational complexity, hence in terms of processing time.

4 Conclusion

Perception based Anti-aliasing for down-sampling of CMYK images has been implemented. Lowest dominancy of "yellow" plane is proved by psycho visual experiments and statistical hypothesis on its result. Consequently, avoidance of yellow plane from anti-aliasing, maintains the same image quality, visually. Contour detection was done separately for each color planes and the horizontal and vertical edges were not taken into consideration for filtering. All these helped to reduce the complexity of the algorithm. Finally, anti-aliasing on the morphologically filtered contour map was done. The final output is observed to have the details and quality of the high resolution input image.

References

1. Franklin C (1977) Crow "The aliasing problem in computer-generated shaded images". Commun ACM 20:799–805
2. Yeh Y-H, Lee C-Y (1999) A new anti-aliasing algorithm for computer graphics images. Proceedings of the international conference on image processing, vol 2. pp 442–446
3. Feiner S, Foley J, van Dam A, Hughes J (1996) Computer graphics: principles and practice, 2nd edn. Addison-Wesley, Reading
4. Oyvind R (2006) Applications of antialiasing in image processing framework setting. Signal Processing Symposium, pp 106–109
5. Ferwerda J, Greenberg D (1988) A psychophysical approach to assessing the quality of anti-aliased images. IEEE Comput Graph Appl 8:85–95
6. Kajiya J, Ullner M (1981) Filtering high quality text for display on raster scan devices. In: Computer graphics (SIGGRAPH'81 Proceedings), vol 15. pp 7–15
7. Wu X (1991) An efficient antialiasing technique. In: Proceedings of the 18th annual conference on computer graphics and interactive techniques, vol 25. pp 143–152
8. Blinn J (1989) Jim Blinn's corner-return of the Jaggy (high frequency filtering). IEEE Comput Graphics Appl 9:82–89
9. http://sipi.usc.edu/database/database.php?volume=misc
10. http://orientalbirdimages.org/birdimages.php
11. http://www.eecs.berkeley.edu/Research/Projects/CS/vision/grouping/BSR/BSR_bsds500.tgz
12. Das A, Parua S (2012) Psycho-visual evaluation of contrast enhancement algorithms by adaptive neuro-fuzzy inference system. Lect Notes Comput Sci 7143:75–83 Springer

13. Spiegel M, Schiller J, Srinivasan A (2004) Theory and problems of probability and statistics (Schaum S Outline Series) 2nd edn. Tata McGraw Hill, New Delhi
14. Gupta S, Sproull RF (1981) Filtering edges for gray-scale displays. Comput Graph 15:1–5
15. Bærentzen J, Nielsen S, Gjøl M, Larsen B (2008) Two methods for anti-aliased wireframe drawing with hidden line removal. Proceedings of the spring conference in computer graphics
16. Ahmad MB, Choi TS (1999) Local threshold and Boolean function based edge detection. IEEE Trans Consum Electron 45:332–333
17. Matthews J (2002) An introduction to edge detection: The Sobel edge detector. Available at http://www.generation5.org/content/2002/im01.asp
18. Burgett SR, Das M (1991) Multiresolution multiplicative auto regression coding of images. In: Proceedings IRRR conference on systems engineering, pp 276–279
19. Gonzalez C, Woods RE (2002) Digital image processing, 2nd edn. Prentice Hall, Englewood Cliffs

Efficient Video Processing Method for Traffic Monitoring Combining Motion Detection and Background Subtraction

Roberts Kadiķis and Kārlis Freivalds

Abstract A computationally efficient video processing algorithm for vehicle detection is presented. The algorithm detects vehicles that arrive on a defined detection line. Inter-frame difference is used to detect the presence of a vehicle and background subtraction is used to find the shape of the object that does not belong to an empty road. If movement is detected, intervals corresponding to the moving objects are created on the detection line. Further processing of intervals allows detecting and counting the objects. The accuracy of the proposed method is analyzed on different roads and in different weather conditions.

Keywords Vehicle detection · Image processing · Movement detection · Background subtraction

1 Introduction

A growing number of vehicles and increasing human demand for mobility leads to a steady increase of traffic congestions and energy consumption. Expansion of roads is an expensive solution; therefore more efficient management of traffic using Intelligent Transportation Systems (ITS) has become an attractive solution. Example of this is a legal framework to accelerate the deployment of innovative

R. Kadiķis (✉) · K. Freivalds
Institute of Electronics and Computer Science, Riga, Latvia
e-mail: r.kadikis@gmail.com

K. Freivalds
e-mail: karlis.freivalds@lumii.lv

Mohan S. and S. S. Kumar (eds.), *Proceedings of the Fourth International Conference on Signal and Image Processing 2012 (ICSIP 2012)*, Lecture Notes in Electrical Engineering 221, DOI: 10.1007/978-81-322-0997-3_12,

transport technologies across Europe, adopted by European Commission on 7 July 2010.

Crucial task of ITS is acquisition of traffic parameters like vehicle count, speed etc. Non-invasive sensors, such as video cameras, are becoming more appealing than the classical approaches using inductive loops. Although research in the field of camera use for monitoring of traffic has taken place for decades [1, 2], there are still efforts to make more effective, robust or capable systems. Task to be resolved is complex due to a continuously changing outdoor environment and the vehicles—objects of interest—may greatly differ from each other. Objects may be moving or static, they may not stand out from the background and they may occlude each other.

A good system for solving this task should be able to detect and correctly count vehicles on the road. The detection algorithm should identify moving and stopped vehicles, and it should achieve this in different seasons and times of the day, so it must be able to adapt to the changing environment. It has to distinguish vehicles from shadows and other background objects and also distinguish vehicles that are close to each other. It is important that detection algorithm is efficient and fast since the most explicit effect of the ITS is achieved if it is able to receive traffic information in real time. Detection of various parameters of vehicle (speed, size, color, motion trajectory) is desirable to facilitate versatile application of the traffic monitoring system.

The method presented in this paper counts vehicles with about 95 % accuracy. The detection algorithm is fast and efficient, because most of the processing operations are carried out on a single row of a video frame. Both moving and stationary vehicles are detected on this row by combining methods of movement detection (inter-frame difference) and object detection (current intensity or color, or edge frame difference with a background reference). The presented algorithm automatically adapts to the changes in lighting and weather conditions. It is achieved by using adaptive background as a reference and automatic detection of thresholds used in the segmentation steps. Optionally, vehicle length and width can be determined. The above-mentioned properties of our algorithm makes it suitable to use in ITS for real-time vehicle detection in a changing environment.

2 Prior Work

Vehicle detection using image processing and computer vision has been a topic of research for more than thirty years. However, challenges of changing outdoor environment, advancements in computer vision and increasing availability of computing power lead to new solutions, which solve the set of aforementioned problems with greater accuracy or speed. A survey paper by Kastrinaki et al. [3] offers a comprehensive overview of different approaches proposed before year 2003.

A widespread method to detect a vehicle in a video stream is to use background subtraction—comparison of the current frame with a reference frame of an empty road. There are different ways how the objects of interest are distinguished from the background. A frequently used parameter is grayscale intensity of the pixels or values of the color components in some color space [4–7]. Edges and corners are robust parameters in a changing lightning or weather conditions [8] where the current edges are compared with the background edges to detect if there are some new objects.

Since vehicle detection systems have to deal with a changing outdoor environment, they use an adaptive background image, which can be estimated in a different ways. There are non-recursive methods where a specified number of frames are stored in a buffer. The value of the background pixels may be then selected as some statistical value of them, such as mean or median. In the recursive methods background is updated using a weighted sum of the background image and the current frame [7], Kalman filter [9], Mixture of Gaussian [10] or a similar technique. In the case of recursive methods, the buffer is not needed; therefore the algorithms are typically more computationally efficient.

When the foreground objects are found, vehicles may be tracked to find their trajectories. Vehicle occlusion is an important problem that has to be addressed appropriately [11].

Since video processing includes manipulation of a large amount of pixels, one of the methods used in the field to make the algorithms faster, is reduction of the number of pixels processed. For vehicle detection, regions of interest are defined within a frame. Several works [1, 12, 13], define one or more lines of interest within a frame, which are perpendicular to the traffic lanes and are also called virtual-lines, detection lines or virtual loop detectors.

In those approaches, lines of background intensity or color components are determined and subtracted from the same lines in the current frame, revealing objects, which do not belong to the background. Adaption to new lighting conditions is not instantaneous, thus any rapid changes in the background may affect such algorithms.

Another approach [14] combines movement detection with background subtraction. Inter-frame difference is used to obtain background. Detection regions are used to count vehicles on each lane. However, background acquisition is carried out on a whole frame and the virtual detectors process already segmented frames. We propose a method that allows processing a single detection line, thus decreasing the required computing power.

In the methods [13] and [14], not only position of the road but also position of the lanes of interest have to be defined, thus these methods are not usable in cases when the number and position of lanes may vary. Our method solves this issue using interval approach that is discussed in the next section.

3 The Proposed Method

Our algorithm detects vehicles that arrive on a defined line in a video frame. This line (a row of pixels) is typically placed perpendicular to the road. All further processing will be carried out on this line only, thus making the detection algorithm computationally efficient. To detect vehicles that cross the line, inter-frame difference is used to detect the presence of the vehicle and background subtraction is used to find the actual objects that do not belong to an empty road.

In the detection process intervals are created, denoting the segments covered by vehicles, on the detection line. Those intervals are stored in a permanent buffer B. Each interval in B is updated with the information acquired from the next frames to reflect the actual overlap of the corresponding vehicle with the detection line. When the vehicle leaves the detection line, its interval is closed and the vehicle parameters are acquired.

The benefit from such approach is that the algorithm is able to detect vehicles on a road where the number of lanes is not clearly defined and may change depending on the traffic flow. The main steps of the presented vehicle detection method are given in the following pseudocode. The algorithm works with the global buffer B of intervals and a global array LG of the processed lines (actually only the previous line is needed).

```
Input: line extracted from the current color video frame LC;
       frame number k;

DETECT-VEHICLES
line LG_k ← convert LC to grayscale
line LD ← medianFilter(|LG_k – LG_k-1|)
TLM ← threshold LD with threshold value thr
interval set IS ← FIND-INTERVALS(LD)
add all contents of IS to B
foreach (intervals I and J in B)
    if I and J overlap
        then merge I and J
line TLI ← INTENSITY-BACKGROUND(B, LG_k)
line TLE ← EDGE-BACKGROUND(B, LG_k)
line TLC ← COLOR-BACKGROUND(B, LC)
foreach (interval I in B)
    int PI ← the number of white pixels of TLI covered by I
    int PE ← the number of white pixels of TLE covered by I
    int PC ← the number of white pixels of TLC covered by I
    if PI+PE+PC < closingConstant and I was not updated for d frames
        then report a detected vehicle
             remove interval I from B
```

Fig. 1 Detection line is perpendicular to the road. A single vehicle has entered the detection line, so an appropriate interval is created and maintained until the vehicle leaves

The input for the proposed algorithm is a line of pixels extracted from a stream of color video frames. Movement detection (steps 1–3) is achieved by using grayscale intensity values of pixels LG. The previous line LG_{k-1} is subtracted from the current one and the absolute value of the difference is taken to obtain the difference line LD. A median filter is applied to it to reduce noise.

A threshold is applied to segment the pixels of line LD into moving objects (white pixels) and static background (black pixels). Automatic acquisition of the threshold value is described in Sect. 3.1.

The detection of a moving vehicle is performed by the method FIND-INTERVALS at step 4 where appropriate intervals are created and added to the buffer B. The detected interval consists of adjacent pixels, which overcome the threshold. The intervals found in the current frame are compared to all already existing intervals. If the new interval overlaps with an existing one, they are merged to correctly identify the full extent of the vehicle. Otherwise, new intervals are simply stored in the buffer B for comparison with future intervals (steps 5–8).

Figure 1 depicts the detection line and an interval that corresponds to the vehicle crossing the detection line.

Intervals are created after detection of the movement; however they must stay existent even when the vehicle has stopped on the detector line. For this purpose, a background line of several parameters—intensity, edges and color is maintained and the difference between the current detection line and the background line is calculated. Segmentation of the difference lines TLI, TLE and TLC of the three parameters is accomplished in steps 9–11 by using a certain threshold; for details, see Sect. 3.1.

Several closing conditions are examined in steps 13–16 in order to suspend the closure of intervals in cases of stopped vehicles. For every interval, pixels of different parameter lines that overcame the threshold (white pixels) are counted. If the sum of pixels counted for each parameter is larger than the value of *closing-Constant* then it is assumed that the vehicle is still on the detection line and the interval is not closed.

Another condition that prevents removal of the interval is appearance of a new interval that is merged with the interval of interest. If no new overlapping intervals are found in the area of already existing interval I for *d* frames and the background parameters do not indicate the presence of vehicle in current frame, the interval is

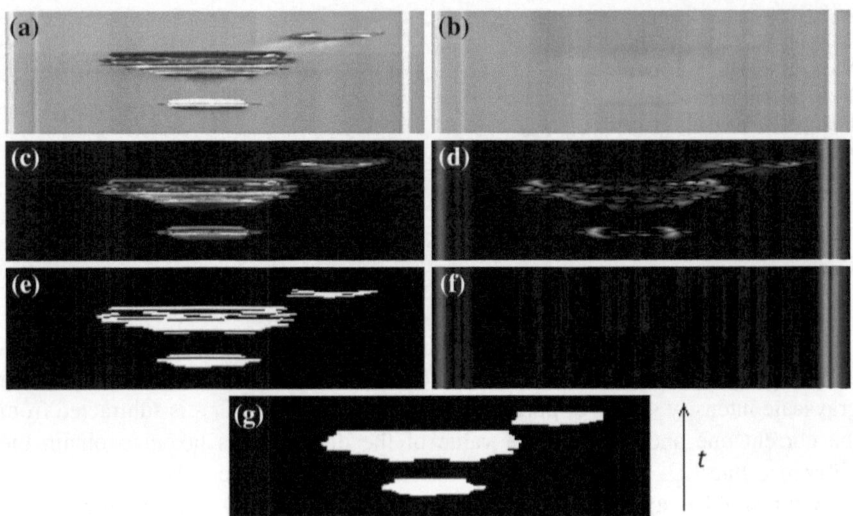

Fig. 2 Spatio-temporal images for some of the steps of the algorithm, **a** Intensity line,**b** Inter-frame difference line, **c** Thresholded inter-frame difference line, **d** Intensity background line, **e** Edge line, **f** Background of edge line, **g** Intervals

removed from the buffer B and a detected vehicle is reported. Later, parameters of the interval such as its width and duration can be examined to discard small intervals that cannot correspond to vehicles. These parameters also may be used to distinguish passenger cars from trucks.

Value of the constant d depends on the frame rate of the input video. Faster frame rates require higher values of d since vehicles move by less amount of pixels in consecutive frames. At the same time the value of d should not be too large, because the interval has to be closed before the next vehicle arrives into the region of interest. For our test videos with 15–25 fps we chose $d = 3$.

Changes of width of an interval may indicate specific occlusion cases thus allowing to improve the counting accuracy. If a few consecutive changes of the interval's width are narrowing and then the interval becomes significantly wider and it has existed long enough to match two possible vehicles, an occlusion is detected. This approach can be extended to detect occlusion of more than two vehicles.

For a high-speed vehicle, some of its important features can skip a single detection line. The movement detection step (step 1–2 of the algorithm) can be improved by using a thicker detection line of width of more than one pixel. Each of these pixels is subtracted from the corresponding pixel in the next frame. Then, the thick line is converted into a thin line LD by taking the maximum of the respective pixels.

In Fig. 2 intermediate results of several steps of the algorithm are depicted in the form of spatio-temporal images. The horizontal lines of these images show the detection lines and the vertical axis of the images corresponds to time with the oldest line at the bottom of the image.

3.1 Background Subtraction

Let us discuss how the background line of edges is obtained and used (algorithm EDGE-BACKGROUND). The algorithm maintains and updates a global background line buffer BEL. There is another global buffer BTL which contain a set of lines collected for the threshold calculation. Initially BTL is empty.

Input: buffer of intervals B;
grayscale input line LG;
frame number k.
Output: thresholded edge line.

```
EDGE-BACKGROUND
1  line LE ← blur line LG with a Gaussian low-pass filter
2  LE ← find edges on LE
3  if B is empty
4      then update the whole background edge line BEL according to (1)
5      else update the regions of BEL outside intervals according to (1) and
6          append LE to BTL
7  LE ← |LE − BEL|
8  int TH ← FIND-THRESHOLD(BTL)
9  line TL ← threshold LE with threshold value TH
10 return TL
```

After blurring the intensity line with a Gaussian low pass filter (step 1), the absolute values of the difference of adjacent pixels are calculated (step 2). This results in a line of edges depicted in Fig. 2e.

In order to find objects that do not belong to an empty road, edge background (Fig. 2f) must be acquired. Every new line contributes to a recursive formation of background (steps 3–5). Every pixel that does not belong to any existing interval is updated as follows:

$$F_k = \alpha \cdot F_{k-1} + (1 - \alpha) \cdot S, \tag{1}$$

where F_k is intensity of the background at the *k-th* frame, F_{k-1} is intensity of the background in the previous frame, S is the intensity value of the pixel in the current frame, α is the background update rate. Smaller value of α gives the current frame a bigger impact on the formation of the background. We use $\alpha = 0.98$ in our tests.

At step 6 those lines are placed in buffer BTL which have at least one interval defined in the input buffer B (some vehicle is detected on this line by the motion detection step). Only a fixed number of lines are kept in this buffer (200 in our case); the oldest lines are discarded.

Then, at step 7, the difference of the current line and the background buffer BEL is calculated. At step 8, a threshold is calculated based on the collected lines in BTL. Threshold calculation is based on the Otsu's method [15]. At the system initiation step, while not enough lines are accumulated, some initial threshold has to be used. As in the case of motion detection, the difference line is thresholded (step 9) giving the result of this method.

Algorithms INTENSITY-BACKGROUND and COLOR-BACKGROUND are similar to EDGE-BACKGROUND, except that object detection from the intensity and color lines do not need the step of edge finding (step 2). The color lines are formed by representing the input line in the YCrCb color space. Both color components Cr and Cb are processed independently and the results are summed in the method COLOR-BACKGROUND before thresholding.

4 Results

The proposed method was implemented in C++ programming language. OpenCV image processing library was used for video input/output and common image processing tasks. The method was tested on several real-life videos obtained in a different weather conditions.

The ideal camera position for the proposed algorithm is directly above the road facing downwards so that vehicles do not occlude each other. However, in such placement, only a small section of road is visible, thus limiting the range of observation. Therefore, a more practical placement is tilting the camera towards horizon. We analyze our method on 6 real life videos for the vehicle counting task. In the first three videos (Fig. 3a, b, c) we analyze the performance of our algorithm

Fig. 3 The test videos

Table 1 Results of the tests

Video	Ground-truth	Detected	Errors			Counting accuracy (%)
			Occlusion	Counted as two	Other	
A	100	105		5		95.2
B	100	105		4	1	95.2
C	100	100	3	3		100
D	77	71	7	3	2	91.5
E	42	44			2	95.5
F	48	46	2			95.7

in various weather conditions. The test (a) was carried out on a day, when movement of shadows caused significant and rapid changes in the scene lighting. The test (b) was carried out after rain, when wet road caused reflections of the headlights. The test (c) was carried out during an intensive rain. The resolution of these videos was 640×480. The traffic intensity in those videos was small, thus the number of occlusions was not significant. The next three videos (Fig. 3d, e, f) show cases with several traffic lanes and fairly intensive traffic. The results of evaluation of the method are shown in Table 1.

The tests show that the proposed method is robust to changes in lighting—moving shadows are not detected as vehicles. Also headlight reflections from a wet road or an intensive rain do not create additional problems.

Most of the errors were caused by two factors. Sometimes, when occluded vehicles crossed the detection line, occlusion was not detected thus the visually overlapping vehicles were counted as one. An opposite problem when a vehicle was counted as two also was present. It occurred when an interval was closed before the vehicle has been fully passed the detection line and a new interval was created for the same vehicle. Only a few errors were caused by other factors such as shadows of vehicles or people or birds crossing the detection line. In the test (c) the number of vehicles counted by the algorithm was correct, but it should be noted that there were 6 errors that offset each other, so there can be about 6 % error rate in similar but not so lucky cases.

Similar counting methods, that also use line or small regions of frame to count vehicles, report the following results: in [14] sfor daytime videos varied from 70.31 % to 98.39 %; the method in [13] had the accuracy of 97.73 %; in [12], the precision of the method was 86 %–96 %, while recall was 89 %–96 %. Since different videos were used to evaluate all the aforementioned methods, the accuracy rates are not directly comparable.

5 Conclusions

This paper introduces an algorithm for vehicle detection and counting that is based on a novel idea of maintaining the interval of crossing of a vehicle with the detection line. The intervals are updated during several frames to merge different parts of the same object, to track changes of the width of the passing object and to detect occlusions of vehicles in specific cases.

An advantage of the proposed method is efficiency with respect to computing power since only the pixels of the detection line have to be considered. In practice, receiving and decoding the video consumes more resources than our method, hence exact performance measurements are of little value. Also the interval approach allows using the method on roads where the number and direction of lanes is not defined or may vary.

The performed tests showed that the proposed method is robust to different lighting and weather conditions. The counting accuracy is promising, but further improvements are needed to correctly identify cases with vehicle occlusion. Detection of shadows cast by other vehicles also must be added to reduce the possibility of merging of intervals that represent different vehicles.

Computational efficiency of algorithm allows further accuracy improvements by using multiple detection lines. It would allow analyzing a broader area of the road. If there are occluded vehicles on some detection lines and the same vehicles are not occluded on other lines, it could be possible to determine the correct answer. Also, use of multiple detection lines would allow detection of speed and lengths of the vehicles, thus allowing classification of the vehicles.

Acknowledgments This work was supported by European Regional Development Fund project Nr.2010/0285/2DP/2.1.1.1.0/10/APIA/VIAA/086

References

1. Michalopoulos PG (1991) Vehicle detection video trough image processing: The autoscope system. IEEE Trans Veh Technol 40(1):21–29
2. Beymer D, McLauchlan P, Coifman B, Malik J (1997) A real-time computer vision system for measuring traffic parameters. In: Proceeding of IEEE computer society conference on computer vision and pattern recognition. 495–501
3. Kastrinaki V, Zervakis M, Kalaitzakis K (2003) A survey of video processing techniques for traffic applications. Image Vis Comput 21:359–381
4. Rodriruez T, Garcia N (2010) An adaptive, real-time, traffic monitoring system. Mach Vis Appl 12:555–576
5. Vibha L, Venkatesha M, Prasanth GR, Suhas N, Shenoy PD, Venugopal KR, Patnaik LM (2008) Moving vehicle identification using back-ground registration technique for traffic surveillance. In: Proceedings of the international multi conference of Engineers and Computer Scientists, vol 1
6. Ji X, Wei Z, Feng Y (2006) Effective vehicle detection technique for traffic surveillance systems. J Vis Commun Image Process 17:6477–6658
7. Gupte S, Masoud O, Martin RFK, Papanikolopoulus NP (2002) Detection and classification of vehicles. IEEE Trans Intell Transp Syst 3(1):37–47
8. Hu Q, Li S, He K, Lin H (2010) A robust fusion method for vehicle detection in road traffic surveillance. In: Proceedings of the advanced intelligent computing theories and applications, and 6th international conference on intelligent computing. pp 180–187
9. Cheung S-CS, Kamath C (2005) Robust background subtraction with fore-ground validation for Urban traffic video. EURASIP J Appl Signal Process 2005(1):2330–2340

10. Stauffer C, Eric W, Grimson L (2000) Learning patterns of activity using real-time tracking. IEEE Trans Pattern Anal Mach Intell 22(8):747–757
11. Koller D, Weber J, Malik J (1994) Robust multiple car tracking with occlusion reasoning. Lect Notes Comput Sci 800:189–196
12. Anan L, Zhaoxuan Y, Jintao L (2007) Video vehicle detection algorithm based on virtual-line group. In: Proceeding of IEEE APCCAS. pp 2854–2858
13. Zhang G, Avery RP, Wang YA (2007) Video-based vehicle detection and classification system for real-time traffic data collection using uncalibrated video cameras. Transportation research record. J Transp Res Board 1993:138–147
14. Lei M, Lefloch D, Gouton P, Madani K (2008) A video-based real-time vehicle counting system using adaptive background method. IEEE international conference on signal image technology and internet based systems. pp 523–528
15. Otsu N (1979) A threshold selection method from gray-level histograms. IEEE Trans Syst Man Cybern 9(1):62–66

Combination of Features for Crosslingual Speaker Identification with the Constraint of Limited Data

B. G. Nagaraja and H. S. Jayanna

Abstract Mel frequency cepstral coefficients (MFCC) has proven to be effective in speaker identification, but does not provide satisfactory performance in limited data condition. This paper presents a combination of features from different languages for Crosslingual speaker identification with the constraint of limited data. However, combined features can increase the complexity of the speaker identification system by doubling the dimensionality of the features. Frame reduction and smoothing are achieved using an adaptive weighted-sum algorithm. Experiment results show that the proposed method gives an average 11 % improved in performance over conventional MFCC method.

Keywords Frame reduction · Crosslingual · MFCC

1 Introduction

Speaker identification is defined as a task of recognizing speakers from their speech [1]. Mankind has an ability to learn more than one language [2]. Some nations (India, Canada etc.) are officially bilingual or multilingual. In India, more than 50 languages are officially recognized by the Govt. of India and the Indian citizens can speak more than one language fluently. Criminals often switch over to

B. G. Nagaraja (✉) · H. S. Jayanna
Department of Information Science and Engineering, Siddaganga Institute of Technology, Tumkur 572103, India
e-mail: nagarajbg@gmail.com

H. S. Jayanna
e-mail: jayannahs@gmail.com

Mohan S. and S. S. Kumar (eds.), *Proceedings of the Fourth International Conference on Signal and Image Processing 2012 (ICSIP 2012)*, Lecture Notes in Electrical Engineering 221, DOI: 10.1007/978-81-322-0997-3_13,

another language, especially after committing a crime. So, training a person's voice in one language and identifying him in some other language or in a multilingual environment is of interesting task especially in the Indian context [3].

In the modern day digital automated world, speaker identification system plays a very important role in the field of fast growing internet based communications/ transactions. Nowadays, more and more users of such systems are using several languages. So, if we have a priori knowledge of languages known by the speaker, then it is possible to build a robust speaker identification system in a language different from the one used during training (Crosslingual system). In Crosslingual speaker identification (A/B), training is done in one language (say A) and testing is done in another language (say B) [3, 4]. Major aspects that can make speech to deviate from one language to other are fluency, intonation and word stress [5]. In [6], it was mentioned that language consistency matters for Crosslingual speaker recognition.

Most of state-of-the-art speaker identification systems work on Monolingual (preferably English) using sufficient data. In non-cooperative scenario such as forensic investigation, speech data may be a few seconds and task is to identify the speaker. In such application, it is required to validate the speaker using limited amount of speech data. Speaker identification in limited data conditions refers to the task of recognizing speakers where both the training and test speech present only for few seconds. In this work, limited data refers to the case of having few seconds (≤ 15 s) of speech data [7]. This paper concentrates on closed-set text-independent Crosslingual speaker identification with the constraint of limited data.

In our previous work [4], an attempt was made to identify speaker in the context of Mono and Cross-lingual with the constraint of limited data using MFCC as feature vectors and Vector Quantization (VQ) as modeling technique. It was observed in Crosslingual study that the use of English language either in training or testing gives better identification performance.

The rest of the paper is organized as follows: Sect. 2 describes the database used for the experiments. Feature extraction using MFCC, frame reduction and smoothing using an adaptive weighted-sum algorithm along with the proposed feature combination are presented in Sect. 3. Discussion on the Crosslingual experimental results is presented in Sect. 4. Section 5 gives Summary of the present work and scope for future work.

2 Speech Database for the Study

Since the standard Multilingual database is not available, experiments are carried out on an our own created database of 30 speakers who can speak the three different languages (E-English, H-Hindi and K-Kannada). The database includes 17-male and 13-female speakers. The voice recording was done in the Engineering college laboratory. The speakers were undergraduate students and faculties in an engineering college. The age of the speakers varied from 18–35 years. The speakers were asked to read small stories in three different languages. The training and testing data were recorded in different sessions with a minimum gap of two days. The approximate training and testing data length is two minutes. Recording was done using free downloadable wave surfer 1.8.8p3 software and Beetel Head phone-250 with a frequency range 20–20 kHz. The speech files are stored in .wav format.

3 Feature Extraction

Features are extracted using MFCC technique. Speech recordings were sampled at the rate of 8 kHz and pre-emphasized (factor 0.97). Frame duration of 20 and a 10 ms of overlapping durations are considered. After framing, windowing (Hamming) method is carried out to minimize the spectral distortion. 35 triangular band pass filters are considered. First 13 coefficients are considered as feature vectors. Since the 0th coefficient can be regarded as a collection of average energies of each frequency bands, it is unreliable [8].

Given a sequence of original feature vectors $F = [f_1 f_2 \ldots f_l]^T$, where the nth vector f_n contains M feature elements. Frame reduction for a sliding window of length 3 is obtained by [9]:

$$F^{(m)}_{reduction} = 1/N \sum_{i=n}^{n+3} w_i^{(m)} f^{(m)}_{2n+i-1} \tag{1}$$

where $f_i^{(m)}$ denotes the mth feature element of the vector f_i and $w_i^{(m)}$ is the weight of such feature element. Weight values are transformed from the Euclidean distance between two feature vector frames within the sliding window [9]. Subscript (2n + i −1) is used for frame skipping of sliding window to reduce 50 % length of the feature sequence. N is a normalization factor and for a sliding window of length 3, N is given by [9]:

$$N = \sum_{i=1}^{3} w_i^{(m)} \tag{2}$$

The following steps are used in speaker identification process. The detailed block diagram is given in Fig. 1.

(a) Choose language 'A' training data.
(b) Extract the features using MFCC.
(c) Perform frame reduction and feature smoothing.
(d) Choose language 'B' training data.
(e) Extract the features using MFCC.
(f) Perform frame reduction and feature smoothing.
(g) Combine the features (languages A and B).

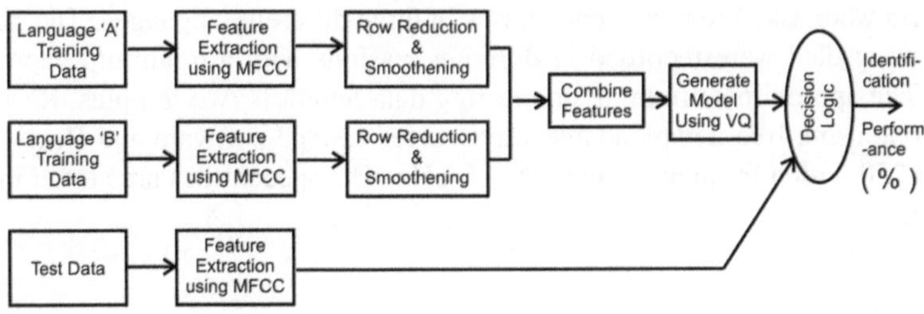

Fig. 1 Block diagram of the proposed method

(h) Generate the speaker model using VQ.
(i) Choose the testing data.
(j) Extract the features using MFCC.
(k) Compare test features with speaker model.
(l) Use the Decision logic to find out the winner.

4 Experimental Results

The Crosslingual experimental results for the 30 speakers for 15 s of training and testing data and for different codebook size are given in Table 1. Note: A/B indicates training with language 'A' and testing with language 'B'.

The results show that the combination of the features perform better than the baseline MFCC technique in all Crosslingual speaker identification experiments. The improvement in the performance may be due to different aspect of speaker-

Table 1 Crosslingual speaker identification performance (%)

Codebooks							
Train/test language	Technique	16	32	64	128	256	P_i
H/E	MFCC	66.66	73.33	76.66	73.33	76.66	**76.66**
(H+E)/E	Proposed	70.00	73.33	73.33	80.00	86.66	**86.66**
K/E	MFCC	66.66	66.66	66.66	76.66	80.00	**80.00**
(K+E)/E	Proposed	83.33	86.66	86.66	90.00	93.33	**93.33**
E/H	MFCC	66.66	60.00	70.00	73.33	73.33	**73.33**
(E+H)/H	Proposed	80.00	80.00	86.66	93.33	90.00	**93.33**
K/H	MFCC	66.66	60.00	70.00	70.00	76.66	**76.66**
(K+H)/H	Proposed	86.66	86.66	90.00	90.00	90.00	**90.00**
E/K	MFCC	70.00	66.66	73.33	73.33	70.00	**73.33**
(E+K)/K	Proposed	76.66	73.33	73.33	83.33	83.33	**83.33**
H/K	MFCC	50.00	63.33	60.00	60.00	63.33	**63.33**
(H+K)/K	Proposed	73.33	76.66	73.33	83.33	76.66	**83.33**

P_i represents the maximum identification performance among the number of codebooks

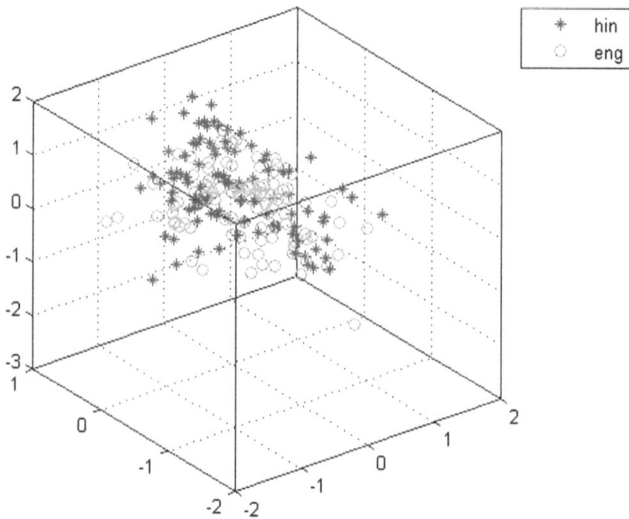

Fig. 2 The first 100 features from Kannada and English training speech data of a speaker in 3-dimensional view

specific information in different languages. The Fig. 2 shows the distribution of the features in the feature space taken from different languages.

5 Conclusion

The main objective of the work is to increase the performance of the Crosslingual speaker identification with the constraint of limited data condition. To achieve this the present work considered combination of row reduced and smoothed features from training and testing languages for creation of speaker specific models. Feature combination is a very effective feature-domain method to improve the performance of Crosslingual speaker identification system. Experimental results showed that the proposed combination method outperforms the baseline MFCC technique in all Crosslingual speaker identification. In order to study the robustness of the system needs to be verified with different languages, different data sizes and large amount of speaker set.

Acknowledgments This work is supported by Visvesvraya Technological University (VTU), Belgaum-590018, Karnataka, India.

References

1. Atal BS (1976) Automatic recognition of speakers from their voices. Proc IEEE 64(4):460–475
2. Halsband U (2006) Bilingual and multilingual language processing. J. Physiol Paris 99:355–369
3. Arjun PH (2005) Speaker recognition in Indian languages: A feature based approach. Ph.D. dissertation, Indian Institute of Technology, Kharagpur
4. Nagaraja BG, Jayanna HS (2012) Mono and cross lingual speaker identification with the constraint of limited data. In: Proceedings of IEEE, PRIME-2012, Periyar University, Salem, pp 439–443
5. Durou G (1999) Multilingual text-independent speaker identification. In: Proceedings of MIST 1999 workshop, Leusden, pp 115–118
6. Arjun PH, Sitaram S, Sharma E (2009) DA-IICT cross-lingual and multilingual corpora for speaker recognition. In: Proceedings of IEEE advances in pattern recognition, Kolkata, pp 187–190
7. Jayanna HS (2009) Limited data speaker recognition. Ph.D. dissertation, Indian Institute of Technology, Guwahati
8. Picone JW (1993) Signal modeling techniques in speech recognition. Proc IEEE 81(9): 1215–1247
9. Nuratch S, Boonpramuk P, Wutiwiwatchai C (2010) Feature smoothing and frame reduction for speaker recognition. In: Proceedings of IEEE international conference on Asian language processing, pp 311–314

A Competitive Performance Analysis of Proactive Routing Protocols of MANET Under Short Time Communication Scenario

A. P. Janani, M. Sakthivel and K. Baskaran

Abstract The performance of proactive routing protocols in a highly Mobile Ad hoc NETwork (MANET) is questionable due to the resulting link failures caused by the high mobility. Since the routing protocols require a certain time to achieve stable performance due to the periodic route discovery and maintenance mechanisms, it is not suitable to use them in a communication scenario such as a very short time military rescue operation that may end in seconds. Here the routing protocol should provide better performance from the very starting instance of the operation. In this work, proactive protocols DSDV and OLSR are evaluated and their behaviour is studied while using TCP and UDP traffic under short term communication scenario in a typical, small mobile ad hoc network.

Keywords DSDV · OLSR · MANET proactive routing protocols · Performance analysis TCP · UDP

A. P. Janani (✉)
Department of Information Technology, Sri Subramanya College of Engineering and Technology, Palani, India
e-mail: janani97@gmail.com

M. Sakthivel
Department of Mechanical Engineering, Anna University of Technology, Coimbatore, India
e-mail: sakthi_vel_m@yahoo.com

K. Baskaran
Department of Computer Science and Engineering, Government College of Technology, Coimbatore, India
e-mail: baski_101@yahoo.com

Mohan S. and S. S. Kumar (eds.), *Proceedings of the Fourth International Conference on Signal and Image Processing 2012 (ICSIP 2012)*, Lecture Notes in Electrical Engineering 221, DOI: 10.1007/978-81-322-0997-3_14, © Springer India 2013

1 Introduction

Mobile ad hoc networks are temporary networks without any definite infrastructure or centralized administration. The topology changes rapidly due to the frequent mobility of the nodes and this dynamic nature make routing a challenging task. Certain areas of application of MANETs include military operations, emergency rescues and civilian environments.

The third generation networking technology used in military networks is based on ad hoc principle which is considered in this work. The routing protocols in ad hoc networks are mainly based either on the distance-vector (DV) routing algorithm or on the link-state (LS) routing algorithm. Though both of them find out the shortest path from the source to destination nodes they differ from each other by the information type the node maintains.

Based on the nature of design and behaviour, the ad hoc routing protocols are categorized into proactive routing protocols, reactive routing protocols, geographical routing protocols and hierarchical routing protocols. The frequently used routing schemes are proactive and reactive routing protocols. The hybrid routing protocols encompasses the advantages of both the proactive and reactive routing protocols.

The proactive protocols namely Destination-Sequenced Distance Vector (DSDV) Routing protocol [1] and the Optimized Link State Routing (OLSR) protocol [2] maintain the routes to every node, including the nodes not involved in the transfer of packets. Hence a significant part of the available bandwidth is occupied.

1.1 Problem Definition

The performance of proactive and reactive routing protocols is questionable if we try to use them during a communication scenario such as a very short time military rescue operation that may end in seconds. In such highly confidential military operations, generally they will maintain radio-silence up to the very starting instance of the rescue operation to avoid the possibility of overhearing or sniffing their communication. So, from the instance of switching-on to the end of rescue operation, the available time will be only in seconds. But, usually the routing protocols will require certain time to achieve stable performance because of the periodic route discovery and maintenance mechanisms in their inherent design. In addition to that, all the existing transport protocols will require certain time to estimate the bandwidth to send and receive packets accordingly. This is the very nature of any transport protocol and the core idea of congestion control mechanisms. So, selection of a transport protocol also becomes very important in such short time communication scenarios. Further, in such highly mobile scenarios, link failures caused by the high mobility will increase the retransmission of data packets as well as the frequent propagation of the routing information through the network and hence rapidly increases the overhead in communication. So if a MANET routing protocol along with a transport protocol is used for the communication scenario such as a very short time military rescue operation that may end in seconds, practically there will not be much time for the protocols to get their optimum level of performance. In such short time applications, the routing protocol in use should provide better performance even from the very starting instance of the operation irrespective of the transport protocol in use. In this work, the performance of two widely used proactive routing protocols are evaluated under short term communication scenarios with TCP as well as UDP and their behaviour under various mobility conditions is studied in a typical, small, short time mobile ad hoc network.

1.2 Related Works

Many researches have been done for the qualitative as well as quantitative analysis of ad hoc routing protocols. The evaluation was based on different metrics and different simulators. The comparison and analysis of protocols is performed under varied workload, mobility and size of the ad hoc networks in many works [3–6]. The performances of AODV, DSDV and DSR are compared in a grid environment [7]. A survey of theoretical analysis of the routing overhead involved has also been conducted [8].

In military applications, a change in the protocol design to increase the level of security was performed [9]. Secure communication algorithms have been proposed to achieve the optimal secrecy throughput in military MANETs [10]. A Network Engineering Design Analytic Toolset (NEDAT) that applies network-science-based approaches to design MANETs for use in military's network-centric warfare (NCW) approach was developed [11]. A new multi-tier adaptive MANET security protocol was proposed for military applications [12].

2 MANET Routing Protocols

In MANETs, sending updates happens either often or as and when needed. But sending updates may increase the overhead since the nodes move frequently. Flat routing protocols are suitable in a small network where each node maintains information about the other nodes in its routing table but become insufficient in large MANETs. In a hierarchical routing protocol the nodes need to maintain smaller routing tables since the nodes only keep track of their levels in the hierarchy. But the maintenance of the hierarchical structure results in overhead. The geographical routing protocol utilizes node locations to organize the MANET structure and fortunately they are scalable and effective than flat routing protocols. Proactive routing protocols are extensively used.

2.1 Pro-active/Table Driven Routing Protocols

Proactive MANET protocols are table-driven and they decide the structure of the network where each node maintains a complete picture of the network. So, only a minimum delay is observed in deciding the route to be taken.

In a MANET the high mobility of the nodes makes the routing information in the routing table invalid. Another drawback is that the increased mobility of nodes causes traffic overhead while generating unnecessary routes. Increased network size still worsens the situation. If the nodes transmit rarely, much of the routing information goes unused. Energy is consumed in continuously updating the unused entries in the routing table. In conclusion, proactive MANET protocols are best suited for low node mobility networks or networks with frequent data transmission.

2.1.1 DSDV

DSDV is a hop-by-hop routing protocol where messages are exchanged between one hop neighbours. Periodical or triggered updates keep the routing table updated continuously. When there is any change in the routing tables in one of the neighbouring nodes, then updates are triggered. The routing queries are sent from a node while a packet for the unknown route is cached until route replies are received from destination.

2.1.2 OLSR

The OLSR being an IP routing protocol can be used in mobile ad-hoc networks and in wireless ad-hoc networks. It uses the HELLO and Topology Control (TC) messages to discover the route and followed by the dissemination of link state information throughout the complete network. Using the topology information each node calculates the shortest next hop forwarding paths. Since OLSR is proactive, power and resources are wasted to propagate information about unused routes. Multipoint relays (MPRs) are used to flood topology information and hence the redundancy of the flooding process is removed to some extent.

3 The Transport Protocols

The widely used transport protocols are TCP and UDP. They are designed to provide reliable peer to peer communication even under worst network conditions. The transport protocol behavior depends on the available network bandwidth and the traffic load of the link.

3.1 User Datagram Protocol

The User Datagram Protocol (UDP) is connection less protocol which sends datagram to other hosts without requiring prior connection establishment. It is simple without any mechanism for maintaining the reliability, ordering and data integrity. So the datagram may be missed, arrive out of order or duplicated. UDP does not concentrate on error detection and correction.

Since UDP will try to deliver the packet in a lesser reliable manner, there may be lot of missing packets, so the application should take care of the missing packets. Under an un-congested network, UDP will not increase the overhead. But in a congested network, there will be lot of undelivered packets.

3.2 Transmission Control Protocol

Transmission Control Protocol (TCP) is a connection oriented protocol which is reliable, providing an ordered delivery of packets between the nodes even under poor network conditions. As the TCP needs a connection establishment phase followed by data transfer, latency is observed during the first phase. This latency is not negligible in case of a very short communication scenario like military rescue operations.

So, selection of a transport protocol is significant in such short time communication scenarios. The link failures caused by the high mobility will increase the frequent propagation of the routing information through the network. So new TCP connections should be established in search of new routes which increase the latency. As a result the underlying routing protocol takes time for an optimum performance.

4 Simulation and Metrics

NS-2 [13] has been used to evaluate the performance of DSDV and OLSR protocols. DSDV protocol is available in the default installation of NS-2. The NS-2 implementation of OLSR is also available as a separate patch. We extended NS-2 with the OLSR extension.

4.1 Parameters of the Simulation

Area of the Network	700 × 700 sq. m
Number of Nodes	30 Mobile nodes
Channel type	Wireless Channel
Radio-propagation model	Two Ray Ground

Antenna type	Omni Antenna
Interface queue type	Drop Tail/PriQueue
MAC type	802_11
Max packet in Queue	50
Mobility Model	Random waypoint
Mobility	0, 10,20,30,40 and 50 m/s
Number of CBR sources	2
Number of Sinks	2
Traffic Rate	512 byte per 0.1 s
Simulation time	50 s

4.2 The Evaluation Metrics

The following four metrics were considered for evaluating the routing protocols. The first two are related with overhead measurements and the other two are PDF and Throughput.

4.2.1 Overhead

Normalized routing load is the number of routing packets transmitted per data packet sent to the destination.

4.2.2 Normalized Routing Load

It is the ratio of the packets sent over the data packets received by the destination.

4.2.3 Packet Delivery Fraction/Ratio (PDF/PDR)

PDF [14] is the ratio of delivered packets to the sink at destination to those packets sent by the Constant Bit Rate (CBR) source, here the application layer. PDF measures the reliability of the protocol. The metric gives the rate of loss faced by the transport protocol, which in turn affects maximum throughput the network supports.

$$P = \frac{1}{C}\sum_{f=1}^{e} \frac{R_f}{N_f} \quad (1)$$

where

P is the ratio of successfully delivered packets

C is the total number of flows or connections

F is the flow id which serves as index
R_f is the number of packets received from flow f
N_f is the number of packets transmitted to f.

4.2.4 Throughput

How far the network is constantly providing data to the sink is measured by the metric throughput. It is given by the number of packets arriving at the sink per second.

$$\text{Throughput (in Kbps)} = \frac{\mathit{Total\,received\,bytes} \times 8 \times 1,000}{\mathit{Total\,duration}} \tag{2}$$

5 Results and Analysis

5.1 The Simulation Results

The performance of the two proactive protocols DSDV and OLSR have been evaluated at different speed of node mobility. As far as overhead and NRL are considered, OLSR is exhibiting poor performance with both UDP as well as TCP as shown in Figs. 1 and 2. As far PDF is considered OLSR performs better under under TCP as in Fig. 3. Throughput is more in OLSR under TCP as the mobility increases as depicted by Fig. 4.

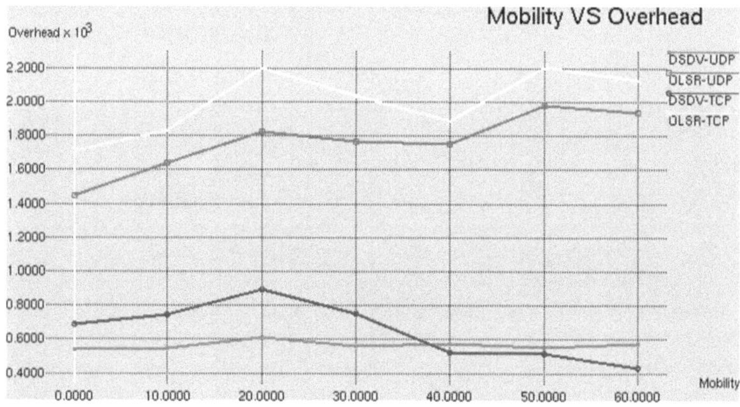

Fig. 1 Mobility vs overhead

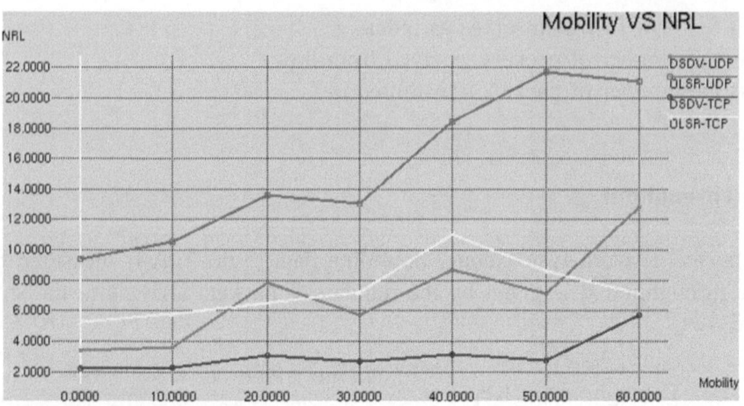

Fig. 2 Mobility vs NRL

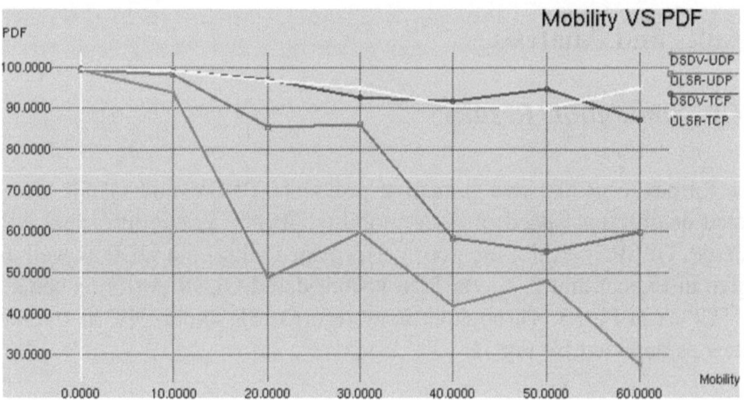

Fig. 3 Mobility vs PDF

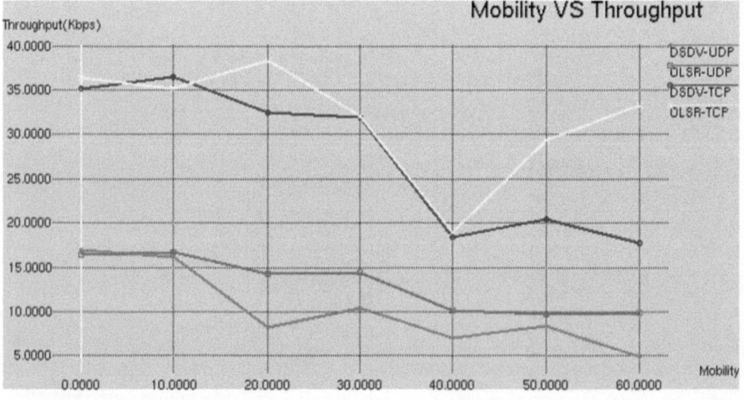

Fig. 4 Mobility vs throughput

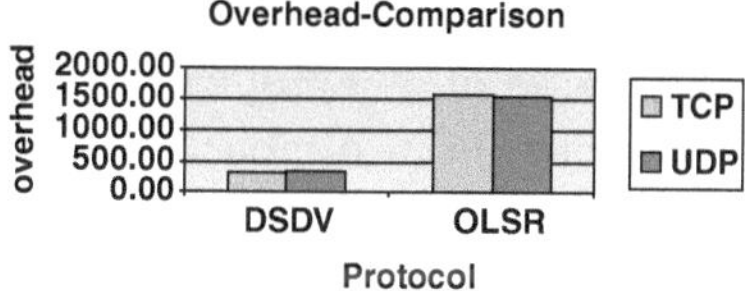

Fig. 5 Comparison of overheads

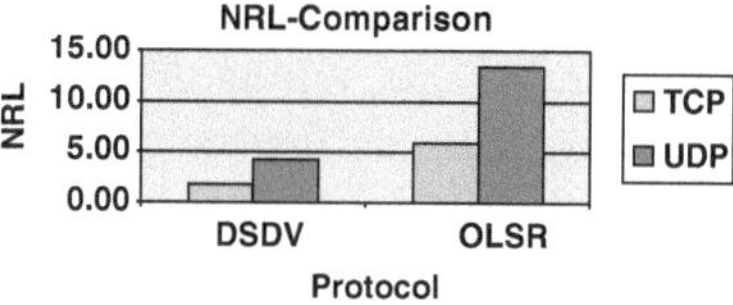

Fig. 6 Comparison of NRL

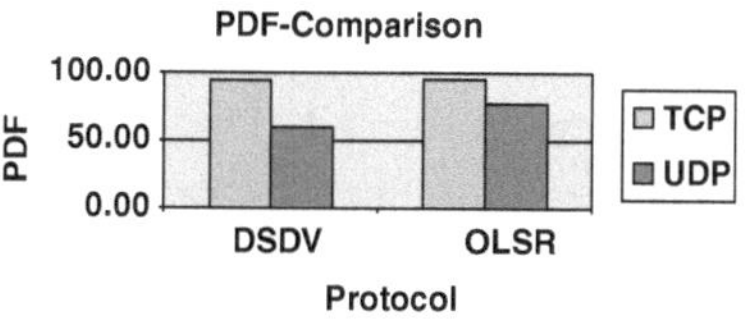

Fig. 7 Comparison of PDF

5.2 The Average Performance

The following graphs show the average performance of routing protocols for which, the average results of all the mobilities were considered.

As shown in Fig. 5, overhead is more in OLSR both under TCP and UDP. From Fig. 6, it is clear that NRL is more in OLSR. As shown in Fig. 7, according to PDF almost both the protocols performed equally but TCP performed better than UDP.

According to throughput, both the protocols gave almost equal result for each of the transport protocol. TCP performed better than UDP as seen in Fig. 8.

5.3 Discussion about the Results

The simulation of the two proactive was repeated several times with different node speeds. As far as the 50 s simulation output is concerned, the following are the observed facts:

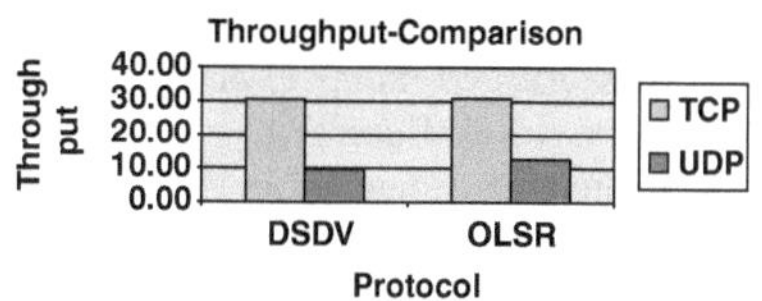

Fig. 8 Comparison of throughput

1. The above results obviously prove that the performance decreases with the increase of mobility.
2. Among the two proactive protocols, even though both of them caused much overhead, the overhead caused by OLSR was much greater than DSDV protocol.

These protocols were evaluated under simple CBR connections without much intense traffic. Although OLSR and DSDV seem to be perform very poor under lower traffic rate, during the repeated simulations, we observed that the deviation in performance during short time communication was very much dependant on the rate of the traffic and type of the transport agent.

6 Conclusion and Future Work

In this paper, the performance analysis of the MANET proactive routing protocols has been performed. The results indicate that the performance of the protocols decreases with the increase of node mobility. DSDV performs well when overhead and NRL are considered. Both the protocols gave equal result as far as PDF and Throughput are considered.

It is not suitable to use the MANET routing protocols in a communication scenario such as a very short time military rescue operation that may end in seconds, since there will not be much time for the routing protocol to get its optimum level of performance. Here the routing protocol should provide better performance from the very starting instance of the operation.

Still the research on military ad hoc networks is an open issue. Though the widely used routing schemes are suitable for military networks, the military environment is entirely different from a typical commercial networking scenario. So the existing MANET routing protocols need to be improved for short term scenario applications. The future work may address the way to improve the performance of any of these routing protocols by innovating efficient route discovery and maintenance mechanisms.

References

1. Perkins CE, Bhagwat P (1994) Highly dynamic destination sequenced distance-vector routing (DSDV) for mobile computers. In: Proceeding of the ACM special interest group on data communications, London, vol 24, pp 234–244
2. Clausen T, Jacquet P (2003) Optimized link state routing protocol (OLSR). RFC 3626, IETF Network Working Group. (online) Available from http://www.ietf.org/rfc/rfc3626.txt. Accessed 20 Dec 2011
3. Rahman A, Islam S, Talevski A (2009) Performance measurement of various routing protocol in ad-hoc network. In: International MultiConference of engineers and computer scientists, vol 1, pp 321–323

4. Boukerche A (2001) Performance comparison and analysis of ad hoc routing algorithms. In: Proceeding of IEEE international conference on performance, computing and communications, pp 171–178
5. Das SR, Perkins CE, Royer EE (2000) Performance comparison of two demand routing protocols for ad hoc networks. In: Proceeding of IEEE INFOCOM, pp 3–12
6. Giannoulis S et al (2007) ZRP versus DSR and TORA: a comprehensive survey on ZRP performance. IEEE Trans Ind Inf 3(1): 63–72
7. Usop NSM et al (2009) Performance evaluation of AODV, DSDV and DSR routing protocol in grid environment. Int J Comput Sci Network Secur 9(7): 261–268
8. Gowrishankar S et al (2008) Theoretical analysis and overhead control mechanisms in MANET: a survey. In: Proceeding of the world congress on engineering London, UK, vol 1, pp 204–210
9. Van Der Merwe J, Dawoud D, McDonald S (2007) A survey on peer-to-peer key management for mobile ad hoc networks. ACM Comput Survey 39(1). doi: 10.1145/1216370.1216371
10. Liang Y, Poor H, Ying L (2009) Secrecy throughput of manets with malicious nodes. In: International symposium information theory, pp 1189–1193
11. Kant L et al (2008) Network science based approaches to design and analyze MANETS for military applications. IEEE Commun Mag 46(11): 56–61
12. Yavuz AA, Alag oz F, Anarim E (2010) A new multi-tier adaptive military manet security protocol using hybrid cryptography and signcryption. Turkish J Electr Eng Comput Sci 18: 1–22
13. Ns-2 network simulator (2012) Available from http://www.isi.edu/nsnam/ns/. Accessed 5 Jan 2012
14. Singh R et al (2011) Performance evaluation of ACO based on demand routing algorithm for mobile ad hoc networks. Int J Eng Sci Technol 3(3): 1809–1815

Content Based Medical Image Coding with Fuzzy Level Set Segmentation Algorithm

Paramveer Kaur Sran, Savita Gupta and Sukhwinder Singh

Abstract Advances in digital medical imaging technologies have resulted in substantial increase in the size of datasets, as a result of improvement in spatial and temporal resolution. In order to reduce the storage cost, diagnostic analysis cost and transmission time without significant reduction of the image quality, a state of the art image compression technique is required. Content based coding is therefore capable of delivering high reconstruction quality over user-specified spatial regions in a limited time, compared to compression of the entire image. Further, CBC coding provides an excellent trade-off between image quality and compression ratio. In this paper a content based compression technique is proposed. The proposed procedure when applied on Computed Tomography (CT) liver image yields significantly better compression rates without loss in the originality of ROI.

Keywords Fuzzy · Image compression · Set Partitioning in Hierarchical Trees (SPIHT) · Region of interest (ROI) · JPEG2000 · Discrete wavelet transform (DWT)

P. K. Sran (✉) · S. Gupta · S. Singh
UIET, Panjab University, Chandigarh, India
e-mail: paramsran@gmail.com

S. Gupta
e-mail: savita2k8@yahoo.com

S. Singh
e-mail: sukhdalip@yahoo.com

Mohan S. and S. S. Kumar (eds.), *Proceedings of the Fourth International Conference on Signal and Image Processing 2012 (ICSIP 2012)*, Lecture Notes in Electrical Engineering 221, DOI: 10.1007/978-81-322-0997-3_15,

1 Introduction

Image coding techniques [1–3] can be classified into two classes: lossless methods and lossy methods. The first class is composed of those methods which reconstruct an image identical to the original; the second comprises compression methods which lose some image details after their application: the reconstruction is an approximation of the original image. In the medical field in particular, the need to conserve the diagnostic validity of the image requires the use of lossless compression methods. The traditional medical image compression techniques (e.g. JPEG, JPEG2K) suffers from low compression rates, low PSNRs, poor reconstructed image quality as well as high MSEs and PRDs (percentage rate distortion). The biggest challenge for the medical image compressors is to retain the diagnostic image quality and to meet the high compression efficiency requirements [4].

Content Based Compression (CBC) combines lossy and lossless compression, together with segmentation. The segmentation process extracts the clinically significant structures (micro-calcifications, masses, ducts, etc.) from the images. In the compression process, important areas are compressed losslessly, achieving high fidelity, while the background regions are compressed lossily, achieving high Compression Ratio(CR) [5]. The proposed content based compression is an excellent method which gives better performance with comparable computational efficiency and the high visual quality of the reconstructed image. Content based compression allows selected parts of an image to be coded with higher quality as compared to the background (BG). It is done using a region of interest (ROI) mask with the priority adjustment. An ROI mask is calculated to determine which wavelet coefficients belong to ROI. After identifying the ROI wavelet coefficients, they are scaled by a certain factor so that they can be coded earlier and placed in the final bitstream before the background coefficients.

2 ROI Extraction

Segmentation is an approach of separating out the region of interest (ROI) and the background (BG) from a medical image. For this purpose, many automatic segmentation techniques [6] like watershed, region growing and split and merge etc., are useful. Here, in order to automatically extract the ROI from a CT liver tissue image, a fuzzy level set segmentation technique [7] is used. The ROI and background area in a CT liver tissue image is shown in Figs. 1a, b.

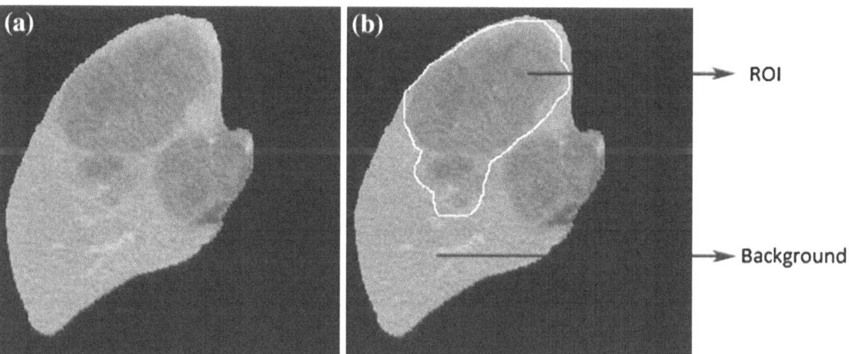

Fig. 1 **a** Original image, **b** with selected ROI

2.1 Fuzzy Based Level Set Segmentation

The fuzzy level set algorithm automates the initialization and parameter configuration of the level set segmentation, using spatial fuzzy clustering. It employs a Fuzzy c-means (FCM) with spatial restrictions to determine the approximate contours of interest in a medical image. One of the biggest challenges in level set segmentation is the intensive computation. The level set function $\emptyset$ converts the 2D image segmentation problem into a 3D problem. There are other constraints for stable level set evolution, too. For instance, the time step and the grid space should comply with the Courant–Friedreichs–Lewy (CFL) condition [8, 9], and the level set function $\emptyset$ should be re-initiated periodically as a signed distance function. To avoid this, a fast level set can be defined as follows:

$$\frac{\partial \emptyset}{\partial t} = \mu\left[\Delta\emptyset - \operatorname{div}\left(\frac{\nabla\emptyset}{|\nabla\emptyset|}\right)\right] + \lambda(\emptyset)\operatorname{div}\left(g\,\frac{\nabla\emptyset}{|\nabla\emptyset|}\right) + vg\delta(\emptyset) \tag{1}$$

Suppose the component of interest in an FCM results is $\mathbf{R}_k : \{r_k = \mu_{nk}, n = x \times N_y + y\}$. Then it is convenient to initiate the level set function as

$$\emptyset_0(x, y) = -4\varepsilon(0.5 - \mathbf{B}_k) \tag{2}$$

Where ε is a constant regulating the Dirac function. The Dirac function in (1) is slightly smoothed as the following function $\delta_\varepsilon(x)$ defined by:

$$\delta_\varepsilon(x) = \begin{cases} 0, |x| > \varepsilon \\ \frac{1}{2\varepsilon}[1+\cos\left(\frac{\pi x}{\varepsilon}\right), |x| \le \varepsilon \end{cases} \tag{3}$$

$\mathbf{B}_k$ is a binary image obtained from

$$\mathbf{B}_k = \mathbf{R}_k \ge b_0 \tag{4}$$

There are several controlling parameters associated with level set methods. Some parameters are determined adaptively for the specific medical image. Given the initial level set function $\emptyset_0$ from spatial fuzzy clustering as in Eq. (2), it is convenient to estimate the length l and the area α by

$$l = \int_1 \delta(\emptyset_0)dxdy; \tag{5}$$

$$\alpha = \int_1 H(\emptyset_0)dxdy, \tag{6}$$

where the Heaviside function $H(\emptyset_0)$ is

$$H(\emptyset_0) = \begin{cases} 1, \emptyset_0 \geq 0 \\ 0, \emptyset_0 < 0 \end{cases} \tag{7}$$

The level set evolution will be faster if the component of interest is large. In this case, the ratio

$$\varsigma = \alpha / l \tag{8}$$

will also be large. It is thereby reasonable to assign the time step τ as ς in this fuzzy level set algorithm. The penalty coefficient μ will be set as

$$\mu = 0.2/\varsigma \tag{9}$$

because their product ($\tau \times \mu$) should be less than 0.25 for stable evolution. The initial level set function $\emptyset_0$ obtained from fuzzy clustering Eq. (2) will approximate the genuine boundaries. Therefore, a comparatively conservative λ

$$\lambda = 0.1\varsigma \tag{10}$$

is used to control topological changes.

The balloon force v undertakes two roles in the level set evolution. First, its sign determines the advancing direction of the level set function: positive for shrinkage and negative for expansion. Second, the larger v is, the faster the level set evolves.

The fuzzy level set algorithm takes the degree of membership of each image pixel μ_k as the distance to the specific component of interest $\mathbf{R}_k$. An enhanced balloon force is applied to pull or push the dynamic interface adaptively towards the object of interest:

$$G(\mathbf{R}_k) = 1 - 2\mathbf{R}_k \tag{11}$$

The resultant balloon force $G(R_k)(\in -1, 1])$ is a matrix with a variable pulling or pushing force at each image pixel. In other words, the level set function will be attracted towards the object of interest regardless its initial position. Then, the evolutionary Eq. (1) is transformed into

$$\xi(g, \emptyset) = = \delta(\emptyset)\mathrm{div}\left(g \frac{\nabla\emptyset}{|\nabla\emptyset|}\right) + gG(R_k)\delta(\emptyset). \tag{12}$$

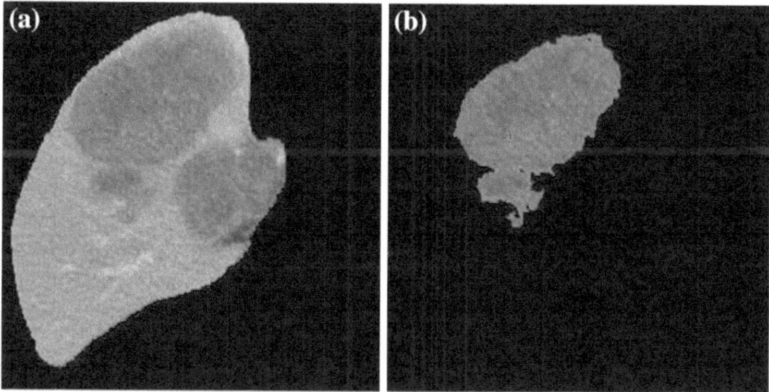

Fig. 2 **a** Original image, **b** extracted ROI

The balloon force can now be derived from spatial fuzzy clustering directly. Moreover level set evolution is now adapted to the distance to the genuine object. Once approaching the object, the level set function will automatically slow the evolution down and will become totally dependent on the smoothing term. Since a conservative λ is adopted here, level set evolution stabilizes automatically. An additional benefit is the flexibility to choose a comparatively large iteration of evolution T for robust segmentation. Without such an enhancement, the operator has to keep an alert to the level set evolution in order to avoid insufficient or excessive segmentation [7]. Figure 2b shows the segmented ROI of original image.

2.2 *Generation of ROI Mask*

When an image is coded with an emphasis of ROI, it is necessary to identify the wavelet coefficients needed for the reconstruction of the ROI. Thus, the ROI mask is introduced to indicate which wavelet coefficients have to be transmitted exactly in order for the receiver to reconstruct the ROI [10]. After image segmentation by fuzzy based level set method, we can attain ROI and BG. If ROI and BG is 0 or 1 respectively, we can attain a ROI mask.

3 Wavelets

Compression techniques based on the wavelet composition of the image have received much attention in the recent literature on medical image compression mainly due to the unique ability of representing the image in such a way that high compression is allowed preserving at the same time fine details of paramount importance [11]. A multiresolution (multiscale) representation of the image that trades off spatial resolution for frequency resolution is provided by the discrete wavelet transform (DWT). The best way to describe DWT is via a filter-bank tree. The input signal is filtered through the low-pass and high-pass analysis filters H_0 and H_1, respectively, and the outputs are subsampled by a factor of 2, that is, every other sample is kept. This sampling rate alteration is justified by the halving of the bandwidth of the original signal. After being quantized and/or entropy coded, the subband signals are combined again to form a full-band signal, by increasing their sampling rate (upsampling by a factor of 2) and filtering with the low-pass and high-pass synthesis filters G_0 and G_1 that interpolate the missing samples. It is possible to design the analysis/synthesis filters in such a way that in the absence of quantization of the subband signals, the reconstructed signal coincides with the original. Such analysis/synthesis systems are said to have the perfect reconstruction (PR) property. A direct extension to the 2D case is to decompose separately the image into low and high frequency bands, in the vertical and horizontal frequency [12]. This is achieved via the separable 2D analysis/synthesis system.

For the 9/7-tap wavelet filter, if L(n) and H(n) denote the low and high subband coefficients respectively, the coefficients necessary to reconstruct X(2n) are L(n-1), L(n), L(n + 1), H(n-2), H(n-1), H(n), H(n + 1). And to reconstruct X(2n + 1), the coefficients L(n-1), L(n), L(n + 1), L(n + 2), H(n-2), H(n-1), H(n), H(n + 1), H(n + 2) are needed as shown in Fig. 3 [13]. The ROI mask generation for CT liver tissue image is shown in Fig. 4.

4 Methodology for Content Based Coding

After separating out the image area ROI and the BG, the ROI-SPIHT compression algorithm is applied. The ROI is determined at each level by simply halving the coordinates of its representation at the previous (finer scale) level. Here, after decomposing the ROI image till level 4, the coefficients which are necessary and sufficient for the lossless reconstruction of the ROI are identified. The corresponding regions in the image subbands are multiplied by the weight factor yielding an amplification of the wavelet coefficients of interest. These same

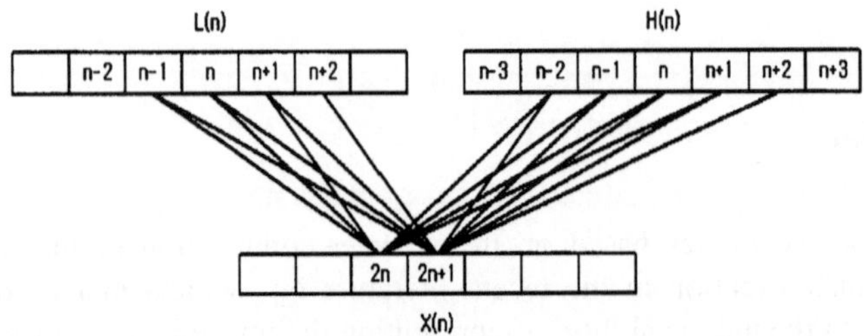

Fig. 3 The 9/7-tap inverse wavelet transform

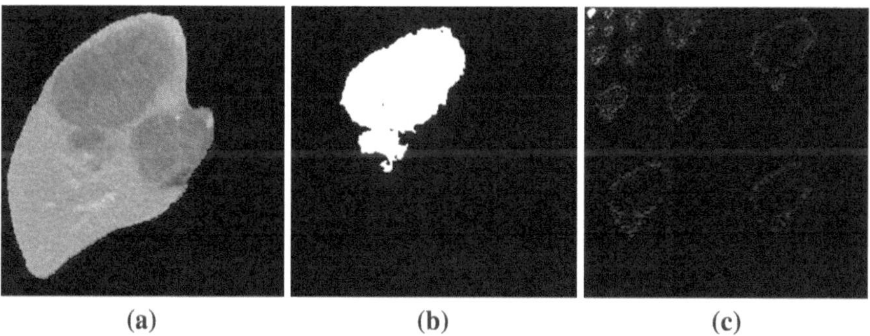

(a) (b) (c)

Fig. 4 **a** Original Image, **b** ROI mask in spatial domain, **c** ROI mask in wavelet domain

coefficients are divided by this factor in the reconstruction phase to undo the emphasis effect. These coefficients are coded first using the ROI-SPIHT encoding algorithm, so the bit budget is mostly spent on the ROI whereas the rest of the image undergoes a coarser quantization, the larger the weight assigned to a ROI, the better its reconstruction with respect to its context.

Algorithm's steps are:

1. Load the image to be compressed.
2. Segment the image into ROI and background images using fuzzy level set segmentation technique respectively.
3. Generate ROI mask to get ready for the selection of region of interest by segmentation method.
4. Calculate wavelet transform of ROI and background images.
5. Identify the ROI coefficients using the ROI mask in wavelet domain.
6. Scale up the ROI coefficients by a certain weight.
7. Perform encoding of the ROI region selectively with the ROI-SPIHT algorithm.
8. Encoding of the BG is also performed with the ROI-SPIHT technique.
9. Now, merge the encoded BG and the ROI bitstreams.
10. The reconstruction of the image is done after decoding the compressed image encoded data and scaling down the ROI coefficients by the same factor as was used during encoding.
11. After reconstruction, the image is correlated with original one and the performance parameters (SSIM, PSNR, CoC) are calculated.

5 Experimental Results

In this work, the results have been compared at various bit rates ranging from 0.0625 to 1.000 bits per pixel (bpp). The compression performance parameters namely bpp, Mean Square Error (MSE), Peak Signal-to-Noise ratio (PSNR) and

Table 1 ROI-SPIHT ROI and entire image parameters for CT liver tissue image

S. No.	Bpp	ROI image area parameters			Entire image parameters		
		MSE	PSNR	CoC	MSE	PSNR	CoC
1.	1.0000	11.9826	37.3453	0.9919	128.8480	27.0300	0.9766
2.	0.5000	12.0174	37.3327	0.9919	136.2143	26.7886	0.9752
3.	0.4000	12.2303	37.2564	0.9917	141.2794	26.6300	0.9743
4.	0.2500	13.3038	36.8910	0.9910	148.8469	26.4034	0.9728
5.	0.1250	17.9396	35.5927	0.9878	166.7910	25.9091	0.9692
6.	0.0625	29.7522	33.3956	0.9797	182.0715	25.5284	0.9660

CoC are calculated and listed in tabular form in Table 1. The comparative graphical results of different compression performance parameters are shown in Fig. 5. The proposed ROI-SPIHT method performs excellent at low bit rates as compared to other methods. The Comparison of PSNR versus bpp for JPEG, JEPEG2 K, SPIHT and proposed ROI-SPIHT algorithm has been shown in Table 2 and is plotted in Fig. 6. The reconstructed image output results of

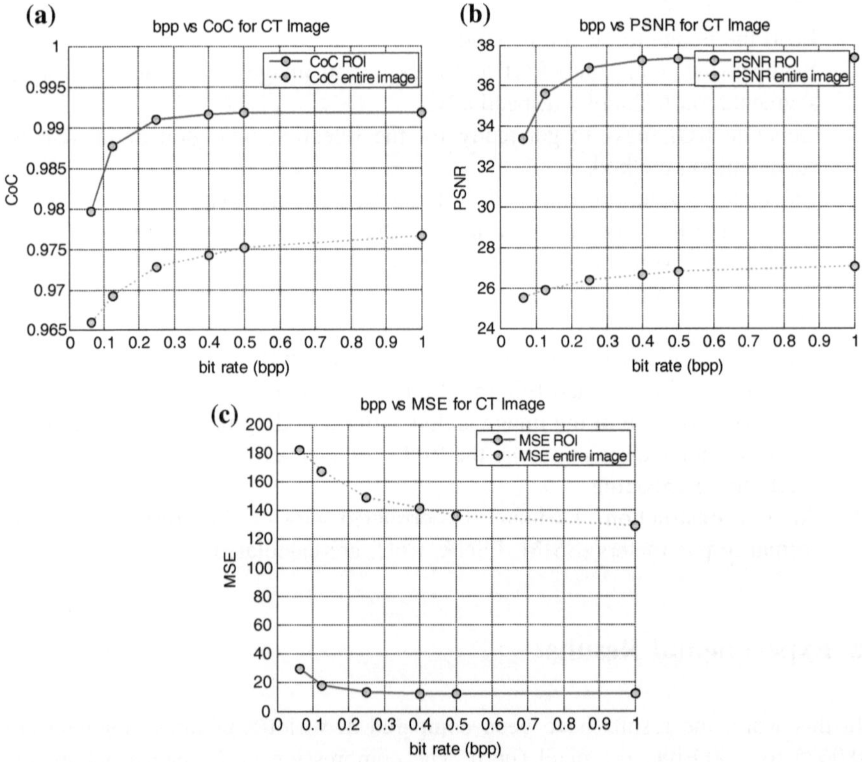

Fig. 5 Graphical results (bpp 'Vs' CoC, PSNR, MSE)

Table 2 Comparison of PSNR (dB) for JPEG, JPEG2K, SPIHT and proposed ROI-SPIHT algorithms

S. No.	Bit rate	JPEG	JPEG2K	SPIHT	Proposed (ROI-SPIHT)
1.	0.0625	21.937	24.951	23.763	33.3956
2.	0.125	22.823	25.021	25.716	35.5927
3.	0.25	24.021	25.264	26.939	36.8910
4.	0.5	25.178	28.587	27.675	37.3327
5.	1.0	27.886	28.639	27.890	37.3453

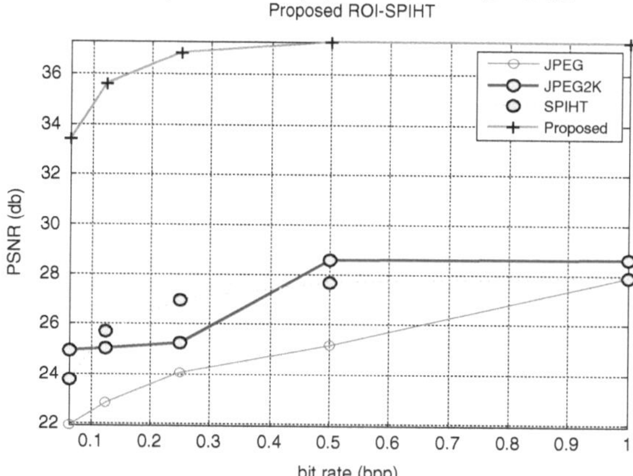

Fig. 6 Comparison of PSNR vs bpp for JPEG, JPEG2K, SPIHT and proposed ROI-SPIHT

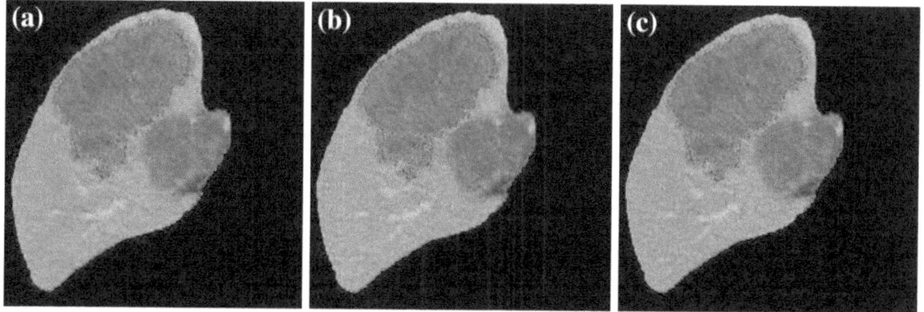

Fig. 7 CT image reconstructed at different bitrates **a** bpp 0.4 **b** bpp 0.6 **c** bpp 0.8

proposed ROI-SPIHT are shown in Fig. 7 for different bitrates. The comparison of SPIHT compressed image and a proposed ROI-SPIHT compressed image is shown in Fig. 8 which shows that the ROI-SPIHT compressed image is better in visual quality as compared to the SPIHT compressed image for the same bpp.

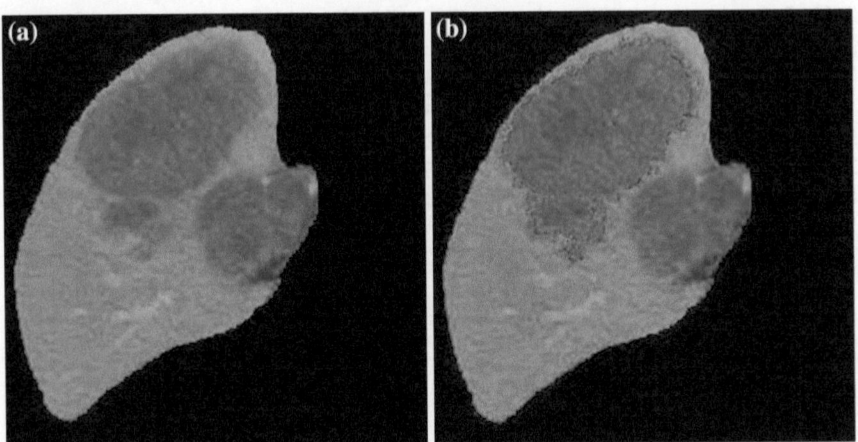

Fig. 8 **a** reconstructed with original SPIHT at bpp.4, **b** reconstructed with proposed ROI-SPIHT at bpp.4

6 Conclusion

ROI coding techniques are of great interest in telemedicine or medical imaging applications with large storage requirements. By comparing the performance of this method on the basis of various parameters, it has been shown that fuzzy level set segmentation integrated with ROI-SPIHT coding gives better rate-distortion performance as compared to SPIHT, JPEG2000 and JPEG.

References

1. Gershikov E, Laviburlak E, Porat M (2007) Correlation-based approach to color image compression. Sig Process Image Commun 22(9):719–733
2. Roterman Y, Porat M (2007) Color image coding using regional correlation of primary colors. Image Vis Comput 25(5):637–651
3. Kotera H, Lanamori K (1990) A novel coding algorithm for representing full color images by a single color image. J Imaging Technol 16:142–152
4. Ansari MA, Anand RS (2009) Context based medical image compression for ultrasound images with contextual set partitioning in hierarchical trees algorithm. Adv Eng Softw 40(7):487–496
5. Chan HY, Grindstead B, Sari-Saraf H, Gleason S (2004): Content-based compression of mammograms with fractal-based segmentation and a modified JPEG2000. Opt Eng 43:2986
6. Oliver A, Freixenet J, Martí J, Pérez E, Pont J, Denton ERE, Zwiggelaar R (2010) A review of automatic mass detection and segmentation in mammographic images. Med Image Anal 14(2):87–110
7. Li BN, Chui CK, Chang S, Ong SH (2011) Integrating spatial fuzzy clustering with level set methods for automated medical image segmentation. Comput Biol Med 41(1):1–10

8. Sethian JA (1999) Level set methods and fast marching methods. In: Press CU (ed) Journal of computing and information technology, vol 11. Cambridge University Press, Cambridge, pp 79–79
9. Osher S, Fedkiw R (2004) Level set methods and dynamic implicit surfaces. Applied Mech Rev(Springer) 57(15)
10. Zhao, hong Li, Ya-nan T, Yong-gang S et al (2007) Lossless Compression of Medical Images by Combining Integer Wavelet Transform with DPCM. J Northeast Univ Nat Sci 28(10):1454–1457
11. Kofidis E, Kolokotronis N, Vassilarakou A, Theodoridis S, Cavouras D (1999) Wavelet–based medical image compression. Future Gener Comput Syst 15:223–243
12. Vetterli M (1984) Multi-dimensional subband coding: some theory and algorithms. Signal Process 6:97–112
13. Park K, Hyeong PK, Park HWPHW (2002) Region-of-interest coding based on set partitioning in hierarchical trees. IEEE Transactions on Circuits and Systems for Video Technology 12:106–113

An Improved IWT Based Algorithm for Transmission and Security of Telemedicine

Aparna Mohanty and Asutosh Kar

Abstract Watermarking is a technique to embed identification codes, called watermarks, into cover media or host media and used for protection of intellectual property as well as authentication and integrity of information. This technique does not restrict access to the information while encryption has the aim of making messages unintelligible to unauthorized persons who might intercept them. A watermark is designed to permanently reside in the host data. In this literature an integer wavelet transform (IWT) based algorithm for transmission and security of medical images has been proposed. The IWT maps integers to integers and allows perfect inevitability with finite precision arithmetic. The extracted image which is obtained should have a high peak signal to noise ratio (PSNR) or else its discarded. At the receiver end the patient digest is cross verified to obtain back the watermarked medical image for further analysis.

Keywords Wavelet · IWT · Bit plane slicing · RSA encryption · RSA decryption

1 Introduction

To Pixels are digital numbers composed of bits. For example, the intensity of each pixel in a 256 level grey scale image is composed of 8 bits (i.e. 1 byte). Each pixel in an image can be represented in terms of its grey scale representation whose

A. Mohanty (✉)
Department of Electronics and Communication Engineering, TITE, Bhubaneswar, India
e-mail: aparna.mhnty@gmail.com

A. Kar
Department of Electronics and Telecommunication Engineering, IIIT, Bhubaneswar, India
e-mail: asutosh@iiit-bh.ac.in

Mohan S. and S. S. Kumar (eds.), *Proceedings of the Fourth International Conference on Signal and Image Processing 2012 (ICSIP 2012)*, Lecture Notes in Electrical Engineering 221, DOI: 10.1007/978-81-322-0997-3_16,

value will vary from 0 to 255 depending on the intensity at the given point [1]. Let the image be composed of 8, 1-bit planes ranging from bit plane 1-0 (LSB) to bit plane 7 (MSB). In terms of 8-bit bytes, plane 0 contains all lowest order bits in the bytes comprising the pixels in the image and plane 7 contains all higher order bits. This is called bit plane slicing [1]. Here instead of highlighting intensity-level ranges, the contribution made to total image appearance by specific bits has been highlighted. Edges characterize boundaries and therefore appear as a problem of fundamental importance in image processing. Edge detecting an image significantly reduces the amount of data and filters out useless information, while preserving the important structural properties in an image [1, 2]. Discontinuities in image brightness are likely to correspond to discontinuities in surface orientation, changes in material properties and variations in scene illumination [3, 4].

In the ideal case, the result of applying an edge detector to an image may lead to a set of connected curves that indicate the boundaries of objects, the boundaries of surface markings as well as curves that correspond to discontinuities in surface orientation. Edges extracted from non-trivial images are often hampered by fragmentation, meaning that the edge curves are not connected, missing edge segments as well as false edges not corresponding to interesting phenomena in the image—thus complicating the subsequent task of interpreting the image data. The edges obtained from natural images are usually not at all ideal step edges [4]. Instead they are normally affected by one or several of the following effects i.e. focal blur caused by a finite depth-of-field and finite point spread function, penumbral blur caused by shadows created by light sources of non-zero radius, shading at a smooth object. A one-dimensional image f which has exactly one edge placed at x = 0 may be modelled as [4]:

$$f(x) = \frac{I_r - I_1}{2}\left(erf\left(\frac{x}{\sqrt{2}\sigma}\right) + 1\right) + I_l \tag{1}$$

At the left side of the edge, the intensity is $I_l = \lim_{x \to -\infty} f(x)$, and right of the edge it is $I_l = \lim_{x \to \infty} f(x)$. The scale parameter σ is called the blur scale of the edge [4, 5].

1.1 Wavelets

Generally, wavelets are purposefully crafted to have specific properties that make them useful for signal processing. Wavelets can be combined, using a "shift, multiply and sum" technique called convolution, with portions of an unknown signal to extract information from the unknown signal [5]. All wavelet transforms may be considered forms of time–frequency representation for continuous-time (analog) signals and so are related to harmonic analysis. Almost all practically useful discrete wavelet transforms use discrete-time filter banks. These filter banks are called the wavelet and scaling coefficients in wavelets nomenclature [2]. These

Table 1 RSA Encryption and Decryption

RSA Encryption	RSA Decryption
For the algorithm to be satisfactory for public-key encryption, the following requirement must be met [4] 1. It is possible to find values of e, d, n such that $T^{\Lambda ed} = T\|n\|$ for all $T < n$ 2. It is relatively easy to calculate $T^{\Lambda e}$ and $C^{\Lambda d}$ for all values of $T < n$ 3. It is infeasible to determine d given e, n	Recipient B does the following [4]:- 1. Uses his private key (n, d) to compute $T = c^d\|n\|$ 2. Extracts the plaintext from the message representative m

filter banks may contain either finite impulse response (FIR) or infinite impulse response (IIR) filters. The wavelets forming a continuous wavelet transform (CWT) are subject to the uncertainty principle of Fourier analysis respective sampling theory. Given a signal with some event in it, one cannot assign simultaneously an exact time and frequency response scale to that event [1]. The product of the uncertainties of time and frequency response scale has a lower bound. Thus, in the scalogram of a continuous wavelet transform of this signal, such an event marks an entire region in the time-scale plane, instead of just one point [5]. Also, discrete wavelet bases may be considered in the context of other forms of the uncertainty principle. Wavelet transforms are broadly divided into three classes: continuous, discrete and multi resolution-based.

1.2 RSA Encryption and Decryption

The RSA algorithm is mainly a public key encryption technique used widely in network communication like in Virtual Private Networks (VPNs). In RSA algorithm encryption and decryption are of following form, for some plain text T and cipher text C [4]:

$$C = T^e|n| \tag{2}$$

$$T = c^d|n| \tag{3}$$

the value of n must be known to both the sender and receiver. The value of e, d is exclusively known to the sender and receiver respectively. Thus, this is a public-key encryption algorithm with a public key of $\{e, n\}$ and private key of $\{d, n\}$ (Table 1), (Table 2).

Table 2 The method of digital signing

Sender A does the following	Signature verification-Recipient B does the following
1. Creates a message digest of the information to be sent	1. Uses sender A's public key (n, e) to compute integer $v = s^e \mid n \mid$ [4]
2. Represents this digest as an integer m between 0 and n-1	2. Extracts the message digest from this integer
3. Uses her private key (n, d) to compute the signature $s = m^d \mid n \mid$	3. Independently computes the message digest of the information that has been signed
4. Sends this signature s to the recipient, B	4. If both message digests are identical, the signature is valid

2 Methodology

2.1 IWT Domain

In order to avoid fractional calculations and saving the computational time for algorithm, we have used the integer wavelet transform which maps integers to integers and reconstruct the original signal without any distortion. Though various wavelet families can be applied for watermarking but in this literature the IWT has been employed as it maps integers to integers and allows for perfect invertibility with finite precision arithmetic. A simple and effective way to construct IWT is to first factor the traditional discrete wavelet transform (DWT) into lifting steps and then to apply a rounding operation at each step [2]. The IWT can thus be used for lossless compression of medical images. One of the main advantages of using the wavelet transform for compression is that it provides a multi resolution representation of the image, which other techniques like spatial-domain prediction cannot offer [6].

IWT (integer wavelet transform) has been used to embed the watermark in frequency domain. IWT performs the transformation of certain set of integers into some other set of integers. The image undergoes the IWT and then the watermark is added in the frequency domain. The data to be entered undergoes arithmetic coding and the data is watermarked [7, 8, 9]. After that the image undergoes inverse integer wavelet transform (IIWT) which gives a watermarked image in spatial domain.

3 Implementation

3.1 Embedding Algorithm

1. First of all the original image is read and loaded into workspace.
2. The image is converted to grayscale if necessary and resized to 256 × 256 size.
3. Now the image undergoes the wavelet decomposition using the liftwave INT2INT method and cdf2.2 matrix.

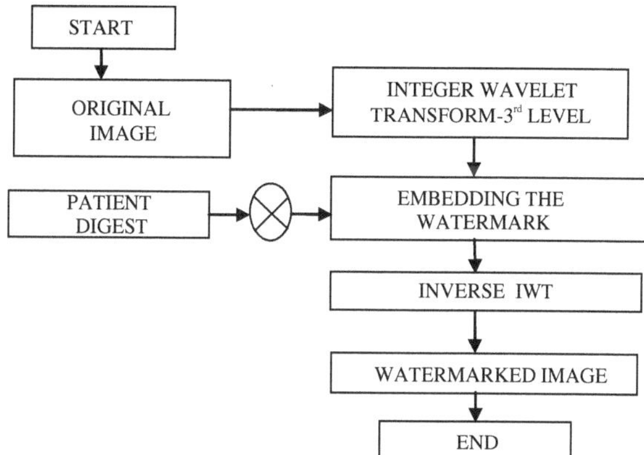

Fig. 1 Block diagram showing the embedding process in IWT domain

4. The image is splinted into CH, CV, CD and CA. Which form a certain set of integers matrix of the image.
5. Now the binary images are created using the fifth bit of CH, CV and CD. After that the data is compressed and how many times one occurs in CH is calculated.
6. Then the sequence is coded using the arithmetic coding. The same is done for the CV and CD.
7. Now the watermark is read and reshaped for insertion. The dimensions of the watermark image should be less than the original image size.
8. Now the compressed data and the watermark are embedded into the original image. The data length is determined and inserted. The value of the each data length is also inserted for decoding [10].
9. For CH, CV and CD the same data is embedded and after this we go for IIWT.
10. The IIWT gives us the watermarked image which is then stored using the imwrite command.
11. The output is a watermarked image whose PSNR value is also determined and thus sent through transmitter to the receiver.

3.2 Extraction Algorithm

1. The watermarked image is received and stored in a variable.
2. Now again the IWT is performed and the image is splitted into CH, CV, CD and CA using the liftwave INT2INT and cdf2.2 matrix.
3. In next step the signal is extracted from the 5th bit of CH, CV and CD. The memory is pre allocated to fasten up the things [11].
4. Again the binary images are constructed from CH, CV and CD. The length of the sequence is also determined for arithmetic decoding.

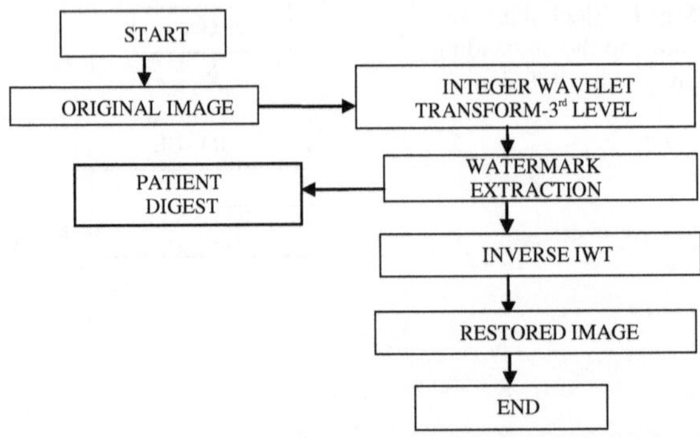

Fig. 2 Block diagram showing the embedding process in IWT domain

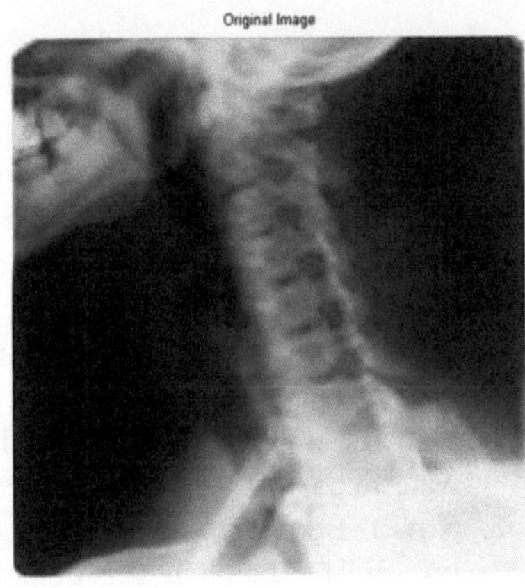

Fig. 3 The original sample medical image that is to be watermarked

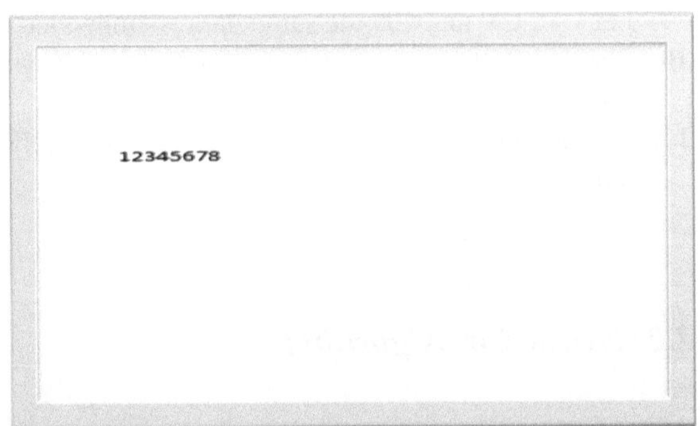

Fig. 4 This shows the watermark image to be used

5. The length of compressed CH, CV, CD and watermark is obtained and they are decoded after that.
6. Now the watermark is removed and the uncompressed data is inserted back into image.

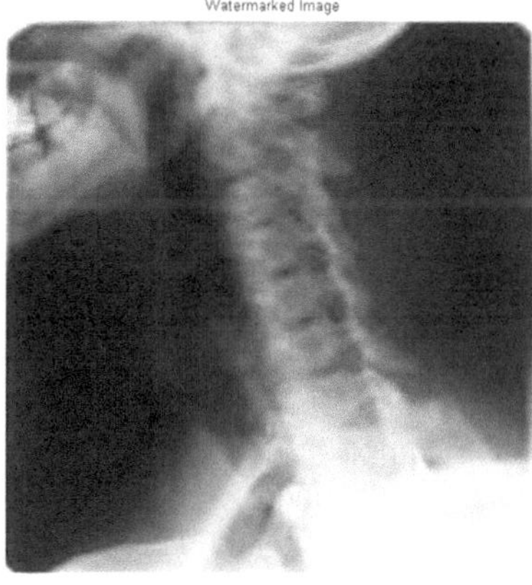

Fig. 5 Block diagram showing the extraction process in IWT domain

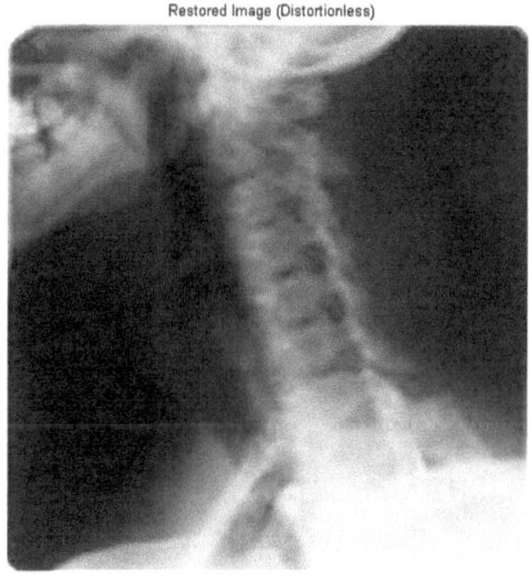

Fig. 6 This is the restored distortion less image at receiver side

Fig. 7 Difference between Restored image and original image

7. After that the inverse integer wavelet transform is done to obtain the distortion less original image back.
8. The PSNR value here comes out to be 38.92 dB which is also acceptable [12–14].

4 Frequency Domain Simulation and Results

The MATLAB 7.7 platform has been chosen for simulation purpose. A sample medical image as shown in Fig. 3 is taken for the processing. The image has been watermarked and IWT transform is employed as explained in Sect. 3.1 and specified in Fig. 1 block diagram. The watermarked image to be used and the watermarked version of Fig. 3 which is to be transmitted after IWT have been shown in Figs. 4 and 2 respectively.

The extraction algorithm as explained in sect. 3.2 and mentioned in Fig. 5 has been applied to the transmitted image i.e. Fig. 2 and the restored distortion less image is obtained as shown in Fig. 6. Figure 7 depicts the difference between restored and original image. The PSNR value was a factor for authentication. If the image is attacked by any kind of noise then the PSNR value will fall to a lower value which is not acceptable. Even if the image is recovered it is still of no use as medical images require high quality n precision. The greater the PSNR value the better is the quality. In our scheme the PSNR value of 39 dB was obtained which is a high value. Thus the quality and the integrity of the images are maintained in our method. Also the patient information can be sent along with the image.

5 Conclusion

The watermark was embedded in the original image by the use of bit plane slicing and RSA encryption. We can embed ample information about patient in the image like patient ID, ERP no. and doctor details. Experimental results show that the proposed method gives a pretty high PSNR value. This method focuses on reducing the degradation of the watermarked image yet providing strong evidence for the authentication of the medical image. Also the PSNR criteria help in obtaining a high quality images at receiver. In general noise with PSNR higher than 30 dB is hardly noticeable.

References

1. Gonzalez RC, Woods RE (eds) (2008) Digital image processing. 3rd edn. Prentice Hall Inc, New Jersey
2. Jain AK (ed) (1989) Fundamentals of digital image processing. Prentice Hall Inc, New Jersey
3. Memon NA, Gilani SAM, Qayoom S (2009) Multiple watermarking of medical images for content authentication and recovery. IEEE proceedings on multitopic conference, INMIC, pp 1–6, Dec 14–15 2009.
4. Aparna M, Asutosh K (2012) A novel approach for spatial domain authentication systems design in telemedicines. In: IEEE proceedings on international conference on computer, communication and informatics, Coimbatore, India, 10 Jan 2012

5. Manasrah T, Al-Haj A (2006) Management of medical images using wavelets-based multi-watermarking algorithm. In: IEEE proceedings on image processing, July 2006
6. Fotopoulos V, Stavrinou ML, Skodras AN (2008) Medical image authentication and self-correction through an adaptive reversible watermarking technique. IEEE Transactions Published on 4 July 2008
7. Feng JB, Lin IC, Tsai CS, Chu YP (2008) Reversible watermarking: Current status an key issues. In: IEEE proceedings on medical images
8. Viswanathan P, Venkata Krishna P (2009) Text fusion watermarking in medical image with semi-reversible for secure transfer and authentication. In: Proceedings on international conference on advances in recent technologies in communication and computing
9. Chemak C, Christophe J, Bouhlel MS (2007) New watermarking scheme for security and transmission of medical images for pocket neuro project. IEEE Radio Eng 16(4)
10. Kong X, Feng R (2001) Watermarking medical signals for Telemedicine. IEEE Trans Inf Technol Biomed 5(3):195–201
11. Garcia E, Guynnette H, Lapayre JC, Moulin T (2005) Adaptive tele-application for remote neurology diagnosis. J Telemedicine E-Health 11(6):692–702
12. Chemak C, Lapayre JC, Bouhlel MS (2007) A new scheme of image watermarking based on 5/3 wavelet decomposition and Turbo-Code. In the Book of abstracts of the (WSEAS) European Computing Conference (ECC'07), Athens, Greece, p 11, Sep 2007
13. Adams MD, Kossentni F (2000) Reversible integer-to-integerwavelet transforms for image compression: performance evaluation and analysis. IEEE Trans Image Process 9(6): 1010–1024
14. Kutter M, Petticolas FAP (1999) A fair benchmark for image watermarking systems. In: Proc of SPIE Security and watermarking of multimedia contents. San Jose, USA, p 226–239

Optimal Color Palette for Error Diffusion Techniques

Mohammed Hassan and Chakravarthy Bhagvati

Abstract Color quantization is an important problem for many applications in graphics and multimedia in which only a limited number of colors can be displayed or printed simultaneously. Reconstruction of an image with a limited number of colors (color palette) leads to highly visible degradations in image quality known as false contours. A way to overcome this problem is to perform dithering techniques. In this paper we propose a color quantization method for the use with color dithering techniques in a way that better results will be obtained after dithering. The results show that our method completely removes false contours as well as prevents color impulses that are common results of applying dithering techniques to color quantized images.

Keywords Color quantization · Error diffusion · Combined quantization · Error diffusion

1 Introduction

Color quantization is the process of reducing the number of colors presented in a digital color image while maintaining the appearance of the original as much as possible. Originally, color quantization has been used to satisfy the display hardware constraints that allow a limited number of colors to be displayed simultaneously. Today the original motivation of color quantization has changed due to availability of inexpensive full color displays. However, color quantization is still

M. Hassan (✉) · C. Bhagvati
Department of Computer and Information Sciences, University of Hyderabad, Hyderabad 500046, India
e-mail: mohdnow@yahoo.com

Mohan S. and S. S. Kumar (eds.), *Proceedings of the Fourth International Conference on Signal and Image Processing 2012 (ICSIP 2012)*, Lecture Notes in Electrical Engineering 221, DOI: 10.1007/978-81-322-0997-3_17,

an important problem in the fields of image processing and computer graphics; it can be used in mobile and hand-held devices where memory is usually small [1], for low-cost display and printing devices where only a small number of colors can be displayed or printed simultaneously [2], it also can be used in lossy compression techniques [3]. Another aspect of importance of color quantization is that the human visual system can't perceive more than 20,000 different colors at any given time [4] while a full color image may contain up to 16 million different colors; this large number of colors makes it difficult to handle a variety of color-based tasks such as computing histograms or other useful statistics. The most commonly used algorithms for color quantization are: median-cut [5], center-cut [6], octree [7], k-means [8], Wu's quantizer [9], fuzzy C-means [10], and SOM [11].

Replacing a larger set of colors by a much smaller set of representative colors (color palette) will cause highly visible degradations in image quality. The most disturbing of these degradations is the appearance of false contours in the quantized image regardless of color quantization algorithm used. False contours appear when batches of colors in a smooth gradient area are mapped into a constant palette colors forming a flat regions. Boundaries between such regions may be visible as false contours. A common way to overcome this problem is the use of dithering techniques that make use of the averaging property of the human eye to the colors in a neighborhood of the point of interest and create the illusion of more colors. A dithering technique called error diffusion achieves this effect by distributing the error encountered in quantizing a pixel to neighboring pixels. This results in an alternation of palette colors between neighboring pixels and is perceived as a new color by the eye. Some well-known error diffusion filters are Floyd–Steinberg [12], Stucki [13], and Jarvice–Judice–Ninke [14].

There is some previous work for jointly quantizing and dithering color images in the literature. Orchard [15] used binary tree splitting (BTS) to create the color palette in a way to minimize the total squared difference between the actual and the quantized images and then modified dithering techniques are combined with quantization process. A modified BTS quantizer of [15] is proposed in [16] in such a way that at leaves the pair of cluster centers are given equal displacements from each other following the split in order to span a larger volume of color space. This creates an illusion of more colors after dithering. In competitive learning and error diffusion algorithm [17], the quantization error is diffused to neighboring pixels during learning process. In [18] three new fuzzy methods to jointly quantize and dither the color images are presented, in two of them a fraction of the quantization error is distributed to neighboring pixels during creating color palettes and then perform quantization and dithering as usual with the resultant palette, while the third method is to enlarge the convex hull of the quantization colors to obtain a color palette which is more suitable for dithering.

In this paper, we propose a quantization method which is designed to obtain better results with minimum visual degradation after the error diffusion technique is applied. In Sect. 2, the CIELAB color space is simply introduced in which we derive our algorithm. Section 3 describes the proposed quantization method. The results of the study are presented in Sect. 4 and are followed by the conclusions.

2 Choice of Color Space

The CIE $L^*a^*b^*$ (CIELAB) and $L^*u^*v^*$ (CIELUV) color spaces [19] are considered to be perceptually uniform and referred to as uniform color spaces in which the Euclidean distances between different colors in the color space correspond approximately to perceived color differences. Gentile, Allebach and Walowit have studied the application of uniform color spaces to the problems of image quantization [20] and halftoning [21], the results indicate that such color spaces can significantly improve quality of the delivered quantized image relative to the standard RGB color space. The CIE recommended using XYZ coordinate system to transform RGB to $L^*a^*b^*$. The following equations are used to transfer RGB to XYZ.

$$\begin{pmatrix} X \\ Y \\ Z \end{pmatrix} = \begin{pmatrix} 0.4124564 & 0.3575761 & 0.1804375 \\ 0.2126729 & 0.7151522 & 0.0721750 \\ 0.0193339 & 0.1191920 & 0.9503041 \end{pmatrix} \begin{pmatrix} R \\ G \\ B \end{pmatrix} \tag{1}$$

And then from XYZ to CIELAB color space

$$L^* = \begin{cases} 116(Y/Y_n)^{1/3} - 16, & \text{if } Y/Y_n > 0.008856 \\ 903.3(Y/Y_n), & \text{if } Y/Y_n \leq 0.008856 \end{cases} \tag{2}$$

$$a^* = 500(f(X/X_n) - f(Y/Y_n)) \tag{3}$$

$$b^* = 200(f(Y/Y_n) - f(Z/Z_n)) \tag{4}$$

where

$$f(t) = \begin{cases} t^{1/3}, & \text{for } t > 0.008856 \\ 7.787 * t + 16/116, & \text{for } t \leq 0.008856 \end{cases} \tag{5}$$

Here $Y_n = 1.0$ is the luminance, and $X_n = 0.950455$, $Z_n = 1.088753$ are the chrominances for the D_{65} white point.

A useful rule of thumb in CIELAB color space is that any two colors can be distinguished if the Euclidean distance between these two colors is greater than threshold 3 [22]. This threshold is known as the Just Noticeable Color Difference (JNCD) threshold. Therefore all the colors within a sphere of radius equal to the

JNCD threshold are perceptually indistinguishable from each other. However, it will be assumed throughout this paper that the original and color quantized images are specified in CIELAB color space.

3 Proposed Color Quantization Method

Usually color quantization and dithering are performed sequentially, the quality of the dithering process depending on the selection of the color palette. Standard methods of color quantization do not take dithering into account. Therefore, the desirable qualities obtained by quantization are often disturbed by the dithering process. Our method of color quantization is designed to obtain better results with minimum visual degradation after the error diffusion technique is applied. It consists of three steps: in the first step, the colors in original image are quantized into a number of colors that is much smaller than the desired number of colors. Then in the second step, the remaining palette colors are selected around each palette color generated in the first step based on the size of the clusters and the CIELAB JNCD threshold, and finally error diffusion technique is incorporated within the color mapping processing.

3.1 Generation of Initial Palette Colors

In this step, the colors in the original image are quantized into a small set of M colors that is much less than the desired K palette colors ($M = 2^{\lfloor \frac{\log_2 K}{2} \rfloor}$). Any quantization algorithm can be used to generate the initial set of palette colors. We have chosen Wu's quantizer [9] for its computational speed and accepted delivered quality to generate the initial set of palette colors. Wu's quantizer is a variance-minimization algorithm that bipartitions the color space in such a way that the cutting plane is chosen perpendicularly to the R, G and B axes separately, and the plane that minimizes the sum of variances at both sides is chosen to cut the cube into two boxes. Next, the box with the larger variance is partitioned into two smaller boxes by the same cutting criterion.

3.2 Selection of the Remaining Palette Colors

After the generation of the initial palette colors, the remaining $(K - M)$ palette colors are selected around each initial palette color based CIELAB JNCD threshold value 3. The use of CIELAB JNCD threshold is to ensure that the selected palette colors will be far from each other by a distance that is enough for the selected palette colors to be distinguishable from each other. The number of

Fig. 1 Floyd–Steinberg vector filter for a single pixel

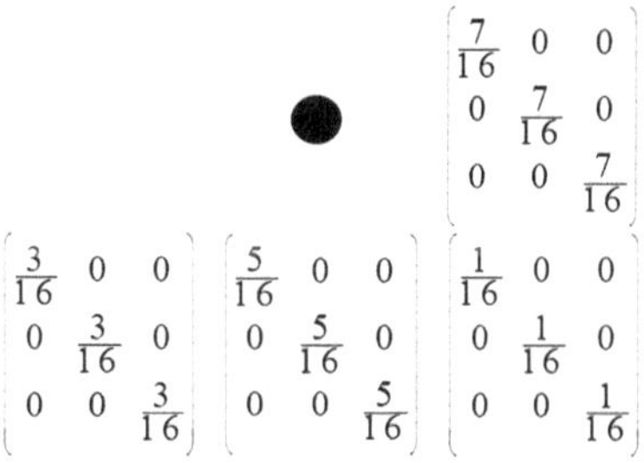

$$\bullet \quad \begin{pmatrix} \frac{7}{16} & 0 & 0 \\ 0 & \frac{7}{16} & 0 \\ 0 & 0 & \frac{7}{16} \end{pmatrix}$$

$$\begin{pmatrix} \frac{3}{16} & 0 & 0 \\ 0 & \frac{3}{16} & 0 \\ 0 & 0 & \frac{3}{16} \end{pmatrix} \begin{pmatrix} \frac{5}{16} & 0 & 0 \\ 0 & \frac{5}{16} & 0 \\ 0 & 0 & \frac{5}{16} \end{pmatrix} \begin{pmatrix} \frac{1}{16} & 0 & 0 \\ 0 & \frac{1}{16} & 0 \\ 0 & 0 & \frac{1}{16} \end{pmatrix}$$

selected palette colors around a given initial palette color is proportional to size of cluster of image colors represented by that initial palette color.

For an initial palette color P_i that represents a cluster $\{C_1, C_2, \ldots, C_{R_i}\}$ of image colors, a set S_i of $\frac{R_i}{N}(K-M)$ distinct image colors will be selected around P_i from $\{C_1, C_2, \ldots, C_{R_i}\}$ and are assigned as new palette colors. N is the total number of distinct colors in the original image. The selection of the S_i of palette colors around P_i will be performed as follows:

for $j = 2, \ldots, |S_i| + 1$ and $P_{i_1} = P_i$

$$P_{i_j} = \min_{P \in \{P_{i_1}, P_{i_2}, \ldots, P_{i_{j-1}}\}} \{d(P, C) > 3 : C \in \{C_1, C_2, \ldots, C_{R_i}\}\} \tag{6}$$

where d is the Euclidian distance between two colors in the CIELAB color space.

3.3 Palette Colors Mapping with Error Diffusion

Color quantization process is considered in two stages: the selection of an optimal color palette and the mapping of each pixel of the image to a color from the color palette. Regardless of the color quantization algorithm used, false contours are among the most disturbing degradations appeared in the delivered color quantized image. A way to overcome this problem is to perform dithering techniques. In this paper we incorporate error diffusion technique within the pixel mapping stage to improve the quality of the displayed image. The pixels are chosen in raster ordering from left to right and top to bottom. Each pixel is quantized and the quantization error is propagated forward to the neighboring pixels that have not yet been quantized. Figure 1 shows Floyd–Steinberg [12] filter that we used to perform the error diffusion.

4 Results and Discussions

Standard color quantization process is generally considered as a clustering problem to find the Kpalette colors that minimize some error criterion for all the colors in an image. However, two problems are associated with performing error diffusion techniques for color quantized images. The first problem is: if a given color is close to the boundary of a color cluster, the quantization errors accumulate and eventually a palette color from a different cluster is produced. This exposes itself as a color impulse that is very disturbing. On the other hand, in smooth regions usually the colors are very close to a palette color with a small dithered error. Therefore those colors will be mapped to the same palette color forming flat regions. This results in less alternation of palette colors between neighbor pixels to give full illusion of smooth gradation. These two problems have been solved in our proposed method (as shown in Figs. 2, 3) because the palette colors are distributed around few initially selected palette colors. Therefore it needs a small diffused error to jump into another close palette color in the same cluster.

Fig. 2 128-colors quantized images followed by Floyd–Steinberg error diffusion. **a** Proposed method, **b** K-means, **c** Median cut, **d** Octree, **e** SOM, **f** Wu's quantizer

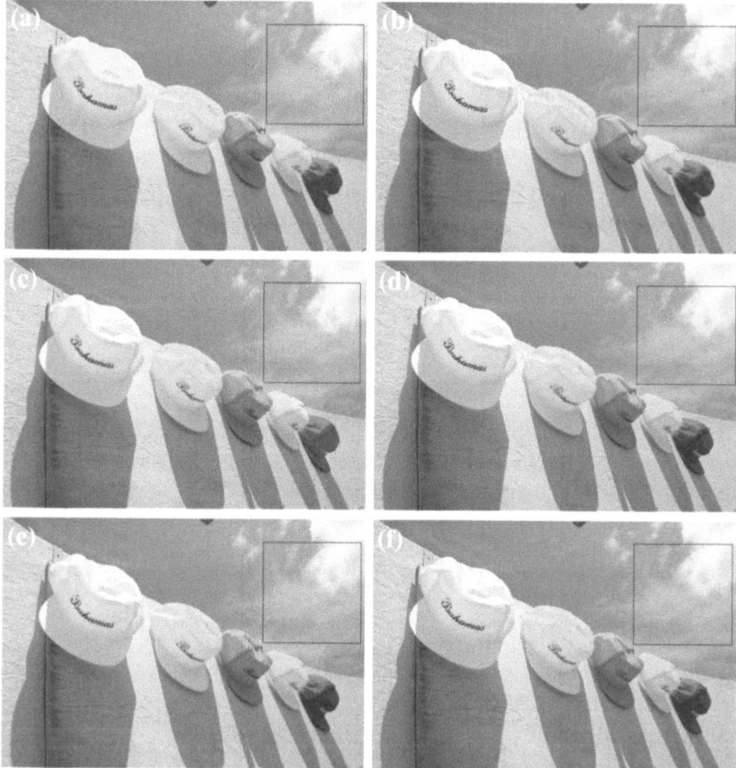

Fig. 3 128-colors quantized images followed by Floyd–Steinberg error diffusion. **a** Proposed method, **b** K-means, **c** Median cut, **d** Octree, **e** SOM, **f** Wu's quantizer

In Figs. 2 and 3 two full color images are quantized by the proposed method. The same two images are also presented to five popular quantization algorithms namely: k-means [8], median-cut [5], octree algorithm [7], SOM [11], Wu's quantizer [9] in which the Floyd–Steinberg error diffusion technique was incorporated within the pixel mapping process. It is observed that applying the Floyd–Steinberg dithering on the other quantization algorithms causes the creation of color impulses. The color impulses are in form of isolated green and dark blue pixels on sky area in Fig. 3b–f. It is also noted that some false contours remain visible in these images on the smooth gradient area in Fig. 2b–f. Figures 2a and 3a are the results of the proposed method that shows the best enhanced performance when compared to the other algorithms, as expected: The appearance of disturbing colored impulses and false contours are completely eliminated.

5 Conclusion

In this paper, we presented a novel method for quantization of color images to be used with error diffusion techniques. The results show that the proposed method gives the best enhanced performance when used with error diffusion techniques compared to the other color quantization algorithms in which it completely removes false contours and prevents color impulses that are common results of applying dithering techniques to color quantized images.

References

1. Rui X, Chang C, Srikanthan T (2002) On the initialization and training methods for Kohonen self-organizing feature maps in color image quantization. In: First IEEE international workshop on electronic design, test and applications, pp 321–325
2. Scheunders P (1997) A genetic C-means clustering algorithm applied to color image quantization. Pattern Recognit 30:859–866
3. Velho L, Gomes J, Sobreiro M (1997) Color image quantization by pairwise clustering. In: Proceedings of the 10th Brazilian symposium on computer graphics and image processing. IEEE Computer Society, New York, pp 203–207
4. Sharma G (1996) Digital color imaging. CRC Press, Boca Raton
5. Heckbert P (1982) Color image quantization for frame buffer display. ACM Trans Comput Graph (SIGGRAPH) 16:297–307
6. Joy G, Xiang Z (1993) Center-cut for color-image quantization. Visual Comput 10:62–66
7. Gervautz M, Purgathofer W (1988) A simple method for color quantization: octree quantization. In: New trends in computer graphics. Springer, Berlin, pp 219–231
8. Lloyd SP (1982) Least squares quantization in PCM. IEEE Trans Inf Theory 28:129–137
9. Wu X (1991) Efficient statistical computations for optimal color quantization. In: Arvo J (ed) Graphics Gems II. Academic press, New York, pp 126–133
10. Lim YW, Lee SU (1990) On the color image segmentation algorithm based on the thresholding and the fuzzy C-means techniques. Pattern Recognit 23:935–952
11. Dekker AH (1994) Kohonen neural networks for optimal colour quantization. Network: Comput Neural Syst 5:351–367
12. Floyd RW, Steinberg L (1976) An adaptive algorithm for spatial grey scale. Proc Soc Inf Disp 17:75–77
13. Stucki P (1981) Mecca-a multiple-error correcting computation algorithm for bilevel image hardcopy reproduction. Research report RZ1060, IBM Res. Lab
14. Jarvis JF, Judice CN, Ninke WH (1976) A survey of techniques for the display of continuous tone pictures on bilevel displays. Comput Graph Image Process 5:13–40
15. Orchard MT, Bouman CA (1991) Color quantization of images. IEEE Trans Signal Process 39:2677–2690
16. Akarun L, Ozdemir D, Yalcin O (1996) A modified quantization algorithm for dithering of color images. Electron Lett 32:1185–1186
17. Scheunders P, De Backer S (1997) Joint quantization and error diffusion of color images using competitive learning. In: International conference on image processing, pp 811–814
18. Ozdemir D, Akarun L (2001) Fuzzy algorithms for combined quantization and dithering. IEEE Trans Image Process 10:923–931
19. Robertson A (1977) The CIE 1976 color-difference formulae. Color Res Appl 2:7–11

20. Gentile R, Allebach J, Walowit E (1990) Quantization of color images based on uniform color spaces. J Imaging Technol 6:11–21
21. Gentile R, Walowit E, Allebach J (1990) Quantization and multilevel halftoning of color images for near-original image quality. J Opt Soc Am A 7:1019–1026
22. Mahy M, Van Eycken L, Oosterlinck A (1994) Evaluation of uniform color spaces developed after the adoption of CIELAB and CIELUV. J Color Res Appl 19:105–121

A New Morphological Filtering Algorithm for Pre-processing of Electrocardiographic Signals

Rishendra Verma, Rini Mehrotra and Vikrant Bhateja

Abstract Pre-Processing of Electrocardiographic (ECG) signals involves the baseline wander elimination and impulse noise filtering to facilitate automated analysis. In this paper a new morphological filtering algorithm using combinations of flat (two dimensional) structuring elements is proposed for pre-processing of ECG signals. Usage of two dimensional structuring elements, (over single dimension) aids in controlling effective inhibition of noise, leading to reconstruction with minimal signal distortion. Signal to noise ratio (SNR) and Root Mean Squared Error (RMSE) are used as quantitative evaluation measures for optimizing the selection of size of the structuring elements. Experimental results show that the proposed algorithm yields effective pre-processing of ECG signals, thereby eliminating the discussed artifacts.

Keywords Baseline wandering · Bottom-hat filtering · ECG · Flat structuring element · Top-hat filtering

R. Verma (✉) · R. Mehrotra · V. Bhateja
Department of Electronics and Communication Engineering, Shri Ramswaroop Memorial Group of Professional Colleges, Faizabad Road, Lucknow 227105, Uttar Pradesh, India
e-mail: verma.rishendra90@gmail.com

R. Mehrotra
e-mail: rini684.mehrotra@gmail.com

V. Bhateja
e-mail: bhateja.vikrant@gmail.com

Mohan S. and S. S. Kumar (eds.), *Proceedings of the Fourth International Conference on Signal and Image Processing 2012 (ICSIP 2012)*, Lecture Notes in Electrical Engineering 221, DOI: 10.1007/978-81-322-0997-3_18,

1 Introduction

Electrocardiogram (ECG) represents the weak non-stationery electrical signal generated by heart. During each heart beat an electrical signal is generated by cells in the sinoatrial node in the upper part of the heart which then travels and spreads all over the ventricles. A single normal cycle of ECG represents the consecutive atrial and ventricular depolarization during every heartbeat which is associated with the peaks and troughs of ECG waveform [1–3]. This is detected as tiny rise and falls in voltage between the electrodes attached to the heart. ECG signals are frequently plagued by impulse noise in diverse forms. Power line interference of 50/60 Hz is a common artifact corrupting the raw ECG which appears as a sinusoidal wave. Another artifact is baseline wander, where the baseline (of ECG waveform) starts to drift up and down in a sinusoidal pattern due to respiration [4]. Therefore, a Pre-processing of ECG signals for providing baseline correction and noise elimination is a necessary pre-requisite in understanding the functioning of heart, diagnosis of cardiovascular diseases. Traditional methods for pre-processing of ECG include: high-pass filtering [5] and band-pass filtering [6]. Since baseline drift has relatively lower frequency; baseline correction is performed by high-pass filtering. Both of these pre-processing techniques possess sharp cut-off frequencies which often distort the signal. Morphological Filtering [7] is also employed for pre-processing of ECG signals. Yan Sun et al. proposed MMF algorithm [8] which retained the characteristics of the ECG signal but could not denoise the signal effectively. In addition, the computational time of this algorithm was high. Wavelet based pre-processing techniques [9, 10] requires numerous experiments for ruling out scales and thresholds. Donoho et al. [11] proposed soft thresholding which has continuity at the threshold critical point. However, there are differences between the processed intrinsic mode function and the original IMF which introduces error when signal is reconstructed. Chu et al. [12] used combination of morphological opening and closing operators for Pre-processing which ultimately distorted the trait points in the signal which is not appropriate for ECG analysis. Liu et al. [13] proposed morphological filtering by using linear structuring element whose size depends on the sampling frequency. This approach successfully removed the distortion due to baseline drift but was not capable in suppressing the noise. Also, other methods like FIR and IIR filters [14] were also not appropriate in improving signal quality as additive noise has same frequency band as ECG signal. She et al. [15] proposed EMD improved thresholding for pre-processing of ECG signals where the reconstructed signal showed the presence of noise at various points in the ECG signal. Yuan Gu et al. [16] used the morphological filters with flat structuring element but was able to reduce ST segment distortion only up to 50 %. Hence, a new morphological filtering algorithm for Pre-processing of ECG signals is introduced using morphological filters. The proposed morphological filtering algorithm uses combinations of flat structuring element(s) which achieves low distortion and high noise suppression. Root Mean Squared Error (RMSE) and Signal-to-noise ratio (SNR) are used as quality parameters to evaluate the performance of proposed algorithm in comparison to other methods. The rest of this paper is organized as follows. The proposed methodology is described in Sect. 2. Section 3 details the results and discussions and lastly, the conclusion is given under Sect. 4.

2 Proposed Methodology

2.1 Morphological Filtering

Morphological filtering is a non-linear transformation technique primarily used for local modification of geometrical features of a signal. This shape information (of features) is extracted by using a structuring element (of appropriate dimensions) to operate on the input signal. Erosion, dilation, opening and closing are the common morphological operators used. Top-hat (signal—opening) and bottom-hat (closing—signal) are derived from opening and closing operators. Top-hat filtering computes the morphological opening of the signal followed by subtracting the result from the original signal. Bottom-hat filtering involves subtracting the input signal from morphological closing operation on the ECG signal [17]. Therefore, the Pre-processing of ECG signals can be achieved by using various combinations of top-hat and bottom-hat filters.

2.2 Two Dimensional Flat Structuring Element

A structuring element plays very significant role in extracting the characteristic information from the targeted signal. An important aspect in baseline wander removal and noise suppression is the selection of optimum size of the structuring element. Improper selection of structuring element may distort the adjacent wave in the ECG signal. Hence, the size of structuring element should be greater than the width of the characteristic wave [7]. Two-dimensional, or flat, structuring elements consist of a matrix of 0's and 1's, typically much smaller than the signal being processed. The center pixel of the structuring element, called the origin, identifies the pixel of interest, the pixel being processed. The pixels in the structuring element containing 1's denote the portion of signal to be retained while 0's denotes the signal which is to be ignored. This enables proper extraction of characteristic wave for baseline correction and denoising without introducing any distortion in ECG waveform. A disk-shaped structuring element is used with the proposed morphological filtering algorithm for the pre-processing of the ECG signal as shown in Fig. 1a, b, c. The dimensions of the structuring element are determined experimentally using the performance evaluation parameters.

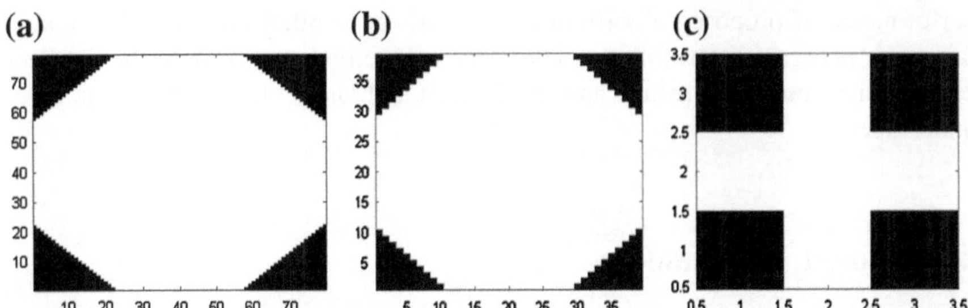

Fig. 1 Structuring Elements (se1-se3) for **a** Baseline Correction (se1) of radius = 40; **b** Baseline Correction (se2) of radius = 20; **c** Noise suppression (se3) of radius = 1

2.3 Proposed Algorithm for Baseline Correction

Baseline wandering being a low frequency artifact is removed by employing morphological operators that have both low pass and high pass filtering characteristics. Therefore, in the proposed algorithm, noisy baseline drifted signal is subjected to top-hat filtering and bottom-hat filtering as shown in Fig. 2. With a flat structuring element of appropriate dimensions, top-hat and bottom-hat filtering can be used to remove baseline wandering signal. The size of the structuring element (se1 > se2) used for top-hat operation is kept larger than that used for bottom-hat filtering. This ensures proper extraction of characteristic wave. The noisy baseline drifted signal (f_o) is subjected to perform two set of operations. Top-hat and bottom-hat filtering on corrupted signal (f_o) yields signal peaks (f_2) and signal valleys (f_1) respectively. The top-hat and bottom-hat filtering are performed

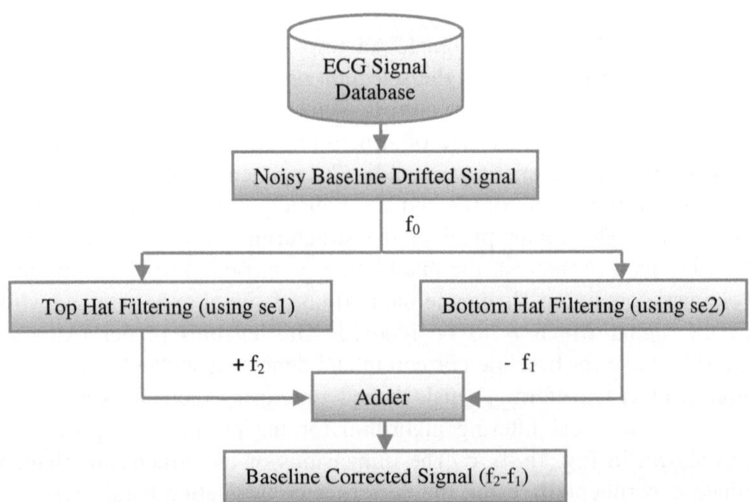

Fig. 2 Block diagram of the proposed baseline correction algorithm

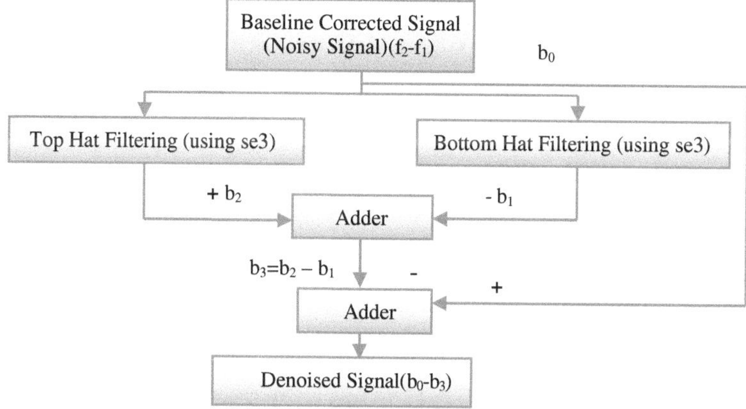

Fig. 3 Block diagram of the proposed noise suppression algorithm

with the structuring elements mentioned in Fig. 1. The signals obtained after top-hat and bottom-hat filtering are subtracted (f_2-f_1) to yield the corrected baseline signal.

2.4 Proposed Algorithm for Noise Suppression

ECG signal is mainly corrupted by high frequency impulse noise. During baseline correction, some of the impulse noise is removed. However, for further noise suppression, the signal obtained after baseline correction (b_0) is made input to the proposed noise suppression algorithm. This involves top-hat and bottom-hat filtering using structuring element (se3) of optimum shape and size as shown in Fig. 1c. Top-hat and bottom-hat operations using se3 yields the high frequency noise which corrupts ECG signal. The top-hat (b_2) and bottom-hat (b_1) filtering yield noisy component which is subtracted from the baseline corrected signal to give denoised ECG signal which is as shown in Fig. 3. The structuring element (se3) is comparable to the width of the noise component present hence noise can be removed without sacrificing the signal quality.

3 Experimental Results and Discussions

3.1 Quantitative Evaluation of Obtained Results

The performance of proposed algorithm is evaluated with the help of two quality parameters: Root Mean Squared Error (MSE) and Signal to Noise Ratio (SNR) in dB which is defined as:

$$\text{RMSE} = \frac{1}{N}\sqrt{\sum_{n=1}^{N}(f - fp)^2} \tag{1}$$

$$\text{SNR(dB)} = 10\log\left(\frac{\mu}{\sigma}\right) \tag{2}$$

where: f is original signal, fp is reconstructed signal, N is total signal length, μ is the mean of the signal and σ is the standard deviation of the signal. The Root Mean Square Deviation (RMSD) or Root Mean Square Error (RMSE) is a measure of the differences between values predicted by a model and the values actually observed. RMSE is a good measure of accuracy of an algorithm hence lower the value of RMSE lesser is the distortion. SNR can be used to quantify the noise level in the ECG signal, thus higher the value of SNR, better is the amount of noise suppression.

3.2 Simulation Results

For simulation purposes in the present work, ECG signals are adapted from internationally accepted MIT-BIH database that consists of recordings which are observed in clinical practice. The ECG signal (record no. 106 and 228 are shown in Figs. 4a and 5a) is taken from MIT-BIH database [18]. Baseline drift and Gaussian noise (10 dB) is added to generate the noisy drifted ECG signal as shown in Figs. 4b and 5b. The distorted ECG signal is first processed through the proposed baseline correction algorithm using disk shaped structuring elements. The baseline corrected signal (shown in Figs. 4c and 5c) is then subjected to the proposed noise suppression algorithm. The finally processed ECG signal is shown in Figs. 4d and 5d. Figures 4c and 5c shows the baseline corrected signal where baseline is adjusted to or brought down to zero reference. As it can be seen that characteristic wave of ECG are preserved therefore proposed baseline correction algorithm produces no distortion. Moreover, during baseline correction some amount of impulse noise is also reduced which occurs in the form of high peaks in the signal. It can be observed from Figs. 4d and 5d that impulse noise is reduced by noise suppression algorithm without distorting the characteristic wave in the ECG signal. The performance of proposed algorithm on the various recordings is shown in Table 1.

3.3 Comparative Analysis

Performance comparison of proposed algorithm is done with the results obtained in the Improved thresholding [15]. The results obtained from various Pre-processing algorithms (on the record no.106 of MIT-BIH after adding 10 dB Gaussian White Noise and Baseline Drift) are evaluated with the help of above mentioned quality parameters and are tabulated in Table 2.

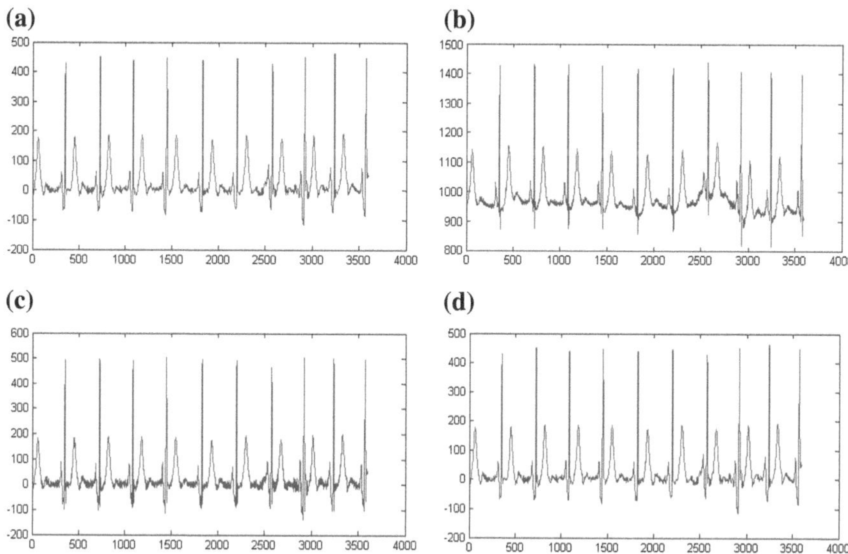

Fig. 4 **a** The original ECG (record no. 106 from MIT-BIH arrhythmia database). **b** The noisy drifted ECG obtained by adding 10 dB Gaussian white noise and baseline drift to the original ECG signal. **c** Baseline corrected signal. **d** De-noised signal

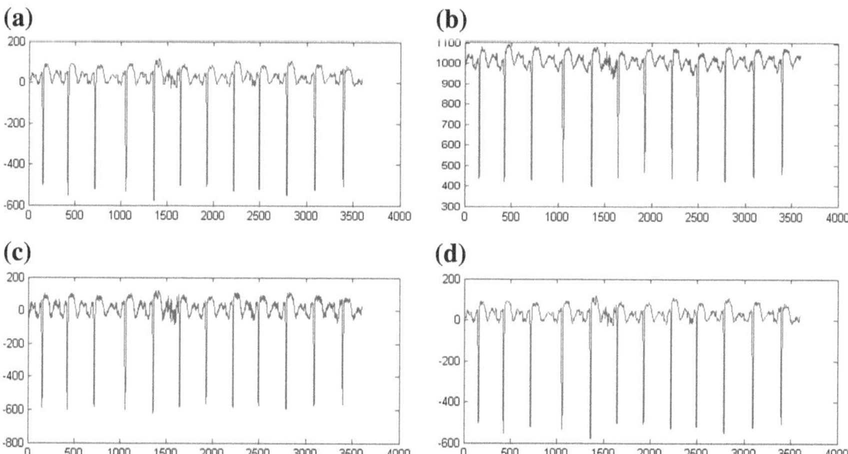

Fig. 5 **a** The original ECG (record no. 228 from MIT-BIH arrhythmia database). **b** The Noisy drifted ECG obtained by adding 10 dB Gaussian white noise and baseline drift into the original ECG signal. **c** Baseline corrected signal. **d** De-noised signal

From above set of tabulations, it can be observed that in case of the proposed algorithm the values of RMSE is lesser than those obtained in by previous algorithms, which shows that the reconstructed ECG signal contains minimum

Table 1 Performance evaluation of the proposed algorithm

Record #	SNR (dB)	RMSE
106	16.325	0.000043
228	16.982	0.000540
230	18.188	0.000136

Table 2 Performance comparison of various pre-processing algorithms

Algorithm	Performance indices	
	SNR (dB)	RMSE
MMO [8]	9.8385	0.0023
Wavelet threshold [10]	14.9427	0.0013
Soft-thresholding [11]	9.4272	0.0024
FIR filter [14]	14.5194	0.0013
Improved-thresholding [15]	15.0427	0.0012
EMD threshold [19]	10.0137	0.0022
Proposed algorithm	16.3254	0.000043

distortion. Moreover, higher values of SNR obtained in the present work depicts better noise suppression in comparison to previous algorithms.

4 Conclusion

In this paper, a morphological filtering algorithm using two dimensional structuring elements is proposed for the pre-processing of ECG signal. Two dimensional flat structuring elements overcomes the disadvantage of 1-D structuring element as it helps in proper extraction of signal features thereby controlling signal distortion effectively along with impulse noise suppression. Moreover, it provides better pre-processing of ECG signal than various Wavelet and thresholding algorithms. The results obtained in the present work gives lower values of error indices and shows significant improvement in SNR in comparison to other algorithms.

References

1. Dupre A, Vincent S, Iaizzo PA (2005) Basic ECG theory, recordings and interpretation. In: Handbook of cardiac anatomy, physiology, and devices, pp 191–201
2. Gupta R, Bera JN, Mitra M (2010) Development of an Embedded System and MATLAB-based GUI for online acquisition and analysis of ECG signal. J Measur 43(9):1119–1126
3. Sayadi O, Shamsollahi MB (2007) ECG baseline correction with adaptive bionic wavelet transform. In: Proceeding of the 9th international symposium on signal processing and its applications, Sharjah, pp 1–4

4. Kumar Y, Malik GK (2010) Performance analysis of different filters for powerline interface reduction in ECG signal. Int J Comput Appl (0975–8887)3:1–6
5. Christov II, Dotsinsky IA, Daskalov IK (1992) High-pass filtering of ECG signals using QRS elimination. Med Biol Eng Comput 30:253–256
6. Pei S, Tseng C (1995) Elimination of AC interference in ECG using IIR notch filter with transient suppression. IEEE Trans Bio-Med Eng 42(11):1128–1132
7. Chu CHH, Delp EJ (1996) Impulsive noise suppression and background normalization of electrocardiogram signal using morphological operators. IEEE Trans Biomed Eng 36:262–273
8. Sun Y, Chan KL, Krishnan SM (2002) ECG signal conditioning by morphological filtering. Comput Biol Med 32(6):465–479
9. Sayadi O, Shamsollahi M (2007) Multiadaptive bionic wavelet transform: application to ECG de-noising and baseline wandering reduction. EURASIP J Adv Sig Process, pp 1–11
10. Alfaouri M, Daqrouq K (2008) ECG signal de-noising by wavelet transform thresholding. Am J Appl Sci 5(3):276–281
11. Donoho DL (1995) De-noising by soft-thresholding. IEEE Tran Inf Theor 41(3):612–627
12. Chu CHH, Delp EJ (1988) Electrocardiogram signal processing by morphological operators. In: Proceeding of the conference on computers in cardiology, Washington, pp 153–156
13. Liu Z, Wang J, Liu B (2011) ECG signal denoising based on morphological filtering. In: Proceeding of the 5th international conference on bioinformatics and biomedical engineering, Wuhan, pp 1–4
14. Jan J, Sch Ilden TS (1985) Removal of baseline wander and powerline interference from the ECG by an efficient FIR filters with the reduced number of taps. IEEE Trans Biomed Eng 32(12):1052–1060
15. She L, Xu Z, Zhang S, Song Y (2010) Denoising of ECG based on EMD improved-thresholding and mathematical morphology operation. In: Proceeding of the 3rd international conference on biomedical engineering and informatics, vol 2. Yantai, pp 838–842
16. Gu Y, Zheng G, Dai M (2011) A morphology algorithm based on 2-dimensional flat structuring element on ECG baseline wander elimination. In: Proceeding of the conference on computing in cardiology, Hangzhou, pp 817–820
17. Zhang F, Lian Y (2007) Electrocardiogram QRS detection using multiscale filtering based on mathematical morphology. In: Proceeding of the 29th annual international conference of the IEEE, Lyon, pp 3196–3199
18. Moody GB, Mark RG (2001) The impact of the MIT-BIH arrhythmia database. IEEE Eng Med Biol Mag 20:45–50
19. Huang NE, Shen Z, Long SR et al (1998) The empirical mode decomposition and the Hilbert spectrum for non-stationary time series analysis. Proc R Soc Lond A 454:903–995

An Adaptive Image Steganography Algorithm Using Successive Pixel Difference Embedding and Integer Wavelet Transform

Anirudh H. Yajurvedi and Sudarshanpatil Kulkarni

Abstract In this paper an adaptive, secure and high capacity image steganography algorithm based on Integer Discrete Wavelet Transform is proposed. The algorithm is adaptive as it chooses the optimum value for threshold based on cover image and secret image through an iterative process, using least possible number of DWT coefficients. The experimental results show that the algorithm chooses the optimum value for the variable threshold for the given cover image and secret image and helps in obtaining the best PSNR for the steganography image.

Keywords Secure image steganography · High embedding Capacity · DWT · Adaptive scheme

1 Introduction

The internet is an important medium for digital data transmission. However, being a fully open medium, it has brought us not only convenience but also some hazards and risks as it is easier for some malicious users to illegally copy, destroy, or change it. Hence information security becomes an essential issue. To make the transactions more secure, many steganography methods are designed. The basic idea is to embed the confidential data into a cover-media viz. texts, voices, images, and videos and then transmit it secretly. The three things to be considered when designing a steganography system are *invisibility, capacity and security.*

The Least significant bit (LSB) substitution method is the most common steganography technique used and it is highly susceptible to statistical steganalysis

A. H. Yajurvedi (✉) · S. Kulkarni
Department of Electronics and Communication Engineering, SJCE, Mysore, India
e-mail: anirudhyajurvedi@gmail.com

Mohan S. and S. S. Kumar (eds.), *Proceedings of the Fourth International Conference on Signal and Image Processing 2012 (ICSIP 2012)*, Lecture Notes in Electrical Engineering 221, DOI: 10.1007/978-81-322-0997-3_19,

attacks. In this method, the quality of steganographed image reduces as the payload increases. Modified LSB substitution techniques such as pixel value differencing (PVD) method [1], Triway PVD [2] and modulus arithmetic based LSB replacement [3] try to enhance the security as well as increase the embedding capacity. The embedding capacity can be increased by embedding lesser information per pixel of the secret image. This is implemented in this work by substituting the successive pixel difference instead of actual pixel value, using modulus arithmetic. This logic leads to achieve a very high embedding capacity along with enhanced security. During the embedding process, the algorithm assigns optimum values for variables threshold and pixel to achieve the best PSNR for the steganography image. Transform domain techniques are more robust, have a higher resilience to attacks against the stego-image such as compression, cropping and image processing [4]. DWT domain steganography is the most preferred transform domain technique [5]. Steganography algorithms based on DWT technique have been discussed in earlier reports [6, 7]. In these studies though DWT technique is used they do not have high embedding capacity. Hence in the present work is to achieve a very high embedding capacity, by selectively choosing to embed the pixels of secret image in the DWT coefficients of cover image. The transform domain technique with the proposed LSB substitution technique provides a highly secure steganography technique along with very high embedding capacity.

Discrete Wavelet Transform: Wavelets are special functions that are used as basis functions for representing signals. A key advantage wavelet transforms has over Fourier transforms is temporal resolution in addition to frequency resolution. It captures both frequency and location information (location in time). For Digital images, applying 1 level 2-D DWT (Discrete Wavelet Transform) separates the image into a lower resolution approximation image or band (LL) as well as horizontal (HL), vertical (LH) and diagonal (HH) detail components as shown in Fig. 1. With the DWT, the significant parts (smooth parts) of the spatial domain image exist in the approximation band that consists of low frequency wavelet coefficients and the edge and texture details usually exist in high frequency sub bands, such as HH, HL, and LH.

Generally, wavelet filters have floating point coefficients, hence, when the input data consist of sequences of integers (as in the case for images), the resulting filtered outputs no longer consist of integers, which doesn't allow perfect reconstruction of the original image. However, with the introduction of wavelet transforms that map integers to integers based on lifting schemes [8], it is possible to

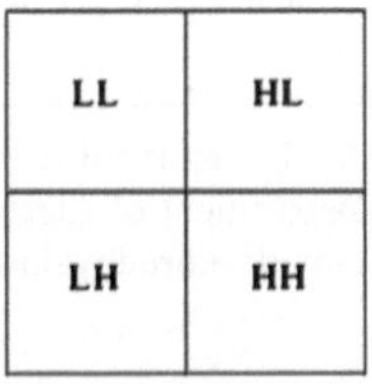

Fig. 1 Components of 1-level 2-dimensional discrete wavelet transform

LL	HL
LH	HH

characterize the output completely with integers [7, 9]. Based on this idea, a method has been described [10] that embeds the message bit stream into the LSB's of the integer wavelet coefficients of a true-colour image. To achieve perfect reconstruction in the present work IDWT has been used.

2 Proposed Method

In this section, both proposed embedding and extraction algorithms will be explained.

2.1 Block Diagram and Brief Explanation of Embedding Process

In the embedding process, a 1 level 2-D discrete wavelet transform operation is performed on the cover image. The first few coefficients in each sub-band are reserved to store the essential information needed to retrieve the image at the receiving end. They include number of pixels embedded in each sub-band, the value of threshold used to embed and the size of secret image. To simplify the embedding process the secret image is converted into a row matrix. The actual value of 1st pixel of secret image to be hidden in a sub-band is embedded by modifying the DWT coefficients by modulus operation. From the next pixel, the difference between the current pixel and the previous pixel embedded is inserted by modulus operation. The exact embedding procedure is explained in the following algorithm. This method helps to improve the embedding capacity. The procedure is repeated by taking the next sub-band, until all the pixels are embedded. Now, the inverse DWT of the modified coefficients is taken to get the steganography image (Fig. 2).

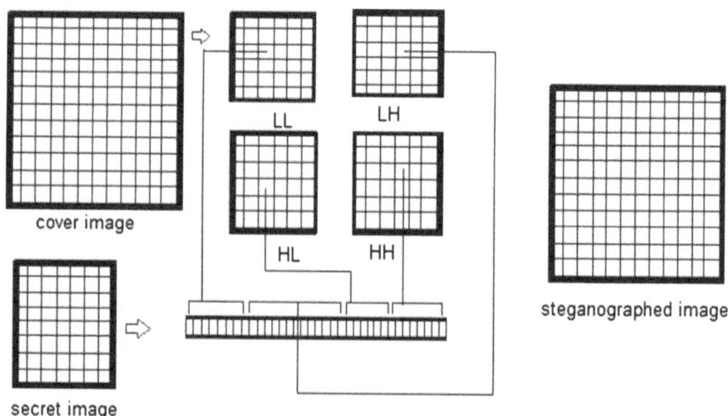

Fig. 2 Block diagram of embedding process

3 Embedding Algorithm

Input: Cover image, Image to be hidden.
Output: Steganography image.
Variables used:

Fs Row matrix, with modified coefficients of each sub-band forming 4 rows
Fh The reshaped row matrix of image to be hidden
Fc The matrix with reshaped sub-band matrices of DWT as its rows
Th Threshold. It specifies the maximum absolute difference of pixel values between 2 successive pixels in image to be hidden.

A large value of threshold provides higher embedding capacity but decreases the PSNR of the steganography image. Hence the algorithm has to choose the value of threshold optimally.

If the absolute difference is less than or equal to the threshold value then the difference is stored using single coefficient in DWT sub-band of cover image.

If the difference lies outside the range then algorithm uses multiple pixels to store the actual difference. The number of coefficients to be used is determined in the following algorithm. Let n_c be the number of coefficients of DWT sub-band of cover image to be utilized to store the actual difference in case the difference lies outside the specified range.

1. The cover image and image to be hidden are taken as inputs and 1 level 2-D. Integer wavelet transform is performed on the cover image, to obtain the 4 sub-bands.
2. Each of the sub-band matrices is reshaped into a row matrix. A new matrix with these row matrices as the rows are defined (*Fc*). The image to be hidden is also reshaped into row matrix (*Fh*).
3. The value of *Th* is initially set to zero and then is set to 4.
4. The first 23 pixels are reserved in each sub band, to store the essential information needed to retrieve the hidden image. This method is followed to spread the essential data evenly across all sub-bands.
    a. In the first 10 pixels of the chosen sub-band of cover image, the total no of pixels of hidden image that are embedded in that sub-band is stored, through LSB substitution.
    b. The 11–20 pixels in 1st sub-band are reserved to store the row size of the hidden image, through LSB substitution.
    c. The 11–20 pixels in 2nd band are used to store the column size of the hidden image, through LSB substitution.
    d. The other 3 pixels in each sub band are used to store the number of sub bands utilized, threshold value used to embed the image, number of pixels used to store the actual pixel value if the difference lies outside the specified threshold range.

5. The pixel value of 1st pixel to be hidden in a new sub-band of cover image is converted into 8 bit binary number. The 4 significant bits are stored in the 24th pixel of the chosen sub-band of cover image by LSB substitution. The least significant 4 bits are stored in 25th pixel by LSB substitution.
6. The divisor for modulus operation is defined as follows

$$N = (2 \times Th) + 3$$

The above equation is used to assign a value to N so that a unique value can be assigned to each difference value obtained in the range specified above and 2 additional values for those which exceed the range, in upper and lower directions.

A modulo operation with N as divisor yields remainder in the range 0 to N−1.

7. 26th coefficient onwards, the coefficients in the chosen sub-band are modified as follows, in order to store the secret image in it.

The difference between 2 successive pixels, starting from the first pixel value in image to be hidden is taken.

$$d = Fh(1, i + 1) - Fh(1, i),$$

where, $1 \leq i \leq (row\_h \times col\_h)$

The difference is then compared with the threshold range.

Case 1 $|d| \leq Th$
N-2 remainders of the modulo operation are reserved to store each of these difference. These assign a unique value v to each of the difference d obtained. This assigned value is embedded into the coefficients of cover image through modulus arithmetic LSB substitution. This difference is stored with modulus subtraction as follows

$$Fs(di, mi) = Fc(di, mi) - (Fc(di, mi)\%N) + v$$

Case 2 $d < (-Th) or d > Th$
The two remainders of the modulo operation are reserved to store each of the above 2 cases. The assigned value indicates whether the difference is greater than the threshold range or it is less than the threshold range. This special value is inserted into the cover image as below

$$Fs(di, mi) = Fc(di, mi) - (Fc(di, mi)\%N) + v$$

Now the exact difference is expressed in 8 bit binary form. If the number of variable coefficients n_c is 4, next 4 coefficients are modified as shown below, to store 2 bits of 8 bit representation in each coefficient. Let the decimal representation of 2 bits to be embedded be represented by Vt.

$$Fs(di, mi + t) = Fc(di, mi + t) - (Fc(di, mi + t)\%4) + Vt$$

where, $1 \leq t \leq 4$

If the number of variable coefficients n_c used is 2, next 2 coefficients modified as shown below, to store 4 bits of 8 bit representation in each coefficient.

Let the decimal representation of 2 bits to be embedded be represented by Vt.

$$Fs(di, mi + t) = Fc(di, mi + t) - (Fc(di, mi + t)\%16) + Vt$$

where, $1 \leq t \leq 2$

8. If the last coefficient in the current sub-band is reached then it is not possible to embed any more pixels in the current sub-band. Thus the total number of pixels embedded in the current sub-band is stored and move to the next band and follow the steps from 5 to 9. This process is repeated until the image is hidden completely or all the coefficients in all sub-bands are used. The process is stopped if the image is hidden completely otherwise step 10 is executed. The threshold is increased by 1 and steps 5–8 are followed.
9. If the threshold crosses a specified value and if the image could not be embedded completely, then variable coefficients used $n_c$is reduced from 4 to 2. Now, steps 5–9 are repeated until the image is completely hidden.
10. Now, the inverse DWT transform is taken to get the steganography image.

4 Block Diagram and Brief Explanation of Extraction Process

The block diagram of the extraction process is shown in Fig. 3. The received image is taken and 1 level 2-D wavelet transform is applied on it. The essential information regarding the value of threshold used, size of hidden image, number of sub-bands used to embed the image and the number of pixels embedded in each sub-band is first extracted from the DWT coefficients of the sub-bands. Now the procedure to extract the actual value of 1st pixel embedded in a sub-band is as

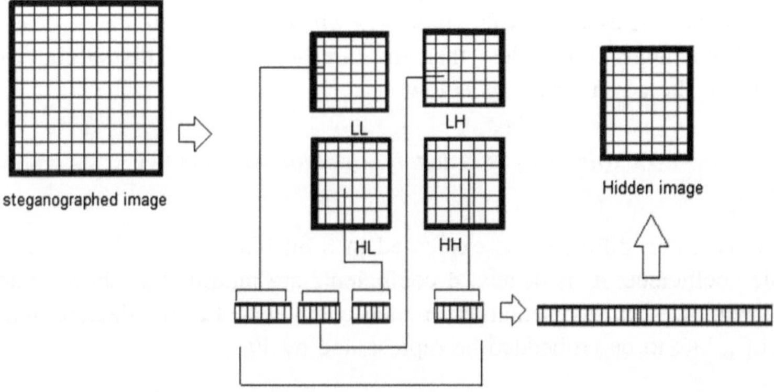

Fig. 3 Block diagram of extraction process

explained in the algorithm. Next, the successive pixel differences are determined and added to the previous pixel value to obtain the actual value of it. This process is repeated on all the sub-bands in which the image is embedded. After all the pixels are successfully extracted, the row matrix is reshaped into a 2-D matrix of the size of secret image.

5 Extraction Algorithm

1. A 1 level 2-D discrete wavelet transform is performed on the steganography image to get the 4 sub-bands and from the first 10 coefficients of each sub-band, the number of pixels of hidden image that are embedded in the sub-band is found by extracting 2 LSB bits from binary representation of each coefficient.
2. From 11–20th coefficients of 1st sub-band, the number of rows present in hidden image is determined. In a similar fashion, from the 11–20th coefficient of 2nd sub-band, the number of columns in hidden image is extracted.
3. From the 21st, 22nd, 23rd coefficients of 1st sub-band, the value of threshold, pixel and number of sub-bands used for embedding is extracted.
4. Let us define variable N, the divisor for modulus operation as follows. $N = (2 \times Th) + 3$, The 1st sub-band (LL band) is chosen.
5. From 24th coefficient of current sub-band, the MSB 4 bits of binary representation of the pixel value of 1st pixel of hidden image, embedded in the current sub-band is found, by extracting rightmost 4 bits. From 25th coefficient the LSB 4 bits are extracted and the 2 nibbles are concatenated to form an 8 bit representation, this gives the actual pixel value of 1st pixel.
6. From the 26th pixel the modulus operation is performed using N as the divisor.
7. Let the remainder obtained from above operation on C_ith coefficient be R. R is compared with the table that was used for embedding the difference.
8. If the difference lies within the range defined by threshold, comparison yields the actual difference between the present pixel and previous pixel of hidden image. Else it would yield 2 special values which indicate that the difference lies out of range and next 4 or 2 coefficients yield the actual difference.
9. Let the difference be d. The value may be positive or negative.
10. This difference is added to the pixel value of the previous pixel extracted to get the actual pixel value of present pixel.

$$Pi + 1 = Pi + d$$

11. Now, the next embedded coefficient is taken and steps 8–10 are repeated, until all the pixels hidden in the current sub-band are extracted.
12. Consider the next sub-band and repeat the steps 6–10. This is repeated until all the pixels of hidden image are extracted.

Table 1 Results with standard cover image and smooth secret image (Test case 1)

Image to be hidden and size	Cover image and size	Relative size of hidden image (%)	Total bits embedded	PSNR along with wavelet used (in db)			
				Haar	Bior 2.2	Rbio 1.1	Cdf 2.4
Apple.png 128 × 128	Living room.tiff 512 × 512	6.25	131,072	55.8123	58.7359	63.1654	58.7048
Apple.png 256 × 256	Living room.tiff 512 × 512	25	524,288	51.1224	53.2269	53.1255	53.0993
Apple.png 360 × 360	Living room.tiff 512 × 512	49.438	1,036,800	48.5716	50.0646	47.4154	49.8113
Apple.png 440 × 440	Living room.tiff 512 × 512	73.8	1,548,800	45.7025	47.0923	46.3949	46.9443
Apple.png 480 × 480	Living room.tiff 512 × 512	87.89	1,843,200	40.8866	42.0102	39.6308	42.0070

Fig. 4 The image to be hidden (Apple.jpg) and cover image (livingroom.tiff)

13. The row matrix of the obtained pixels is reshaped into a matrix of size of the hidden image, to obtain the complete hidden image.

6 Experimental Results

The simulation experiments of the algorithm were conducted by choosing two different images which were to be hidden. One of the images chosen is a smooth image (Image to be hidden in Fig. 4 and Table 1) and another is a standard image (Image to be hidden in Fig. 5 and Table 2). We chose to hide them using the same cover image in order to compare their experimental results. The experiment has been carried out for different sizes of hidden image and also using different

Table 2 Results with standard cover image and standard secret image (Test case 2)

Image to be hidden and size	Cover image and size	Relative size of hidden image (%)	Total bits embedded	PSNR along with wavelet used (in db)			
				Haar	Bior 2.2	Rbio 1.1	Cdf 2.4
Lena.tiff 128 × 128	Living room.tiff 512 × 512	6.25	131,072	50.8814	53.7851	57.9665	53.7495
Lena.tiff 256 × 256	Living room.tiff 512 × 512	25	524,288	44.2741	45.6731	44.6373	45.7054
Lena.tiff 360 × 360	Living room.tiff 512 × 512	49.438	1,036,800	37.7879	38.7762	37.3058	38.8450
Lena.tiff 440 × 440	Living room.tiff 512 × 512	73.8	1,548,800	34.0194	35.1316	33.2750	35.3176
Lena.tiff 480 × 480	Living room.tiff 512 × 512	87.89	1,843,200	31.0694	31.4431	30.0181	31.4659

Fig. 5 The image to be hidden (Lena.tiff) and cover image (livingroom.tiff)

wavelet bases, with cover image remaining the same. The resulting PSNR values are noted down for each case.

7 Observations and Conclusions

Following observations can be made from the above results,

1. The PSNR is high when a less detailed image is embedded. This is because the successive pixel values in case of such images are nearly the same. This shows that the algorithm is more effective for smooth images.
2. The PSNR of the steganography image decreases as the size of payload is increased and as noticed from the PSNR values and the decrease is steady.

3. Also, there is no significant difference between the PSNR of steganography images obtained using wavelets chosen from different wavelet families, for the same payload.
4. The algorithm goes from providing best invisibility (measured in terms of PSNR) for small sized payloads to best embedding capacity for larger payload, which is not evident in earlier reports. This is achieved through the iterative selection of values for threshold and pixel. Thus the results were found to be most optimal for the given cover image and hidden image.

References

1. Wu DC, Tsai WH (2003) A steganography method for images by pixel-value differencing. Pattern Recogn Lett 24:1613–1626
2. Chang KC, Chang CP, Huang PS, Tua TM (2008) A novel image steganography method using tri-way pixel-value differencing. J Multimedia 3(2):37–44
3. Jaafar S, Manaf AA, Zeki AM (2007) Steganography technique using modulus arithmetic. In: 9th international symposium on signal processing and its applications, pp 1–4
4. Katzenbeisser S, Petitcolas FAP (1999) Information hiding techniques for steganography and digital watermarking. Artech House, Norwood, pp 56–92
5. Sutaone MS, Khandare MV (2008) Image based steganography using LSB insertion. In proceedings of IET conference on wireless, mobile and multimedia networks, pp 146–151
6. Banoci V, Bugar G, Levicky D (2011) A novel method of image steganography in DWT domain. In: 21st international conference radioelektronika (RADIOELEKTRONIKA), pp 19–20
7. Calderbank AR, Daubechies I, Sweldens W, Yeo BL (1996) Wavelet transforms that map integers to integers. Appl Comput Harmonic Anal (ACHA)
8. Jinna SK, Ganesan L (2011) Reversible image data hiding using lifting wavelet transform and histogram shifting. Int J Comput Sci Inf Secur 7(3):283–289
9. Bilgin A, Sementilli PJ, Sheng F, Marcellin MW (2000) Scalable image coding using reversible integer wavelet transforms. IEEE Trans Image Process 9(11):1972–1977
10. Tolba MF, Ghonemy MA, Taha IA, Khalifa AS (2004) Using integer wavelet transforms in colour image-steganography. IJICIS 4(2):75–85
11. Nag A, Biswas S, Sarkar D, Sarkar PP (2011) A novel technique for image steganography based on DWT and Huffman encoding. Int J Comput Sci Secur 4(6):561–570
12. Reddy HSM, Raja KB (2009) High capacity and security steganography using discrete wavelet transform. Int J Comput Sci Secur 3(6):462–472
13. Kumar KBS, Raja KB, Pattnaik S (2011) Hybrid domain in LSB steganography. Int J Comput Appl 19(7):35–40
14. Masud Karim SM, Rahman MS, Hossain MI (2011) A new approach for LSB based image steganography using secret key. In: 14th international conference on computer and information technology, pp 286–291
15. Zhiweil K, Jing L, Yigang H (2007) Steganography based on wavelet transform and modulus function. J Syst Eng Electron 18(3):628–632

On Generation of Silhouette of Moving Objects from Video

Soharab Hossain Shaikh, Sugam Kumar Bhunia and Nabendu Chaki

Abstract Object classification from video is a well known topic of research in the context of computer vision. Video processing for the purpose of real time object classification and action recognition has great importance in building an intelligent surveillance system. This paper proposes a novel technique for extracting human silhouette from video in real time. Identification of the moving human objects is performed first using frame differencing technique followed by a number of steps for extraction of the human silhouette. The proposed method has been tested on a good number of videos having varying textured background with noise related to illumination change. The proposed method can also be applied for extraction of silhouettes of other types of animate and inanimate moving objects from a video with the view of object classification and recognition. The experimental results as documented in the paper establish the effectiveness of the proposed method.

Keywords Moving object detection · Frame differencing · Human silhouette extraction · Online video processing · Surveillance video

S. H. Shaikh (✉) · S. K. Bhunia
A.K.Choudhury School of Information Technology, University of Calcutta, Kolkata, India
e-mail: soharab.hossain@gmail.com

S. K. Bhunia
e-mail: sug9am@gmail.com

N. Chaki
Department of Computer Science and Engineering, University of Calcutta, Kolkata, India
e-mail: nabendu@ieee.org

Mohan S. and S. S. Kumar (eds.), *Proceedings of the Fourth International Conference on Signal and Image Processing 2012 (ICSIP 2012)*, Lecture Notes in Electrical Engineering 221, DOI: 10.1007/978-81-322-0997-3_20,

1 Introduction

Real time object classification and action recognition from video stream is very important for building an intelligent surveillance system. Object classification from video is a well known area of research in pattern recognition and computer vision. Over last two decades, several researchers are working on finding proper object classification techniques in dynamic background scenario. In this context, classification of objects from the object silhouettes is another important area of research. Several attempts have been made in this context as found in [1, 2]. These techniques use the silhouette of objects and their templates for the classification task. The success rate of classification depends on how accurately the input and template object silhouettes have been extracted. The extraction of object silhouettes from a real time video is itself a challenging task. The accuracy of silhouette generation of objects is important. Moreover the extraction technique should be cost effective in terms of computation and dynamic in the sense that it must be done on real time video without the necessity of offline video processing.

This paper proposes a technique for extraction of human silhouette from a video in real time. In this proposed technique it is not necessary to store the video and perform offline processing. So, this method can readily be used for real time surveillance purposes. This paper presents the experimental results found on generation of only human silhouette from video. However, the proposed methodology can be applied for generating other types of objects (e.g. moving car, animals etc.) as well. The interesting applications of the proposed method include object classification from silhouettes, human gait recognition and action identification, video surveillance to mention a few.

2 Previous Works

Moving object detection from video is a well known arena of research for past few years [3–5]. In [3] a method for low-cost moving object detection has been proposed using frame differencing technique and boundary tracking. Paper [4] focuses on moving object detection based on background subtraction technique. In [5] the authors have considered robust background subtraction method for changing background. Attempts have also been made on object classification from object silhouettes as found in [1, 2, 6–9]. In [6, 9] methods have been presented on silhouette analysis based gait recognition for human identification. In [10] the authors design a morphological moving object signature for the purpose of human identification. This paper proposes a new inter-frame signature of binary silhouettes. It captures the dynamics of moving objects and generates a robust set of features suitable for classification. In [2] a silhouette based method for object classification and human action recognition has been presented. In [8] a method has been presented for automated video surveillance using silhouette based human motion detection and analysis.

Most of the papers which address the issue of object identification and action recognition in real time video processing require object silhouette as the most fundamental input. In [1] an approach based on the graph cuts techniques has been used for silhouette extraction. However, the proposed technique is expensive in terms of computation and for this reason a separate graphics processing unit (GPU) has been used to make this computation possible in real time. This incurs higher hardware cost which may not be affordable for all kind of surveillance requirements. In [4] several background modeling techniques have been discussed which compute suitable adaptive model by observing the nature of every pixel in the background over several frames. This may not be suitable for real time processing. Also in most of the surveillance applications the semantics between background and foreground movements are known in advance. This information can be used to design an algorithm that reflects the most important change to track the foreground object without modeling each pixel individually.

3 Proposed Method

In the present work the following assumptions have been made; the camera is stationery, there is no sudden change of illumination in the video frames, the motion of the human objects have been considered across the camera frame from left-to-right, right-to-left or any angular direction, the moving object is not present in the frame from the start of the video. Even through all these restrictions the proposed method is suitable for surveillance video as all the assumptions are satisfied in case of most of the surveillance videos in general.

The major challenges needed to overcome are the following: (1) *White background*: As the background subtraction is done by frame differencing technique which applies pixel-by-pixel computation (Present-Frame—Reference-Frame). Due to this if the Reference-Frame is completely white or its intensity is greater than the moving object in the Present-Frame then the moving object can not be detected using frame differencing technique. (2) *Multicolor background*: If the color of the background is not uniform, such that, some part of the background in the reference frame may have intensity greater than the present frame and some other part may have intensity less than the foreground moving object making the detection of the moving object difficult using frame differencing technique. (3) *Movements in the background*: There might always be little movements of objects present in the background. For example, a video is taken where several trees are present in the background; so there are movements of the leaves of different trees in the video frames. (4) *Illumination change*: There always are some illumination changes on the sharp edges present in the background (e.g. edges of the wall, door, window etc., static car, statue, pole or any fixed object present in the background). Moreover the videos suffer from this type of illumination related problems due to different qualities of the camera sensors. Figure 1 shows the major functional steps of the proposed method. In the following subsections detailed description of each step is given.

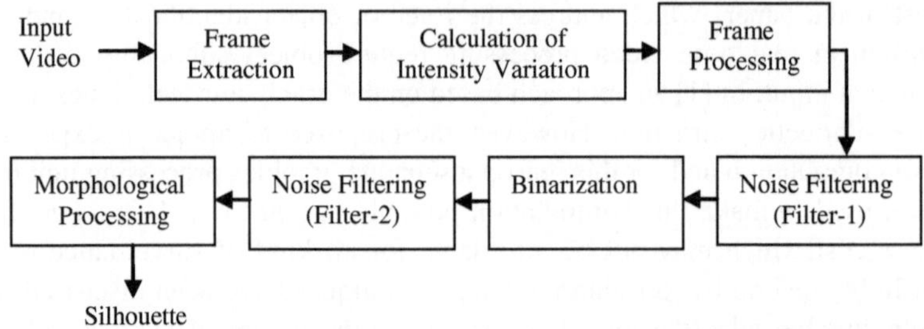

Fig. 1 Major function steps of the proposed method

3.1 Frame Extraction

In this step frames are extracted from the video stream. All the frames are converted from RGB to gray-scale for further processing. The height (h) and width (w) of the frames are noted.

3.2 Calculation of Intensity Variation

In this step the mean intensity value and maximum intensity value of each pixel across all the frames (until the frame where the moving object enters into the frame) have been calculated. A metric has been calculated which consists of the difference between these two quantities corresponding to each pixel position.

$$\mathrm{C(i,j)} = \mathrm{Max(PF(i,j)\ for\ all\ the\ frames)} - \mathrm{Mean(PF(i,j)\ for\ all\ the\ frames)}.$$

where PF (i,j) is the present frame intensity value at spatial pixel position (i,j). This metric C(i,j) is a measure of maximum intensity variation of the background with respect to their mean at each pixel position (i,j).

3.3 Frame Processing

The first frame of the video is considered to be the reference (Fig 2b). Gradually other frames from the second frame onwards have been considered as the present frame e.g. (Fig 2a). The reference frame has been subtracted from the present frame. Figure 2 shows the problem discussed in Sect. 3 (White background).

Figure 2a is the original RGB frame extracted from a test video. It is seen from this figure as well as from the reference frame (Fig 2b) that the background wall is almost white, so it has a larger intensity value than the foreground moving object (here it is a man). If the reference frame (Fig. 2b) is subtracted from the present

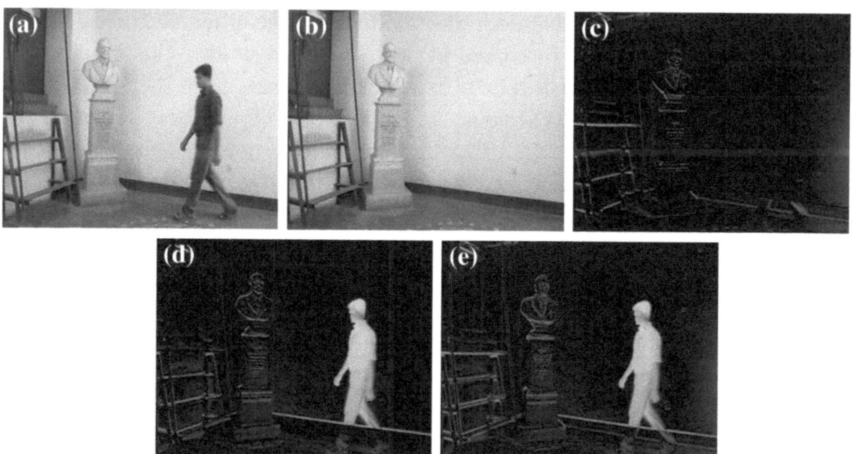

Fig. 2 Frame Processing. **a** PF = Present Frame **b** RF = Reference Frame **c** I1 = PF–RF **d** I2 = RF–PF **e** I = I1 + I2

frame (Fig. 2a), the frame shown in Fig. 2c is produced which shows that only a portion of the leg of the human being is observed. In Fig. 2d, the present frame (Fig. 2a) is subtracted from the reference frame (Fig 2b) and most of the moving object part is distinguished at this time. But at the bottom the leg part is not fully shown. Figure 2c and d are added (taking set union) to get the content of Fig 2e where most of the moving object part is distinguishable.

3.4 Noise Filtering (Filter-1)

Noise filtering task is divided into two parts named as Filter 1 and Filter 2, which are two filtering actions applied successively to eliminate illumination noises from the subtracted frames. The results of the noise processing are shown in Sect. 4 on Experimental Results.

PF(i,j) represents the present frame intensity value at position (i,j) for a particular frame. Mean [PF(i,j)] represents the mean intensity value of the pixel at position (i,j). Max [PF(i,j)] represents the maximum intensity at pixel position (i,j). Mean [PF(i,j)] and max[PF(i,j)] have been computed over all the frames starting from the first reference frame to the frame until the moving object enters the camera window. These two metrics are known as the global mean and global maximum intensity values respectively. The boundary tracking and frame differencing technique as proposed in [3] has been used for detecting the presence of the moving object in a frame.

The noise filtering action (Filter-1) starts with the assumption that the pixels which are not suffered greatly by noise should have the global maximum intensity value (max(PF(i,j))) very close to its global mean intensity value (mean(PF(i,j))) until the moving object enters into frame window; i.e, the difference between these intensity values (i.e. C (i,j) discussed in Sect. 3.2) is very low. Consequently the difference between the gray levels of each of these pixels from the Global Mean Intensity value is very large. For noise-suffered pixels C(i,j) value is larger than the corresponding magnitude of the pixels not suffered by noise. These assumptions regarding noise-suffered pixels and those not suffered by noise have been found experimentally and have been shown in Figs. 3 and 4 respectively. From the observations of these figures (Figs. 3 and 4), the following things can be interpreted: as shown in Fig. 4, the absolute difference between C(i,j) and Global Mean Intensity value of each of these pixels is comparatively much smaller than the pattern shown in Fig. 3 (with few exceptions).As the mean values of the pixels in Fig. 3 are quiet large so it may happen that some of these pixels may not have the present frame intensity value (PF(i,j)) greater than their computed Global Mean Intensity value (mean(PF(i,j)). The intensity of a pixel (belonging to the category of Fig. 3) at subtracted frame (Say, (I(i,j)) is much greater than C(i,j). The pixels corresponding to Fig. 4, may have the subtracted frame intensity value (I(i,j)) greater than C(i,j) value. Because the analysis of Fig. 4 shows that a greater percentage of pixels have Global Mean Intensity value larger than their corresponding C(i,j) value.

The quantity (PF(i,j)–C(i,j)), i.e., the difference between the present frame intensity value and the C(i,j) value of a pixel has produced a value larger than the previously discussed C(i,j) value for both of the cases of Fig. 3 (all the time) and Fig. 4 (most of the times). When a moving object passes through a pixel, its intensity value changes by a larger extent which can be reflected as the intensity (I(i,j)) of that pixel in the corresponding subtracted frame.

This intensity value should have a larger value than (PF(i,j)–C(i,j)), when a moving object passes through that pixel according to Fig. 3. On the other hand, I(i,j) for a noise-suffered pixel e.g. (Fig. 4) should have a smaller value than the quantity

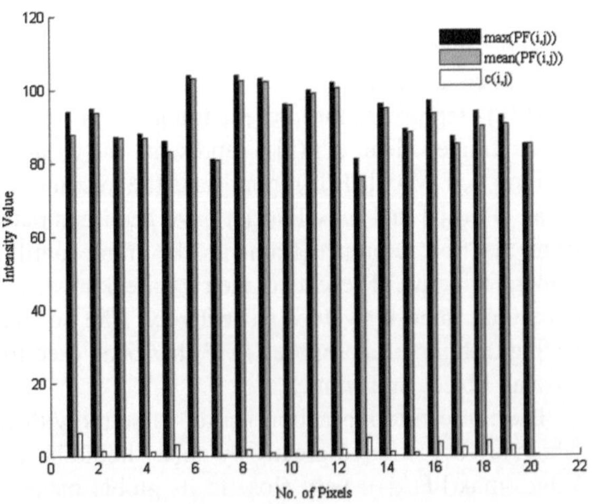

Fig. 3 Intensity variation plot for pixels not-suffered by noise

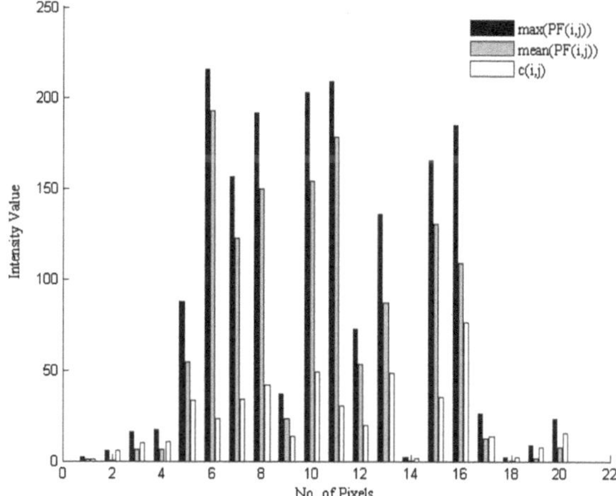

Fig. 4 Intensity variation plot for noise suffered pixels

(PF(i,j)–C(i,j)), most of the times. So the quantity (PF(i,j)–C(i,j)), acts as noise filter which passes the pixels containing the moving objects and blocks most of the noise-suffered pixels. Filter-1 can be represented as:

$$K(i,j) = (PF(i,j) - C(i,j))$$

3.5 Binarization

The next step is Binarization. If the intensity of any pixel in the subtracted frame (I(i,j)) has a intensity value larger than the quantity K(i,j) then the corresponding pixel is set to one (*white*) otherwise to zero (*black*).

3.6 Noise Filtering (Filter 2)

Further noise filtration can be achieved with by observing the face that the noises remain in a clustered way in the pixels of the frames. This type of noise is created by intensity fluctuation of the pixels due to illumination change over the background of the frames. The edges of different objects (e.g. building edges, window edges, car edges) are sensitive to illumination problems. Movements of tree leaves cause continuous illumination changes in video frames. An example video (scattered frames) is shown in Fig. 5 depicting the fact (these frames have been obtained after applying Filter-1). Presence of this type of noise will cause an inaccurate generation of object silhouette, so this noise is to be removed. This leads to the development of Filer-2 for further noise filtering.

Filer-2 can be defined as follows: Compute the maximum intensity at each pixel calculated over all the frames until the frame where the moving object enters the

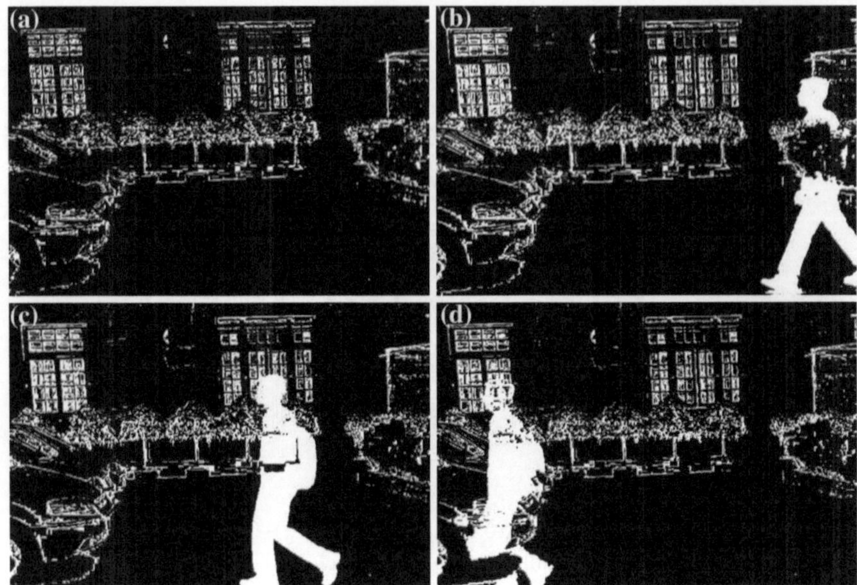

Fig. 5 Presence of noise (*After Filter-1*). **a** Frame No 5. **b** Frame No 5. **c** Frame No 67. **d** Frame No 91

frame window and store in a matrix (say E(i,j)) having the same dimension (hxw) as that of a frame. E (i,j) = Max (I (i,j)) for all the frames at spatial pixel position (i,j). Compute the difference between the subtracted frame intensity at each pixel (Say I(i,j)) and the value stored in E(i,j). This eliminates the noise present in the subtracted frame. D(i,j) = I(i,j)–E(i,j). This quantity D(i,j) acts as Filter-2.

3.7 Morphological Processing

The output of the previous step is gone through morphological opening and closing operations [11] to remove any bridges or gulfs or isolated holes or joins some narrow breaks or long thin gulfs to find out the proper silhouette without noise.

4 Experimental Results

A number of video samples have been used for the experimentation purpose. The authors have considered both indoor and outdoor videos. Some of the video samples were taken where there is insufficient light over the moving objects making it difficult to detect during frame differencing. Results of some of the test samples have been produced in this section.

Figure 6 shows the results of an indoor video where a moving object enters the frame from right to left in angular direction. It is observed in Fig. 6e that the background edges are prominent after background differencing and application of Filter-1. After applying Filter-2 and morphological processing those background noises have been removed (Fig. 6f).

Figure 7 shows the results of a video taken in outdoor environment where a moving object enters the frame from right to left laterally. The background of the frames of this video is highly textured due to the presence of window, trees and existence of stationery vehicles. It is observed in Fig. 7d that a lot of noise is present in the frame. Fig. 7e shows the frame after removal of most of the noise by filtering operation (Filter-1). But this frame still contains noise as there are continuous illumination changes on the edges of different objects. This noise is removed in the next processing step when Filer-2 is applied on the frame. After applying Filter-2 and morphological processing the moving object silhouette is obtained.

Figure 8 shows another video with a specific frame at different processing levels. In this figure it is observed that the moving object enters the frame from left to right laterally.

The generated silhouettes can be used for creating a database of objects to be used for template matching based object classification. Generation of human silhouette from video is also important for the purpose of human gait recognition and action identification. Identification of backpack is another important aspect. As shown in Figs. 7 and 8 the persons are carrying backpacks which is evident from the generated silhouettes.

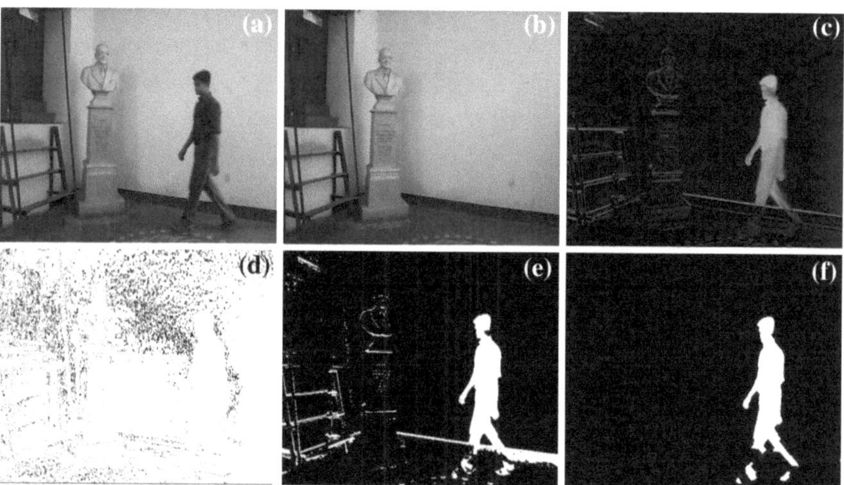

Fig. 6 Experimental Results (*right to left angular*). **a** Present Frame **b** Reference Frame **c** Subtracted Framed **d** Noise in frame **e** Binarization and Filer-1 **f** Filter-2 and Morphological Processing

Fig. 7 Experimental Results (*right to left lateral*). **a** Present Frame **b** Reference Frame **c** Subtracted Frame **d** Noise in frame **e** Binarization and Filer-1 **f** Filter-2 and Morphological Processing

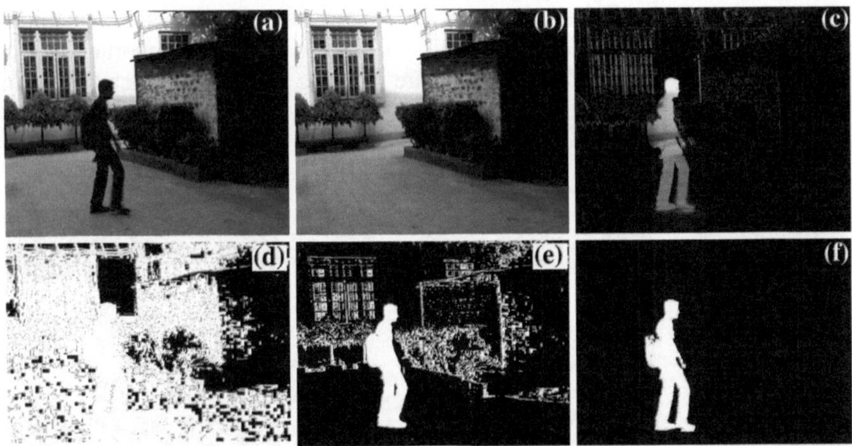

Fig. 8 Experimental Results (*left to right lateral*). **a** Present Frame **b** Reference Frame **c** Subtracted Frame **d** Noise in frame **e** Binarization and Filer-1 **f** Filter-2 and Morphological Processing

5 Conclusion

This paper proposes a novel approach for generating human silhouette from video. This method can be applied for preparing a template database of human silhouettes at different steps of motion. This database can be used for human action recognition and further classification purpose. The proposed method can be extended for generating the silhouettes of other moving objects (e.g. car, animals etc.) in a video. Boundary of the objects can be extracted from the generated silhouettes from which a shape signature can be found which can act as an object description. Such object descriptions can be used as feature for classifying different types of moving objects present in a video (e.g. human being, car, animals etc.). The proposed method works in real time.

References

1. Garrett AZ, Satio H (2009) Real-time online video object silhouette extraction using graph cuts on the GPU, In: Proceedings of the 15th international conference on image analysis and processing, Springer, Heidelberg, 985–994
2. Dedeoğlu Y, Töreyin UB, Güdükbay U, Çetin EA (2006) Silhouette-based method for object classification and human action recognition in video, HCI/ECCV, LNCS 3979, 64–77
3. Shaikh SH, Chaki N (2011) A low cost moving object detection method using boundary tracking, springer proceedings of the international conference on computer science engineering and applications (ICCSEA)
4. Shireen YE, Khaled ME, Sumaya HA (2008) Moving object detection in spatial domain using background removal techniques–state-of-art. Recent Patents Comput Sci 1:32–54
5. Seki M, Fujiwara H, Sumi K (2000) A robust background subtraction method for changing background. In: proceedings IEEE workshop applications of computer vision, (2000) 207–213
6. DeCann B, Ross A (2010) Gait curves for human recognition, backpack detection and silhouette correction in a night time environment. In: Proceedings of SPIE conference on biometric technology for human identification 7, April, Orlando, USA
7. Barnich O, Jodogne S, Droogenbroeck VM (2006) Robust analysis of silhouettes by morphological size distributions, LNCS, vol 4179. Springer, Verlag, 734–745
8. Ekinci M, Gedikli E (2005) Silhouette based human motion detection and analysis for real-time automated video surveillance. Turk J Elec Engin 13(2):199–229
9. Wang L, Tan T, Ning H, Hu W (2003) Silhouette analysis-based gait recognition for human identification, vol 25(12). IEEE transactions on pattern analysis and machine intelligence, December, 1505–1518
10. Barnich O, Droogenbroeck VM (2009) Design of a morphological moving object signature and application to human identification, IEEE international conference on acoustics, speech and signal processing (ICASSP), 853–856
11. Gonzalez CR, Woods ER (1997) Digital image processing, second edition. Pearson Education North Asia Ltd and Publishing House of Electronics Industry

Fast Brain Abnormality Detection Method for Magnetic Resonance Images (MRI) of Human Head Scans Using K-Means Clustering Technique

T. Kalaiselvi, K. Somasundaram and M. Rajeswari

Abstract This paper proposes a rapid method to classify the brain MRI slices into normal or abnormal using brain extraction algorithm (BEA), K-means and knowledge based techniques. BEA is used to extract the brain part from the original magnetic resonance images (MRI) of head scans. K-means is a simple and quicker segmentation process used to segment the brain into known brain regions, like white matter (WM), gray matter (GM) and cerebro-spinal fluid (CSF). Any abnormalities of brain usually affect the normal brain tissues (BT). At times, their intensity characteristics are identical to CSF class. This knowledge is used to analyze the segmented classes of brain by K-Means and thus identify the abnormal slices and location of abnormality within the slices. Experiments were done with datasets collected from medical schools. The results were compared with existing method. The proposed work took only 2 s to produce the results where as the existing requires 12 s per brain extracted slices. The proposed method never produced wrong classification but sometimes missed the abnormal slices. But the existing method had mixed possibilities. This proposed method could be used as a preprocessing technique in brain related studies and thus saves radiologist's time, increases accuracy and yield of diagnosis.

Keywords Brain extraction · K-means · Brain segmentation · Abnormality detection · Location finding

T. Kalaiselvi (✉) · K. Somasundaram · M. Rajeswari
Image Processing Lab, Department of Computer Science and Applications, Gandhigram Rural Institute (Deemed University), Gandhigram 624302 Dindigul, Tamil Nadu, India
e-mail: kalaivpd@gmail.com

K. Somasundaram
e-mail: ka.somasundaram@gmail.com

M. Rajeswari
e-mail: mraji.1986@gmail.com

Mohan S. and S. S. Kumar (eds.), *Proceedings of the Fourth International Conference on Signal and Image Processing 2012 (ICSIP 2012)*, Lecture Notes in Electrical Engineering 221, DOI: 10.1007/978-81-322-0997-3_21, © Springer India 2013

1 Introduction

Automatic defects detection in MR images is very important in many diagnostic and therapeutic applications. Magnetic Resonance Imaging (MRI) is one of the safe and best imaging technique to analyze the brain abnormalities. MRI usually produces a stack of 2D slices of human brain and thus considered to be time-consuming diagnostic process. Several methods have been developed to automate the brain abnormality detection and location identification of the brain tumors. They used several popular techniques like feature extraction, deformable methods, neural networks, image segmentation operations, neural networks and genetic algorithms.

Some of the related existing processes are discussed here. A method proposed by Tuhin et al. [1] is used for detection and segmentation of brain tumors in MRI using K-means algorithms. It is a very time-consuming task that requires to be completed manually by medical experts. A work done by AmirEhsan et al. [2] is automatic defects detection in MR images using feature extraction technique by neural networks. This work required all types of MRI images like T1, T2 and PD images. A work done by Noor Elaiza et al. [3] proposes a method of k-nearest Neighbor (k-NN) analysis in abnormalities segmentation of MRI brain images. The segmentation is done by executing five steps of k-NN, which are, determination of k value, calculation of Euclidian distance objective function, sortation of minimum distance, assignment of majority class and determination of class based on majority ranking.

A method developed by Somasundaram and Kalaiselvi [4] is used to analyze the MRI head scans and detect abnormality in brain using Fuzzy c-means technique. This method is based on bilateral symmetry of the human brain. Finally a fuzzy symmetric measure (FSM) is calculated to discriminate between normal and abnormal scans. An extension of this work is done [5] for locating tumors and thus separating tumor regions from normal CSF region using extended maxima transform. These location details provide an initial contour to activate the deformable models to fix the exact boundary of tumors [6].

The framework of all the existing methods is commonly divided into two stages: (1) decision making, (2) diseases detection. In stage 1, all the existing methods identify the abnormal slices from the patient MRI volume and roughly finding the location of abnormality. In stage 2, they use some advanced techniques to find the exact boundary, size and volume of affected brain region.

The purpose of this proposed work is to classify the brain slices into normal and abnormal automatically and thus reduce the medical experts' time on decision making in stage-1. K-Means is one of the simple and fast segmentation techniques and thus quickens the decision process [7]. This proposed work is developed using brain extraction algorithm (BEA), K-Means clustering technique and a knowledge based technique. Any structural abnormality of brain usually affects the normal brain tissues (BT) and their symmetry property. BT is a combination of white matter (WM) and gray matter (GM) of the brain. The abnormality intensities are mostly identical to cerebro-spinal fluid (CSF) intensity in MRI of head scans. This brain image knowledge is used to develop this proposed work. From the original MRI of head scans the brain part is initially extracted using BEA by removing the non-brain tissues. This helps to segment the BT and CSF from brain using K-Means techniques. The BT and CSF classes produced by K-Means are further processed to detect the abnormal slices and the location of abnormality within the slices. Datasets collected from popular medical school websites were used for experimental study. The results produced by the proposed method were compared with the existing method [4].

The remaining part of this paper is organized as follows. In Sect. 2, an overview of the method used is discussed. In Sect. 3 and 4, the evaluation parameter and materials used in the experiment are given. The results and discussions are presented in Sect. 5. The conclusion is given in Sect. 6.

2 Method

The proposed method has four stages as given in Fig. 1. In stage-1, the brain portion is extracted from the head region by eliminating the non-brain regions. Stage 2 uses the K-means clustering technique to segment the extracted brain portion into known regions like BT and CSF. Stage 3 uses a procedure to check the presence of abnormality by using the BT and CSF images. If a slice is identified as an abnormal slice then the location will be detected using region searching algorithm.

2.1 Stage 1: Brain Portion Extraction

numerous brain extraction algorithms (BEA) are available in literature [8–10]. The T2-BEA [7] makes use of anisotropic diffusion process, optimal thresholding and morphological processes to separate the brain from non-brain portions. The diffusion process is used to highlight the brain from T2 head scan. Then an intensity threshold is computed using which a rough binary brain portion is generated. The morphological operations, erosion and dilation, and connected component analysis are then performed on the rough brain portion to produce the brain mask. Finally the brain mask is used to extract the brain from T2 scans. In Fig. 2, the extracted brain portions are shown in column 2.

2.2 Stage 2: Brain Regions Segmentation

The extracted brain is segmented into four major classes WM, GM, CSF and background using K-means clustering algorithm. K-means algorithm is an unsupervised clustering algorithm that partition (or cluster) n data points into k disjoint

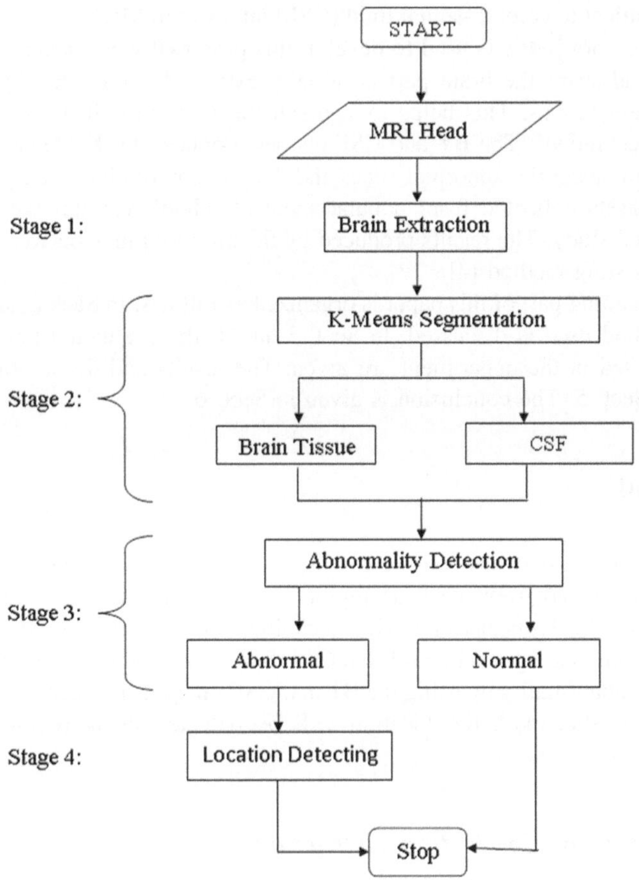

Fig. 1 The flowchart of Proposed Work

subsets c_j containing p_i data points [11]. The input data points were classified into multiple classes based on their inherent distance from each other. The centroid of the cluster is the mean value of all data points with respect to k. K-means uses an iterative algorithm that minimizes the sum of distances from each object to its cluster centroid, over all clusters. K-Means algorithm partitions the data into k mutually exclusive clusters, and returns a vector of indices indicating to which of the k clusters it has assigned.

This algorithm is composed of the following steps.

Step1: Initialize the centroids c_j, $j = 1, 2, \ldots k$
Step2: Assign each data point to the group that has the closest centroid.
Step3: When all points have been assigned, calculate the positions of the k centroids.

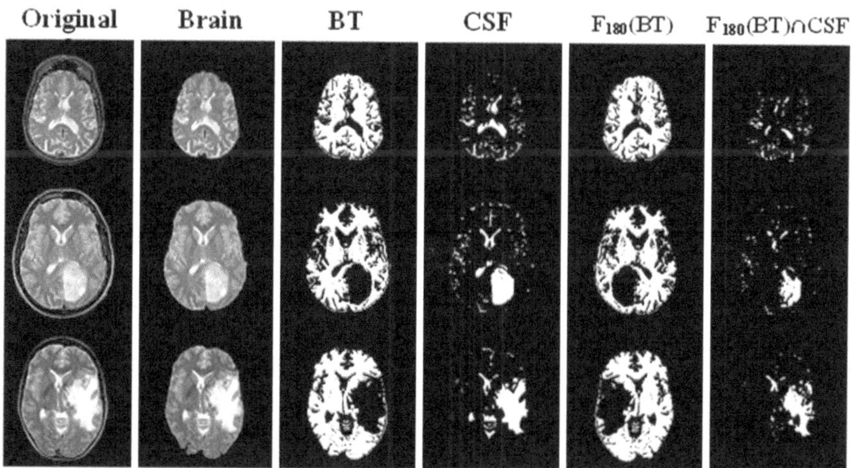

Fig. 2 Results of proposed method at different stage for three slices taken from three volumes respectively. Column one shows the original slices, column two shows the brain extracted at stage 1, column three and four shows the BT and CSF classes produced by K-means method at stage 2 and column five and six are the images used by stage 3 for abnormality detection

Step4: Repeat steps 2 and 3 until the centroids no longer move. This produces a separation of the data points into groups from which the metric to be minimized can be calculated.

The clusters are found by minimizing the objective function,

$$J = \sum_{j=1}^{k} \sum_{i=1}^{n} ||p_i^{(j)} - c_j||^2 \quad (1)$$

where $||p_i^{(j)} - c_j||^2$ is a measure of intensity distance between a data point $p_i^{(j)}$ and the cluster center c_j.

Here the initial centroids of the intensity range 0–255 for the gray image are taken by using the formula

$$\mathrm{m_u} = ((1 : \mathrm{k}) * \mathrm{m}) / (\mathrm{k} + 1) \quad (2)$$

where $\mathrm{m_u}$ is initial centroid vector, k is number of clusters, m is maximum value of the gray image.

The four region based K-means algorithm produces the image of WM, GM, CSF and background. A brain tissue (BT) region will be created by combining the WM and GM regions together as shown in column three of Fig. 2 and CSF class is given at column four.

2.3 Stage 3: Abnormality Detection

The segmented images help to detect the abnormal region in the brain image, as the brain diseases occupy the locations of normal tissues and their intensity characteristics differ from the surrounding normal tissue. Using these segmented images the abnormal region is found by using flipped BT and CSF regions. To classify the slices, the following procedure should be executed.

Step1: The brain tissue (BT) is flipped by 180° known as F_{180} (BT).
Step2: Then the F_{180} (BT) and the CSF images will be compared to get the common area.
Step3: If the common area is > 1,000 then it is known that the slice was affected by tumor.

The results of step 1 and 2 are given in column five and six of Fig. 2. At step 3 the threshold value is set as 1,000 to classify the slices into normal or abnormal after doing an experimental study with several MRI scan images.

2.4 Stage 4: Location Finding

Once the slice is classified as abnormal, the location of abnormality should be detected. Abnormal regions usually have CSF like intensities. Hence in clustered CSF region, the hemispheres that are having largest area are considered to be tumor affected area.

3 Evaluation Parameters

Both qualitative and quantitative validations were used for the performance evaluation. The qualitative evaluation is simply the visual inspection of the result done by the experts in the field. For quantitative study, fuzzy symmetric measure (FSM), false alarm (FA) and missed alarm (MA) parameters were used to check the results of abnormality detection procedure. They also used to compare the results with existing method [4].

FSM is used to check the symmetry property of cerebral hemisphere about the longitudinal fissure. FSM is given by

$$\mathrm{FSM} = 1/1 + \left((\mathrm{n_L} - \mathrm{n_R}/100)^2\right) \tag{3}$$

where n_L and n_R are the number of foreground (white) pixels in the left and right half of the image present at either side of the hemisphere. The symmetry values calculated from normal classes are generally much larger than 0.1, and the values

Table 1 Dataset used

Sl no	Volume identity	Gender	Age	Clinical	Total slices
1	V1	Female	76	Normal	43
2	V2	Male	35	Astrocytoma	29
3	V3	Female	42	Metastatic bronchogenic carcinoma	24

for abnormal classes are much smaller than 0.1. The existing method [4] fixed 0.05 as a threshold value to FSM. If FSM of selected region is much < 0.05 then the slice is abnormal.

False alarm is an indication when the input scan without tumor is marked as abnormal, during analysis. Missed alarm is an indication when an abnormal image is not marked during the analysis. The percentage of such parameter for every dataset is calculated and compared with existing method.

4 Materials

Three dataset of normal and abnormal subjects were used in this experiment. The dataset were taken from the website 'The Whole Brain Atlas' maintained by the Department of Radiology and Neurology at Brigham and Women's Hospital, Harvard Medical School. The dimension of each slice in volume V1–V3 is 256 $\times$ 256 pixels and slice thickness varying from 2 to 5 mm with 260 mm field of view. So the pixel dimension is fitted to 1 $\times$ 1 mm. The Table 1 shows the dataset used for this proposed method.

5 Results and Discussion

The images in the dataset are initially processed by the existing BEA algorithm at stage 1. Then the respective extracted brain images are used for further processing with stage 2.

At stage 2, the K-Means algorithm is applied on the BEA images. Using K-Means technique, the extracted brain portion is divided into two sub regions, BT regions and CSF regions. This algorithm was tested on many images, in order to find an optimal solution in terms of time taken for processing the images. The processing time to segment the image is approximate to 0.0013 s and iteration is approximate to 10.

At stage 3, the proposed method imposed the knowledge based technique to classify the slice into normal or abnormal. This stage produces a common area using the F_{180} (BT) and CSF regions. Here the abnormal images are to be identified by the area that is greater than or approximate to 1,000 in terms of pixel.

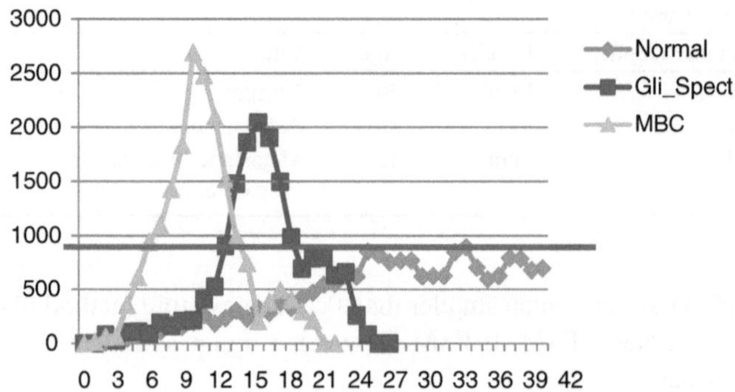

Fig. 3 A Graph plotted against the slice numbers and common area for three volumes

A graph is plotted against the slice number and the respective area of three volumes as given in Fig. 3. In this graph, the entire line graph of normal volume V1 is displayed below 1,000 and thus proves that no abnormality is presented in V1. A part of other two line graphs lies above 1,000. There are corresponding to V2 and V3 volumes, which confirm that the V2 and V3 have some abnormal slices that may be affected by brain diseases or tumors. The slices 13–19 in V2 and slices 7–14 in V3 were detected as abnormal slices. This is again proved by FSM parameter calculation for the CSF class as given in Table 2.

In the proposed system, the symmetry value FSM is calculated for CSF classes of detected abnormal slices and found to be $\ll 0.05$. Hence it is confirmed that the slices are having abnormal regions. Then, this proposed method aimed to detect the location of abnormality within the slice. This again depends on the CSF class of abnormal slices. In stage 4, the location detection procedure compares the area of hemispheres of CSF class and fixed the largest hemisphere as abnormal region. In V2 and V3, almost the right side hemispheres of CSF class are larger and given in Table 2. So the results confirmed that the abnormality is present in the right side hemisphere of the volume. Thus the abnormality location analysis is further narrowed down to cerebral hemispheres processing.

For qualitative evaluation study, FA and MA were used to compare the results with existing method for the selected three volumes. Table 3 shows the values of FA and MA of both proposed and existing methods for three datasets, normal, Gli_Spect and MBC. In normal volume V1, nothing is to be detected as abnormal so that FA and MA were found to be zero in both methods. In second volume V2, out of 29, the actually affected slices are 11–20 and the proposed method detected 13–19 as abnormal. So there is no FA but three slices 11, 12 and 20 were marked as MA. In third volume V3, out of 24, the actually affected slices are 4–15 and the proposed algorithm detected slices 6–14 as abnormal. So there is again no FA but three slices were marked as MA.

The existing method produced both FA (13 % to V2 and V3) and MA (4 % to V3). This proved the existing method had chances for both false as well as missing alarms. But there is no FA in the proposed method. It has only MA that may be due

Table 2 FSM of CSF class generated by K-means method for abnormal slices

Volume	Slice	BT class			CSF class			FSM
		Full area	Left	Right	Full area	Left	Right	
V2	13	12894	7137	5757	1862	484	1378	0.01236
	14	12682	7522	5160	2824	648	2176	0.00426
	15	12306	6639	5667	3391	944	2447	0.00441
	16	14205	7516	6689	3204	867	2337	0.00461
	17	15105	8171	6934	3320	1003	2317	0.00576
	18	14760	7968	6792	2529	543	1986	0.00478
	19	13865	7054	6811	2390	776	1614	0.01404
V3	6	7790	4666	3124	1787	382	1405	0.00946
	7	9925	6279	3646	2393	467	1926	0.00468
	8	11681	6639	5042	2658	343	2315	0.00256
	9	12639	7551	5088	3057	281	2776	0.0016
	10	13102	7986	5116	3927	590	3337	0.00132
	11	14326	8396	5930	3710	906	2804	0.00277
	12	15675	8726	6949	3061	1006	2055	0.00901
	13	13163	7015	6148	3395	1288	2107	0.01469
	14	14235	7403	6832	2087	761	1326	0.03037

Table 3 Performance of proposed method in terms of FA and MA

No	Volume	Total slice	Abnormal slices		Proposed FA		Proposed MA		Existing method	
			Actual affected	Detected abnormal	Slices	%	Slices	%	FA %	MA %
1	V1	43	–	–	–	0	–	0		
2	V2	29	11–20	13–19	–	0	11,12,20	10	13	0
3	V3	24	4–15	6–14	–	0	4,5,15	13	13	4

to hard segmentation nature of K-means method at stage 2. But this MA will not affect the further analysis of brain abnormalities than that of FA.

6 Conclusions

This paper proposed knowledge based automatic method to classify the brain slices into normal and abnormal. This method used the BET and K-means segmentation technique to produce the required regions like BT and CFS classes quickly. Then the brain anatomy based knowledge is used to analyze the classes to arrive to a conclusion regarding the type of a slice. Finally the location of abnormality is detected for the affected slices. The results produced by the proposed method were compared and found to be better than the existing method. This can be used to help the radiologists to quicken their diagnostic process.

References

1. Paul TU, Randhyopadhyay SK (2012) Segmentation of brain tumor from MRI images reintroducing K-means with advanced dual localization method. Int J Eng Res Appl 2: 226–227
2. Lashkari AE (2010) A neural network based method for brain abnormality detection in MR images using Gabor wavelets. Int J Comput Appl 4:9–10
3. Khalid NEA, Ibrahim S (2011) MRI brain abnormalities segmentation using K-nearest neighbors (K-nn). Int J Comput Sci Eng, 980–983
4. Somasundaram K, Kalaiselvi T (2010) Fully automatic method to identify abnormal MRI head scans using fuzzy segmentation and fuzzy symmetric measure. ICGST-GVIP J 10:1–6
5. Somasundaram K, Kalaiselvi T (2010) Automatic detection of brain tumor from MRI scans using maxima transform. UGC Sponsored National Conference on Image Processing-NCIMP, 136–140
6. Kalaiselvi T, Somasundaram K, Vijayalakshmi S (2012) A novel self initialization technique for tumor boundary detection from MRI, ICMMSC12, CCIS 283, Springer, Berlin, NY, pp 464–470
7. Selvy PT, Palanisamy V, Purusothaman T (2011) Performance analysis of clustering algorithms in brain tumor detection of MRI. Eur J Sci Res 62:321–330
8. Somasundaram K, Kalaiselvi T (2010) Fully automatic brain extraction algorithm for axial T2-weighted magnetic resonance images. Comput Biol med 40:816–821
9. Somasundaram K, Kalaiselvi T (2011) Automatic brain extraction methods for T1 magnetic resonance images using region labeling and morphological operations. Comput Biol Med 41(2011):716–725
10. Somasundaram K, Kalaiselvi T (2010) Brain extraction method for T1 magnetic resonance scans, presented in IEEE sponsored international conference on signal processing and communication (IEEE-SPCOM2010), July, Indian institute of sciences (IISc), Bangalore, India, 18–21
11. Somasundaram K, Kalaiselvi T (2009) A comparative study of segmentation techniques used for MR brain images, presented in international conference on image processing, computer vision, and pattern recognition—IPCV'09, WORLDCOMP'09, Las Vegas, Nevada, 597–603

A New Shape Feature Extraction Method for Leaf Image Retrieval

B. Vijayalakshmi

Abstract In this paper we present an algorithm for extracting features for leaf identification. Contour-based shape feature extraction is one of the important research contents in content based image retrieval. We introduce the hybrid method which is a combination of threshold and Sobel segmentation algorithm and extract the feature. The experimental results show that the proposed method has better performance.

Keywords Feature extraction · Edge detection · Threshold

1 Introduction

The recognition and identification of plant has great significance to explore genetic relationship of plants and explain the evolution law of plant system. However, it is a very time consuming task, which is usually done by botanist currently. When recognizing and identifying plant, people usually observe leaf, stem, fruit and so on to extract discriminating feature. But most of the people who don't have sufficient knowledge to discriminate features. so they need computer based system to recognition and identification of plant using leaf images. This is need for the development of automatic plant identification system.

Several researches in leaf identification have been explored, but there are still many challenges for researches to try other approaches for better performance of

B. Vijayalakshmi (✉)
Assistant Professor, Department of MCA, K. L. N. College of Engineering, Sivagangai District, TamilNadu, India
e-mail: bviji1977@yahoo.com

Mohan S. and S. S. Kumar (eds.), *Proceedings of the Fourth International Conference on Signal and Image Processing 2012 (ICSIP 2012)*, Lecture Notes in Electrical Engineering 221, DOI: 10.1007/978-81-322-0997-3_22,

the identification system. A certain method may give good performance in specific samples of leaves, but doest not guarantee to perform good result for other ones.

Present research work in leaf identification is involved with combination of multiple methods. In this paper a one such multiple feature methods with a new shape feature extraction process is proposed and then analysis performance of the proposed method empirically.

First, a preliminary segmentation based on the gray color conversion, image resizing. At the second level edge detection using Sobel operator and single thresholding method. Finally fine feature extraction method is proposed.

The rest of this paper is organized as follows. Section 2 discusses the general architecture. Section 3 discuss the related works. Section 4 explains proposed multiple method. Section 5 presents result and discussion. In Sect. 6 conclusion is given.

2 General Architecture for Leaf Recognition

There are four stages involved in recognition of the plant using leaf image. In the first stage, the leaf image is captures. In the second stage, preprocessing and segmentation is carryout for obtain the gray image and detect edge of image. In the third stage, feature extracted. In the last stage, extracted feature of image is compared with standard database data of trained leaf image. The details of the four stage is discussed as follows in Fig. [1].

In this paper a new feature extraction process is introduced in stage three. The complete discussion on the multi method process for plant identification is presented in sect. 4.

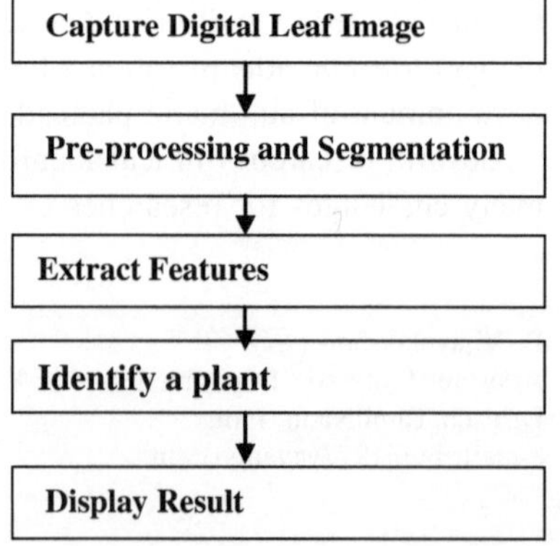

Fig. 1 General Architecture for leaf recognition

3 Related works

A new approach that combines a thresholding method and an artificial neural network (ANN) classifier is proposed to extract leaf veins by [1]. The Comparative of five types of threshold methods such as Mean method, P-tile method, Histogram dependent Technique (HDT), edge maximization technique (EMT) and Visual Techniques are discussed and used three satellite images of testing. In this paper, they proposed HDT and EMT are best techniques in [2].Many types of Thresholding including Shape based algorithm, Histogram Shape-Based Thresholding Methods, Clustering-Based Thresholding Methods, Entropy-Based Thresholding Methods, and Thresholding Based on Attribute Similarity, Spatial Thresholding Methods, and Locally Adaptive Thresholding and based on that nearly 40 types of thresholding are compared [3]. Several other issues remains also to be addressed.

Sobel operator and histogram method are used to detecting edge and proposed the Field Programmable Gate Array (FPGA) which is used to as alternative for the implementation of software algorithm to i mage processing is implemented by [4].

The Comparative of seven edge detection techniques such as Sobel operator technique, Prewitt technique, Kiresh technique, Laplacian technique, Canny technique, Roberts technique and Edge Maximization Technique (EMT) are discussed to choose the best technique for edge detection. They are used satellite images of testing. It reports that Kiresh, EMT and Perwitt techniques respectively are the best techniques for edge detection in those images by [5].

Extended Sobel, Prewitt and Kirsch edge operators are proposed for image segmentation of mammographic images are implemented by [6]. They are using Mammography Images. Edge direction histogram texture feature extracted by using sobel operator to capture the spatial distribution of edge in the six directions in [7].

Fast edge detection algorithm which runs in $\Theta(hw)$, where h and w are the height and width of the source image by using Sobel edge operator and Canny edge detector by [8]. Different types of features are identified in [9], they are using white pixels.

Four types of shape features (seed boundary, seed edge, seed roundness, seed ripples) are discussed in [10] to seed identification. They counts the white pixels in the boundary of a seed, but our proposed system counts the black pixels which will takes less time.

4 The proposed Multiple Method

The detail along with the step by step procedure involved in the proposed method is discussed as follows

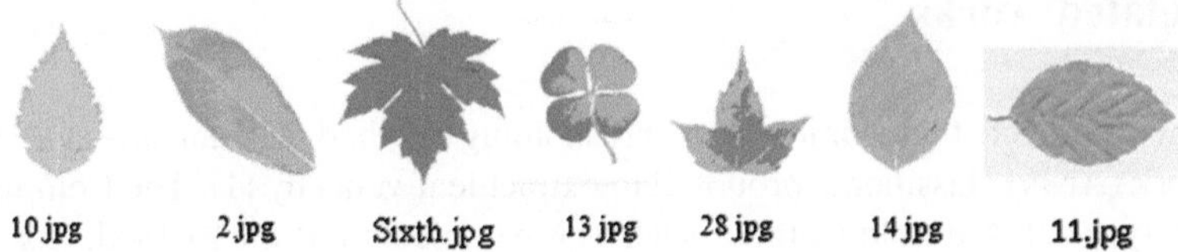

Fig. 2 Sample leaf images

4.1 Stage One: Capturing Leaf Image

In this paper the leaves samples are collected in the forest and leaves are full-grown. The images of leaves of plants were obtained from a five mega pixel CCD camera. The distance between camera and the leaf was maintained to be 15 cms and the image was taken from the any direction. All the images were taken in natural day light with white background and stored on JPEG format with 800 × 600 resolutions. Some the sample images are shown in Fig. [2].

4.2 Stage two- Pre-processing and Segmentation

The leaf image pre-processing refers to the initial processing of input leaf image to correct the geometric distortions and eliminate the noise. These operations are called pre-processing because they are used before the real analysis and manipulations of the image data in order to extract any specific information.

In the pre-processing two steps are involved. The first indicates, convert RGB image to gray image and second stage resizing of the image.

4.2.1 Converting RGB image to Gray image

For converting RGB image to Gray image the formula [1] Presented in [11] is used and formula is as follows,

$$\text{Gray} = 0.2989 * \text{R} + 0.5870 * \text{G} + 0.1140 * \text{B} \tag{1}$$

where R, G, B correspond to the colour of the pixel, respectively.

The following Fig. 3 shows the sample Gray image.

4.2.2 Resizing the Gray image

The input images may have different sizes which can affect the results. The resizing of the image can be defined in two different ways: (1) Resize the every

Fig. 3 Transform RGB image into Gray image

image to predefined size. (2) compare the trained image size (called first image) and input image (called second image) size. In this system we have used second type of resizing.

4.2.3 Image Segmentation

Image segmentation is the process of extracting features or regions of interest from an acquired image for further intelligent computer analysis. The image is sliced into multiple regions based on some property of the pixels [12]. These properties are intensity, texture, position or some local or global statistical parameters.

There are three main board categories in image segmentation: (1) Edge-based segmentation (2) Region—based segmentation (3) special theory- based segmentation. And some sub-classes are included in the main categories too. In this paper, we follows the edge based segmentation

4.2.4 Edge-based Segmentation

Contour-based segmentation approach assumes that the different objects in an image can be segmented by detecting their boundaries. Whereas region-based techniques attempt to capitalize on homogeneity properties within regions in an image, boundary-based techniques rely on the gradient features near an object boundary as a guide. Hence, contour based segmentation methods that rely on detecting edges in the image is inherently more prone to noise and image artifacts. Sophisticated pre- and post-processing is often needed to achieve a satisfactory segmentation result. The simplest contour- based technique is edge detection [13]. Edge-based segmentation is described in [14] also. There are nine operators in the edge based segmentation. In this paper, sobel operator is used.

The Sobel operator performs a 2D spatial gradient measurement on an image and so emphasizes regions of high spatial frequency that correspond to edges [15]. Typically it is used to find the approximate absolute gradient magnitude at each point in an input grayscale image. In theory at least, the operator consists of a pair of 3×3 convolution kernels as shown in Fig. 1 (Fig. 4).

Fig. 4 Sobel Convolution Kernel

-1	-2	-1
0	0	0
1	2	1
Gy		

-1	-2	-1
0	0	0
1	2	1
Gx		

These kernels are designed to respond maximally to edges running vertically and horizontally relative to the pixel grid, one kernel for each of the two perpendicular orientations. The kernels can be applied separately to the input image, to produce separate measurements of the gradient component in each orientation (call these *Gx* and *Gy*). These can then be combined together to find the absolute magnitude of the gradient at each point and the orientation of that gradient [16].

The gradient magnitude is given by Eq. (2)

$$|G| = \sqrt{Gx^2 + Gyb^2} \tag{2}$$

Typically an approximate magnitude is computed using Eq. (3):

$$|G| = |GX| + |GY| \tag{3}$$

The Sobel operator has the advantage of simplicity in calculation.

Thresholding

Thresholding approaches segment scalar images by creating a binary partitioning of the image intensities. A thresholding procedure attempts to determine an intensity value, called the *threshold*, which separates the desired classes [17]. The segmentation is then achieved by grouping all pixels with intensity greater than the threshold into one class, and all other pixels into another class. We can use single threshold or more than one threshold which is called multiple thresholding. In this paper single thresholding are used. When there is only a single threshold T, any point (x, y) for which $f(x, y) > T$ is called an object point; and a point (x, y) is called a background point if $f(x, y) < T$.

Thresholding techniques can be mainly divided into global, local, and dynamic thresholding techniques. In this paper global thresholding is used.

4.2.5 Global thresholding

When T depends only on $f(x, y)$ (in other words, only on gray-level values) and the value of T solely relates to the character of pixels, this thresholding technique is called global thresholding technique [1]. There are a number of global thresholding techniques such as: minimum thresholding, Otsu, optimal thresholding, histogram concave analysis, iterative thresholding, entropy-based thresholding, MoM-keeping thresholding and so on.

4.3 Stage Three: Feature Extraction

Feature extraction involves simplifying the amount of resources required to describe a large set of data accurately. Transforming the input data into the set of features is called *feature extraction.*

In this section we presented new method proposed for feature extraction. The step by step details of proposed method is as follows.

Step1: Let determine the Global threshold value. (In this algorithm, threshold value is 250)
Step2: Evaluate contour of the image by applying the Sobel edge detection operator and calculate the gradient magnitude (GM) to each pixel.
Step3: IF GM > threshold, assign zero (black) to that pixel. Otherwise assign 255 (white).
Step4: After completion of step 3, we will get contour of the image.
Step5: Then extract the features that are count the black pixel in the edge of the leaf.

4.4 Stage Four: Identify a Plant

The Feature extraction method proposed in this paper provides the required data to understand its plants. In order to correctly identify the plant of the leaf data extracted from the feature extraction method is compared with trained data related to various plants which already stored in database.

The Complete discussion in the 4.1–4.4 is present in the form of algorithm (Fig. 5) and this completed method is implemented in Matlab version 7.4.

5 Result and Discussion

Fifty leaf images are taken as benchmark images to test the accuracy of our algorithm (that is 50 leaf images are trained). To test the accuracy of the proposed method in identification of plant using leaf image, the twenty-five primary date (leaf image) and twenty from secondary data (image of leaf) available in the [18]

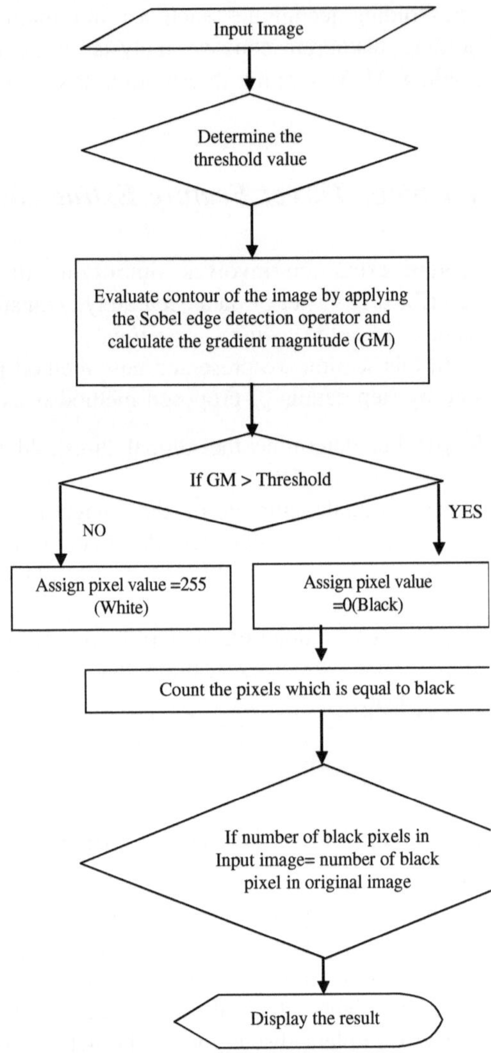

Fig. 5 A Flowchart for identification of plant using leaf image

are given as input (testing images). The result obtain from the proposed method are given in the table [1].

The first three (Leaf 1–Leaf 3) sample result from secondary images and last three (Leaf 4–Leaf 6) sample result from primary images.

The correctly identified plant and its associated trained images and input image are shown in Fig. 6. Out of the fifty, thirty-seven leafs are matched with trained data and identified the plant. The remaining 13 leaves the correctly identified plant and its associated trained images and input image are shown in Fig. 6. Out of the fifty, thirty-seven leafs are matched with trained data and identified the plant. The

Table 1 Shows the no of black pixels in trained image and test image

	No of black pixels in the Original image	No of black pixels in the test image	Result
Leaf 1	3,724	3,724	Plant is identified
Leaf 2	3,022	3,022	Plant is identified
Leaf 3	1,528	1,528	Plant is identified
Leaf 4	4,437	4,437	Plant is identified
Leaf 5	5,368	5,368	Plant is identified
Leaf 6	5,004	5,004	Plant is identified

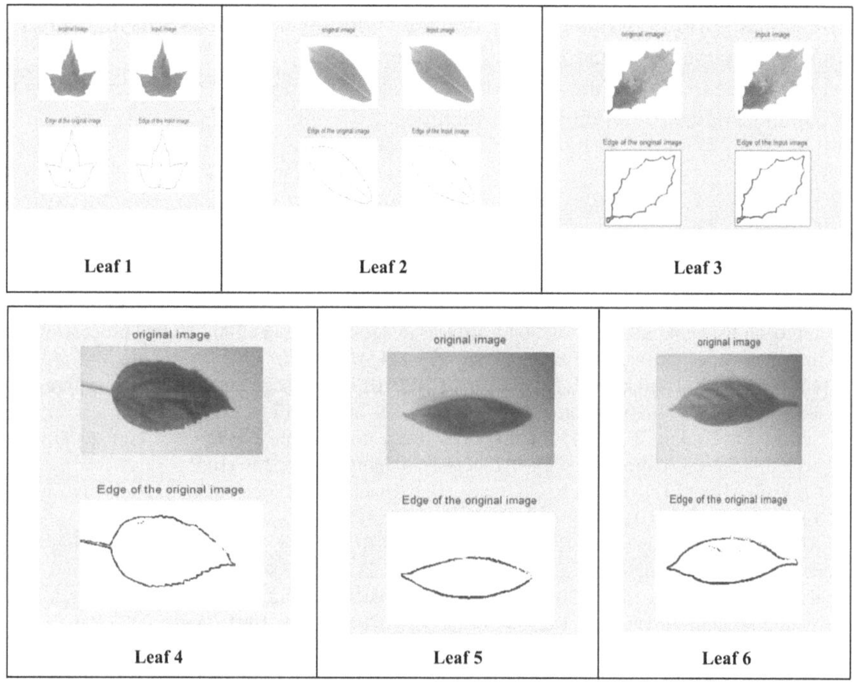

Fig. 6 correctly identified plant leaf

remaining 13 leaves are not identified, and this could be the fact that the training data correcting to the input leaf are missing in the database.

6 Conclusion

The Computer aided identification of plants has gained more attention and is proving to be a very important tool in many areas including agriculture, forestry and pharmacological science. In addition, with the deterioration of environments, many of the rare plants have died out, and so, the investigation of plant recognition can contribute to environmental protection.

The ultimate goal of this project is to develop a system where a user in the field can take a picture of an unknown plant leaf, feed it to the system carried on a portable computer, and have the system classify the species and display information about that leaf within a couple of seconds. We provide the one feature extraction algorithm in this paper. Compared with other methods, this algorithm is efficient in recognition and easy to implementation.

Future research directions include train data corresponding to medical plants to be capture and test with additional features.

References

1. Cham KM, Hui J, Voorde PV, Fu HS (2007) Combined thresholding and neural network approach for vein pattern extraction from leaf images, December 2007, Signal processing and information technology, IEEE Int Symp, 978-1-4244-1835-0
2. Al-amri SS, Kalyankar NV, Khamitkar SD (2010) Image segmentation by using threshold techniques, May 2010. J Comput, Vol 2(5), ISSN 2151-9617
3. Sezgin M, Sankur B (2004) Survey over image thresholding techniques and quantitative performance evaluation, January. J Electron Imaging, 13(1), 146–165
4. Chikkali MPS (2011) FPGA based image edge detection and segmentation. Int J Adv Eng Sci Technol 9(2):187–192
5. Al-amri SS, Kalyankar NV, Khamitkar SD (2010) Image segmentation by using edge detection. Int J Comput Sci Eng 02(03):804–807 ISSN: 0975-3397
6. Kekre HB, Gharge SM (2010) Image segmentation using extended edge operator for mammographic images. Int J Comput Sci Eng, 02(04) 1086–1091, ISSN: 0975-3397
7. Anami BS, Nandyal SS, Govardhan A (2010) A combined color, texture and edge features based approach for identification and classification of indian medicinal plants, September. Int J Comput Appl (0975–8887), 6(12)
8. Sleit A, Dalhoum ALA, Al-Dhamari I, Tareef A, An edge detection algorithm for online image analysis. Recent Adv Appl Math, ISBN: 978-960-474-150-2
9. Lurstwut B, Pornpanomchai C (2011) Plant seed image recognition system, December, IACSIT Int J Eng Technol, 3(6)
10. Lurstwnt B, Pornpanomchai C (2011) Plant seed image recognition system, December. Int J Eng Technol, 3(6)
11. Valliammal N, Geethalakshmi SN (2011) Automatic recognition system using preferential image segmentation for leaf and flower images, October. Comput Sci Eng Int J, 1(4)
12. Gonzalez RC, Woods RE (2002) Digital image processing, 2nd edn. Prentice Hall, NJ
13. Liew AWC, Hong Y (2006) Current methods in the automatic tissue segmentation of 3D magnetic resonance brain images, Curr Med Imaging Rev, 2, 000-000, 1573-4056/06
14. Singh KK, Singh A (2010) A study of image segmentation algorithms for different types of images, September. IJCSI Int J Comput Sci Issues, 7(5), ISSN (Online): 1694-0784

15. Jayaraman S, Esakkirajan S, Veerakumar T, Digital image processing, Mc Graw Hill, NY
16. Matthews J (2002) An introduction to edge detection: The sobel edge detector, Available at http://www.generation5.org/content/2002/im01.asp
17. Liu P (2004) A survey on threshold selection of image segmentation. J Image Graphics, pp 86–92
18. http://www.imageprocessingplace.com/downloads_V3/root_downloads/image_databases/leaf%20shape%20database/leaf_shapes_downloads.htm

A Review of Defect Detection on Electrical Components Using Image Processing Technology

Geoffrey O. Asiegbu, Ahmed M. A. Haidar and Kamarul Hawari

Abstract Image processing technology in the recent years has gain lots of recognitions in the fields of electrical power system engineering. It has been widely used in detection of anomalies on electrical component parts. It has also been effectively applied during testing, inspection and preventive maintenance works. Current researches in thermal imaging technology have shown the interest in development of unsupervised computer aided scrutiny system. This is because of robustness and speed of defect detection analysis compared to conventional or traditional method of testing and inspection. Numerous methods have been used to detect and analyze abnormalities in electrical components such as infrared thermal image, x-ray image, binary, and gray scale images. Procedures normally used in scrutinizing defective components can be classified into five stages thus image acquisition, preprocessing, segmentation, classification and decision-making. This paper presents the review of electrical equipments defect detection techniques using different forms of image analysis approach in detecting and classifying the severity of defects in electrical components. Some advantages and disadvantages of these approaches are also elaborated.

Keywords Image processing · Anomaly detection · Thermal inspection · Preventive maintenance

Geoffrey and Haidar: Proceedings of the 4th International Conference on Signal and Image, ICSIP2012, pp. 1– 8. springerlink.com © Springer India 2012.

G. O. Asiegbu · A. M. A. Haidar (✉) · K. Hawari
Faculty of Electrical and Electronics Engineering, University Malaysia, Pahang, Malaysia
e-mail: ahaidar67@yahoo.com

Mohan S. and S. S. Kumar (eds.), *Proceedings of the Fourth International Conference on Signal and Image Processing 2012 (ICSIP 2012)*, Lecture Notes in Electrical Engineering 221, DOI: 10.1007/978-81-322-0997-3_23, © Springer India 2013

1 Introduction

Devastation on electrical components caused by little or no attention to the thermal radiations has imposed some serious issues to heat management in electrical power network engineering systems. Electrical power components, which have a temperature of operating condition above absolute zero, emit infrared radiation. This radiation can be measured on the infrared spectral band of the electromagnetic spectrum using thermal imaging technology. Attention to thermal radiations will go a long way in attenuating the alarming rate of defects on electrical components. This on the other hand plays a vital role in stack performance, durability and overall system efficiency [1]. The electrical components failures can be avoided if the temperature threshold is detected in order to take timely corrective action. Generally, heat in electrical equipment mainly occur due to the current increasing and in some cases in the form of free convection during a temperature or current dependent components variation on the internal and external surfaces of the stack fibers under different electrical loads and environment conditions [2].

Experiment on infrared radiation was first conducted by William Hershel in 1800; it was proved that there were forms of lights that are invisible in the eyes of human beings [3, 4]. This kind of radiation can be measured on the thermal infrared image spectral band of the electromagnetic spectrum ranging from about 1 mm down to 750 nm. Any equipment, operating at a temperature above absolute zero, certainly emits infrared radiation energy proportional to its surface temperature. Images within this range of electromagnetic spectrum are not visible to human eyes. Therefore a specially designed thermal camera is required the creates pictures of heat essentially measuring infrared energy and convert the data into corresponding images of temperatures permissibly compactable to human eyes to understand the internal status of objects. Thermal camera is characterized by its ability to capture an image of the thermal pattern and measures the emissive power of a surface in an area at various thermal ranges. Thermogram is the digital image produced from infrared thermal camera. One pixel of a thermogram has a specific temperature value, and contrast of the image is derived from the surface temperature gradient.

X-ray is another type of electromagnetic wave having a wavelength shorter than infrared wave. X-ray imaging technology is a diagnostic tool popularly used in medical fields for detecting abnormalities in human system. Recent study showed the application of X-ray image in detecting anomalies on electrical power components base on digital silhouette technology and digital subtraction angiograph. The combination of this two techniques is used in detecting defects in the scanned X-ray image of electrical power components. Its principle is that it can get the X-rays before the vascular injection in the check parts, which is called mask image denoted by r(x, y). Then after the injection vascular in the same parts of the X-ray image, another image is produced called surplus pieces also known as live image. The digitalized live image produces an output image represented as d(x, y). The available silhouette obtained between the mask and the live image, contains only the characteristics of digital silhouette image of vascular [5].

This shows that digital silhouette and digital subtraction angiograph technology are useful during the preprocessing/enhancement stage of image analysis to eliminate background interference between two images. These methods neither show the cause of defect nor identify the location of affected components in electrical power network system. Therefore, these techniques are not effective method of defect detection on electrical equipments. Secondly, X-ray imager presents its captured image in gray scale level only and tells nothing about equipments or component temperature. Since nearly every components or equipment gets hot before it fails [6] therefore, X-ray imager cannot be used to detect defects on equipments that have thermal problem.

2 Defect Detection

In many manufacturing operations, electrical power systems engineering have been the fundamental pillars whose contributions cannot be over emphasized. An unexpected shutdown of even a minor piece of component could have a major impact on production. Since nearly every component or equipment gets hot before it fails, non-destructive thermal inspection is a valuable and cost-effective diagnostic tool with many industrial applications [7–9]. Provision of proper maintenance for many components has been the key objective for every power utility company, but such a target has always been a challenging one. Repairs and maintenance of equipments are classified under these conditions thus: when equipment malfunctions, time–dependent, and status-dependent maintenance [10]. The most popular one is status-dependent maintenance, also known as preventive maintenance [11]. Using infrared thermography of Electrical equipments and components as a non-invasive adjunctive diagnostic methodology for defect detection on electrical equipment have given thermal imaging technology an edge over other fault detection technology. Other advantages of thermal imaging technology are contact-less, easy interpretation of thermal data, wide coverage of inspection, free from dangerous radiation, and robustness.

More applications are in accurate measurements of objects that are moving, very hot, difficult to reach, impossible to shut off, dangerous to contact, also obtain images in situations where contact to damaged or contaminated object, or change in temperature of the object is very difficult. Above all, this technology has been successfully used to solve many industrial problems.

Infrared thermography technology is a robust non-destructive inspection technique [12]. The inspection can be conducted efficiently by keeping a distance from the inspected equipment considering the specified minimum distance for focus of 61 cm or 24 inches [13]. There is no need to halt equipment under operation while an inspection is going on, because the collection of information for inspection is by telemetry, hazardous operations can be avoided [14]. For these reasons, Infrared thermography is widely used for many applications involving preventive maintenance [15].

Typical Fault Localizations are given below:

- Connectors, Relays, and Switches: Find poorly secured, corroded, or current overloaded hardware.
- Semiconductors: Find poorly bonded, die attached, open, shorted, or leaky active devices.
- Circuit Boards: Find overstressed components, plated through holes, poor heat sinks, and bad solder joints.
- Discrete Components: Find overstressed transformers, capacitors, and resistors.

More often than not, loose contacts or connections and misalignment have been the most reoccurring faults found in electrical power system transmission and distribution lines. In fact, such problems are very difficult to detect by mere eye looking especially on steady state or no-load conditions. Facts and figures have shown that, thermographically between 1999 and 2005, 48 % power distribution system problems that were found in conductors were on lose contact parts and bolted connections, in addition to that, 45 % of thermal faults were as a result of disconnecters and misalignments, rust and rough or dirty contact joints, while, 7 % were on other electrical faults [16].

3 Thermal Image Operation

Image segmentation in its simple definition is a process of subdividing an image into its constituent parts. Subdivision of image should stop after extraction and identification or classification of image components. However, the level of subdivision is application dependant. Image segmentation is an important task in machine vision applications. The success of image vision analysis or machine vision operations is highly dependant open the success of the autonomous segmentation of object or segmentation of an image. There are two approaches of image segmentation: discontinuity based image segmentation and similarity based image segmentation. In discontinuity based image segmentation approach, the subdivision or partition is based on some abrupt changes in the intensity level of an image or gray levels of an image. Discontinuity based image segmentation approach involves identification of isolated points, lines, and edges. Edge is a boundary between two regions in an image having distinct intensity gray levels. Similarity based image segmentation approach involves an idea of grouping those pixels in an image which are similar in some sense. There are three approaches under similarity based image segmentation: thresholding, Region growing, and Region splitting and merging techniques. Under the thresholding technique, there are four different types of thresholding operations: global, dynamic or adaptive, optimal, and local thresholding operations. The simplest approach under similarity based image segmentation approach is thresholding technique because of its simplicity and fast computational result; the application of similarity based image segmentation approach becomes more useful [17].

Many researchers like [18–20] have applied Otsu thresholding methods on defect detection using image segmentation process. Otsu's global thresholding technique has been used because of its simplicity and low computational cost. Segmentation using Otsu method is base on calculation of first and second order moments. However, Otsu method is insensitive in detecting objects in the neighborhood of the processed image so it cannot give detailed analysis of defective equipment. Sauvola's local thresholding technique attempt to solve the problem associated with Otsu's technique by calculating the local threshold t(x, y) from local average m(x, y) and the standard deviation s(x, y) [21–23]. This method could have offered a solution but, it takes a longer computational time. In the both thresholding techniques (Otsu's global and Sauvola's local), the input image is in gray scale level and the output is in the binary scale which is not very intuitive compared with RGB color space input and output result. K-means clustering algorithm was used for color image segmentation and the feature extraction was unsupervised. Experimental result have shown that K-means algorithm is insensitive to number of regions segmented and blocks in each clusters does not have to be neighboring blocks [24].

Here segmentation involves converting from RGB color space to CIELAB enables quantification of visual color differences. The L*a*b* color space is derived from the CIE XYZ tristimulus values. The L*a*b* space consists of a luminosity layer 'L*', chromaticity-layer 'a*' indicating where color falls along the red-green axis, and chromaticity-layer 'b*' indicating where the color falls along the blue-yellow axis. All of the color information is in the 'a*' and 'b*' layers [25]. Difference between two colors is measured using the Euclidean distance metric. This algorithm is sensitive to situations where attention to specific regions of interest is needed. The extracted regions are sensitive to classified string labels with respect to specific display positions of targeted regions. Suprathreshold Stochastic Resonance tends to solve the problem of K—means clustering algorithm by been insensitive to object size and position. The multi-object, multi-color image, input RGB color image with different color background were operated on using bit-wise logical OR and AND operators. This operation provides the maximum connected regions of the object in the input noisy, blurred image corresponding to noise standard deviation one. Logical bit-wise ANDing of all R, G, and B frames together provides common segmented regions [26].

However, the bit-wise OR and AND logical operator used for RGB image segmentation produced segmented binary output images. There are many improvements in defect detection using imaging processing technology. Nowadays, some articulated researchers have automated the approach towards ensuring easy and fast means of finding faults on electrical power systems using. Ying-Chieh et al. [10] did some good work using thermographic technology to automate defect detection on electrical equipments. Their proposed system is based on the principle of Otsu's statistical threshold selection algorithm using gray-level histograms [27]. The morphological algorithm adopted in their analysis was based on

Fig. 1 Thermal visible and gray scale level based reference point

reference temperature of equipment having similar load and environment condition as shown in Fig. 1.

Certainly, this system will find it difficult to operate in situations where there is no reference temperature. Therefore, defect detection using this algorithm is partial. On the other hand, there are other approaches used in determining the thermal severity of electrical equipment. Qualitative analysis method is one of techniques applied to estimate the maintenance priorities. This method can directly interpret and evaluate the severity of equipment hotspot temperature based on ΔT criteria [28]. The disadvantage of qualitative method is that ΔT criterion does not indicate the exceeded equipment's temperature limits.

However, in quantitative method, as the reference temperature has to be measured, it requires a greater understanding of the variables influencing the radiometric measurement, as well as its limitations [29]. Calculating the histogram distance is another method for finding the similarity between two objects. In this case, the histogram for each evaluated region is compared with other region in order to obtain the ΔT value. Another method is by analyzing the gradient of the segmented region. One of the advantages of utilizing the gradient analysis is that the source of the hotspot that occurred in electrical equipment can be identified [30]. Meanwhile, non-destructive defect detection on electrical equipments using thermal imaging technology was an improved system developed to detect defects on electrical components irrespective of reference temperatures. This technique applied RGB Color image and optimal threshold algorithm for detection of defective parts. The input data is taking as thermogram of electrical equipment with regions of thermal gradients in RGB color space and the output result is in RGB color space that makes this technique more intuitive as shown in Fig. 2. A normalized RGB IRT image is used to estimate a parametric statistical model consisting mixtures of Gaussian Probability Distribution. Regions of electrical thermogram are segmented through the optimal threshold algorithm ascertained using Receiver Operating Characteristic curve and area under convex hull [31].

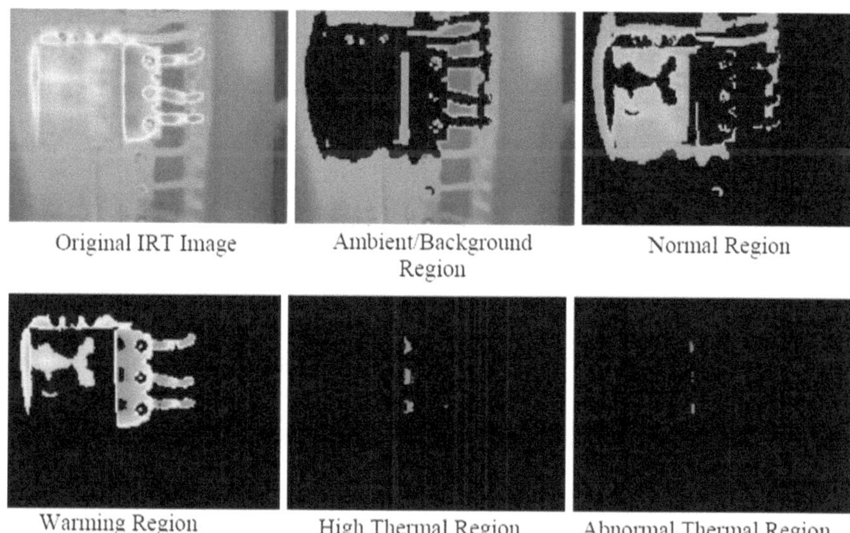

Fig. 2 Overloaded circuit breaker anomaly detection segmentation (http://www.ispyndt.co.nz/infrared+thermal+imaging/infrared+gallery.html)

4 Conclusion

Application of imaging technology to analyze electrical faults has been proved as one of the safe means of testing, inspecting and preventive maintenance checks. Defect detection at early stage on electrical power components is very imperative in order to save cost, equipments and life. Some defect detection methods using different kinds of imagers have been reviewed in this paper: x-ray imager, thermal imager and other imagers. In addition, various techniques have been applied on the produced equipments images to find anomalies that might occur. Results have shown that thermal imager got more advantage over x-ray imager. Automatic or unsupervised defect detection algorithm have been found ineffective to situation where region of interest is so required. However, some good improvements have been recorded to tackle the backdrop of defect detection based on reference temperature. Generally, thermal imaging technology and optimal thresholding technique is considered as robust, non-destructive and contactless methodology with fast computational result in terms of testing, inspection and preventive maintenance work. There is still room for more improvement such as applying artificial neural network system in order to increase the sensitivity of the scrutiny system.

References

1. Matian M, Matian AJ, Brandon NP (2010) Application of thermal imaging to validate a heat transfer model for polymer electrolyte fuel cells. Int J Hydrogen Energy 35:12308–12316
2. Rada N, Triplett G, Graham S, Kovaleski S (2008) High-speed thermal analysis of high power diode arrays. J Solid-State Electron 52:1602–1605
3. Braunovic M, Myshkin NK, Konchits VV (2007) Electrical contacts fundamentals, applications and technology, CRC Press
4. Chuck H (2001) Handbook of nondestructive evaluation, 1st edn. McGraw-Hill Professional
5. Li J, Wang P, Hong Liang P, Jie Weip P, Wang D, Hong Yu P (2012) X-ray image defects detection algorithm based on digital silhouette. Int J Digit Content Technol Appl (JDCTA) 6(7)
6. www.ndted.org/EducationResources/CommunityCollege/OtherMethods/IRT/IR_Applications.htm Techniques and select industrial applications of thermal imaging sighted on 18th April 2011
7. Techniques and select industrial applications of thermal imaging. Available from http://www.ndted.org/EducationResources/CommunityCollege/Other%20Methods/IRT/IR_Applications.htm, sighted on 18th April 2011
8. Lee S (2011) Color image-based defect detection method and steel bridge coating. ASC Annu Int Conf Proc
9. Grys S (2012) New thermal contrast definition for defect characterization by active thermography. J Publ Meas, Elsevier, 30 March 2012
10. Ying-Chieh H, Yao CL (2009) Automatic diagnosis system of electrical equipment using infrared thermography. Int Conf Soft Comput Pattern Recogn
11. Aksyonov YP, Golubev A, Mochortov A, Rodionov V, Minein V, Romanov B, Churtin C, Ignatushin A, (1999) On-line and offline diagnostics for power station HV equipment. IEEE Int Conf Electr Insul Electr Manuf Coil Winding, pp. 637–643, 26–28 October
12. Jabri M, Jerbi H, Braiek NB (2010) Fault detection of synchronous generator based on moving horizon parameter estimation and evolutionary computation. Int Rev Comput Softw (IRECOS) 3(1):38–47
13. Stojcevski B, Kalam A (2010) Fault location in overhead power lines using the IEC61850. Int Protoc Int Rev Comput Softw (IRECOS) 3(5):888–899
14. Chieh Y, Yao CL (2009) Automatic diagnosis system of electrical equipment using infrared thermography, 2009 international conference of soft computing and pattern recognition
15. Fluke, available from http://thermalimagingblog.com/index.php/tag/ifovmeasurement/
16. Martínez J, Lagioia R, Edenor SA (2007) Experience performing infrared thermography in the maintenance of a distribution utility. 19th international conference on electricity distribution Vienna
17. Ning J, Zhang L, Zhang D, Chengke W (2010) Interactive image segmentation by maximal similarity based region merging. Pattern Recogn 43:445–456
18. Thiruganam M, Anouncia SM, Kantipudi S (2010) Automatic defect detection and counting in radiographic weldment images. Int J Comput Appl (0975–8887) 10(2)
19. Du Y, Chang C, Thouin PD (2004) Unsupervised approach to color video thresholding. Opt Eng 43(2)
20. Boberg J (2008) Early fault detection with model-based testing. In: Proceedings of the 2008 sigplan workshop on Erlang
21. Liu Y (2009) A new method of threshold and gradient optimization using class uncertainty theory and its quantitative analysis, Thesis, University of Iowa
22. Shafaita F, Keysersa D, Breuelb TM (2008) Efficient implementation of local adaptive thresholding techniques using integral image, http://code.google.com/p/ocropus/, Article
23. Som HM, Zain JM, Ghazali AJ (2011) Application of threshold techniques for readability improvement of jawi historical manuscript Images. Adv Comput Int J (ACIJ) 2(2)

24. Li J, Wang JZ, Wiederhold G (2000) IRM: integrated region matching for image retrieval. In: Proceedings of the eighth ACM international conference on multimedia, Marina del Rey, California, United States, pp 147–156, October 2000
25. Jha RK, Biswas PK, Chatterji BN (2010) Image segmentation using suprathreshold stochastic resonance. World Acad Sci Eng Technol 72
26. Chitade AZ, Katiyar SK (2010) Color based image segmentation using K–means Clustering. Int J Eng Sci Technol 2(10):5319–5325
27. Ying-Chieh H, Yao CL (2009) Automatic diagnosis system of electrical equipment using infrared thermography. Int Conf Soft Comput Pattern Recogn
28. Haidar AMA, Asiegbu GO, Hawari K, Ibrahim FAF (2012) Electrical defect detection thermal image. Adv Mater Res 433–440:3366–3370
29. Shawal M, Taib S, Kabir S, Ansor M (2011) Image processing methods for evaluating infrared thermographic image of electrical equipments, progress in electromagnetic research symposium proceedings, Morocco, pp 1299–1302
30. Wretman D (2006) Finding regions of interest in a decision support system for analysis of infrared images, Master Thesis, Royal Institute of Technology
31. Soriano M, Huovinen S, Martinkauppi B, Laaksonen M (2000) Skin detection in video under changing illumination conditions. In: Proceedings of the 15th international conference on pattern recognition, vol 1. pp 839–842

Optical Flow in Onboard Applications: A Study on the Relationship Between Accuracy and Scene Texture

Naveen Onkarappa, Sujay M. Veerabhadrappa and Angel D. Sappa

Abstract Optical flow has got a major role in making advanced driver assistance systems (ADAS) a reality. ADAS applications are expected to perform efficiently in all kinds of environments, those are highly probable, that one can drive the vehicle in different kinds of roads, times and seasons. In this work, we study the relationship of optical flow with different roads, that is by analyzing optical flow accuracy on different road textures. Texture measures such as *contrast*, *correlation* and *homogeneity* are evaluated for this purpose. Further, the relation of regularization weight to the flow accuracy in the presence of different textures is also analyzed. Additionally, we present a framework to generate synthetic sequences of different textures in ADAS scenarios with ground-truth optical flow.

Keywords Optical flow accuracy · Texture metrics · Ground-truth optical flow

N. Onkarappa (✉) · A. D. Sappa
Computer Vision Center, Edifici O, Campus UAB 08193 Bellaterra, Barcelona, Spain
e-mail: naveen@cvc.uab.es
URL: www.cvc.uab.es

A. D. Sappa
e-mail: asappa@cvc.uab.es

S. M. Veerabhadrappa
Department of Electrical and Electronics, PES Institute of Technology and Management, Shivamogga 577204, India
e-mail: sujay.veerabhadrappa@gmail.com

Mohan S. and S. S. Kumar (eds.), *Proceedings of the Fourth International Conference on Signal and Image Processing 2012 (ICSIP 2012)*, Lecture Notes in Electrical Engineering 221, DOI: 10.1007/978-81-322-0997-3_24,

1 Introduction

Computer vision has got many applications in human life safety, assistance and comfort. Among them, the safety of automotives and people has got relevant importance in the present world. With the advance in computing performance, computer vision plays a major role in making these assistance and safety applications a reality. The information about the visual motion is very important for ADAS applications such as egomotion, moving object detection, autonomous navigation etc. (e.g., [1, 2]). A well known visual motion estimation instrument is optical flow. Optical flow is the apparent displacement vector field between two image frames. The seminal work on optical flow dates back to 1981. Horn and Schunck proposed a variational formulation to estimate dense optical flow in [3]. At the same time, Lucas and Kanade [4] proposed another approach that computes sparse flow field. In general, optical flow methods can be classified as global and local. Global methods produce dense, whereas local methods give sparse flow fields. A huge amount of work has been proposed during the last three decades; interesting surveys can be found in [5] and [6]. An empirical evaluation of optical flow methods on complex image sequences is presented in [5]. In [7], Galvin et al. evaluate eight different optical flow algorithms. Later evaluation of optical flow algorithms with benchmarking suite of image sequences and tools are proposed by McCane et al. in [8].

The research on optical flow has been getting lots of interests in recent years.[1] Most of the approaches are variational methods [3, 9, 10], which produce dense flow fields. These works concentrate on robust edge-preserving regularization (e.g., [11–13]), and sophisticated data terms [14]. The developments in respective parts of optical flow estimation are discussed in detail in [6]. Recently, in [15], the concepts such as pre-processing, coarse-to-fine warping, graduated non-convexity, interpolation, derivatives, robustness of penalty functions, median filtering are explored and the best formulation out of variants of all these is discussed.

A typical variational method involves a data term and a regularization term that makes the problem well posed. As per our knowledge almost all methods fix the regularization weight empirically. Even though there is a large amount of work on optical flow, there are no considerable efforts to adapt the regularization weight based on some features of the given sequence. In [13], an automatic selection approach based on optimal prediction principle is presented. It predicts the regularization weight based on the computed flow field and data constancy error. But it also involves a brute-force method of selecting the weight empirically. In ADAS domain, the vehicle can be driven in any kind of environment such as urban, highway and countryside, and at different times in a day and in different seasons. Hence, it is impossible to compute accurate optical flow in all the scenarios with a fixed regularization weight. Note that for an ADAS involving visual motion perception, it is very important to estimate accurate optical flow in all such scenarios.

[1] http://vision.middlebury.edu/flow/

In that direction, it is clear that it is necessary to adapt the motion perception algorithms for the occurring scenario directly using some features of the image frames being captured at times. Different scenarios can be regarded as different textures and different structural scenes. In the current work, we study the effect of different textures on optical flow in ADAS domain. Specifically, the change in accuracy of flow field with different regularization weights in the presence of different textures is analyzed; trying to find correlations between accuracy, texture information, and regularization weights.

For the work planned above, it is required to have datasets of image sequences with ground-truth optical flow. Few sequences with ground-truth optical flow have been proposed in [16, 17]. But none of them are suitable for our intended study. It is difficult to obtain ground-truth optical flow for real-world scenarios unless it is done in controlled environments. Particularly for the work in this paper, it is not possible to have or create a real scenario, neither possible to generate dense ground-truth flow field. Therefore, we present a framework to generate synthetic image sequences where the same geometrical scene is used but with different textures; furthermore this framework allows to compute the ground-truths.

The paper is organized as follows. An overview of the basic variational optical flow estimation is presented in Sect. 2. Section 3 describes the texture measures. The framework for generating the sequences and ground-truth is described in Sect. 4. Section 5 presents the experimental analysis. Finally, the work is concluded in Sect. 6.

2 Optical Flow Overview

The classical variational method of Horn and Schunck [3] assumes the constancy of brightness, which is also called *optical flow constraint* (OFC). The OFC can be formulated as: $I_1(\boldsymbol{x}+\boldsymbol{u}) - I_0(\boldsymbol{x}) = 0$, where I_0 and I_1 are two images, $\boldsymbol{x} = (x_1, x_2)$ is the pixel location within the image space $\Omega \subseteq \mathbf{R}^2$; $\boldsymbol{u} = (u_1(\boldsymbol{x}), u_2(\boldsymbol{x}))$ is the two-dimensional flow vector. Linearizing the above equation using first-order Taylor expansion we get OFC as: $(I_{x_1}u_1 + I_{x_2}u_2 + I_t)^2 = 0$, where subscripts denote the partial derivatives. Using only local intensity constraints do not provide enough information to infer meaningful flow fields, make the problem ill-posed. In particular, optical flow computation suffers from two problems: first, no information is available in un-textured regions. Second, one can only compute the normal flow, i.e., the motion perpendicular to the edges. This problem is generally known as the *aperture problem*. In order to solve this problem it is clear that some kind of regularization is needed. The Horn and Schunk [3] method overcomes this by assuming the resulting flow field globally smooth all over the image, that can be realized as penalizing large flow gradients. Combining OFC and regularization in a single variational framework and squaring both constraints yields the following energy function:

$$E(\mathbf{u}) = \int_{\Omega} \{ \underbrace{(I_{x_1}u_1 + I_{x_2}u_2 + I_t)^2}_{Data\ Term} + \alpha \underbrace{(|\nabla u_1|^2 + |\nabla u_2|^2)}_{Regularization} \} \, d\mathbf{x}, \tag{1}$$

where α is the regularization weight. This energy function can be minimized by solving the corresponding Euler-Lagrange equations. Another well known approach to minimize variational energies is by dual formulation [18].

3 Texture Measures

It is necessary to quantify texture properties to study the influence of different textures on the accuracy of optical flow. These texture properties are used to correlate with optical flow accuracy. In this direction, we use basic texture measures such as *contrast*, *correlation* and *homogeneity*. These textural measurements are computed over Gray-Level Co-occurrence Matrix (GLCM) [19] of an image. A GLCM is a matrix that is defined over an image as the distribution of co-occurring values at a given offset. A co-occurrence matrix P is defined over an image I of size $m_1 \times m_2$, parameterized by an offset $(\Delta x, \Delta y)$ as:

$$P_{\Delta x,\Delta y}(i,j) = \sum_{k=1}^{m_1}\sum_{l=1}^{m_2}\begin{cases} 1, & \text{if } I(k,l) = i \text{ and } I(k+\Delta x, l+\Delta y) = j. \\ 0, & \text{otherwise.} \end{cases} \tag{2}$$

The most widely used, computed on the texture metrics on the normalized GLCM, are the following:

$$Contrast = \sum_{n=0}^{N_g-1} n^2 \{\sum_{i=1}^{N_g}\sum_{j=1}^{N_g} p(i,j)\}; \ |i-j| = n \tag{3}$$

$$Correlation = \frac{\sum_{i=1}^{N_g}\sum_{j=1}^{N_g}(ij)p(i,j) - \mu_x\mu_y}{\sigma_x\sigma_y} \tag{4}$$

$$Homogeneity = \sum_{i=1}^{N_g}\sum_{j=1}^{N_g}\frac{p(i,j)}{1+|i-j|} \tag{5}$$

where $p(i,j)$ is the (i,j)th entry in normalized GLCM; N_g is the number of distinct gray levels in the quantized image; μ_x, μ_y, σ_x, and σ_y are the means and standard deviations of p_x and p_y: $p_x(i) = \sum_{j=1}^{N_g} p(i,j)$ and $p_y(j) = \sum_{i=1}^{N_g} p(i,j)$.

4 Framework for Synthetic Dataset

The specific goal of this work is to study the effect of different road textures on the accuracy of optical flow computation in the ADAS scenario and to unveil their relationship. For that, it is required to have images with the same structural scene but with different textures. It is impossible to find such a scenario in the real world. Moreover, it is not possible to obtain ground-truth optical flow from such real-world scenarios unless it is done in sophisticated controlled laboratory environments. So, in the current work, we present a framework to generate synthetic image sequences. The framework can generate image sequences with exactly similar scene structures and with exactly the same acquisition condition (i.e., vehicle speed, camera pose, et.,), but with different textures. There are some works to generate synthetic datasets with ground-truth data in the literature (e.g., [16, 17, 20] and [21]). In [16], authors provide ground-truth information for several synthetic as well as real sequences, whereas in [17], synthetic sequences for ADAS are provided. Later Aodha et al. present a framework for synthetic dataset generation in [20]. Following the framework proposed in [20] a set of synthetic sequences for different speeds of the vehicle are presented in [21]. The datasets provided in these previous works can be used to compare different optical flow techniques, but do not provide sequences with different textures. For the intended study, we generate our own dataset and make it available for the whole community for further research through our website.[2] We use a framework similar to [20] and construct a 3D urban model, which contains road, sky and several buildings using Maya.[3] A camera is made to move in the model along the road, mimicking the camera fit inside a moving vehicle. The image frames are rendered for the camera movement timeline. Similarly, several different image sequences are generated by changing road textures in the model. Also, the ground-truth flow vectors are generated for the camera movement using ray-tracing technique on 3D Maya model. In ADAS, the road surface covers major area in the visibility of the vehicle camera and the structures in the sides of the road vary a lot depending on the environment (urban, highway, hilly etc.). So the flow vectors on the road surface are generally preferred in ADAS applications since they are more reliable and

[2] www.cvc.uab.es/adas

[3] www.autodesk.com/maya

cover a large part of the image. Hence, using the framework described in this section we generate several sequences of the same structural scene with different road textures. The generated images and ground-truth flow fields are of resolution 640×480. Figure 1 shows images from sequences of different textures, but having the same structural scene. The image in Fig. 1 (*bottom-left*) shows the ground-truth flow field. The ground-truth flow field is the same for all the sequences shown there, as the ray-tracing technique used gives the same displacement information based on the structural geometry of the scene irrespective of the texture.

5 Experimental Analysis

In order to make the analysis and conclusion precise and easier, we have considered three sequences of different textures. The textures are different at least on the road surface, but the structures of the scene are exactly the same in all the three sequences. Let us refer to these sequences as T1, T2 and T3. These sequences are selected in the increasing order of the textural property, *contrast*. For further experimentation, we have selected several pairs of images from the sequences at specific locations and present the results as average values of all these selected pairs. Hereinafter, the results obtained from these sets of image pairs are referred to as the results obtained from the corresponding sequences.

Now the task is to find the relation between the accuracy of optical flow, textural property of images and the regularization weight involved in optical flow estimation. For computing the optical flow, we use the method proposed in [15], that involves weighted median filtering. First, optical flow is computed on the sets of image pairs from different sequences. Then, errors (both average angular error (AAE) and average end-point error (AEPE)) are computed between the estimated flow fields and the ground-truth flow fields. The ground-truth flow fields for all the three sequences are the same.

To study the effect of regularization weight on the accuracy, first we empirically analyzed optical flow error on different pairs of images for a wide range of values of regularization weights and determined that the range from 1 to 22 at intervals of 1 or 2 would be fine for further analysis. Then, we estimated optical flow for values of regularization weight within the most suitable range of values (i.e., 1–22). Figure 2 (*left*) shows the curves of AAEs corresponding to sequences T1, T2 and T3 for different values of regularization weight. The minimum AAE for each sequence is marked as $*$ on the curve. Similarly, AEPE error curves are shown in Fig. 2 (*right*).

Further, textural measures explained in Sect. 3 are computed on one of the images of each sequence. Since road is the major and reliable surface in ADAS scenarios, we compute texture metrics on a Region Of Interest (ROI) on the road

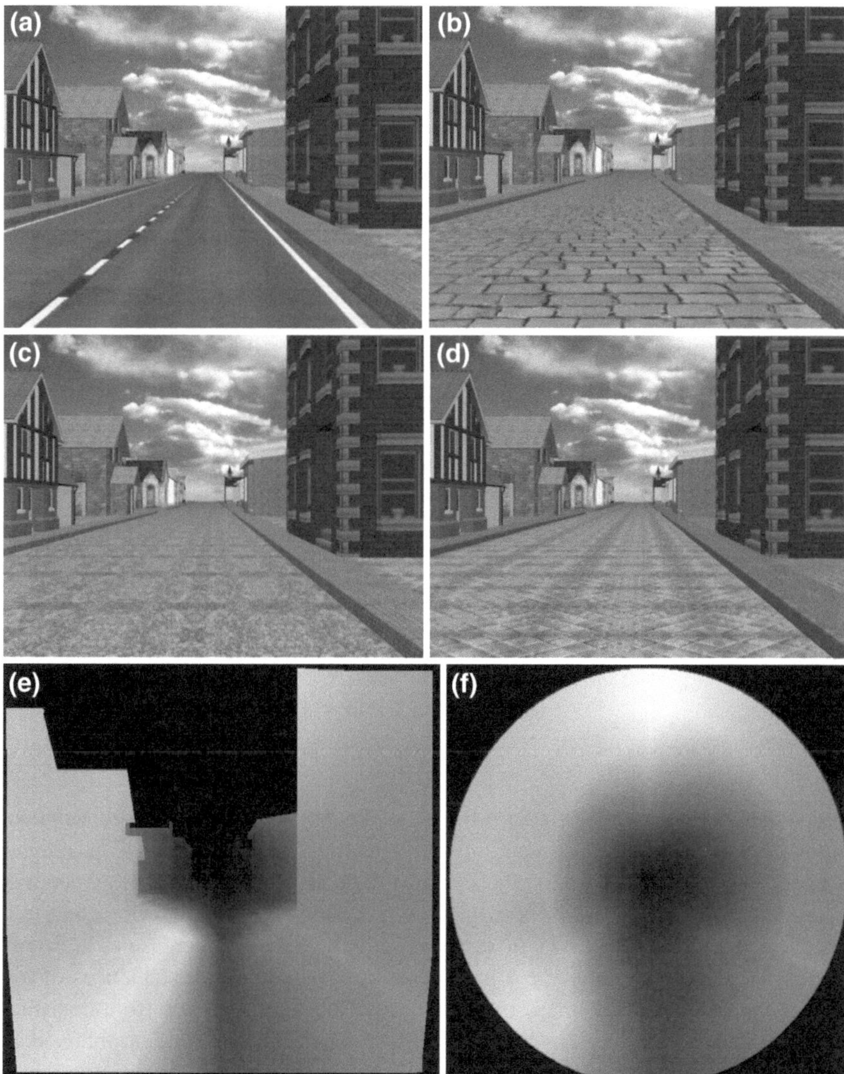

Fig. 1 *Top-left* consecutive frames of a sequence; *top-right* and *middle* frames of different texture; *bottom-left* ground-truth flow field from images in *top-left*; *bottom-right* colormap used to show the flow field

surface. The texture metrics, minimum AAEs, minimum AEPEs, and the regularization weights corresponding to the minimum errors are depicted in Table 1.

The curves in Fig. 2 and texture metrics in Table 1 indicate that image sequences with higher texture *contrast* produce smaller AAE and AEPE,

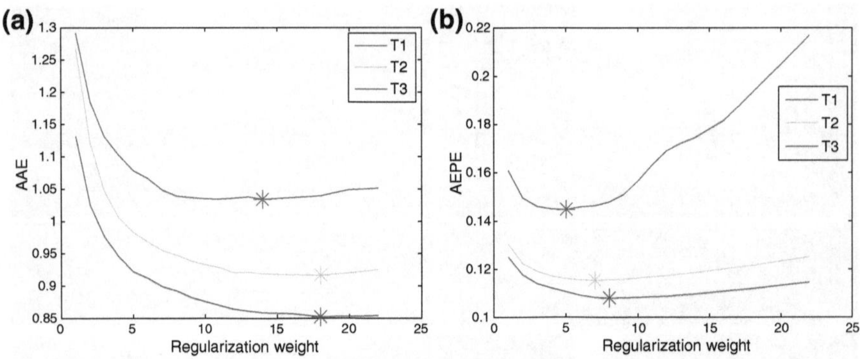

Fig. 2 *Left* AAEs for three sequences for different regularization weights; *right* similarly AEPEs

Table 1 Texture metrics, minimum AAEs, minimum AEPEs, regularization wights for the minimum errors for T1, T2 and T3

Sequence	T1	T2	T3
min AAE	1.0339	0.9159	0.8515
α of min AAE	14	18	18
Contrast	0.0488	0.0850	0.1255
Correlation	0.9386	0.8611	0.7168
Homogeneity	0.9765	0.9575	0.9372
min AEPE	0.1447	0.1152	0.1079
α of min AEPE	5	7	8

independently of the regularization factor α. We can see in Fig. 2, both (*left*) and (*right*), that curve of T1 is above T2, and T2 is above T3. In Table 1, T3 has higher *contrast* than T2, and T2 has higher *contrast* than T1. Similarly, the sequence with lower values of *correlation* or *homogeneity* produce smaller errors. One can see in Fig. 2 (*right*), the AEPE of T1 increases drastically with the increase in α. So one should be careful in tuning the α parameter when dealing with sequences of low texture *contrast*. Another interesting conclusion that can be drawn from this study is that, for higher texture *contrast* sequences, the minimum error is obtained by increasing the regularization factor α. This can be observed in Fig. 2 marked by $*$ and also in the Table 1, the α values of minimum errors. Although in the plots only three curves have been depicted, this conclusion has been extracted from a larger set up using nine sequences of different textures. Figure 3 presents four additional ROIs with the textures from the road surface used to validate the results from our studies. In these four cases, similar behavior like the ones presented in Fig. 2 and Table 1 were obtained.

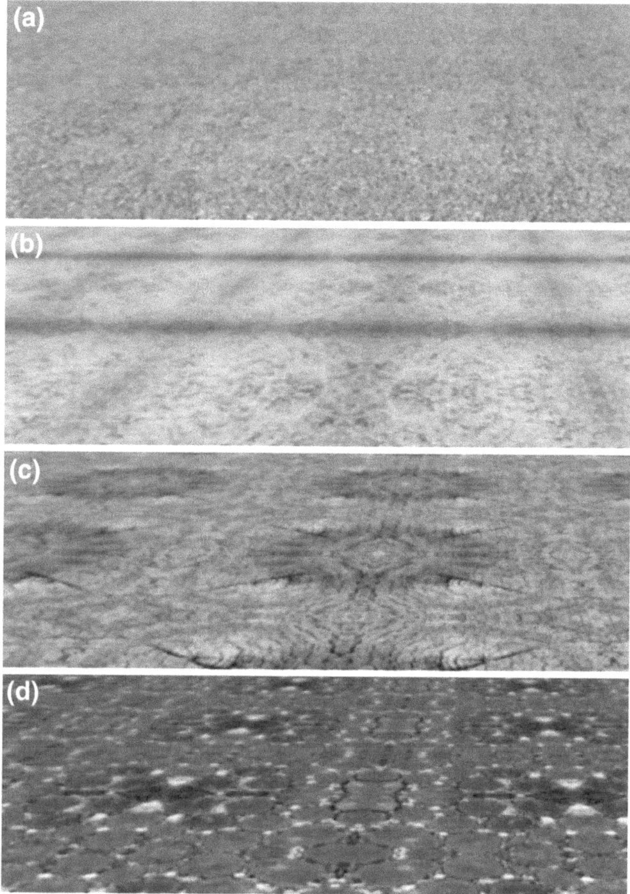

Fig. 3 ROIs of road surface from different sequences with different textures

6 Conclusions

In this work, the behavior of optical flow accuracy for different textural properties and their dependency on the regularization factor in ADAS scenarios is studied. It is evident that the sequence with higher textural *contrast* gives more accurate flow estimation and a more special care should be taken while tuning the regularization factor for sequences with low contrast more than those with higher contrast. It is also observed that, increasing the regularization factor gives better results with increase in textural *contrast* of sequences. A framework to generate synthetic sequences of different textures with the ground-truth optical flow is also presented along with the dataset generated. The presented framework motivates deeper study of the relationship between the optical flow accuracy and scene texture using more sophisticated texture metrics.

Acknowledgments This work has been partially supported by the Spanish Government under Research Program Consolider Ingenio 2010: MIPRCV (CSD2007-00018) and Project TIN2011-25606. Naveen Onkarappa is supported by FI grant of AGAUR, Catalan Government. The authors would like to thank Oisin Mac Aodha for providing the Python code for raytracing with Maya.

References

1. Giachetti A, Campani M, Torre V (February 1998) The use of optical flow for road navigation. Robot Autom IEEE Trans 14(1):34–48
2. Onkarappa N, Sappa AD (2010) On-board monocular vision system pose estimation through a dense optical flow. In: 7th international conference on image analysis and recognition, vol 1, pp 230–239
3. Horn BKP, Schunk BG (1981) Determining optical flow. Artif Intell 17:185–203
4. Lucas BD, Kanade T (1981) An iterative image registration technique with an application to stereo vision (DARPA). In: DARPA image understanding workshop, pp 121–130
5. Barron JL, Fleet DJ, Beauchemin SS (1994) Performance of optical flow techniques. Intern J Comput Vis 12(1):43–77
6. Bruhn A (2006) Variational optic flow computation: accurate modelling and efficient numerics. PhD thesis, Department of Mathematics and Computer Science, Saarland University, Saarbrücken
7. Galvin B, Mccane B, Novins K, Mason D, Mills S (1998) Recovering motion fields: an evaluation of eight optical flow algorithms. In: British machine vision conference, pp 195–204
8. McCane B, Novins K, Crannitch D, Galvin B (October 2001) On benchmarking optical flow. Comput Vis Image Underst 84(1):126–143
9. Brox T, Bruhn A, Papenberg N, Weickert J (2004) High accuracy optical flow estimation based on a theory for warping. In: European conference on computer vision, vol 3024 of LNCS. Springer, pp 25–36
10. Wedel A, Pock T, Zach C, Cremers D, Bischof H (2008) An improved algorithm for TV-L1 optical flow. In: Dagstuhl motion workshop, Dagstuhl Castle, pp 23–45
11. Weickert J, Schnörr C (December 2001) A theoretical framework for convex regularizers in pde-based computation of image motion. Intern J Comput Vis 45(3):245–264
12. Wedel A, Cremers D, Pock T, Bischof H (2009) Structure- and motion-adaptive regularization for high accuracy optic flow. In: International IEEE conference of computer vision, Kyoto, pp 1663–1668
13. Zimmer H, Bruhn A, Weickert J (2011) Optic flow in harmony. Intern J Comput Vis 93(3):368–388
14. Steinbruecker F, Pock T, Cremers D (2009) Advanced data terms for variational optic flow estimation. In: Modeling vision and visualization workshop, Braunschweig, pp 155–164
15. Sun D, Roth S, Black MJ (June 2010) Secrets of optical flow estimation and their principles. In: Conference IEEE on computer vision and pattern recognition, San Francisco, pp 2432–2439

16. Baker S, Scharstein D, Lewis JP, Roth S, Black MJ, Szeliski R (2007) A database and evaluation methodology for optical flow. In: IEEE international conference of computer vision, pp 1–8
17. Vaudrey T, Rabe C, Klette R, Milburn J (2008) Differences between stereo and motion behaviour on synthetic and real-world stereo sequences. In: Image and vision computing New Zealand, Christchurch, pp 1–6
18. Zach C, Pock T, Bischof H (September 2007) A duality based approach for realtime TV-L^1 optical flow. In: Symposium DAGM on pattern recognition, Heidelberg, pp 214–223
19. Haralick RM, Shanmugam K, Dinstein I (1973) Textural features for image classification. IEEE Trans Syst Man Cybern 3(6):610–621
20. Mac Aodha O, Brostow GJ, Pollefeys M (2010) Segmenting video into classes of algorithm-suitability. In: IEEE conference on computer vision and pattern recognition, San Francisco, pp 1054–1061
21. Onkarappa N, Sappa A (2012) An empirical study on optical flow accuracy depending on vehicle speed. In: IEEE intelligent vehicles symposium, pp 1138–1143

Character Segmentation for Cursive Handwritten Text Using Ligature Classification and Transition Feature

S. Gomathi Rohini, R. S. Uma Devi and S. Mohanavel

Abstract Splitting of touching characters in cursive handwritten text is a critical task in segmentation process. A perfect segmentation of character is required to reduce the error rate of recognition. This paper proposes an approach to segment touching/overlapping and shadow characters in the handwritten text using ligature classification. It falls under the category of dissection method, but does not over segment 'm', 'n' and 'u', where the existing methods do. Binarization is the pre processing step for segmentation, which is performed by global or local thresholding. Sauvola's method of threshold calculation is employed in this approach to binarize the gray scale image. The skew of the image is corrected by MATLAB code. Statistical analysis of ligature is done, in order to classify the inter-letter links and intra-letter links for evaluating the segmentation points. The Possible Segmentation Points (PSP) is generated based on the transition feature, followed by removal of invalid PSP by incorporating ligature extraction. The integration of transition feature in dissection method avoids unnecessary segmentation points without any attempt of classification and consequently reduces computational cost. A benchmark database IAM is used for fair comparison. The paper exhibits many examples with challenging and normal cases. The experimental results show that the proposed method achieves the segmentation rate of 92 %.

S. Gomathi Rohini (✉)
Sri Ramakrishna Engineering College, Coimbatore, India
e-mail: rohinismohan@yahoo.com

R. S. Uma Devi
GR Govindarajulu School of Applied Computer Technology, Coimbatore, India
e-mail: umapragaladh@gmail.com

S. Mohanavel
Dr. N.G.P. Institute of Technology, Coimbatore, India
e-mail: mohanavels@gmail.com

Mohan S. and S. S. Kumar (eds.), *Proceedings of the Fourth International Conference on Signal and Image Processing 2012 (ICSIP 2012)*, Lecture Notes in Electrical Engineering 221, DOI: 10.1007/978-81-322-0997-3_25, © Springer India 2013

Keywords Ligature detection and classification · Core detection · Stroke height analysis · Inter-letter links · Intra-letter links

1 Introduction

Segmentation of cursive handwritten words into isolated characters remains an open problem in handwritten document recognition. It is a process, which decides whether the isolated sub images are characters, parts of characters or noises. A perfect segmentation of character is required to reduce the error rate of recognition. There are three approaches for character segmentation:

(1) **Classical/Dissection Approach:** It identifies segments based on 'character like' properties. The word image is divided into segments, each representing a character. It precedes classification and recognition and has no feedback information.
(2) **Recognition based Approach:** It searches the word image for components that matches classes in its alphabet. It splits words into segments and passes each segment to a classifier. If the classification is not satisfactory, again segmentation is called with feedback information.

The classical approach and recognition based approach together is called as analytical approach [1, 2].

(3) **Holistic Approach:** It seeks to recognize words as a whole to avoid the need of segmenting into characters. It is based on the analysis of the order of ascender, descender, loops, vertical strokes and their count. It is heavily on dictionary searching that is costly and prone to mislead by spelling errors.

One of the most challenging problems in the classical approach is segmentation on ambiguous touching characters such as 'm', 'n' and 'u'. They may be misclassified, because the inter-letter link is similar to intra-letter link. The existing techniques in the classical/dissection approach are projection profile and contour analysis.

In vertical histogram based segmentation, number of foreground pixels found in each column is taken into account. The threshold might be based on mean line width or percentage of a mean histogram value. Several shortcomings such as dividing loops and characters containing valleys ('v', 'u', 'm', 'n', and 'k') might occur. This method is very sensitive to threshold value and not suitable for slanted cursive writing.

In segmentation based on contour analysis, the upper profile minima leads to cut in valleys of letters, separating the upper short line of 'r' and dissecting vertical strokes from the body of a character in letters 'b', 'd', 'h', 'k' and 'p'. When maxima is used as segmentation point, then it would intersect single letters and separate hooks from 'a', 'u' and 'd'.

This paper is structured as follows: Sect. 2 surveys different segmentation algorithms reported in the literature. Section 3 describes the pre-processing and character segmentation. Section 4 reports all the experimental results and analysis. Finally Sect. 5 concludes this work.

2 Related Work

Algorithms dealing with character segmentation in the literature are based primarily on analysis of geometric relationship amongst adjacent components. The researchers' contribution on character segmentation is highly remarkable. They are highlighted in the survey papers [3–5] and several segmentation techniques have been proposed [2, 6–9]. Eric Lecolinet et al. [10] used a grapheme-based segmentation technique. It used a recognition procedure, which distinguished the meaningful entities from the non-significant parts within the words. To segment touching characters, Liang et al. [1] presented a technique based on contour analysis and dynamic time warping. In this method, splitting paths could be verified by incorporating character recognition and contextual information. Le Kang et al. [11] proposed another contour based shape decomposition approach that provided local segmentation of touching characters. The shape contour was linearized into edgelets and edgelets were merged into boundary fragments. The method proposed by Ghazali et al. [12] used a hybrid strategy for recognition of strings of characters. It used two-stage Artificial Neural Networks (ANN) based dynamic programming—lexicon driven stage followed by verification stage. Tripathy and Pal [13] proposed a water reservoir concept-based scheme for segmentation of unconstrained Oriya handwritten text into individual characters. Using structural, topological and water reservoir concept-based features, touching characters of the word were segmented. Nicchiotti et al. [14] used a simple procedure for cursive word over segmentation, which was based on the analysis of the handwritten profiles and on the extraction of "white holes".

3 Character Segmentation

In this section, we describe the motivation for extracting transition feature from an input gray scale image and discuss the statistical analysis used in our approach. Without loss of generality, we assume that the character pixels are black, represented as $1'$s in a binary image, while background pixels are white, represented as $0'$s in a binary image. We also assume that the orientations of words are locally uniform and a skew correction method normalizes the word to orient horizontally. The image requires preprocessing in the proposed approach. The segmentation approach has the following steps as mentioned in Fig. 1 below.

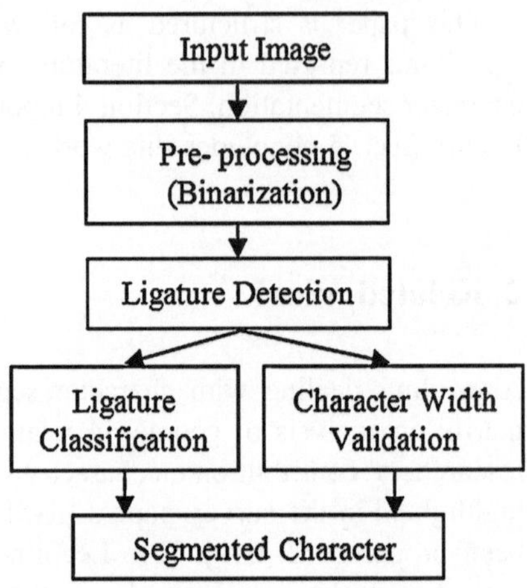

Fig. 1 Block diagram of the proposed segmentation method

3.1 Binarization

It is the starting step in most of the segmentation techniques. Binarization can be performed by global or local thresholding. A global binarization method gives a single threshold value for the entire word image. Its results are good for document images with uniform illumination. Images with non-uniform illumination may suffer from loss, when employing global thresholding methods. In local thresholding, the input image is fragmented into windows and threshold for each window is determined by using simple mean. We have employed Sauvola's method of threshold calculation, which is based on mean and standard deviation [15] to binarize the gray scale image. The skew of the image is corrected by using the MATLAB code given in Fig. 2 below.

Fig. 2 MATLAB code for skew correction

```
h2 = figure; imshow (B);
A = imread ('5cr.jpg');
h1= figure; imshow (A); title ('Original Image');
orange = '[255 255 255]';
a  = 0.45;
T  = maketform ('affine', [1 0 0; a 1 0; 0 0 1]);
B  = imtransform (A, T,'cubic','FillValues', orange);
T  = maketform ('affine', [1 0 0; a 1 0; 0 0 1]);
R  = makeresampler ({'cubic','nearest'},'fill');
B  = imtransform (A, T, R, 'FillValues', orange);
h2 = figure; imshow (B);
```

3.2 Ligature Detection and Classification

A certain pattern naturally takes place when human beings write words. For example, the movement of hand make slower while writing character and faster while writing connector (ligature) between characters. The length of the ligature differs between characters and within characters. This motivates us to consider the ligature classification. When character segmentation is extended to recognition of text images captured through mobile phone, it is not possible to store large amount of database and features for segmentation. This motivates us to consider the extraction of simple transition features that classifies ligature into inter-letter link or intra-letter link. Instead of segmenting directly on the binary image, we read the binary image as two-dimensional array. This reduces the complexity in segmenting. We can avoid merging of character pieces, which arises before validation of segmentation points.

3.3 Extraction of Transition Feature

Given an input binary image I(x, y), for x = 1, 2... width and y = 1, 2... height, we can extract transitions from white to black and black to white from the observation, as indicated by green box in the following Fig. 3.

Vertical scan over this two-dimensional array or matrix, the number of transitions from white to black and black to white is recorded. If the number of transition (NT) is greater than 2, it indicates the presence of a loop or semi-loop. With this information, Possible Segmentation Points (PSP) can be avoided in that area. In order to prevent the dissection from the body of the character, with NT equal to 2, stroke height (SH) is added as second constraint. The SH estimation is a two scan procedure. The first scan on each column of the histogram of the binary image calculates SH of the column by counting the number of black pixel runs in vertical direction. Then mean of the SH of all columns is taken as upper bound

1	1	0	0	0	0	1	1	0	0	0	1	1
1	1	1	0	0	1	1	1	0	0	1	1	1
1	1	1	0	0	1	1	1	0	0	1	1	1
1	1	1	0	0	1	1	1	1	0	1	1	1
1	1	1	1	1	1	1	1	1	1	1	1	1
0	1	1	1	1	1	1	0	1	1	1	1	1
0	0	0	1	1	0	0	0	1	1	0	0	0
0	0	0	0	0	0	0	0	0	0	0	0	0
0	0	0	0	0	0	0	0	0	0	0	0	0

Fig. 3 Binary image with transition features marked

(maximum height) for the run length of the strokes. The second scan discards those strokes, whose run length is greater than the maximum height. Finally SH of the input word image is estimated as the average height of the strokes in the second scan. The columns that do not satisfy those constraints above are converted to 1′s as shown by black lines in Fig. 4.

The black line in the image indicates PSP and the consecutive lines are grouped as segmentation blocks (SB) as shown in Fig. 4. These blocks indicate the ligature and are stored as SB array. In order to classify the ligatures as inter-letter link or intra-letter link, inter block space (IBS) is calculated. The IBS is the distance between consecutive blocks representing foreground pixel. These distances are stored in an array called IBS array. If the width in the IBS array is less than one third of the height of the image, it implies that a character is unlikely to fit in less than this width. So the immediate blocks are removed and the IBS array is altered accordingly. IBS array holds only the valid character width which is greater than one third of the height of the image. Those columns are extracted from IBS array and passed as parameters to segment the characters using the standard Java function, subimage (). Thus the proposed approach is unique by preventing over-segmentation of character such as ‘u’, ‘v’, ‘n’ and ‘k’ without recognition or training.

To avoid over-segmentation of ‘w’ and ‘m’ in which the width of the character is greater than or equal to 1.5 times of height of the image, ligature classification is done. Here, the inverse of IBS is taken. Threshold to classify the ligature is calculated as arithmetic mean of all the elements in SB array. If the block width is less than the threshold value, then it is inter-letter link, otherwise it might be intra-letter link.

The proposed approach works well for normal characters, without any alteration in region of interest as shown in Fig. 5a and b. There are cases which has touching boundaries as denoted by red dotted lines in Fig. 5d and dominance of ascender/descender as shown in the above Fig. 5c denoted by red solid lines. They may miss some valid segmentation points. For these cases, certain processing has to be done prior to ligature detection. Some challenging cases and the processing to be done are listed below:

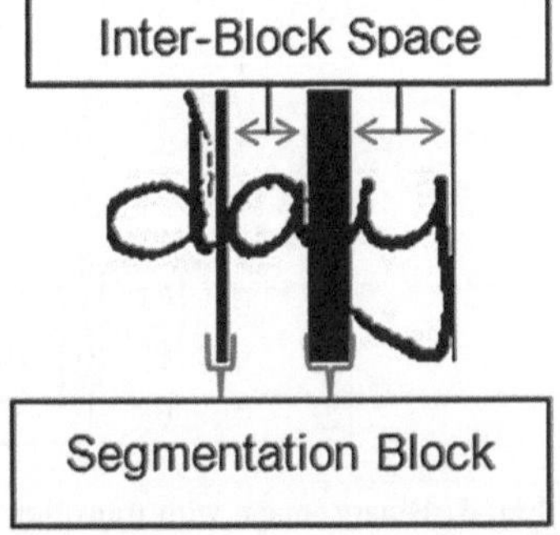

Fig. 4 Segmentation blocks and space between them

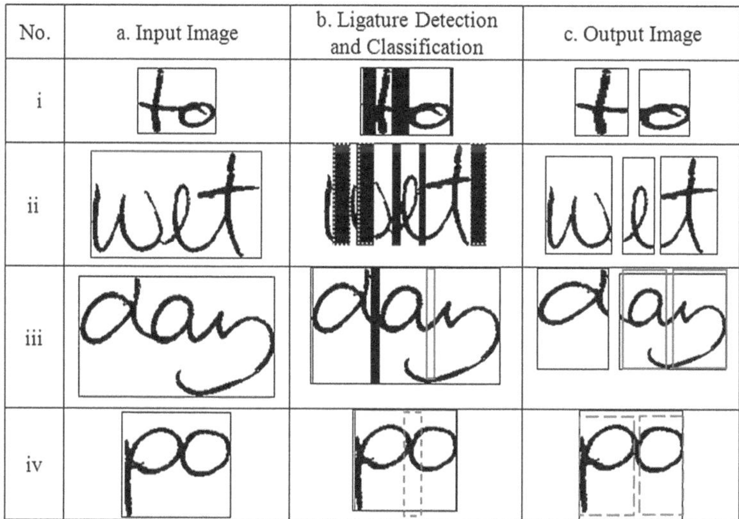

Fig. 5 Result of ligature detection and classification

(1) **Shadow Character Segmentation:** Shadow characters are characters, over which the extension of ascender and descender might share the columns of the other character. Supremacy of ascender and descender in the words might hide the valid segmentation points. To segment the shadow characters and locate hidden valid segmentation points, this prompts us to go for detection of core region. For the given input image, horizontal projection profile is constructed. Core region is plotted in the constructed profile, as the area in which the pixel distribution is greater than the threshold. Here the trimmed mean of pixel distribution is used as the threshold value. Now, the region of interest is restricted to core region, in which ligature detection and classification is done to capture the valid segmentation points hidden under ascender or descender.

(2) **Characters with Touching Boundaries:** Some characters may have touching boundaries, where the width of the character will be greater than the average character width and cannot be segmented directly using ligature classification. Here modified histogram is constructed for those characters. It is constructed by calculating the distance between the top and bottom foreground pixels for each column of the word image. Three fourth of average character width is subtracted from both the ends of the image and the low density region in this area is denoted as character boundary.

(3) **Broken Characters:** Broken characters are characters in which there might be absence of black pixels throughout the column. This occurs often in printed documents due to misprint, but rarely in handwritten documents. The system will consider this area as character boundary. In order to avoid such invalid segmentation points, explicitly the following condition is set—if the number of consecutive white runs is greater than 2, then it is marked as valid segmentation point, else as invalid segmentation point.

Table 1 Statistics about IAM database 3.0

IAM database 3.0	Writers	Pages	Sentences	Text lines	Words
	657	1,539	5,685	13,353	115,320

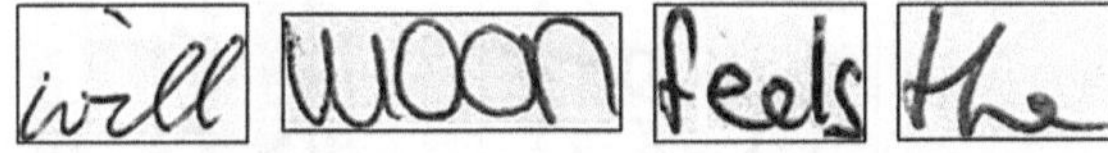

Fig. 6 Handwriting samples in IAM Database

4 Experiments and Results

4.1 Dataset

In this section, we present a quantitative analysis of our algorithm over a large database. The IAM handwriting database contains forms of handwritten English text which can be used to train and test handwritten text recognizers and to perform writer identification and verification experiments. The database contains forms of unconstrained handwritten text, which were scanned at a resolution of 300 dots per inch and saved as PNG images with 256 gray levels. Table 1 below shows some statistics of IAM Database 3.0. We have randomly selected 200 samples for our experiments. The Fig. 6 below shows some samples of words.

4.2 Implementation

We have implemented the proposed approach in Java on a PC with an Intel core 2 duo, 1.6 GHz, 2 GB RAM and Windows XP and evaluated its performance with 200 word images. The proposed approach for segmenting touching characters has to successfully solve single and multi touching problems. It shows that, even when dealing with broken characters and characters that do not contain loop or semi-loop like 'm', 'u', 'n', 'w' and 'v', the proposed approach can segment properly. The extraction of transition feature is not affected by stroke thickness; hence it is used for ligature classification. Some results of pre-processing stage are presented in Fig. 7.

After extracting the transition feature, the columns with NT less than or equal to 2 are marked as PSP. By validating the PSP with the second constraint, that is, the number of pixels in the column less than average SH, some of the PSP are removed. PSPs are generated based on transition feature along with SH analysis. Afterwards, all PSPs are validated using ligature classification. Numerous incorrect segmentation points are rejected accurately and correct segmentation points are left as validated segmentation points as shown in Fig. 8. Ligature classification shows promising accuracy in normal writing style.

Fig. 7 Word images after pre-processing stage

Fig. 8 Sample output of the approach for normal cases

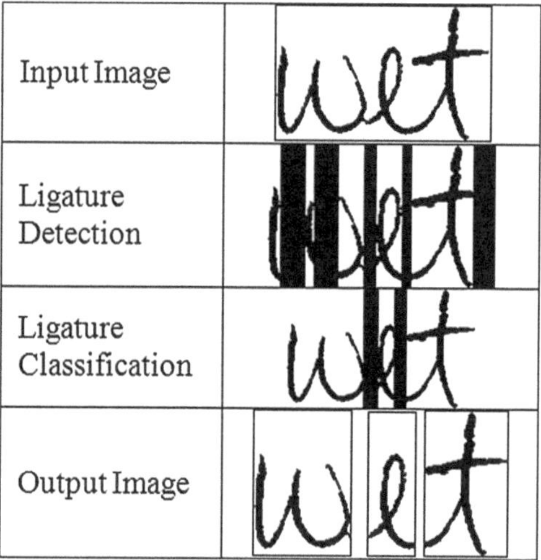

Figure 9 shows ligature classification for challenging cases. Here, the dominance of descender from the letter 'z' hides the valid segmentation point between letter 'a' and 'z'. Hence, the region of interest is restricted to core region, followed by ligature detection and classification to capture the hidden valid segmentation points.

4.3 Analysis and Discussions of Result

Segmentation algorithm based on transition feature in conjunction with ligature detection and classification achieved an accuracy rate of 92 %. It is shown that using simple transition feature, the segmentation rate is better. Proposed ligature detection can provide enough possible segmentation points for further feed to ligature classification to validate those points. Table 2 shows the comparison with another classical approach.

Table 2 Comparison of results

Method	No. of words	No. of words segmented correctly	Percentage of accuracy (%)
Projection profile	200	162	81
Proposed method	200	184	92

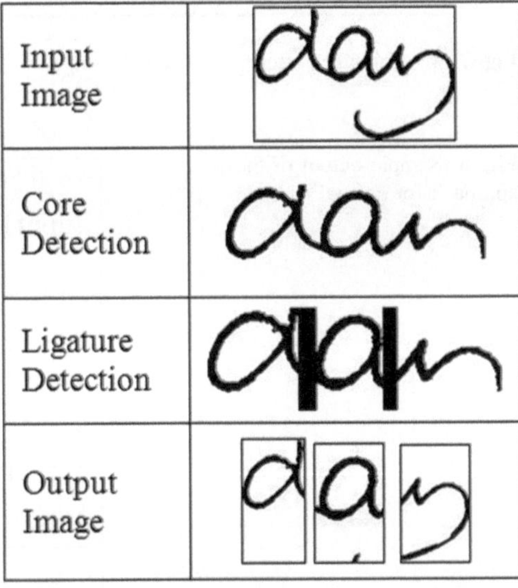

Fig. 9 Sample output of the approach for challenging cases

Three problems are found during segmentation. Firstly, as seen in Fig. 5b segmentation based on transition feature over-segmented the character ‘m’, ‘n’ and ‘u’. This problem is common, since classical segmentation approach cannot distinguish its context. It can be solved using intelligent technique to validate segmentation points as drawn above. Secondly, touched/overlapped characters caused missing of segmentation points. This type of problem is very hard to deal with. When two characters are tied together, ligatures cannot be found and therefore they cannot be segmented. In such cases, considering those two characters alone, the modified histogram is used to locate the ligature that is shared by both the characters. Finally, shadow characters caused missing of valid segmentation points due to domination of ascender and descender. Shadow character problem is an extraordinary case in handwriting domain. Conversely, if it exists in a handwritten word, then it might disturb the character segmentation method and lead to incorrect segmentation of ascender/descender, which has less impact in recognition. But there is no significant loss in core region. Hence overall correct segmentation results are not affected.

Figure 10 depicts the result of ligature detections on word samples with shadow characters. Hidden core region is denoted by the rectangle box. Ligature detection is also carried out in Fig. 10. It shows that the proposed approach leads to some

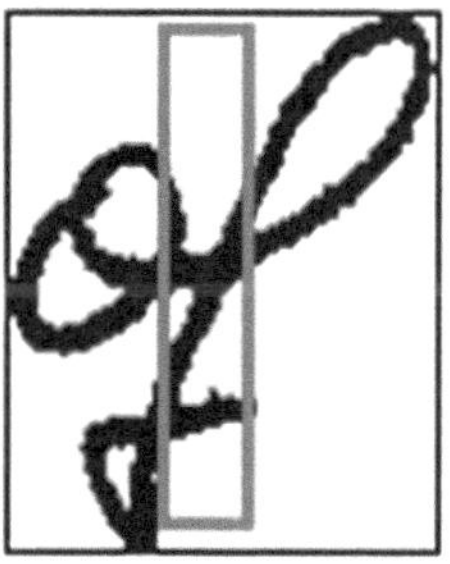

Fig. 10 Ligature detection on shadow characters

insignificant loss in ascender/descender. We are actively investigating to improve the result.

5 Conclusion and Future Works

We presented a classical approach based on ligature detection and classification. Transition feature is used for locating the ligature. Since the extraction of transition feature has less computational complexity, the overall complexity is reduced. The statistical analysis on width of the ligature classifies it into inter-letter link and intra-letter link. Comparative experimental results using IAM database demonstrate the efficiency of the approach. In the future work, we plan to extend our methodology for better segmentation accuracy by getting feedback from recognition phase.

References

1. Liu Z, Cai J, Buse R (2003) Handwritten recognition, soft computing and probabilistic approaches. Springer, Berlin
2. Madhavanath S, Govindaraju V (2001) The role of holistic paradigms in handwritten word recognition. IEEE Trans Pattern Anal and Mach Intell 23(2):1433–1446
3. Casey RG, Lecolinet E (1996) A survey of methods and strategies in character segmentation. IEEE Trans Pattern Anal Mach Intell 18(7):609–706
4. Dzulkifli BM, Fajri K Off-line character segmentation technique for handwritten cursive word—A survey
5. Ribas FC, Oliverira LS, Britto AS, Jr Sabourin R (2012) Handwritten digit segmentation: A comparative study. IJDAR. doi: 10.1007/s10032-012-0185-9
6. Xu L, Yin F, Liu C-L (2010) Touching character splitting of Chinese handwriting using contour analysis and DTW. IEEE 978-1-4244-7210-9/10
7. Louloudis G, Tamatopolous N, Gatos B (2009) A novel two stage evaluation methodology for word segmentation techniques. In: Proceedings of the 10th international conference on document analysis and recognition IEEE. doi: 10.1109
8. Makridis M, Nikolaou N, Gatos B (2007) An efficient word segmentation technique for historical and regarded machine-printed documents. In: 9th international conference on document analysis and recognition IEEE

9. Markus F, Klaus D (2003) Word segmentation of handwritten dates in historical documents by combining semantic a priori-knowledge with local features. In: Proceedings of the 7th international conference on document analysis and recognition IEEE
10. Lecolinet E, Jean-Pierre C A grapheme-based segmentation technique for cursive script recognition
11. Kang L, David D Local segmentation of touching character using contour based shape decomposition
12. Ghazali S, Amjad R, Tanzila S (2010) Improved offline connected script recognition based on hybrid strategy. Int Eng Sci Technol 2(6):1603–1611
13. Tripathy N, Pal U (2006) Handwriting segmentation of unconstrained oriya text. Sadhana 31(6):755–769
14. Nicchiotti G, Scagliola C, Rimassa S (2000) A simple and effective cursive word segmentation method. In: Proceedings of the 7th international workshop on frontiers in handwriting recognition, pp 499–504
15. Sauvola J, Pietikainen M (2000) Adaptive document image binarization. Pattern Recogn 33:225–236

Detection of Digital Photo Image Forgery Using Copy-Create Techniques

S. Murali, Govndraj B. Chittapur and H. S. Prabhakara

Abstract Picture says thousands words and believe million of people in trusted media like television, news paper and Internet where every second photo images are published in social networks. Unfortunately what we seeing are not believable due to the advance photo editing and manipulation tools are available in digital media art world. We propose trust model to detect popular copy-create and copy-move forgery techniques in common image formats available in digital camera. We succeeded to detect image tampering in given known and unknown image reference source for detectable anomalies in photo image.

Keywords Picture · Copy-create · Copy-move · Image tampering unknown source · And known source

1 Introduction to Image Forgery

Photographic forgeries aren't new. People have doctored images since the beginning of photography inception. In society today, we're now seeing doctored images

S. Murali · H. S. Prabhakara
Maharaja Research Foudation, Maharaja Institute of Technology, Mysore, India
e-mail: nymurali@yahoo.com

H. S. Prabhakara
e-mail: prabhassan@gmail.com

G. B. Chittapur (✉)
Basaveshwar Engineering College, Bagalkot, Karnatka State, India
e-mail: govindraj_bec@yahoo.com

Mohan S. and S. S. Kumar (eds.), *Proceedings of the Fourth International Conference on Signal and Image Processing 2012 (ICSIP 2012)*, Lecture Notes in Electrical Engineering 221, DOI: 10.1007/978-81-322-0997-3_26,

Fig. 1 Pakistan's press information department distributed a photo of Prime Minister Yousaf Raza Gilani on January 2012 photo image (**b**) from original image from (**a**)

regularly. While these images might tarnish the public opinion of a celebrity, cases involving manipulated images with more serious implications have arisen in science and law. The art of making an image forgery is as old as photography itself. In its early years, photography quickly became the chosen method for making portraits, and portrait photographers learned that they could improve sales by retouching their photographs to please the purchaser [1]. Though photo manipulation has become more common in the age of digital cameras and image editing software, it actually dates back almost as far as the invention of photography. For recent years, an exhaustive inventory of every photo manipulation would be nearly impossible, so we focus here on the instances that have been most controversial or notorious, or ones that raise the most interesting ethical questions[1]. The photographers of the era also experimented with composting, i.e., combining multiple images into one. An early example of composting appears in the top panel of Figs. 1 [9] and 2 [9].

Recently Pakistan press information Department distributed a photo of Prime Minister Yousaf Raza Gilani speaking with two generals in which his fingers were extended in an awkward pose, and partially clipped. Further investigation determined that the original photo showed him holding a teacup, which had been removed prior to publication of the photo. The department eventually replaced the modified photo with the original.

Digital images offer many attributes for a tamper detection algorithm to take advantage of, specifically the color and brightness of individual pixels as well as an image's resolution and format. These properties allow for analysis and comparison between the fundamentals of digital forgeries in an effort to develop an algorithm for detecting image tampering. This paper focuses on images which are

[1] www.fourandsix.com/photo-tampering-history/category/2012

Fig. 2 Recent forgery photo published in popular news media and magazine in respective months of year 2012

created with known source of origin and unknown source of origin with making forgery of composition of multiple images known as copy-create and part of image is pasted in the photo image known as copy-move images.

2 Related Work on Image Forgery

This section introduces the techniques and methods currently available in the area of digital image forgery detection. A survey of the current research is presented as well as an analysis of the current techniques and methods available to detect image tampering.

Currently, most acquisition and manipulation tools use the JPEG standard for image compression. As a result, one of the Standard approaches is to use the blocking fingerprints Introduced by JPEG compression, as reliable indicators of possible image tampering. Not only do these inconsistencies Help determine possible forgery, but they can also be used to Shed light into what method of forgery was used. Many passive Schemes have been developed based on these fingerprints to detect re-sampling [2], copy paste [3, 4], Luminance level techniques [5, 10], Double Compression JPEG image [6], ANN [7], Wavelet Transformation Coefficient [8].

Above mentioned methodologies are diverted from one technique to another and those are all contain constrain with implementing results, more-ever they are not considered digital photo image as input and most of the digital images are photos in the real life. We propose and concentrating on media photo images to develop an effective algorithm for detecting the forgery region in most popular image format JPEG and other digital camera supported Image formats.

3 Methodologies and Implementation of Detecting Digital Photo Image Forgery

Photo Image forgery is classified into two categories, The first class of image forgeries includes images tampered by means of copying one area within an image and pasting it onto another it is called as Copy-Move Forgery, The second class of forged images deals with creating the forgery by using more than just the single image for copying and pasting. This is done by taking one or more images and copying and pasting from various areas within each to form a forged image. The image processing community formally refers to this type of image as an image "composition," which is defined as the "digitally manipulated combination of at least two source images to produce an integrated result" it is called as Copy-Create Image Forgery. Now we are proposing Copy-create image forgery detection. Many algorithms are there for detecting forgery in copy-move type of image forgeries.

3.1 Pixel Substraction

Initially the original image is read and viewed; fallowing which is the forged image, which is again read and viewed. Once both the images are viewed further both the images undergo pixel subtraction where in a subtraction operation is performed on the images. The subtraction is performed in pixel-by-pixel fashion i.e. each and every pixel undergoes subtraction. On subtraction of images if the pixel value of both the images will be equal then that area is masked in black. If there is any difference in the pixel value between the two images then that area/ region can be identified to be tampered. The result viewed will show a black masked area where there is no difference between the two images i.e. no difference between the pixel value of original and forged images. And the area where there is pixel difference will be highlighted which was a tampered area is been suspected; mathematical model for pixel substraction is given as: Let f (x) is the original image and let f (y) is the forged image. On using pixel subtraction we obtain

$$f(z) = f(x) - f(y) \quad (1)$$

where
f (z) is the suspected and identified forged region.

3.2 Block Matching

The input image of size an x b is divided into 'n' blocks of size m x m pixels by moving the block point to point on the image. Each block is iteratively compared to every other block in the image. In case of complete match both blocks are marked as copied. In case of copy detection, the adjacent neighbors of the marked blocks are then compared. The algorithm confirms the manipulation if at least three blocks in the adjacent neighborhood of the both marked blocks is exact match of each other. In case of colored images, the image is first converted into a matrix of size 3 times an x b by appending the RGB values of each pixel. The copy forge detection approach is subsequently applied on this modified matrix and copied blocks are marked.

3.3 Region Masking

Initially the original 'n' forged image are read and viewed, the forged image is assigned to another variable for security purpose. Further on giving both the original and forged image as input, we keep on checking the image or comparing the image for each region i and j until it reaches the size of the image. If any part of i or j of original image is not equal to i or j of forged image then that area is masked i.e. that area is said to be tampered and hence forgery is detected. If there is no difference found between the two images then there is no forgery taken place in the image.

3.4 Lighting Variation

The first step is to convert an input image of a digital to a grayscale for detecting the lighting variation in the image. The grayscale image can be analyzed by a relationship between a white area and a black for the input image, and made to faster processing than a color image. The second step is to use the filter in order to accurately find the lighting difference from the digital image and remove a noise of the image. The filter is to adequately apply a mask of (e.g., $3 \times 3, 5 \times 5, 7 \times 7, 9 \times 9$)

according to a texture of a variety of images. The image could be eliminated to a steep intensity transition of the object except the lighting gradation formed naturally the objects from the image.

The third step is to adapt a size high contrast from the image. The contrast should be clearly the highlights and shadows formed by the light from the image. It is therefore easy to detect the lighting variation in the image. The fourth step is to convert 256 levels to 10 levels for the 8-bits image for level segment purpose. A simplification of level might be made a change an abrupt tone reproduction from an infinitesimal tone of the object. The fifth step is a process to make a 3-dimensional graph of a virtual image converting the brightness values to a height of the image plane for each pixel from the image.

4 System Boundaries and Experimental Design

All of the methods presented are tailored to the JPEG and other image formats like BMP (Window Bitmap), PNG (Portable Network Graphics) image format. The best way to detect image tampering, in general, is to perform a wide range of tests on the image in question. The methodology presented in this chapter attempts to accomplish this by including a broad range of detection techniques. Table 1 includes a listing of the proposed detection methods and those image formats that work with each.

Table 2 includes the tests conducted with a description for each of the images in the blind image test-bed. The images in this test-bed represent expert level forgeries.

This paper focuses on methods to detect digital forgeries created from multiple images. *Copy-create* image forgeries are Images that result from portions copied and moved within the same image to "cover-up" something are classified as *copy-move* forgeries. Therefore, the experimental design and analysis herein focuses on *copy-create* image forgeries with pixel Substraction, Region masking and lighting variation although copy move forgeries with Block matching Technique.

Table 1 Comparison between proposed detection methods applied for various tested image format

Forgery detection method	RGB color	Grayscale	JPEG	PNG	BMP
Known image source: *successful execution					
Pixel Subtraction	*	*	*	*	*
Region masking	*	*	*	*	*
Block matching	*	*	*	*	*
Unknown image source: *successful execution					
Lighting variation	*	*	*	*	*

Table 2 Comparison of proposed method with the context of design of metric measurements

Experiment	Detection method	Metric
(1)Pixel subtraction	Image subtraction method	Measure and compare the pixel by pixel in test image to identify the difference pixel with original test image and identify any discrepancy, if any exist.
(2) Lighting variation	Luminance and lighting variant level	Measure perceived variability in luminance levels in test image. Identity and document areas of luminance discrepancy, if any exist.
(3)Block matching and region masking	Block matching and region masking	Measure perceived variability in BLOCK and REGION in test image. Identity and document areas of similar block or region discrepancies, if any exist.
(4)Blind image test	All	Execute each detection method on test image to Identity and document abnormal areas, if any exist.

5 Result

The proposed approach has been evaluated using datasets containing different types of tampered images. The whole test data consists of 300 natural scene original images. Images are taken from digital cameras and collected from web. Programs are executed in Matlab7.8 with dual core processor machine. Image tampering is done with advance photo editing software like Adobe Photoshop, Microsoft paint program and gimp and other photo editing tools (Figs. 3, 4, 5, 6).

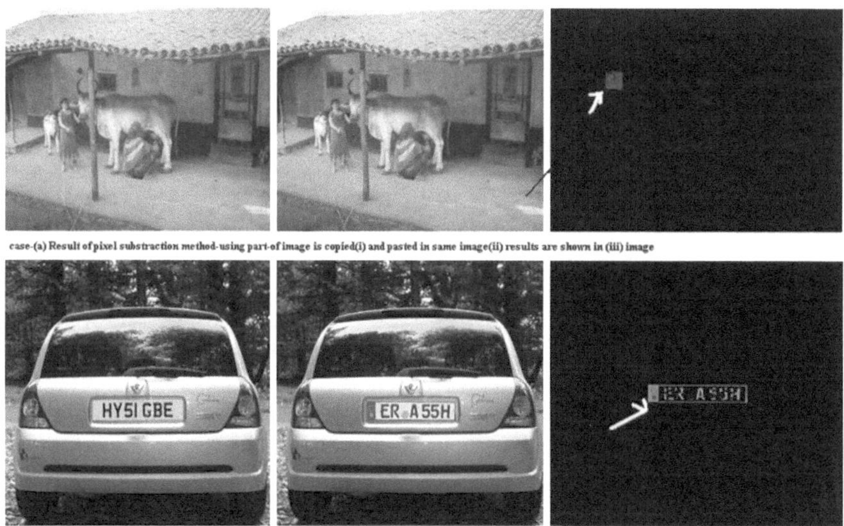

Fig. 3 Result of pixel substraction

Fig. 4 Result of block matching

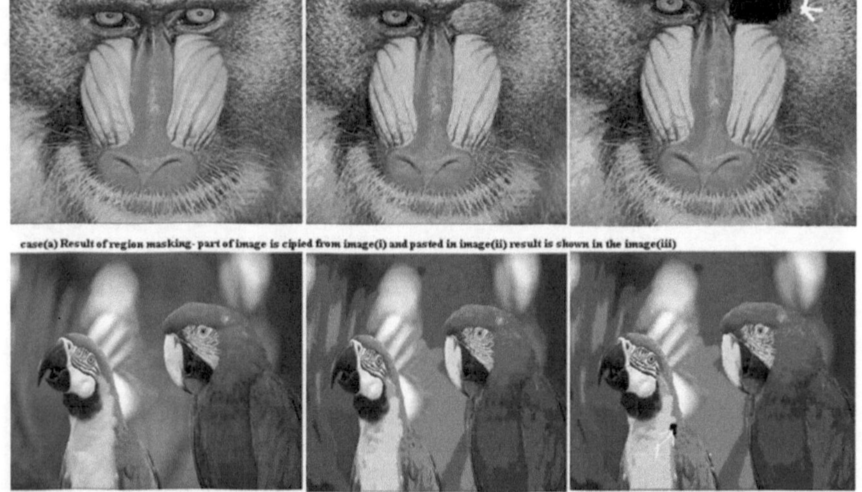

Fig. 5 Result of region masking

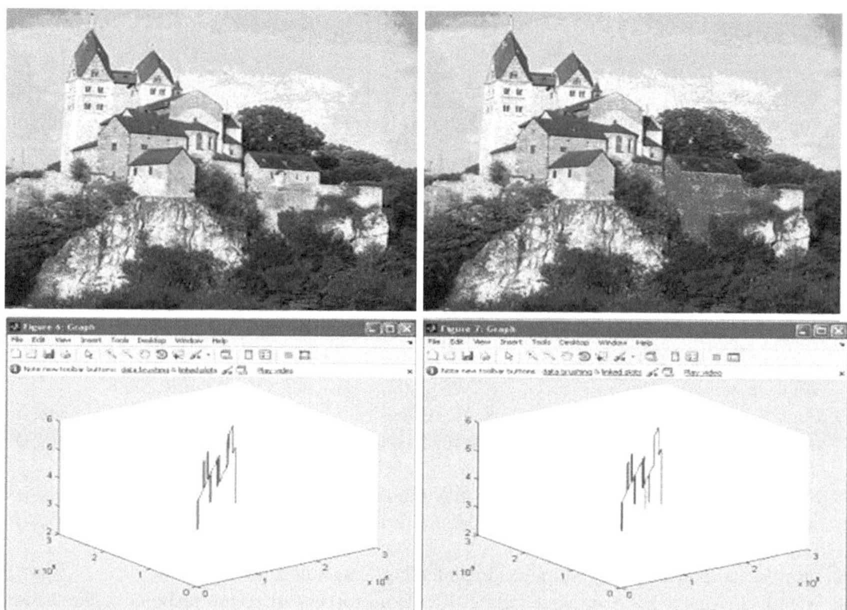

Fig. 6 Result of lighting vitiation

6 Conclusion

This Paper outlined the methodology used in detecting digital image tampering from images with known and unknown origin. While it is difficult to predict exactly how a malicious person will forge an image, a wide range of techniques have been presented to account for tamper methods. The detection system proposed includes methods based on Pixel subtraction, Region masking with the help of known image source, and Block matching each pixels Luminance and Brightness levels of an image, lighting variation mainly used with unknown image source. An experiment testing this method has been set-up that will help in determining the accuracy and correctness of the proposed tamper detection techniques, as well as when each fails. It has been conjectured that these methods will help in detecting various types of image forgeries, but one has to acknowledge that no "silver-bullet" exists to account for every type of forgery imaginable. To wrap up testing, an independent experiment is presented to help analyze the correctness of this system of techniques at accurately identifying blind images as authentic or forged.

References

1. Baxes GA (1994) Digital image processing: principles and applications. Wiley, New York
2. Popescu AC, Farid H (2005) Exposing digital forgeries by detecting traces of re-sampling. IEEE Trans Sig Process 53(2):758–767
3. Ng T, Chang SF, Sun Q (2004) Blind detection of photomontage using higher order statistics. In: IEEE international symposium on circuits and systems, Canada, May 2004
4. Popescu AC, Farid H (2005) Exposing digital forgeries by detecting traces of re-sampling. IEEE Trans Sig Process 53(2):758–767
5. Murali S, Anami Basavaraj S, Chittapur Govindraj B (2011) Detection of digital imager forgery using luminance level techniques. In: IEEE third national conference of computer vision, pattern recognition, image processing and graphics, NCVPRIPG, p 215
6. Fredric J, Lukas J (2003) Estimation of primary quantization matrix in double compressed JPEG images. In: Proceedings of DFRWS 2003, Cleveland, OH, August 2003
7. Gopi ES, Lakshmanan N, Gokul T, KumarGanesh S (2006) Digital image forgery detection using artificial neural network and auto regressive coefficients EEE/CCCGET, Ottawa, May 2006 1-4244-0038-4
8. Sutch Y, Coskum B, Sencar HT, Memon N (2003) Temper detection based on regularities of wavelet transform coefficients. Polytechnic University, Electrical Computer Engineering Department, Brooklyn Ny 11201, USA
9. http://ivms.stanford.edu/~varodayan/mentorship/GargHailuSridharan.pdf
10. Murali S, Anami BS, Chittapur GB (2012) Photo forgery detection techniques. Int J Mach Intell 4(1):405. ISSN 0975-2927

An Enhanced Intrinsic Biometric in Identifying People by Photopleythsmography Signal

N. S. Girish Rao Salanke, N. Maheswari and Andrews Samraj

Abstract In the area of secure authentication, the fusion of Photopleythsmography (PPG) signals for biometric identification is a novel technique. Researchers suggested the use of PPG along with other biometric components for augmenting the biometric robustness. PPG signals have great potential to serve as biometric identification appliance and can be easily obtained with low cost. Use of PPG signals for personnel identification is very appropriate during field operations in day or night. While building a large scale identification system the feature selection from PPG is a critical activity. To have the identification system more accurate, the set of features that deemed to be the most effective attributes are extracted in order to build robust identification system. Applying Kernel Principal Component analysis (KPCA) an efficient supervised learning method for dimensionality reduction and feature extraction in this experiment results in precise classification.

Keywords PPG signal · Authentication · Kernel principal component analysis (KPCA) · Mahalanobis distance

N. S. Girish Rao Salanke (✉)
Research Scholar, School of Computing Science and Engineering,
VIT University, Chennai, India
e-mail: girishraosalanke@gmail.com

N. Maheswari
School of Computing Science and Engineering,
VIT University, Chennai, India
e-mail: maheswari.n@vit.ac.in

A. Samraj
Advance Science and Technology Research Center, Salem, India
e-mail: andrewsmalacca@gmail.com

Mohan S. and S. S. Kumar (eds.), *Proceedings of the Fourth International Conference on Signal and Image Processing 2012 (ICSIP 2012)*, Lecture Notes in Electrical Engineering 221, DOI: 10.1007/978-81-322-0997-3_27, © Springer India 2013

1 Introduction

In search of robust biometrics that is not compromised, researchers started to try with unseen internal organs rather than conventional outer organs with all their varying dimension and activity. They seems to be more reliable types of biometrics in human authentication and identification systems than the simple ones used in applications like personnel security, hospital, finance, digital rights etc. conventional biometric system [1] using behavioral and physiological characteristics to allow recognition of an individual, like face recognition [2], fingerprint [3], iris and retinal patterns [4] and voice [5] are becoming very popular. The common weakness of conventional biometric system is their vulnerability and the possibility to falsity of their features. In 1977, Forsen listed the possible features of a human body that could be used for biometrics were ECG [6] and EEG [7]. The basic approach is to extract the signals from the user during the enrollment phase, the extracted signals from the user then attempts to log in, the particulars are computed, which determines the decision outcome. This is really no different than any other behavioral biometric, the novelty here is the data that are acquired. In order to acquire biological signal data, specialized hardware is required. Especially the EEG and ECG acquisition hardware is complex and tedious processes is involved. In this paper, we examine the use of Photoplethysmograph (PPG) signals, which is acquired using a very simple hardware interface for human identification [8]. PPG has distinct advantages including low development cost, easy to use without any complicated procedure and conveniently accessible to various sites of human body, such as finger, wrist or arm. Moreover, PPG signals provide a noninvasive and accurate methodology to obtain valuable physiological information such as blood oxygen saturation, heart rate and blood flow.

The blood in human body is being pumped from the heart to all parts in the body by blood vessels called arteries. Blood pressure is the force of blood pushing against the walls of the arteries. Each time the heart beats it pumps out a considerable volume of blood to the arteries. Systolic pressure which is the highest blood pressure occurs when heart is in pumping motion. Diastolic pressure is lowest blood pressure when heart is in resting [9]. Since blood pressure are an indirect measurement of heart beats and the blood pressure tend to change according to time and emotion. For instance, blood pressure will rise when a subject is awaken and excited. The unit for measurement of blood pressure is in mmHg and the notation will be systolic followed by diastolic pressure.

The Photoplethysmograph (PPG) signals reflect the change in blood volume caused by blood vessel expansion and contraction, which can be detected by photodiode if external light is illuminated into tissue. This method is based on the idea that if an externally applied pressure in the cuff is equal to the arterial pressure instantaneously, the arterial walls will be unloaded (zero transmural pressure) and the arteries will not change in size. In this condition, the blood volume will not change.

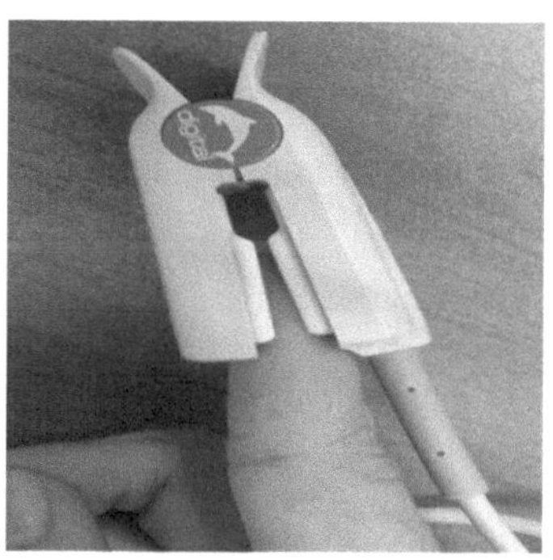

Fig. 1 Finger nail sensor to measure PPG signal

Digital Photoplethysmographic sensor used to read a PPG signal has an infrared emitter and photodiode detector. The intensity of light from infrared emitter which reaches the photodiode detector in either reflection or transmission will be measured to determine the blood volume changes. The Photoplethysmographic sensor will be placed below the tip of the finger and pressure will be applied on the proximal phalanx. Since the cuff is wrapped on the proximal phalanx of the finger rather than arm, it makes less discomfort for prolonged used. The blood volume changes on the finger will be notified by the sensor and transmitted to the system by wired or bluetooth transmitter. There will be a timer attached on the device to time the cuff pressure applied time and transmitter so that it is semi-continuous measurement. This is because to decrease the energy consumption of Bluetooth transmission and to avoid discomfort of the wearer's finger due to continuous pressure applied.

The idea of the blood pressure sensor is much like the Photoplethysmograph Fingernail Sensor for measurement of finger force which is mentioned by Stephen A. Mascaro and H. Harry Asada [10]. The finger force is measuring colouration of fingernail by using reflectance Photoplethysmography. The Fig. 1 shows the fingernail sensor used in this research.

2 Related Work

A brief survey of the related work in the area of identification using PPG signals is presented in this section. Earlier work by Y. Y Gu et al. [11, 12] on biometrics with PPG signals used to represent the pulse using four quantities; the peak number, the upward slope, the downward slope and the time interval from the bottom to peak. Jianchu Yao et al. [13] examines two important issues in applying derivatives of PPG signals as discriminants to identify and verify subjects. Andrews Samraj et al. [8] demonstrated the intra trial variability factor which enhanced the signature verification system that uses PPG Signal as one of its combination. Singular value decomposition (SVD) was used as a dimension reduction tool. In this paper, we propose a technique for human identification by applying kernel principal component analysis for dimension reduction.

3 Methodology

The methodology of this experiment is carried out in four stages as follows

(i) Experimental Setup
(ii) Preprocessing
(iii) Feature Extraction
(iv) Classification

3.1 Experimental Setup

The PPG signals were recorded using a Pulse Oximetry module (from Dolphin Medical, Inc) that was connected to a computer using a data acquisition software and captured with a sampling frequency of 37 Hz [14]. The main purpose of sampling the signal at this particular frequency is to highlight that even at this lowest frequency we were able to capture the signal. Sampling at a higher frequency is very fine and accurate and may be considered later on further progress. Four led's of Plethysmogram were fixed on each subject while recording the data. A total of 9 healthy subjects participated in this study. All the 9 subjects selected were our co- scholars at VIT University and the average age of the subjects is 34. Two PPG recordings (relaxed and stressed) were obtained from each individual with duration of 60 s. Total of 8 signals were recorded for each subject, 4 in relaxed sate and remaining 4 in the stressed state. Before recorded the stressed PPG signal the subjects were made to go up and down the staircase for a couple of times in order to increase their level of stress. There was about 30 min gap between the time of recording the relaxed and the stressed PPG signal. Figures 2, 3 shows PPG signal recorded during the relaxed and the stressed state. The aim for recording the relaxed and the stressed state signal is to detect whether there is significant fluctuation in the PPG signal features of a person in both the states. The relaxed PPG signals of all subjects shows identical waveforms but stressed PPG signals shows unidentical waveforms due to motion artifact. The same signal was used for training and the testing purpose. Both the recordings were taken separately, once when the subject was very much relaxed and the other time when he was stressed.

3.2 Preprocessing

The Preprocessing of the PPG signal consists of Peak detection, Segmentation and finally Scaling. In peak detection, we find the maximum peak of the recorded PPG

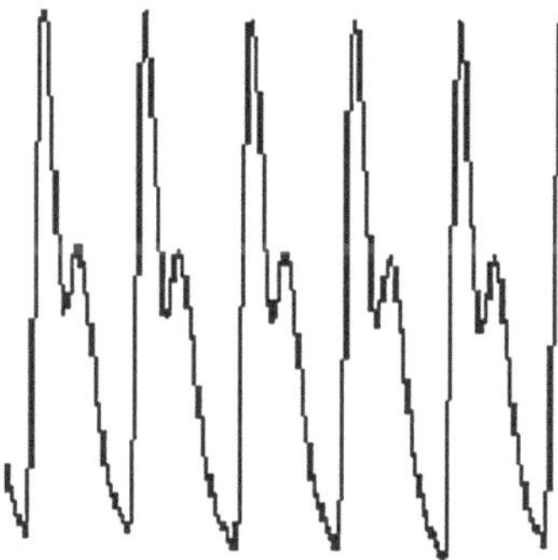

Fig. 2 Relaxed PPG Waveform sampled at 39 Hz

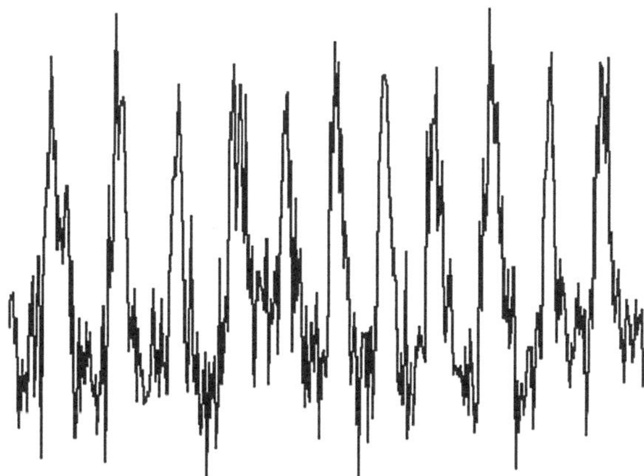

Fig. 3 Stressed PPG Waveform sampled at 37 Hz

signal. A point is considered a maximum peak if it has the local maximal value and is preceded by another peak with their amplitude difference larger than a preset threshold. In Segmentation we define one PPG cycle with the help of the maximum and minimum points obtained through peak detection. In Scaling since one PPG cycle is nothing but a heart rate, and generally heart rate can potentially increase intra-subject variability, the time duration is scaled to 100 samples.

3.3 Feature Extraction

Kernel Principal Component Analysis (KPCA) is used as dimension reduction technique which is used for feature extraction. Kernel PCA is the reformulation of traditional linear PCA in a high-dimensional space that is constructed using a kernel function. In recent years, the reformulation of linear techniques using the 'kernel trick' has led to the proposal of successful techniques such as kernel ridge regression and Support Vector Machines. Kernel PCA computes the principal eigenvectors of the kernel matrix, rather than those of the covariance matrix. The reformulation of PCA in kernel space is straightforward, since a kernel matrix is similar to the in product of the data points in the high-dimensional space that is constructed using the kernel function. The application of PCA in the kernel space provides Kernel PCA the property of constructing nonlinear mappings. Kernel PCA computes the kernel matrix K of the datapoints $\boldsymbol{x}_i$. The entries in the kernel matrix are defined by

$$k_{ij} = \kappa(x_i, x_j) \tag{1}$$

where $\boldsymbol{\kappa}$ is a kernel function, which may be any function that gives rise to a positive-semi definite kernel K. Subsequently, the kernel matrix K is double-centered using the following modification of the entries

$$k_{ij} = -\frac{1}{2}(k_{ij} - \frac{1}{n}\sum_l k_{il} - \frac{1}{n}\sum_l k_{jl} + \frac{1}{n^2}\sum_{lm} k_{lm}) \tag{2}$$

The centering operation corresponds to subtracting the mean of the features in traditional PCA. It subtracts the mean of the data in the feature space defined by the kernel function $\boldsymbol{\kappa}$. As a result, the data in the features space defined by the kernel function will have zero-mean. Subsequently, the principal eigenvectors $\mathbf{v}_i$, of the centered kernel matrix are computed. The eigenvectors of the covariance matrix $\mathbf{a_i}$ (in the feature space constructed by $\boldsymbol{\kappa}$) can now be computed, since they are related to the eigenvectors of the kernel matrix $\mathbf{v}_i$ and eigenvalue $\boldsymbol{\lambda_i}$ through

$$a_i = \frac{1}{\sqrt{\lambda_i}} v_i \tag{3}$$

In order to obtain the low-dimensional data representation, the data is projected onto the eigenvectors of the covariance matrix $\mathbf{a}_i$. The result of the projection (i.e., the low-dimensional data representation Y) is given by

$$y_i = \left\{ \sum_{j=1}^{n} a_1^{(j)} \kappa(x_i, x_j), \ldots, \sum_{j=1}^{n} a_d^{(j)} \kappa(x_i, x_j) \right\} \tag{4}$$

where $\mathbf{a}_i^{(j)}$ indicates the jth value in the vector, $\boldsymbol{a_1}$ and $\boldsymbol{\kappa}$ is the linear kernel function that was also used in the computation of the kernel matrix.

3.4 Classification

Mahalanobis distance based classifier is used to identify the individuals. The use of the Mahalanobis metric removes several of the limitations of the Euclidean metric like automatically accounts for the scaling of the coordinate axes, corrects for correlation between the different features and provides curved as well as linear decision boundaries. The value of r in the given below equation

Table 1 The intra subject variation in terms of Mahalanobis distance

Mahalanobis distance			
Subjects	Relaxed state	Stressed state	Intra variation
1	98.30	114.50	16.20
2	594.70	606.70	12.00
3	1119.20	1136.10	16.90
4	2116.80	2132.40	15.60
5	1009.10	1029.50	20.40
6	3110.60	3142.30	31.70
7	5099.20	5124.90	25.70
8	7130.20	7157.80	27.60
9	8094.10	8112.70	18.60

$$r^2 = (x - m_x)^I C_x^{-I} (x - m_x) \tag{5}$$

is called the **Mahalanobis distance** from the feature vector **x** to the mean vector $\mathbf{m}_x$, where C_x is the covariance matrix for **x**. It can be shown that the surfaces on which r is constant are ellipsoids that are centered about the mean $\mathbf{m}_x$. In the special case where the features are uncorrelated and the variances in all directions are the same, these surfaces are spheres, and the Mahalanobis distance becomes equivalent to the Euclidean distance.

We first generate a matching score by the selected feature compared with the stored template. For each PPG Cycle, the mahalabonis distance is calculated and is compared with the set of samples stored in the system is computed. The template resulting in the smallest distance is considered to be the match.

4 Results and Discussion

The difference in the Mahalanobis distance between the relaxed and the stressed state (Intra Subject Variation) for each subject is calculated. We observe that the mahalanobis distance of the relaxed state is less when compared to the stressed state. The table given below shows the distance for each subject for the maximum value obtained out of 30 samples extracted during the relaxed and the stressed state (Table 1).

We observed that the Intra Variation of all the subjects is very less and the average Inter Variation is 20.52. Figure 4 shows the Intra Subject Variation in Mahalanobis distance for one particular subject across 10 different samples. The Inter subject variation (calculated from the average values between the relaxed and the stressed state), the difference between the mahalanobis distance across different subjects is shown in Figs. 5, 6 reveals the wide gap between the Intra trial variability, this factor can be used as a novel method of identifying people using Photoplethysmographic signal.

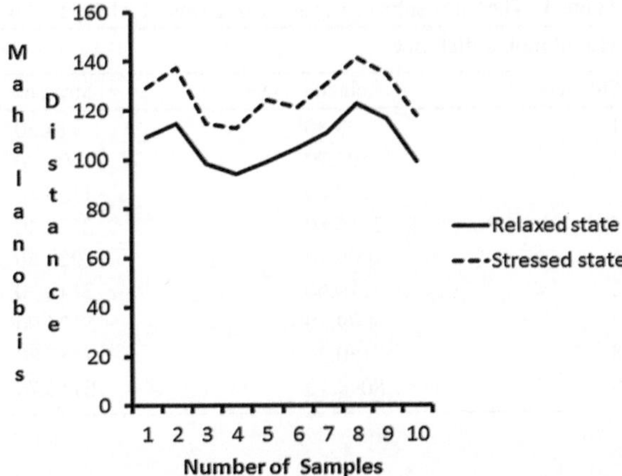

Fig. 4 Intra subject variation in terms of Mahalanobis distance

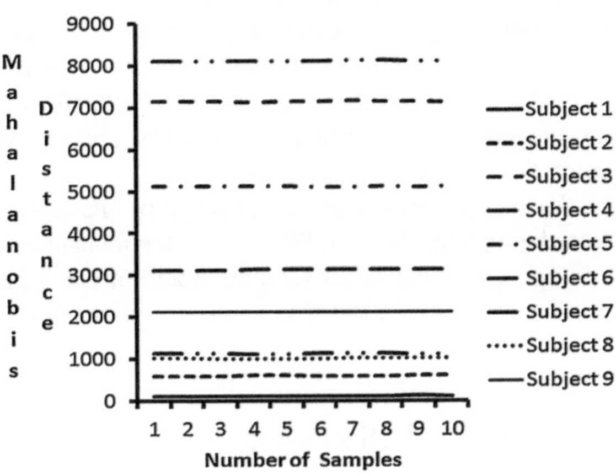

Fig. 5 Inter subject variation in terms of Mahalanobis distance

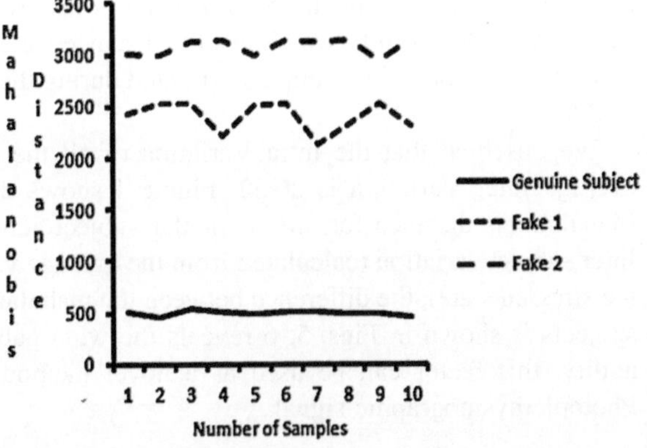

Fig. 6 Mahalanobis distance between a Genuine and Fake subject

5 Conclusion

We have presented a novel approach in this paper with the possibility of using Photoplethysmographic signal in the human identification. We observe that the stressed PPG signal from all the subjects shows distinct waveform when compared to their relaxed state that results in the difference in the mahalonobis distance, which can be used for the verification purpose. The most salient feature in using the Photoplethysmographic signal as a biometric is that it cannot be easily copied or simulated, as compared to other biometrics such as fingerprint, face or even voice.

References

1. Jain AK, Ross A, Prabhakar S (2004) An introduction to biometric recognition. IEEE Trans Circ Syst 14(1):4–20
2. Samal A, Iyengar P (1992) Automatic recognition and analysis of human faces and facial expressions: a survey, pattern recognition. In: Pattern recognition, 25(1):65–77
3. Samraj A, Islam MR, Sayeed MS (2008) A secured fingerprint authentication system. J Appl Sci 8:2939–2948
4. Boles W (1977) A security system based on human iris identification using wavelet transform. In: Proceedings of the 1st international conference knowledge-based intelligent electron systems
5. Dumm D (1993) Using a multilayer pereptron neural for human voice identification. In: Proceedings of the 4th international conference signal processing and application technologies
6. Palaniappan R, Krishnan SM (2004) Identifying individuals using ECG beats. In: proceedings at international conference on signal processing and communications
7. Poulas M (2002) Person identification from the EEG using nonlinear signal classification. In: Methods of information in medicine 41:64–75
8. Samraj A, Sayeed S, Kiong LC, Mastorokis NE (2010) Eliminating forgers based on intra trial variability in online signature verification using handglove and photometric signals. J Inf Secur 1:23–28
9. Allen J (2007) Photoplethysmography and its application in clinical physiological measurement, IOP publishing. Physiol Meas 28(3):1–39
10. Mascaro SA, Asada HH (2001) Photoplethysmograph fingernail sensors for measuring finger forces without haptic obstruction, IEEE transactions on. Robot Autom 17(5):698–708
11. Gu YY, Zhang Y, Zhang YT (2003) A novel biometric approach in human identification by Photoplethysmographic signals. In: J Inf Technol Appl Biomed
12. Gu YY, Zhang YT (2003) Photoplethysmographic authentication through fuzzy logic. In: J Biomed Eng
13. Yao J, Sun X, Wan Y (2007) A pilot study on using derivatives of Photoplethysmographic signals as a biometric identifier. In: J Eng Med Biol Soc
14. Waseem O (2009) Microcontroller design and bluetooth signal transmission for the non-invasive health monitoring system. In: EE B416 Electrical Engineering Biomedical Capstones 17–29

Image Denoising Using Multiple Wavelet Decomposition with Bicubic Interpolation

R. Vijaya Arjunan

Abstract With the advent of better computers and high computing speeds use of images and videos have drastically increased. Today images are a very integral part of our lives from the entertainment industry to medical sciences. In the field of medicine, image processing plays a very important role when it comes to medical imaging. Image processing is utilized to get a clear denoised image for clear and easy diagnostic study. With increase in the usage, need to compress the images and store as many as possible in limited spaces have thus become a necessity. The emphasis being on the ability to convert them back into clear crisp image with minimum noise when the need be. Compressing an image is significantly different than compressing raw binary data. General purpose compression programs can be used to compress images, but the result is less than optimal. This is because images have certain statistical properties which can be exploited by encoders specifically designed for them. Also, some of the finer details in the image can be sacrificed for the sake of saving a little more bandwidth or storage space. This also means that lossy compression techniques can be used in this area. In this work, noisy image is decomposed using three different DWT transforms (haar/db6/coif5). Input images are processed with various noise levels (20, 50 and 80 %) with both salt and pepper and Additive White Gaussian Noise (AWGN). Various error metrics such as Mean Squared Error (MSE), Peak Signal to Noise Ratio (PSNR), Root Mean Squared Error (RMSD), Mean Absolute Error (MAE), and Structural Similarity index Measure (SSIM) are computed and compared with other state of art methods for stability performance.

R. Vijaya Arjunan (✉)
Department of CSE, Manipal University, Manipal, India
e-mail: profrvijayaarjunan@gmail.com

Mohan S. and S. S. Kumar (eds.), *Proceedings of the Fourth International Conference on Signal and Image Processing 2012 (ICSIP 2012)*, Lecture Notes in Electrical Engineering 221, DOI: 10.1007/978-81-322-0997-3_28,

Keywords Peak signal to noise ratio · Root mean squared error · Mean absolute error · Structural similarity index measure · Additive white gaussian noise · Image denoising

1 Introduction

Image or video denoising refers to the recovery of a digital image or video that has been affected by noise. Digital images are corrupted by various types of noise during acquisition and transmission. The discrete wavelet is essentially sub band–coding system and sub band coders have been quite successful in speech and image compression. In this work, we have compared different DWT wavelets and a newer approach for video/image frame denoising based on discrete wavelet transforms (Haar, DB6 and Coif5) has been deduced. The main scheme of de-noising is how to keep the poise between degrading the image/video noise at the same time preserving the edge information. So we aim to construct a de-noising algorithm which not only eliminates noise but also preserves signal image. The findings can be extensively used to improve image quality in various fields such as medical sciences. From literature survey we found DWT (Discrete Wavelet Transform) is one of the suitable tools, which can be applied for image processing application. The paper is organized as follows, Sect. 1 gives the general work done with introduction, Sect. 2 gives the methods adopted, Sect. 3 details the proposed methodology and Sect. 4 details the experimental results followed by conclusion and references.

2 Proposed Method

a. Discrete Wavelet Transform

A transform represents the pixel values in some other, but equivalent form. Transforms allow for some very efficient and powerful algorithms in which the entire image is processed as a single large block. 2D wavelet decomposition of an image is performed by applying 1D DWT along the rows of the image first, and, then, the results are decomposed along the columns. When DWT is applied to an image it results in decomposition of the image into 4 sub band images referred to as low–low (LL), low–high (LH), high–low (HL), and high–high (HH). A schematic sketch of level 3 DWT decomposition is shown in Fig. 1.

b. Inverse Discrete Wavelet Transform

Inverse Discrete Wavelet Transform is a process by which components can be assembled back into the original signal without loss of information. IDWT

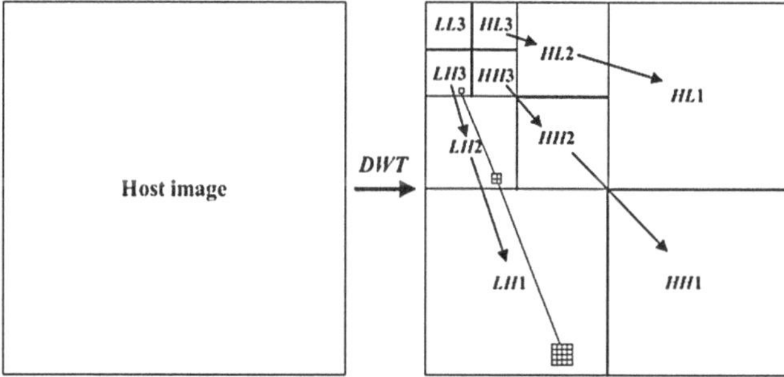

Fig. 1 Result of 3 level DWT (Haar/COIF5/DB6)

reconstructs a signal from the approximation and detail coefficients derived from decomposition. The IDWT differs from the Discrete Wavelet Transform (DWT) in that it requires up sampling and filtering, in that order. Up sampling, also known as interpolating means the insertion of zeros between samples in a signal.

3 Proposed Methodology

A three second high definition video stream (Open heart surgery video) is taken for verification of proposed methodology. Using FFMPEG, the video stream is converted into image frames (Approximately 180 frames per 1 s video is obtained). Samples of 3 images are taken into consideration for processing. Varying noise levels of 20, 50 and 80 % are impinged onto the randomly selected image frames. The proposed method of image denoising is depicted in Fig. 2.

DWT (Haar, DB6 and Coif5) are applied on the processed image at level 1. Similarly DWT is applied to the horizontal, vertical and diagonal coefficients resulted from level 1 at level 2, and inverse DWT is applied on the modified subbands. The resultant horizontal coefficients are added subsequently with the approximate coefficient of level 2 decomposition. Similarly, results are obtained for vertical and diagonal coefficients. Bicubic interpolation is applied to the resultant signal to form the new horizontal, vertical and diagonal coefficients for level 1 reconstruction.

3.1 Functional Process

Discrete Wavelet Transform on the noisy images is applied using Matlab. Functions from the Image Processing Toolbar of Matlab are used to do so. Application of DWT is done in two levels.

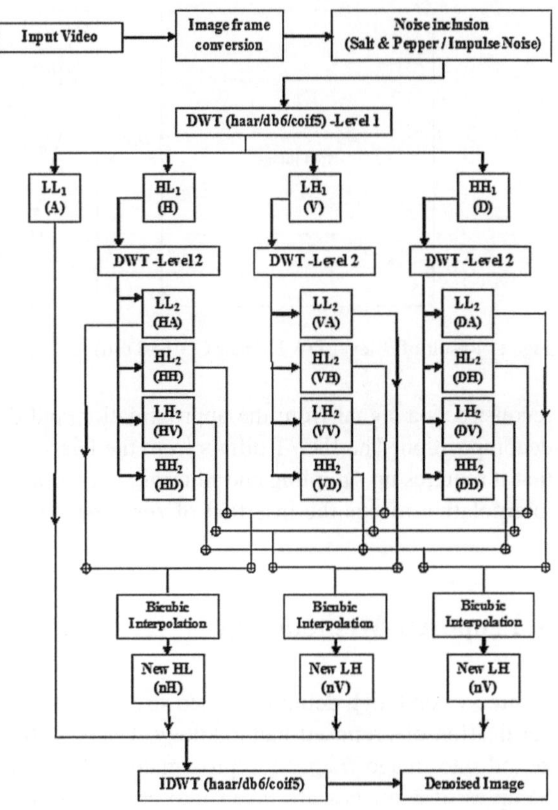

Fig. 2 Proposed algorithm for image denoising

Level 1: DWT (haar/db6/coif5) is applied to noisy image and the decomposed image are stored as a1 = approximate coefficient, h1 = horizontal coefficient, v1 = vertical coefficient and d1 = diagonal coefficient. It computes the approximation co-efficient cA and the detail co-efficient (cH, cV, cD) for the input image. Level 1 output is shown in Fig. 3.

To write the image file to disk "imwrite ()" function is used imwrite (A, colormap, filename, fmt) writes the image A to the file specified by filename in the format specified by fmt. Colormap is applied to the image before writing it to the disk.

Level 2: DWT (haar/db6/coif5) is applied again to the horizontal (h1.bmp), vertical (v1.bmp), diagonal (d1.bmp) images we get after level 1 decomposition, the resultant files are stored as shown in Fig 4.

Level 3: The coefficients we get after level 2 decomposition are added in a particular fashion to form a new resultant image. The proposed method of adding the coefficients is shown below in Fig. 5.

Level 4: IDWT is applied to get the denoised image as shown in Fig. 6. An important condition to be satisfied when applying this function is that all the coefficients should be of the same size.

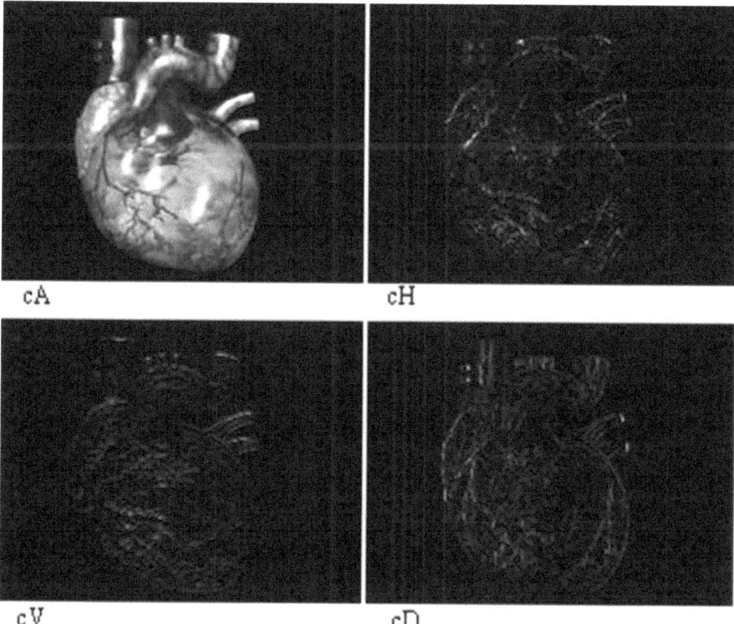

Fig. 3 Output of 1 level DWT

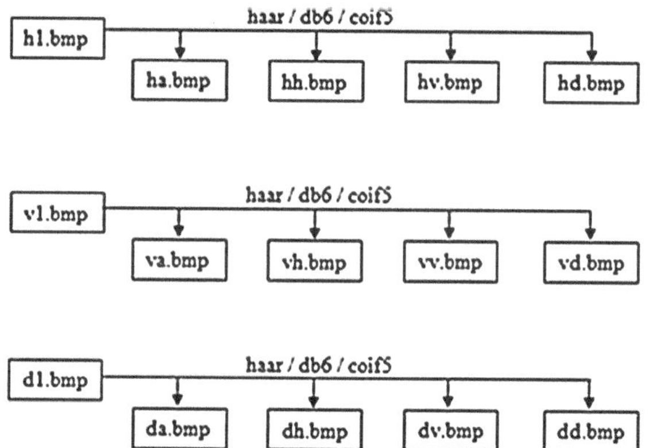

Fig. 4 Output of 2 level DWT

hh.bmp + vh.bmp + dh.bmp + ha.bmp = nh.bmp

hv.bmp + vv.bmp + dv.bmp + va.bmp = nv.bmp

hd.bmp + vd.bmp + dd.bmp + da.bmp = nd.bmp

Fig. 5 Addition of co-efficients

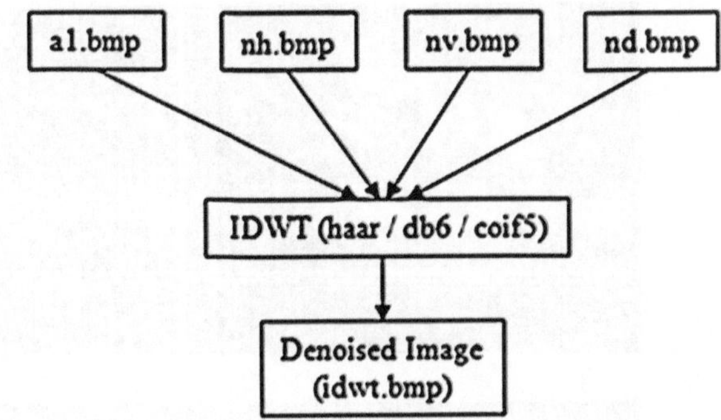

Fig. 6 IDWT process

4 Performance Metrics

Digital images are subject to a wide variety of distortions during acquisition, processing, compression, storage, transmission and reproduction, any of which may result in a degradation of visual quality. The simplest and most widely used full-reference quality metric is the Mean Square Error (MSE), computed by averaging the squared intensity differences of distorted and reference image pixels, along with the related quantity of Peak Signal to Noise Ratio (PSNR). The other error metrics used to compare Mean Absolute Error (MAE), Root Mean Square Error (RMSE) and Structural Similarity Index Measure (SSIM).

4.1 Peak Signal to Noise Ratio

The PSNR is the ratio between the maximum possible power of a signal and the power of corrupting noise that affects the fidelity of its representation. It is most easily defined via the Mean Squared Error (MSE) as,

$$MSE = \frac{1}{mn}\sum_{i=0}^{m-1}\sum_{j=0}^{n-1}[(I(i,j) - K(i,j)]^2 \tag{1}$$

The PSNR is defined by Eqn. (3) as,

$$PSNR = 10.log_{10}\left(\frac{MAX^2i}{MSE}\right) \tag{2}$$

$$= 20log_{10}\left(\frac{MAX_i}{\sqrt{MSE}}\right) \tag{3}$$

Here, MAXi is the maximum possible pixel value of the image (255).

4.2 Root Mean Square Deviation

The Root-Mean-Square Deviation (RMSD) or Root-Mean-Square Error (RMSE) measures the difference between values predicted by a model and the values actually observed. These individual differences are called as residuals and the RMSD serves to aggregate them into a single measure of predictive power. The RMSD of an estimator $\hat{\theta}$ with respect to the estimated parameter θ is defined as the square root of the mean square error, given in Eqn. (5).

$$\mathrm{RMSD}\left(\hat{\theta}\right) = \sqrt{\mathrm{MSE}\left(\hat{\theta}\right)} = \sqrt{\mathrm{E}\left(\left(\hat{\theta} - \theta\right)^2\right)} \tag{4}$$

$$\mathrm{RMSD}(\theta 1, \theta 2) = \sqrt{\mathrm{MSE}(\theta 1, \theta 2)} = \sqrt{\frac{\sum_{i=1}^{n}(x1, i - x2, i)^2}{n}} \tag{5}$$

4.3 Mean Absolute Error

The Mean Absolute Error is given in Eqn. (6).

$$\mathrm{MAE} = \frac{1}{n}\sum_{i=1}^{n}|fi - yi| = \frac{1}{n}\sum_{i=1}^{n}|ei| \tag{6}$$

MAE is an average of the absolute errors $e_i = |f_i - y_i|$, where f_i is the prediction and y_i the true value.

4.4 Structural Similarity Index Measure

SSIM is a full reference metric, in other words, the measuring of image quality based on initial uncompressed or distortion free image as reference. SSIM is designed to improve on traditional methods like PSNR and MSE, which have proved to be inconsistent with human eye perception. The difference with respect to other techniques is that these approaches estimate perceived errors on the other hand SSIM considers image degradation as image change in structural information. The SSIM is given in Eqn. (7).

$$\mathrm{SSIM(X,Y)} = \frac{(2\mu x\mu y)(2\sigma xy + C2)}{(\mu x2 + \mu y2 + C1)(\sigma x2 + \sigma y2 + C2)} \tag{7}$$

Table 1 Performance of proposed system with HAAR wavelet

Input image	Noise % level	Error metrics					
		PSNR		MSE	RMSD	MAE	SSIM
		DWT	Proposed method				
	20	10.5078	7.7527	1.0910 e + 004	104.4493	36.842	0.1291
	50	6.5115	7.6796	1.1095 e + 004	105.3321	57.176	0.0699
	80	4.4684	6.976	1.303 e + 004	114.219	62.156	0.0268
	20	13.1758	8.4989	9.1874 e + 003	95.8507	63.205	0.087
	50	6.1592	7.5745	1.1367 e + 004	106.6141	67.341	0.0423
	80	2.4286	7.01	12974 e + 004	113.9033	73.071	0.0174
	20	10.4726	8.5146	9.1541 e + 003	95.6772	58.6097	0.1075
	50	6.4725	7.566	1.1389 e + 004	106.7186	62.699	0.0537
	80	4.4269	6.9831	1.3025 e + 004	114.1257	68.4348	0.0217

Table 2 Performance of proposed system with DB6 wavelet

Input image	Noise % level	Error metrics					
		PSNR		MSE	RMSD	MAE	SSIM
		DWT	Proposed method				
	20	10.5078	7.2124	1.2355 e + 004	111.1522	56.7367	0.0625
	50	6.5115	7.1337	1.2581 e + 004	112.647	48.4145	0.0413
	80	4.4684	7.0134	1.2394 e + 004	113.7291	52.0952	0.0186
	20	13.1758	7.2573	12228 e + 004	110.5795	66.7277	0.0407
	50	6.1592	7.2182	12333 e + 004	111.0785	58.9267	0.0261
	80	2.4286	7.0559	1.2808 e + 004	113.1732	63.4947	0.0124
	20	10.4726	7.2691	1.2195 e + 004	110.4294	62.236	0.0512
	50	6.4725	7.1628	1.2497 e + 004	111.789	54.8628	0.0329
	80	4.4269	6.9821	1.3028 e + 004	114.1387	59.5274	0.015

Table 3 Performance of proposed system with Coif5 wavelet

Input image	Noise % level	Error metrics					
		PSNR		MSE	RilSD	MAE	SSIM
		DWT	Proposed method				
	20	10.5078	7.1742	1.2464 e + 004	111.643	62.0539	0.0662
	50	6.5115	7.2266	1.2315e + 004	110.9711	45.87235	0.0453
	80	4.4684	7.0134	1.2394 e + 004	113.7291	52.0952	0.0186
	20	13.1758	7.1709	1.2474 e + 904	111.6851	71.8065	0.0427
	50	6.1592	7.3205	1.2051 e + 004	109.7784	56.4142	0.0299
	80	2.4286	7.0669	1.2776 e + 904	113.0302	65.3691	0.0142
	20	10.4726	7.0764	1.2748 e + 004	112.9069	69.1297	0.0531
	50	6.4725	7.2953	1.2121 e + 004	110.0972	51.8728	0.0362
	80	4.4269	6.9985	1.2979 e + 004	113.9234	61.409	0.0162

5 Experimental Results

The performance analysis of the proposed method of image denoising is qualitatively analyzed with PSNR, MSE, RMSD, MAE and SSIM with Haar, DB6 and Coif5 wavelets. Also, the PSNR is compared with classical DWT method for confirming stability of the proposed system. Results are presented in Tables 1, 2, 3.

6 Conclusions

The goal of this project was to compare different Discrete Wavelet Transform and analyze their denoising capability for an image having varying percentages of "Salt and Pepper" noise. The comparison was performed with the help of various error metrics such as MSE, MAE, RMSD, PSNR and SSIM. From those results following conclusions were drawn:

Inference 1: For low level noise (Up to 20 %), classical DWT is the best choice in denoising as it exhibits more than 2 dB increase when compared to the proposed system.

Inference 2: For moderate level noise (From 30 to 50 %), the proposed system is found better as it exhibits 1 dB increase with PSNR when compared to classical DWT.

Inference 3: For high level noise (More than 60 %), the proposed system is found supreme as it exhibits more 2.5 dB increase with PSNR when compared to classical DWT.

Proposed algorithm inference: For low of noise (up to 20 %), Haar wavelet is found better with PSNR and for moderate level of noise (From 30 to 50 %), DB6 wavelet is found better and at higher noise level (more than 60 %) Coif5 wavelet is better but very closely followed by DB6.

7 Future Enhancement

There is a lot of scope for future work to be done in this project domain. Some enhancements that can be made to the current project are, the proposed system is tested only with three different 2 D wavelet transforms i.e. Haar/Db6/Coif5. A comparative study be done, further covering various other wavelet transforms. More over the scheme is verified only with salt and pepper noise which could be further extended for other type of noise such as, Impulse noise, Additive White Gaussian Noise etc. For signal interpolation, bicubic interpolation technique was utilized; other, different type of interpolation methods like bilinear interpolation etc. could be used for algorithmic stability.

References

1. Raviraj P Sanavullah MY (2007) The modified 2D-Haar wavelet transformation in image compression. Middle East J Sci Res 2(2):73–78, ISSN 1990-9233
2. Talukder KH, Harada K (2010) Haar wavelet based approach for image compression and quality assessment of compressed image. IAENG Int J Appl Math 36:1 IJAM_36_1_9
3. Vetterli M, Kovacevic J (1995) Wavelets and sub-band coding. Prentice Hall, Englewood Cliffs, NJ
4. Sethi N, Krishna R, Arora RP (2011) Image compression using haar wavelet transform. ISSN 2222-1719 paper ISSN 2222-2863 (Online)
5. Fahmy SA (2008) Generalized parallel bilinear interpolation architecture for vision systems. International conference on reconfigurable computing and FPGAs
6. McAndrew A (2004) An Introduction to Digital Image processing with matlab
7. Pratap R (2003) Getting started with MATLAB a quick introduction for scientists and engineers, Oxford, ISBN-0-19-515014-7
8. Wang Z (2004) Image quality assessment: from error visibility to structural similarity. IEEE Trans Image Process 13(4):600–612

FPGA Synthesis of SIRM Fuzzy System-Classification of Diabetic Epilepsy Risk Levels from EEG Signal Parameters and CBF

N. B. Balamurugan, M. Jothi and R. Harikumar

Abstract Now a day the epilepsy risk level classification is one of the most important thing in diabetic patient's treatment. That risk level classification is proposed in this paper. SIRM fuzzy processor is synthesized through the FPGA. FUZZY rules are the best way to deal with natural conditions. In this paper epilepsy classification is performed through minimum no of rules. Cerebral blood flow level and EEG signals are used as input parameters. SIRM fuzzy processor with the tuned and untuned conditions is checked for various input values. The better the fuzzy system is identified based on performance and quality values. The tuned SIRM system with five rules is selected which has the performance of 98.58 % and quality value of 36.56. The SIRM system is simulates through VHDL and synthesized by FPGA which has performance value of 98.28. This SIRM fuzzy model is compared with other techniques like homogeneous system, heterogeneous system.

Keywords Epilepsy risk level · Cerebral blood flow · VLSI design and simulation · SIRM fuzzy system

N. B. Balamurugan (✉)
Department of ECE, Thiagarajar College of Engineering, Madurai, India
e-mail: nbbalamurugan@tce.edu

M. Jothi
Department of IT, K.L.N. College of Engineering, Madurai, India
e-mail: sangeethjothi2@gmail.com

R. Harikumar
Department of ECE, Bannari Amman Institute of Technology, Sathyamangalam, India
e-mail: harikumar_rajaguru@gmail.com

Mohan S. and S. S. Kumar (eds.), *Proceedings of the Fourth International Conference on Signal and Image Processing 2012 (ICSIP 2012)*, Lecture Notes in Electrical Engineering 221, DOI: 10.1007/978-81-322-0997-3_29,

1 Introduction

This paper elucidates the design and synthesis of SIRM fuzzy processor for epilepsy risk level classification. The Cerebral Blood Flow (CBF) measurements, Electroencephalogram (EEG) signals are the input parameters and aggregation operators are used to find the slope of Bell shaped fuzzy membership functions. The sequences of operation involved are

- Measurement of CBF using tetra polar method which is explained in ref [1, 2].
- Recording of EEG signals with extracted features such as energy, positive and negative peaks, Events and sharp wave, Clinical data index of convulsions, duration of seizure and total body depression time after seizure production.

1.1 Back Ground

In clinical neurobiological practice detection of abnormal activity plays an important role in diagnosis epilepsy [3, 4]. Analysis of EEG includes detection of patterns and features characteristic abnormal conditions. A severity in disease can be detected by Quantitative abnormalities. EEG easily detects the oxygen delivery and utilization in brain [5]. x1, x2, x3, x4 are variables used in aggregation operator application. y1, y2, y3, y4 are derived from the clinical parameters and EEG signals respectively [6]. Nonlinear fashion is used for the normalization [7] with the secant and sigmoid function.

Now consider the general aggregation operator f(x) used in our approach [8] is given by Eq. 1.

$$f(x) = \alpha 1(x1 + x2 + x3) + 1 - \alpha 1(y1 + y2 + y3 + y4) \tag{1}$$

2 Single Input Rule Model (SIRM) Fuzzy System

In the conventional if–then fuzzy inference model [9], all the input items of systems are put into antecedent part of each fuzzy rule. To implement the fuzzy logic expert knowledge is usually expressed as [10],

IF < a set of conditions are satisfied>
THEN < a set of Consequence can be inferred>

Both the numbers of fuzzy rules and parameters increases exponentially, and defining fuzzy rules becomes difficult for large scale systems. To solve the problems, the single input rule modules (SIRM) fixed important degree for each input items. The SIRM fuzzy model is shown in Fig. 1. The SIRM fuzzy model is designed with five rules in the rule base

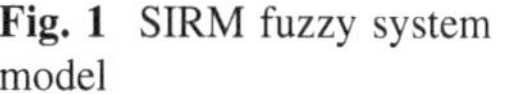
Fig. 1 SIRM fuzzy system model

SFR (1) IF CBF is very low THEN output epilepsy risk is normal
SFR (2) IF CBF is low THEN output epilepsy risk is low risk
SFR (3) IF CBF is medium THEN output epilepsy risk is medium risk
SFR (4) IF CBF is high THEN output epilepsy risk is high risk
SFR (5) IF CBF is very high THEN output epilepsy risk is very high risk.

2.1 Membership Function of SIRM Fuzzy System

The membership function of measured CBF is bell shaped and slope function is embedded with EEG signal information. The output epilepsy risk level of this system uses triangular membership function. The input membership of the SIRM fuzzy system is shown in Fig. 2.

A Group of 200 diabetic patients is used for calibration, training and testing the SIRM fuzzy system. Result shows that the missed classification is fully removed.

The performance [11] of the system is given in Eq. (2)

$$\text{Performance} = \{(\text{PI} - \text{MI} - \text{FI})/\text{PI}\} * 100 \quad (2)$$

PI: Perfect classification
MI: Missed classification
FI: False alarm.

3 Analysis of SIRM Fuzzy System

The performance obtained for the SIRM fuzzy system is low as 86.9 %. After doing literature survey and analysis [11, 12 and 13] to tune the SIRM fuzzy system the following decisions have been taken

- Both the input and output will have bell shape membership function.
- The edges of input and output membership function will start at unity.

3.1 Analysis of Various Fuzzy Systems

To validate the advantages of SIRM system it must be compared with the other approaches. Performance comparison is given in Fig. 3. Two input homogeneous system miserably fails in the all risk level classification. The Heterogeneous system needs additional EEG signal parameter. Thus the SIRM fuzzy system gives good results.

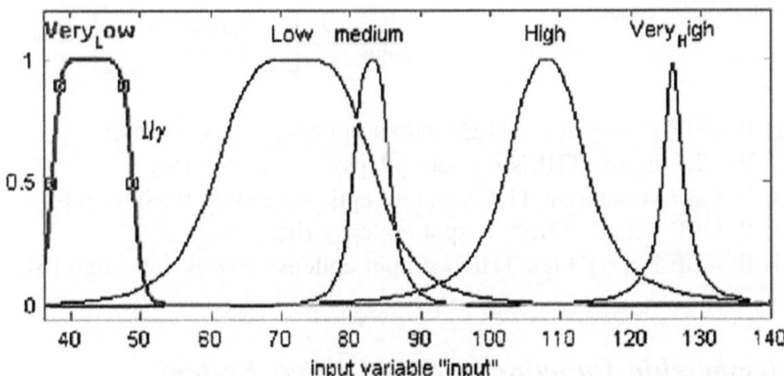

Fig. 2 Input membership function of SIRM fuzzy system

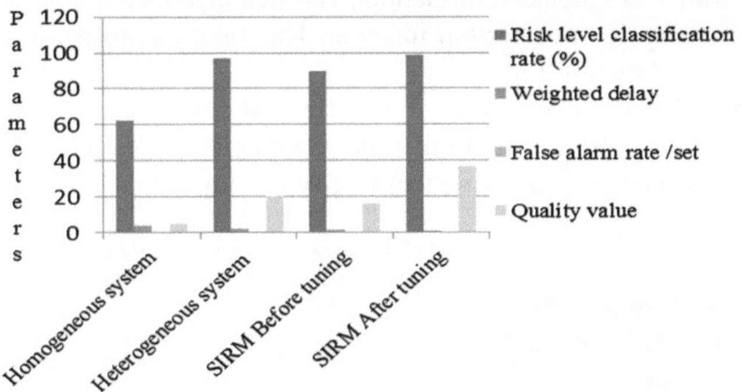

Fig. 3 Performance of different fuzzy systems

3.2 Quality Value

Overall quality of the classifier [14] is determined by three factors

1. Classification rate
2. Classification delay
3. False alarm rate

The quality value is defined in Eq. (3)

$$Qv = C/\left(R_{fa} + 0.2\right) * \left(T_{dly} * P_{dct} + 6 * P_{msd}\right) \tag{3}$$

where C—scaling constant
R_{fa}—The number of false alarm per set
T_{dly}—Avg delay of onset classification
P_{msd}—Percentage of perfect risk level missed
P_{dct}—Percentage of perfect risk level classification.

Justification for Qv are

- Qv monotonically increases when R_{fa} decreases
- The lower false alarm rate is better the classifier performance. In order to classify a perfect risk level with the shortest possible delay, the classification window lasts 2.0 for EEG signal based fuzzy system Quality value obtained for four different fuzzy system are shown in Fig. 4.

4 VLSI Design and Synthesis of SIRM Fuzzy Processor

The sustained growth in VLSI technology is fueled by continuous shrinking of transistor to ever small. The key parameters for the superior performance are

Miniaturization of size
Higher packing density
Higher circuit speeds
Lower power dissipation.

The VLSI design and simulation using Hardware Description language (VHDL) [8, 15] is undertaken for the tuned SIRM fuzzy system for epilepsy risk level classification. Fuzzy system [16] is modeled with input and output ports. The bell shape membership functions are evaluated from the functional parameters like deviation slope and center value of CBF is modeled [5].

System involves with floating point numbers. So VLSI implementation needs [16] floating point operators. To avoid the complexity, the system is developed with the scaling factor either in the multiplication or division. The simulation results are shown in Fig. 5. Design Flow Steps of VHDL consists of various stages to promote to FPGA. FPGA can accept only the BIT-File. So all the processes are converted into the related Bit format.

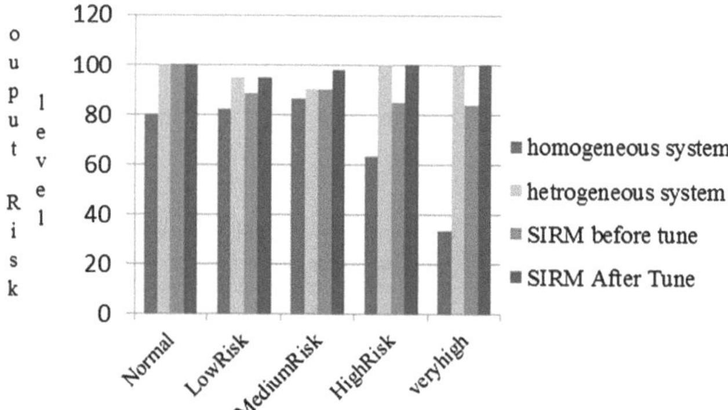

Fig. 4 Quality value for four fuzzy system

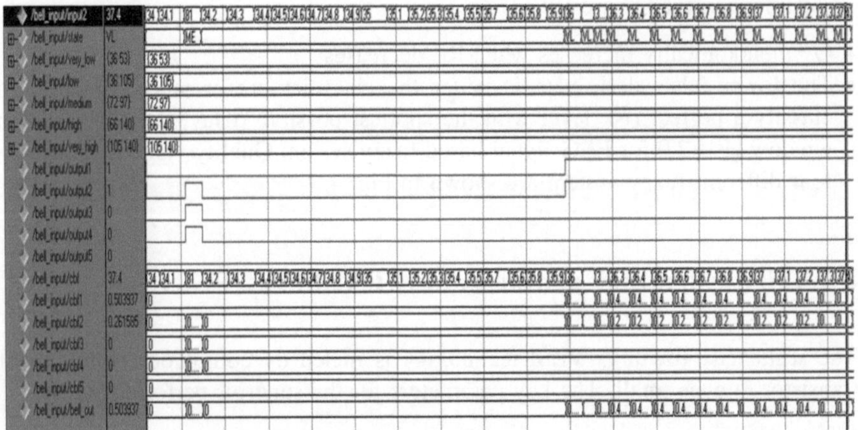

Fig. 5 Simulation window for bell membership function

The simulated results for SIRM fuzzy processor is synthesized using Xilinx ise .9.1i [17]. Generally Synthesis is the process of physical implementation of the circuit which reads the Hardware Description Language. The system contains the following elements.

- Logic gates
- Register transfer logic
- Arithmetic Logic units
- Multiplexer
- Comparators.

The FPGA block diagram consists of Configurable Logic Block as a main logic cells. Lookup tables or truth tables are used to implement a single logic function by storing the correct logic state in a memory location. Various memory types are available. Depends on the application memory types can be chosen [18].

Synthesis of Bell shape membership function is shown in Fig. 7. Various CBF values for Bell membership function is applied in VLSI simulation and it is synthesized through Xilinx integrated tool environment (Fig. 6).

Xilinx FPGA editor for Bell shape membership function is shown in Fig. 8. To specify the pins to which the inputs and outputs to be connected is done by the

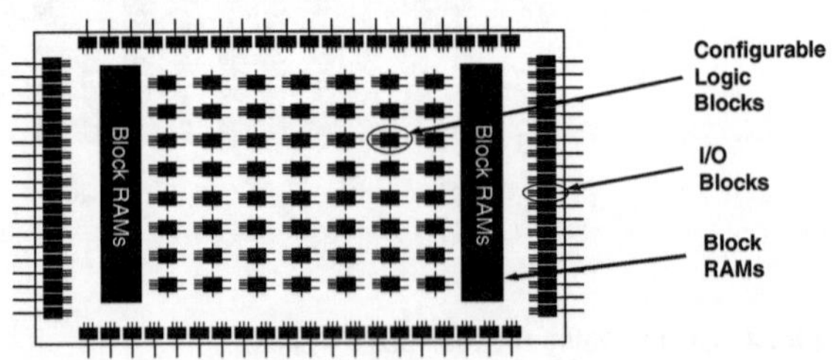

Fig. 6 Basic block diagram of FPGA

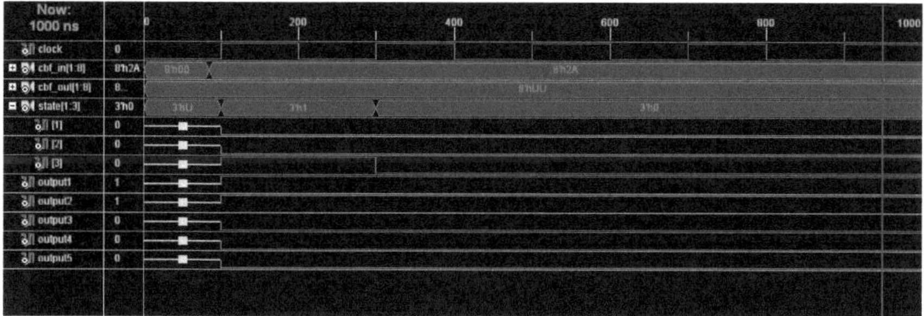

Fig. 7 Synthesis window for bell membership function

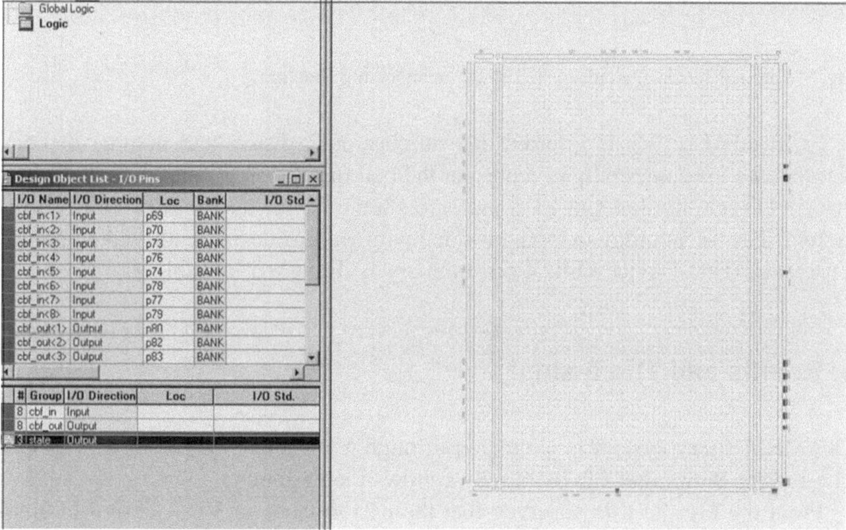

Fig. 8 Xilinx FPGA editor

"ASSIGN PACKAGE PIN" in the "process" pane in the left of the editor window. First the "Pace Editor" is loaded. Here we can have two options

(1) Package view
(2) Architecture view

SIRM Bell input system can be synthesized for both the cases

(1) Bell input-Bell output
(2) Bell input-Triangle output

Figure 9 shows the internal diagram of Bell shape membership function which gives RTL schematic.

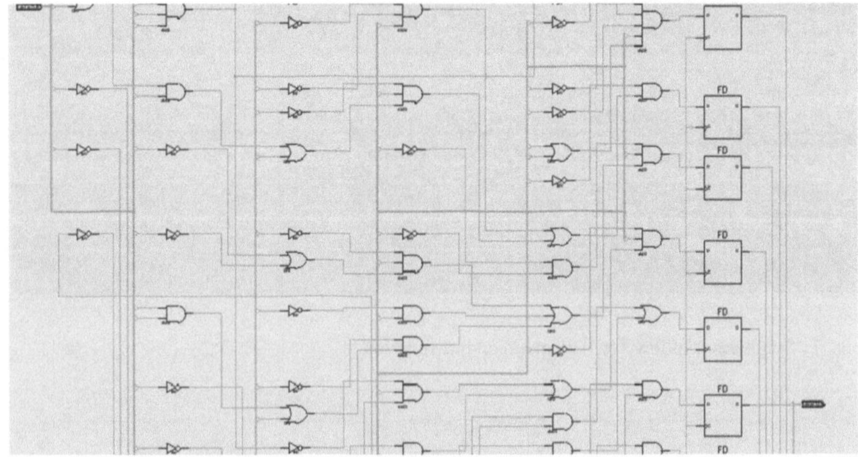

Fig. 9 Internal block diagram of bell shape membership function

From this Fig. 9 it is inferred that number of Registers and number of comparators are used internally to represent the bell membership function it is inferred that 53 D-type flip flop and 13 comparators are used for the measured CBF fuzzy system. The Technology mapping is done with various device family like Spartan, Virtex etc. That Device utilization summary is discussed in Table 1.

5 Results and Discussion

The SIRM fuzzy system is simulated through VHDL and synthesized in FPGA. The results shows that FPGA closely follows the system.

From the Fig. 10 it is observed that the performance of VLSI design in tuned SIRM fuzzy system is as close to the performance of traditional MATLAB based tuned SIRM fuzzy system. This system suffers from false alarm in the low risk level and high risk level regions.

Table 1 Device utilization summary

Logic utilization	Spartan2	Virtex2	Virtex2P
No of 4 input LUTs	18/384	18/1536	18/2816
Logic distribution			
No of occupied slices	10/192	10/68	10/1408
No of slices containing related logic	10/10	10/10	10/10
Total no of 4 input LUT's	18/384	18/1536	18/1536
No of bonded IOB's	24/36	24/94	25/140
No of GCLK's	1/4	1/4	1/16
Total equivalent gate count for design	154		157

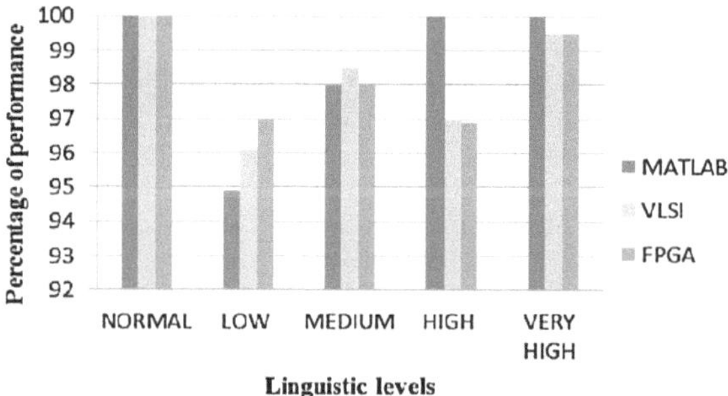

Fig. 10 Performance comparison of MATLAB, VLSI and FPGA

6 Conclusion

A SIRM fuzzy processor is simulated for various C.B.F values. The result shows that FPGA closely follows the MATLAB. The simulation is done with arithmetic operations for scaled up parameter values which is actually floating point value. Synthesis of this design is difficult because of the scaled up values. This will increase the simulation time and the processing time in synthesized hardware. For on-line diagnosis, the same design can be synthesized by incorporating floating point arithmetic operations which gives more accurate output value and produce same performance as that of MATLAB simulation.

References

1. Harikumar R, Selvan S (2002) Fuzzy based classification of patient state in diabetic neuropathy using cerebral blood flow. J Syst Soc India Paritantra 7(1):37–41
2. Harikumar R, Selvan S (1997) Analysis of cerebral blood flow in diabetic Neuropathy using impedance Technique. In: Proceedings Of NCBME'97 Anna University, Chennai, pp 3.11–3.14
3. Clement C, Pang et al. (2003) A comparison of algorithms for detection of spikes in the EEG. IEEE Trans Biomed Eng 50(4):521–526
4. Adlassnig KA (1986) Fuzzy set theory in medical diagnosis. IEEE Trans Syst Man Cybern 16(3):260–265
5. Muro et al. (1989) A mathematical model of cerebral blood flow chemical regulation -part II. IEEE Trans Biomed Eng 36(2):192–201
6. Yager's RR (1998) On ordered weighted averaging aggregation operators in multi criteria decision making. IEEE Trans Syst Man Cybern 18(1):183–190
7. Harikumar R, Sukanesh R, Sabarish Narayanan B (2003) Application of aggregation operators in fuzzy logic based Classification Of diabetic epilepsy risk level. In: Proceedings of annual convention and exhibition (ACE) IEEE India Council ACE, Pune
8. Paramasivam K, Harikumar R, Gunavathi K (2003) Simulation of VLSI design using parallel architecture for epilepsy risk level diagnosis in diabetic neuropathy. In: Proceedings of national conference on VLSI design and testing, Coimbatore, India, Febuary 21st and 22nd

9. Yi JQ et al. (2002) A proposal of SIRMs dynamically connected fuzzy inference model for plural input fuzzy control. Fuzzy Sets Syst 125(1):79–92)
10. Sukanesh R, Harikumar R, Shanmugam Jothi M (2009) FPGA synthesis of heterogeneous and SIRM fuzzy system for classification of diabetic epilepsy risk levels. IE India, 90
11. Leo P, Karall MdJoslin (1989) Diabetes manual, LEA and FEBIGER. Philadelphia London, Chapter 16
12. Mathews JNS et al (1991) Statistical method for the estimation of cerebral blood flow using the kety-schmidt technique. Clin Sci 97:485–492
13. Gayton AC (1996) Textbook of medical physiology: prism books Private Limited. 9th edn. Bangalore
14. Alison A, Dingle et al. (1993) A multi stage system to detect epileptic from activity in the EEG. IEEE Trans Biomed Eng 40(12):1260–1268
15. Verilog HDL Language Reference Manual. IEEE (2001)
16. Mahamoud A, Jayabharathi MD (1995) FPGA for fuzzy controllers. IEEE Tran Syst Man Cybern 25(1)
17. Xilinx ISE web pack (2009)
18. Basic concepts from en.wikipedia.org/wiki/Very-large- scale integration
19. Kim YD, Hyung LK (1997) High-speed flexible fuzzy hardware for fuzzy information processing. IEEE Trans Syst Man Cybern A 27:45–56

Hierarchical Clustering Algorithm for Intensity Based Cluster Merging and Edge Detection in Medical Images

R. Harikumar, B. Vinoth Kumar, G. Karthick, L. K. Chand and C. Navin Kumar

Abstract Edge detection in medical images is an intrinsic difficult problem as the gray value intensity images may show different edges through Improved Mountain Clustering based medical image. We present a clustering based indexing technique, where the images in the database are grouped into clusters of images with similar color content using a hierarchical clustering algorithm. An initial comparative study of various medical datasets shows the differences and properties of these approaches and makes clear that the proposal has interesting properties.

Keywords Hierarchical clustering · Edge detection · Cluster merging · Canny edge detector

1 Introduction

Image is well-known for its applications in exploratory pattern analysis [1], grouping, decision making and machine-learning situations for medical images. Many of the image retrieval systems extract specific features from a query image and compare these features with the corresponding pre-computed features of all the images in the database. The search time, therefore, increases linearly with the size of the database. Efficient feature representations and similarity measures have

R. Harikumar (✉)
Professor, ECE, Bannari Amman Institute of Technology, Sathyamangalam, India

B. Vinoth Kumar
Assistant Professor, EEE, Bannari Amman Institute of Technology, Sathyamangalam, India

G. Karthick · L. K. Chand · C. Navin Kumar
UG Student, ECE, Bannari Amman Institute of Technology, Sathyamangalam, India

Mohan S. and S. S. Kumar (eds.), *Proceedings of the Fourth International Conference on Signal and Image Processing 2012 (ICSIP 2012)*, Lecture Notes in Electrical Engineering 221, DOI: 10.1007/978-81-322-0997-3_30,

been used to speed-up the search process. The growing size of the database results in long search delays that may be unacceptable in many practical situations. Even if the time required to compare two images is very short, the cumulative time needed to compare the query image with all database images is rather long and is probably longer than the time an average user wants to wait. A part of the goal is to group similar images into clusters and to compute the cluster centres, so that during retrieval, the query image need not be compared exhaustively with all the images in the database. To retrieve similar images for a given query, the query image is initially compared with all the cluster centres. Then a subset of clusters that have the largest similarity to the query image is chosen and all the images in these clusters are compared with the query image. The clustering technique is not directly dependent on the nature of similarity measure used to compare the images, so this technique can be used for any general similarity measure. The experimental evaluation shows that this clustering based indexing technique offers high retrieval accuracy with a considerable reduction in the number of similarity comparisons required.

Edge detection plays an important role in image processing and analyzing systems. For medical images, in which channels may provide different or even conflicting information, edge detection becomes more important and essential. Not only it helps to visualize colour images, but also it allows to roughly localizing objects. Edge detection in gray-scale images has been thoroughly studied and is well established. Two main approaches to detect edges in colour/gray images based on monochromatic [2] and cluster techniques [3, 4] have been published. The first one detects edges for each individual band, and then combines the results over all bands. The latter treats each pixel in a medical image as a vector in the spectral domain, and then performs edge detection in this domain. This approach is more efficient than the first one since it does not suffer from the localization variability of edge detection result in the individual cluster. Therefore, in the scope of this paper, we only focus on edge detection methods based on cluster techniques.

A method to extend the edge detection for gray-scale images to multi-channel images [3]. The main idea is to find the direction for a point x for which its vector in the spectral domain has the maximum rate of change. Therefore, the largest eigenvalue of the covariance matrix of the set of partial derivatives is selected as the edge magnitude. Then, a thresholding method could be applied to reveal the edges. However, the problem with this method is how to determine the scale for each channel since the derivatives taken for different channels are often scaled differently. Variations of this approach have been used in [5–7]. A comparison of these techniques is presented in [8]. One typical problem in analyzing the multi-dimensional images is that the size of samples is very large while the number of samples is small. Therefore, it is necessary to employ available information in both spatial and spectral domain to make use of the properties of different parts, or objects in medical images.

Vector- valued ranking operators to detect edges in color images [4]. First, the image is divided into small windows. Then, for each window, the vector-valued data of pixels are ordered increasingly based on the R-ordering algorithm [9]. Finally, "outliers" are determined and considered as edges. The disadvantage of this method is due to the difficulty in determining window's size. If window's size is set too small, it will cause the problem of discontinuous edges between neighboring windows become more serious. If window's size is too big, some pixels are located far from the others, and therefore, uncorrelated from the others. As a consequence, ordering vector-valued data may not gain a good result. A statistically based method for edge detection in color images [10]. They considered each pixel as a point in the feature space. A clustering algorithm is applied for a fuzzy segmentation of the image and then outliers of the clusters are considered as edges. Actually, this method performs image segmentation rather than edge detection and often procedures multiple responds for a single edge. Pixel intensity is good for measuring the similarity among pixels, and therefore it is good for the purpose of image segmentation. But it is not good for measuring the abrupt changes to find edges. Pixel gradient value is much more appropriate for that. Therefore, in our approach, we first consider each pixel as a point in the spectral space composed of gradient values in all image bands, instead of intensity values. Then, a clustering algorithm is applied in the spectral space to classify edge and non-edge pixels in the image. Finally, a thresholding strategy similar to the Canny edge detection method [11] is used to refine the results.

2 Image Clustering

Searching large databases of images is a challenging task especially for retrieval by content. Most search engines calculate the similarity between the query image and all the images in the database and rank the images by sorting their similarities. One problem with this exhaustive search approach is that it does not scale up for large databases. The retrieval time for exhaustive search is the sum of two times: *Tsim* and *Tsort*. *Tsim* is the time to calculate the similarity between the query and every image in the database, and *Tsort* is the time to rank all the images in the database according to their similarity to the query.

$$T_{exhaustive} = nT_{1sim} + O(n\log n) \quad (1)$$

where n is the number of images in the database, $T1sim$ is the time to calculate the similarity between two images, and $O(n\log n)$ is the time to sort n elements.

When the images in the database are clustered, the retrieval time is the sum of three times, the time to calculate the similarity between the query and the cluster centers, the time to calculate the similarity between the query and the images in the nearest clusters and the time to rank these images. Therefore the total search time is:

$$T_{cluster} = kT_{1sim} + lT_{1sim} + Ol\log l \quad (2)$$

Here k is the number of clusters, l is the number of images in the clusters nearest to the query. Since $k \ll n$ and $l \ll n$, $T_{cluster} \ll T_{exhaustive}$.

2.1 Challenges of Image Retrieval

What use is the sum of human knowledge if nothing can be found? Although significant advances have been made in text searching, only preliminary work has been done in finding images and videos in large digital collections. In fact, if we examine the most frequently used image and video retrieval systems (i.e. www.google.com) we find that they are typically oriented around text searches where manual annotation was already performed.

Image and video retrieval is a young field which has its genealogy rooted in artificial intelligence, digital signal processing, statistics, natural language understanding, databases, psychology, computer vision, and pattern recognition. However, none of these parental fields alone has been able to directly solve the retrieval problem. Indeed, image and video retrieval lies at the intersections and crossroads between the parental fields. It is these curious intersections which appear to be the most promising.

What are the main challenges in image and video retrieval? We think the paramount challenge is bridging the semantic gap. By this we mean that low level features are easily measured and computed, but the starting point of the retrieval process is typically the high level query from a human. Translating or converting the question posed by a human to the low level features seen by the computer illustrates the problem in bridging the semantic gap.

However, the semantic gap is not merely translating high level features to low level features. The essence of a semantic query understands the meaning behind the query. This can involve understanding both the intellectual and emotional sides of the human, not merely the distilled logical portion of the query but also the personal preferences and emotional subtones of the query and the preferential form of the results.

3 Image representation and Similarity Measure

3.1 Image Representation

Several histogram-based approaches have been proposed for image retrieval by colour. These approaches are based on calculating a similarity measure between the colour histogram of the query image and the images in the database. The difference between these approaches is mainly in their choice of the colour space and the similarity measure. Since these approaches use a single histogram to calculate similarities, the results are expected to reflect only global similarity. For example, if a user submits a query image with a sky at the top and sand at the bottom, the retrieved results would have a mix of blue and beige, but not necessarily with blue at the top and beige at the bottom. This can be achieved only if the image representation reflects the local colour information.

In this paper, we use a scheme2 that allows retrieval based on local colour features. Images in the database are divided into rectangular regions. Then every image is represented by the set of normalized histograms corresponding to these rectangular regions. It should be noted here that the choice of the rectangular region size is important. In one extreme, the whole image is considered as a single region which reflects the global colour information. As the size of the region becomes smaller, the local variations of colour information are captured by the histograms. The size of the region should be small enough to emphasize the local colour and large enough to offer a statistically valid histogram. In the experiments, images were partitioned into 16 rectangular regions. The number of partitions is represented by P in the following sections.

3.2 Similarity Measures

The similarity between two images is measured by calculating the similarity between the histograms of the corresponding rectangular regions. Then a single measure of similarity between the two images is calculated by adding the individual similarities between the corresponding regions. We have used the histogram intersection measure to compare the individual histograms (probabilities). Given two *normalized* histograms, $p = \{p1, p2,\ldots, pm\}$, $q = \{q1, q2,\ldots, qm\}$, the similarity measure is defined by

$$s_{p,q} = \sum_{i} \min(p_i, q_i) \tag{3}$$

4 Hierarchical Clustering

Let n be the number of images in the database, the similarity between all pairs of images is precomputed. The hierarchical clustering 6 is performed as follows:

The n images in the database are placed in n distinct clusters. These clusters are indexed by $\{C_1, C_2,\ldots, Cn\}$. For the kth cluster, the set E_k contains all the images contained in that cluster and N_k denote the number of images in the cluster.

$$E_k = \{k\} \text{ and } N_k = 1 \text{ for } k = 1, 2, \ldots, n$$

Two clusters C_k and C_l are picked such that the similarity measure $S_{k,l}$ is the largest. (The similarity measure between two clusters is defined in the following subsection). These two clusters are *merged* into a new cluster C_{n+1}. This reduces the total number of unmerged clusters by one. $En + 1$ is updated to $E_{n+1} = \{E_k U E_l\}$ and N_{n+1} is updated to $N_{n+1} = N_k + N_l$. Out of the two children of C_{n+1}, one is referred to as the right child $RC_{n+1}=k$ and the other as the left child $LC_{n+1}=l$. The similarity measures between the new cluster C_{n+1} and all the other unmerged clusters are computed as discussed below.

Steps 2 and 3 are repeated until the number of clusters has reduced a required number or the largest similarity measure between clusters has dropped to some lower threshold.

Figure 1 shows a simple example of hierarchical clustering with 8 images. The clustering is stopped after reaching two clusters. For this example, $N_{14} = 5$, $N_{12} = 3$, $E_{14} = \{1,2,3,4,5\}$ and $E_{12} = \{6,7,8\}$. Also, $RC_{14} = 5$ and $LC_{14} = 13$. Each cluster has an associated tree. The clustering algorithm presented here does not directly depend on the nature of similarity measure. We have presented a performance evaluation of the clustering algorithm for different similarity measures elsewhere.

4.1 Cluster Center Computation

After clustering, an appropriate center has to be obtained for each cluster. Since, every image is represented by P histograms corresponding to P partitions; it is apt to use a similar representation for cluster centers. A simple representation would be the average of histograms of all the images in the cluster. Since the number of images in each cluster can be large, averaging over all the images may be computationally expensive. One solution is to find a smaller number of representative images and use the averages of their corresponding histograms to represent the

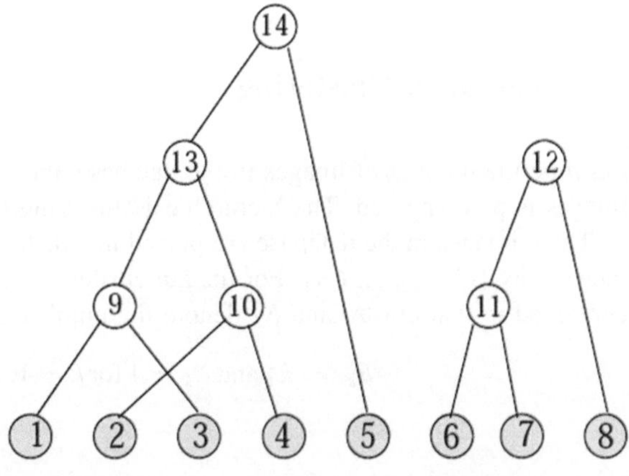

Fig. 1 A sample of cluster merging process with hierarchical clustering

cluster center. In the following discussion, the maximum number of representative images for a cluster is limited to R. These representative images have to be chosen carefully so that the cluster center computed from them is close to all the images in the cluster.

The tree structure that is obtained as a by-product of the clustering algorithm can be effectively used to select the representative set of images. In Fig. 1, let us consider selecting the representative images for cluster C_{14}. From the tree structure it can be inferred that the images 1 and 3 belong to cluster 9 and images 2 and 4 belong to cluster 10. Hence a good selection of representative images, for $R = 3$, is to select one from $\{1, 3\}$, another from $\{2, 4\}$ and 5. If $R = 2$, then it is apt to select one from $\{1, 2, 3, 4\}$ and 5 as representatives. Similarly for C_{12}, it is better to select 6 and 8 or 7 and 8 instead of 6 and 7. Such a selection would result in a representative set that captures the diversity of images present in the cluster.

Let Ci be a cluster for which a set of representative images is to be selected. A set of R nodes is chosen from the tree associated with Ci and from each of these nodes a representative image is selected, resulting in R representative images. The following steps explain this procedure:

1. Set $n = 0$ and form a set $n = \{\mathrm{i}\}$. If $R = 1$, then go to Step 5.
2. Each element in n is an index of one of the nodes in the tree associated with C_i. Find an element k such that N_k is the largest.
3. Form a new set $n + 1$ by copying all the elements of n except k and adding RC_k and LC_k.
4. Repeat steps 2 and 3 until the number of elements contained in n is equal R.
5. Now n contains R nodes from the tree associated with Ci. From each of these nodes a representative image is chosen. If k=n, and $N_k = 1$ and the selection is straightforward, i.e. the associated image is selected. If $N_k > 1$, then it is necessary to select a single image representative from E_k. This is done by selecting the one that has the maximum average similarity measure with the other N_{k-1} images of E_k.

After selecting a set of R representative images, the averages of their P corresponding histograms are used to represent the cluster center. In the example presented above, the cluster center for C_{14} is represented by P histograms which are obtained by averaging the corresponding histograms of partitions of images 2 and 5.

4.2 Cluster Center Optimization

The technique presented in the previous subsection enables us to compute a cluster center for each unmerged cluster. After computing the cluster center, it is necessary to evaluate the optimality of the clustering and the computation of cluster centers. Cluster centers are optimal, if for each image contained in a cluster, the similarity measure between the image and that cluster center is larger than the similarity measure between the image and all the other cluster centers. However this may not be true, especially given the fact that we have only used a representative set and not all the images in the cluster to compute the cluster centers. As a result, an image may have a larger similarity measure with other cluster centers than with its own cluster center. To optimize the cluster centers, it is necessary to move all such images to their closest clusters. Note that the similarity measure between an image and a cluster center is computed in exactly the same manner as presented in Sect. 3 for computing the similarity measure between two images.

5 Result of Cluster Merging

The sample MRI brain image taken based on the intensity value classified into six cluster and merged these cluster got the original image. Brain image classified into six clusters and merge all cluster got the original image (Figs. 2, 3, 4, 5, 6, 7, 8).

6 Clustering Based Edge Detection in Medical Images

First, the spatial derivatives of each channel in a medical image are determined. From [8, 12], it is well-known that the use of convolution masks of 3 × 3 fixed size pixels is not suitable for the complex problem of determining discontinuities in image functions. Therefore, we use the 2-D Gaussian blur convolution to determine the partial derivatives. Another advantage of using the Gaussian function is that we could smooth the image to reduce the effect of noise, which commonly occurs in medical images. After the spatial derivatives of each channel are determined, the gradient magnitude of each pixel is calculated using the hypotenuse functions. Then each pixel can be considered as a point in the spectral

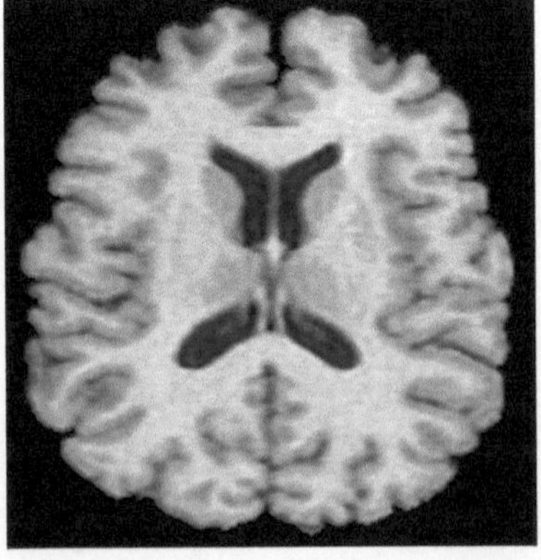

Fig. 2 Original image

Fig. 3 $k = 1$ cluster

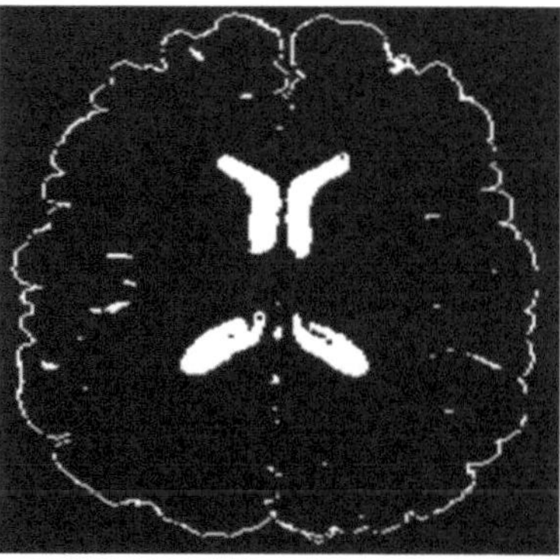

Fig. 4 $k = 2$ cluster

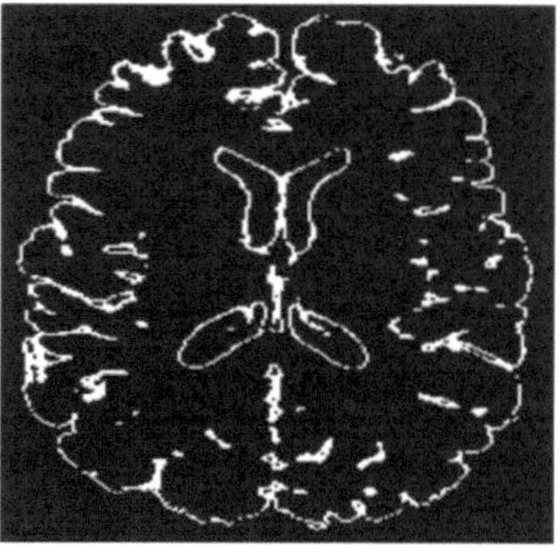

Fig. 5 $k = 3$ cluster

space, which includes gradient magnitudes over all channels of the medical images. The problem of finding edges in the medical images could be considered as the same problem as classifying points in a spectral space into two classes: edge and non-edge points. We then use a clustering method based on the k-means algorithm for this classification purpose. One important factor in designing the

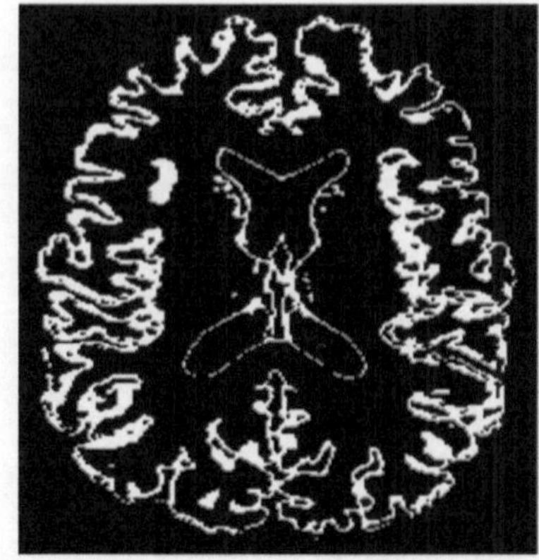

Fig. 6 $k = 7$ cluster

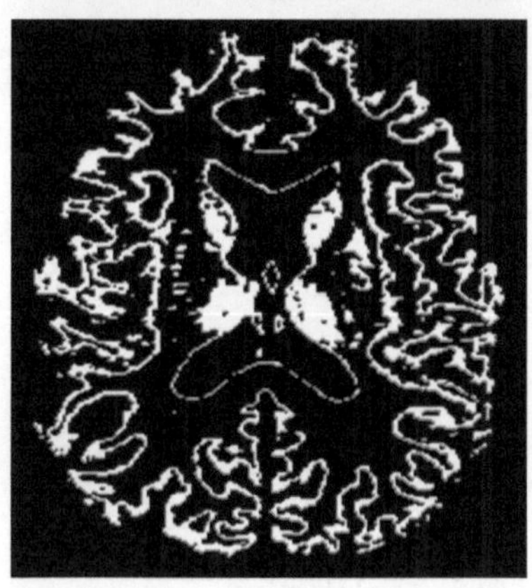

Fig. 7 $k = 5$ cluster

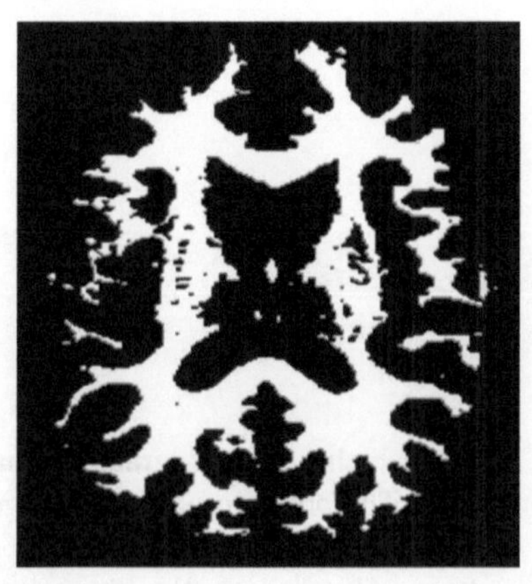

Fig. 8 $k = 6$ cluster

k-means algorithm is to determine the number of clusters N. Formally; N should be two as we distinguish edges and non-edges. However, in fact, the number of non-edge pixels often dominates the pixel population (from 75 to 95 %). Therefore, setting the number of clusters to two often results in losing edges since points in spectral space tend to belong to non-edge clusters rather than edge clusters. In practice, N should be set to be larger than two. In this case, the cluster with the highest population is considered as non-edge cluster. The remaining $N-1$ cluster are merged together and considered as edge cluster. Apparently, detected edges depend on the number of clusters. Therefore, setting number of clusters should be investigated carefully based on the goal of specific application. Experimental results show that [4, 9] is a reasonable range of N for our application. After applying the k-means algorithm to classify each point in spectral space into one of N clusters, a combined classifier method is applied to remove noise as well as isolated edges [13]. The main idea of this method is to combine the results of two separate classifiers in spectral domain and spatial domain. This combining process is repeated until stable results are archived. In the proposed method, the results of two classifiers are combined using the maximum combination rule.

A thresholding algorithm as in the Canny edge detection method [11] is then applied to refine results from the previous step, i.e. to make the edges thinner. There are two different threshold values in the thresholding algorithm: a lower threshold and a higher threshold. Different from Canny's method, in which threshold values are based on gradient intensity, the proposed threshold values are determined based on the confidence of a pixel belonging to the non-edge cluster. A pixel in the edge cluster is considered as a "true" edge pixel if its confidence to the non-edge cluster is smaller than the low threshold. A pixel is also considered as an edge pixel if it satisfies two criteria: its confidence to the non-edge cluster is in a range between the two thresholds and it has a spatial connection with an already established edge pixel. The remaining pixels are considered as non-edge pixels. Confidence of a pixel belonging to a cluster used in this step is obtained from the clustering step.

The proposed algorithm is briefly described as follows:

Algorithm 1 Edge detection for medical images

Input: A medical image I, number of clusters N. Output: Detected edges of the image.

Step 1: Smoothing the medical image using Gaussian blur convolution.

Calculating pixel gradient value in each image channel.
Forming pixel as a point composed of gradient values over all bands.

Step 2: Applying k-means algorithm to classify points into N clusters.
Step 3: Refining the clustering result using the combined classifier method.
Step 4: Selecting the highest population cluster as non-edge cluster, merge other clusters as edge cluster.
Step 5: Applying the thresholding algorithm to refine results from Step 4.

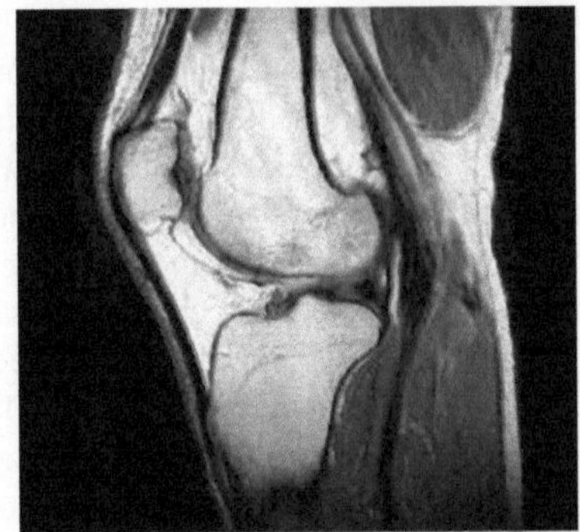

Fig. 9 Original image

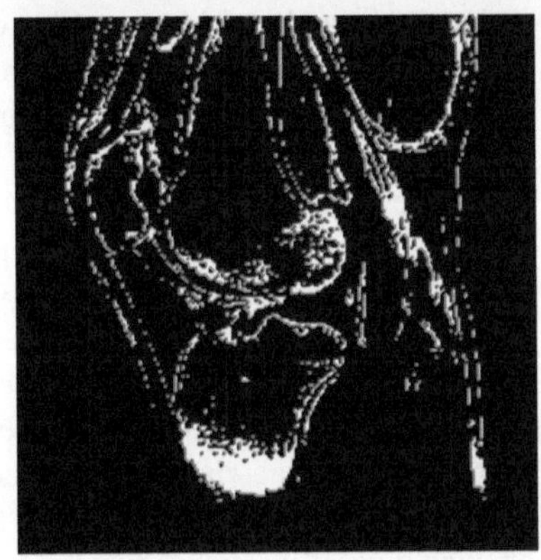

Fig. 10 Intensity clustering based edge detection

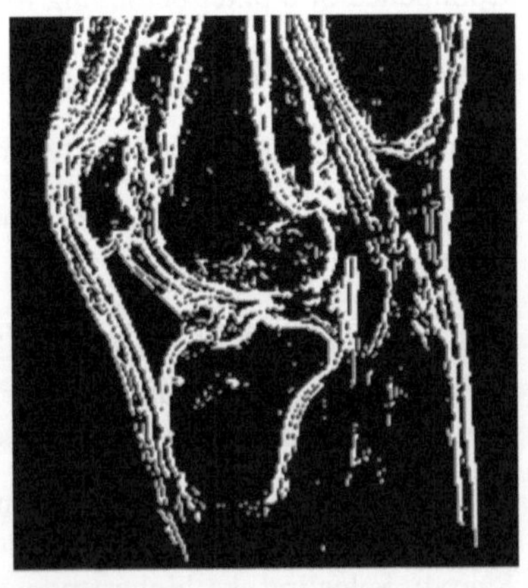

Fig. 11 Canny edge detector

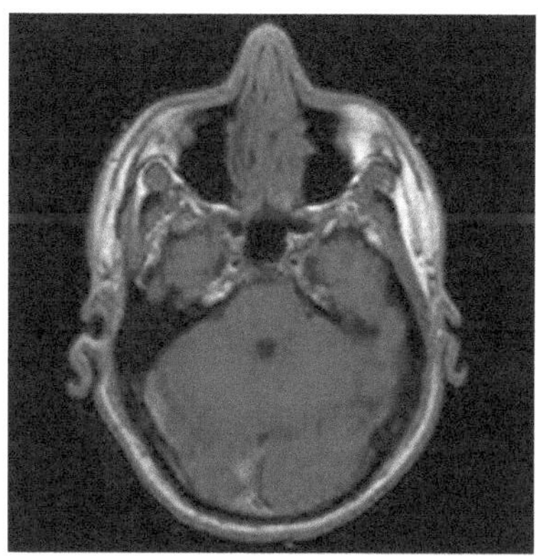

Fig. 12 Original image

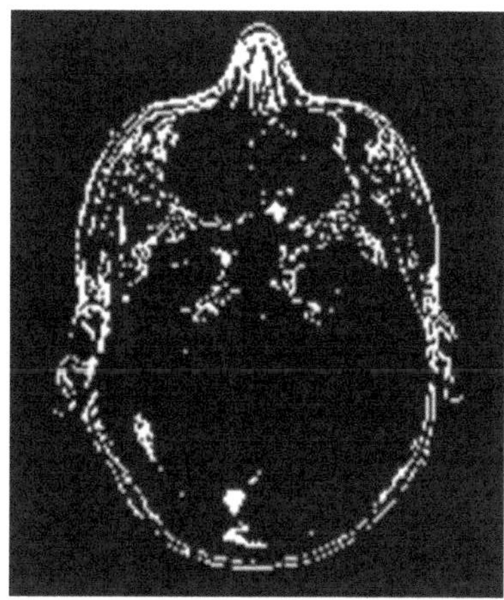

Fig. 13 Intensity clustering based edge detection

7 Result of Edge Detection

In order to evaluate the effectiveness of the proposed method, we have compared it with two typical edge detection methods: Huntsberger method [10], an intensity clustering based method and canny edge detection method. From the Figs. 9, 10, 11, 12, 13, 14 as shown below which infers that Hunts Berger method performs worst: losing edges and creating discontinuous edges. Therefore, the focus will be on the performance between Canny method and the proposed method.

Fig. 14 Canny edge detector

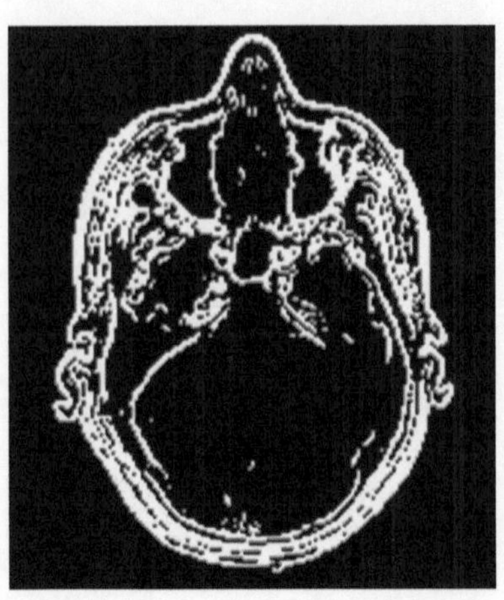

8 Conclusion

In this paper we have presented an algorithm for scalable image retrieval by color, where images are represented by local color histograms and edge detection is proposed. The similarity between images is calculated based on local color properties. Images in the database are clustered into groups having similar color contents. This grouping enables searching only images that are relevant to the query image. The proposed edge detection method enables the use of multivariate statistical information in multi-dimensional space. Based on pixel values, it also provides a better representation of edges comparing to those based on intensity values, e.g. Huntsberger method [10]. As the results, the method reduces the affect of noise and preserves more edge information in the images. More thorough investigation in determining the number of clusters N and the values for the high and low thresholds must be further invested to improve the results.

Acknowledgments The authors express their sincere thanks to the Management and the Principal of Bannari Amman Institute of Technology, Sathyamangalam for providing the necessary facilities for the completion of this paper.

References

1. Swain MJ, Ballard DH (1991) Color indexing. Intl J Comput Vision 7(1):11–32
2. Robinson GS (1977) Colour edge detection. Opt Eng 16(5):479–484
3. Di Zenzo S (1986) A note on the gradient of a multiimage. Comput Vision Graph Image Process 33(116–125):1986
4. Trahanias PW, Venetsanopoulos AN (1993) Colour edge detection using vector statistics. IEEE Trans Image Process 2:259–264

5. Cumani A (1991) Edge detection in multispectral images. Graph Models Image Process 53:40–51
6. Chapron M (2000) A color edge detector based on statistical rupture tests. IEEE Int Conf Image Process II:820–823
7. Lambert P (1993) Using eigenvectors of a vector field for deriving a second directional derivative operator for color images. Int Conf Comput Anal Images Patterns 719:149–156
8. Koschan A, Abidi M (2005) Detection and classification of edges in colour images. Signal Process Magazine Special Issue Color Image Process 22:67–73
9. Barnett V (1976) The ordering of multivariate data. J Royal Statist 139(3):318–343
10. Huntsberger TL, Descalzi MF (1985) Color edge detection. Pattern Recogn Lett 3:205–209
11. Canny J (1986) A computational approach to edge detection. IEEE Trans Pattern Anal Mach Intell 8:679–698
12. Marr D, Hildreth E (1980) Theory of edge detection. In: Proceedings of royal society of London, pp 187–217
13. Paclik P, Duin RPW, van Kempen GMP, Kohlus R (2005) Segmentation of multi-spectral images using the combined classifier approach. Image Vision Comput J 21:473–482

5. [illegible] (1998) [illegible] multidimensional [illegible] Image Process [illegible]
6. [illegible] (2009) A [illegible] based [illegible] for [illegible]
7. [illegible] (1991) Using [illegible] of a [illegible] for deriving [illegible] directional [illegible] images. Int Conf Computational Image [illegible]
8. [illegible] (2003) [illegible] Magazine [illegible] Signal Process [illegible]
9. [illegible]
10. [illegible] (1985) [illegible] Pattern Recogn Lett 3:[illegible]
11. Canny J (1986) A computational approach to edge detection. IEEE Trans Pattern Anal Mach Intell [illegible]
12. Marr D, Hildreth E (1980) Theory of edge detection. In: Proceedings of royal society of London [illegible]
13. [illegible] (2005) Segmentation of [illegible] using the combined [illegible] Image Vision Comput [illegible]

A Comparative Analysis of Image Fusion Methods Using Texture

Jharna Majumdar and Bhuvaneshwari S. Patil

Abstract Image fusion produces a single image from a set of input images such that the fused image have more complete information useful for human or machine perception. In the proposed paper, authors have used feature based image fusion, where textures of the image are used as feature. Image Texture is a process that can be applied to the pixel of an image in order to generate a measure (feature) related to the texture pattern, to which that pixel and its neighbors belong. Authors have used five different texture feature extraction methods for fusion of multi sensor, multi focal, multi temporal and multi spectral imagery. The methods are: GLCM, Runlength, Statistical, Tamura and Texture Spectrum. The performance of fusion algorithm is measured using a number of nonreference quality assessments metric. A meaningful comparison of results and analysis show the suitability of various texture features for fusion of images from multiple modalities.

Keywords GLCM · Runlength · Statistical · Tamura and texture spectrum · Image fusion · Quality parameters

1 Introduction

Image fusion is an emerging technology for many military, surveillance and medical applications. Image fusion combines the information of the same scene of two or more images from identical and different type of sensors to generate

J. Majumdar (✉)
Department of CSE(PG), Nitte Institute of Technology, Bangalore, India
e-mail: jharna.majumdar@gmail.com

B. S. Patil
Department of CSE(PG), Nitte Institute of Technology, Bangalore, India

Mohan S. and S. S. Kumar (eds.), *Proceedings of the Fourth International Conference on Signal and Image Processing 2012 (ICSIP 2012)*, Lecture Notes in Electrical Engineering 221, DOI: 10.1007/978-81-322-0997-3_31,

precise, comprehensive and reliable image description or interpretation of the scene, which is more suitable for human visual perception or subsequent higher level processing such as segmentation, feature extraction, target identification etc. A review of image fusion techniques is available in [1, 2].

With the advancement of technology today, large number of sensors is available at affordable cost. Each of these sensors, works on different band of electromagnetic spectrum, provides information of the same scene at different spatial, temporal and spectral resolutions. This provides multi sensor, multi focal, multi temporal and multi spectral image data. For the full exploitation of increasingly sophisticated data from multiple sources, advance image fusion techniques [3, 4] and analysis tools to find their suitability for a given scenario is important [5, 6]. Effective contribution of such sensor with variety of features could extend the value of the final composite image which will have more complete and detail information content. Fused image may provide increase interpretation capabilities and bring more reliable results since data with different characteristics are combined. Based on the above argument, we have used multi sensor, multi focal, multi temporal and multi spectral imagery for our study.

Texture is a key component for human visual perception [7, 8]. Texture is the change and repetition of image grey in space and is defined as a function of the spatial variation in pixel intensities. Image texture can be quantitatively evaluated as having one or more of the properties such as fineness, coarseness, smoothness, granulation, randomness or regular. Texture features are one kind of features which can reflect the visual property of homogenous performance in images and are independent of color and brightness. A texture feature extraction method is a process that can be applied to a pixel of a given image in order to generate a measure (feature) related to the texture pattern to which that pixel and its neighbors belong. The performance of the different families of texture methods basically depends on the type of processing they apply, the neighborhood of pixels over which they are evaluated and the texture content. In the present work, instead of using image intensity we have used texture as a 'cue' for image fusion.

In several application areas, image fusion is an introductory stage to many important tasks like human monitoring, activity tracking, and situation awareness [9, 10]. In those situations, performance of fusion algorithm must be measured in terms of improvement in the subsequent tasks. For many applications, human perception of the resultant image after fusion is of fundamental importance and as a consequence fusion results are evaluated by subjective measurement. Objective measurement for evaluation of image fusion performance is not easy, since most of them require an ideal fused image, called reference image for the purpose of comparison. Mean squared error (MSE) based metrics are widely used which needs a reference image. Recently researchers have come up with several objective performance metrics for image fusion, which do not need the knowledge of ground truth. Some of them are used in this work.

In this paper, our basic goal is to study a large number of texture feature extraction methods for fusion of multi modal imagery and conclude using quality metric which features are most effective for fusing a given kind of imagery. To start with, authors have made a database containing images from multiple modalities. A study is made for the following texture feature extraction methods such as, GLCM Method, Run Length Method, Statistical Method, Tamura Method and Texture Spectrum methods. After features are extracted and three different strategies for fusion such as average, weighted average and maximum based fusion are used. A set of objective non reference quality parameters one used to analyze the fusion results are used. The final goal of this paper is to find out for a given set of input image, the most suitable texture feature which would give a fused image with more information content compared to other features.

2 Methodology

2.1 Texture Feature Extraction Methods

In this section, the texture feature extraction methods used in the present study is discussed in brief.

2.1.1 Gray Level Co-occurrence Matrix

The co-occurrence matrix defines the spatial relationship between any two pixels in an image as a function of distance and direction [11–13]. It gives the frequency of occurrence of a pair of intensity values in the given image. The texture features extracted to generate feature images using co-occurrence method are shown below. Here P (i, j) denotes the probability of the co-occurrence of a pair of pixels.

$$\text{Angular second moment} = \sum_{i,j=0}^{N-1} P_{i,j}^2 \quad (1)$$

$$\text{Entropy} = \sum_{i,j=0}^{N-1} P_{i,j}(-\ln P_{i,j}) \quad (2)$$

$$\text{Contrast} \sum_{i,j=0}^{N-1} P_{i,j}(i-j)^2 \quad (3)$$

$$\text{Inv. Diff. Moment} \sum_{i,j=0}^{N-1} \frac{P_{i,j}}{1+(i-j)^2} \quad (4)$$

$$\text{Maximum probability}\, Max\,(P_{i,j}) \quad (5)$$

$$\text{Variance}\ \sigma_i^2 = \sum_{i,j=0}^{N-1} P_{i,j}(i - \mu_i)^2 \tag{6}$$

The features that characterize the content of the co-occurrence matrix are the following: Angular Second Moment or Energy, a measure of textual uniformity of an image; Entropy, a measure of the disorder in an image; Contrast, a measure of the amount of local variance in an image; Inverse Difference Moment, a measure of Homogeneity.

2.1.2 Run Length Method

A run length is a continuous set of maximum number of pixels in the same direction that have the same gray level. Each run is defined by its grey level, length and direction. Let B (a, r) is the number of runs in all directions having length r and grey level a. Let M, N be image dimensions, L the number of grey levels, Nr the maximum run length in the images, and K the total number of runs given by

$$\sum_{a=1}^{L}\sum_{r=1}^{N_r} B(a,r), \tag{7}$$

then we can define the following features defining image texture[14]. Six Run Length features are No of primitives; Short run emphasis, divides each run length value by the length of the run squared; Long run emphasis, multiples each run length value by the length of the run squared; Entropy, squares the number of run lengths for each grey level, When runs are equally distributed throughout the grey levels, the function takes on its lowest values; Run length non uniformity, squares the number of runs for each length and if the runs are equally distributed throughout the lengths, the function will have a low value; and Run Percentage, a ratio of the total number of runs to the total number of possible runs if all runs had a length of one and it should have its lowest value for pictures with the most linear structure.

$$\text{Short Run Emphasis} = \frac{1}{K}\sum_{a=1}^{L}\sum_{r=1}^{Nr}\frac{B(a,r)}{r^2} \tag{8}$$

$$\text{Long Run Emphasis} = \frac{1}{K}\sum_{a=1}^{L}\sum_{r=1}^{Nr} B(a,r)r^2 \tag{9}$$

$$\text{Entropy} = \frac{1}{K}\sum_{a=1}^{L}(\sum_{r=1}^{Nr} B(a,r))^2 \tag{10}$$

$$\text{Run Length uniformity} = \frac{1}{K}\sum_{r=1}^{Nr}\left(\sum_{a=1}^{L} B(a,r)\right)^2 \tag{11}$$

$$\text{Run Percentage} = \frac{K}{\sum_{a=1}^{Nr}\sum_{r=1}^{L} rB(a,r)} \tag{12}$$

In the above equation K is given by $K = \sum_{a=1}^{L}\sum_{r=1}^{Nr} B(a,r)$.

2.1.3 Statistical Method

In the Statistical method, texture features are extracted based on the statistics of geometrical properties of connected regions in a sequence of binary images [15]. The first step of the approach is to decompose a texture image into a set of binary images. For each binary image, geometrical attributes such as the number of connected regions and their irregularity are statistically considered. These sets of binary images are termed as binary stack.

For each binary image f b(x, y; α) where α is the threshold, all 1-valued pixels in the binary stack are grouped into a set of connected pixels termed as NOC1(α) and all 0-valued pixels in the same stack are grouped as NOC0(α). Statistical method gives four features such as 0-Mean, 0-Variance, 1-Mean and 1-Variance

$$Mean = \frac{1}{\sum_{\alpha=1}^{n_l-1} g(\alpha)}\sum_{\alpha=1}^{n_l-1} \alpha.g(\alpha) \tag{13}$$

$$Variance = \frac{1}{\sum_{\alpha=1}^{n_l-1} g(\alpha)}\sum_{\alpha=1}^{n_l-1} (\alpha - sample\_mean)^2.g(\alpha) \tag{14}$$

where g(α) is one of the two functions NOC1(α) and NOC0(α).

2.1.4 Tamura Texture Method

Tamura et al. [16] explored texture representation from different angle and proposed a computational approximation on visual Properties like coarseness, contrast, line-likeness, regularity and roughness. These features correspond more closely to human visual perception [7, 8].

$$\text{Coarseness, } F_{crs} = \frac{1}{m \times n} \sum_{i}^{m} \sum_{j}^{n} S_{best}(i,j) \tag{15}$$

$$\text{Contrast, } F_{con} = \sigma / (\alpha_4)^n \tag{16}$$

$$\text{Roughness, } F_{rgh} = F_{crs} + F_{con} \tag{17}$$

$$\text{Line likeness, } F_{lin} = \sum_{i}^{n} \sum_{j}^{n} P_{Dd}(i,j) \cos\left[(i-j)\frac{2\pi}{n}\right] \Big/ \sum_{i}^{n} \sum_{j}^{n} P_{Dd}(i,j) \tag{18}$$

$$\text{Regularity, } F_{reg} = 1 - r(\sigma_{crs} + \sigma_{con} + \sigma_{rgh} + \sigma_{lin}) \tag{19}$$

Coarseness relates to distances of notable spatial variations of gray-levels; Contrast measures how the gray-levels vary in the image and to what extent their distribution is biased to black or white; Roughness, gives sum of coarseness and contrast measures; Line likeness is defined as an average coincidence of the edge directions that co-occurred in the pair of pixels separated by a distance 'd' along the edge direction in every pixel; regularity, regular in a wide range of viewing angles.

2.1.5 Texture Spectrum Method

A complete definition of the texture spectrum employs the determination of values as the texture unit, the texture unit number (NTU) and finally the texture spectrum [17, 18]. The texture units (TU) describe the local texture of a given pixel whereas the texture spectrum gives the global texture characteristics of an image. Texture Spectrum is the frequency distribution of all the TU, with x-axis indicating texture unity and y-axis indicating the frequency occurrence of the TU. In the texture spectrum, the increase in percentage of texture components in an image will result in a tendency to form a particular distribution of peaks.

Texture Spectrum approach gives four texture features other than TU and NTU. These parameters are the texture features. They are defined as: Black-White symmetry, Central Symmetry, Geometric Symmetry and Degree of Direction

$$\text{Black - White symmetry} \quad BWS = \left[1 - \frac{\sum_{i=0}^{3279} |S(i) - S(3281+i)|}{\sum_{i=1}^{6560} S(i)}\right] \times 100 \tag{20}$$

$$\text{Central Symmetry } CS = \sum_{i=0}^{6560} S(i) \times [K(i)]^2 \tag{21}$$

Geometric Symmetry $$GS = \left[1 - \frac{1}{4}\sum_{j=1}^{4}\frac{\sum_{i=0}^{6560}|S_j(i) - S_{j+4}(i)|}{2 \times \sum_{i=1}^{6560} S_j(i)}\right] \times 100 \quad (22)$$

Degree of Direction $$DD = \left[1 - \frac{1}{6}\sum_{m=1}^{3}\sum_{n=m+1}^{4}\frac{\sum_{i=0}^{6560}|S_m(i) - S_n(i)|}{2 \times \sum_{i=1}^{6560} S_m(i)}\right] \times 100 \quad (23)$$

2.2 Quality Metric

The quality metric represents how well the important information in the source images is represented in the fused image. Identification of suitable quality metric for measuring the performance of image processing algorithms is a challenging task.

In this paper, we have used objective measurement of quality parameters to assess the performance of image fusion [19–22]. The parameters are Cross Entropy, measures the similarity in information content between the source and the fused image; Wang and Bovik measures UQI that measures image similarity across distortion types; Gemma Piella Variance, Gemma Piella Entropy, Gemma Piella Spatial Frequency, all of which are used to measure the quality of the fused image with respect to the original images.

$$Cross\ \text{Entropy},\ CE(X;Z) = \sum_{i=0}^{L} h_x(t)\log_2\frac{h_x(t)}{h_z(t)} \quad (24)$$

Wang and Bovik,

$$Q0 = 4\sigma_{xy}\bar{x}\bar{y}/(\bar{x}^2 + \bar{y}^2)(\sigma_x^2 + \sigma_y^2) \quad (25)$$

or

$$Qo = \frac{\sigma_{xy}}{\sigma_x\sigma_y}\cdot\frac{2\bar{x}\bar{y}}{x^2 + \bar{y}^2}\cdot\frac{2\sigma_x\sigma_y}{\sigma_x^2 + \sigma_y^2} \quad (26)$$

Gemma Piella Fusion Quality Index,

$$Q\,(a,b,f) = \frac{1}{|W|}\sum_{w\in W}(\lambda_a(w)Q_o(a,f|w) + \lambda_b(w)Q_o(b,f|w)) \quad (27)$$

3 Experimental Results and Analysis

In the proposed work, we have studied the performance of image fusion algorithms for fusion of multi sensor, multi focal, multi temporal and multi spectral imagery.

Five different texture features extraction methods are used for 3 sets of images from each category as shown in Fig. 1. Although there are large amount of research work on fusion algorithms, no universally accepted standard has emerged for evaluating image fusion performance. In our experiments, after a study of large number of quality parameters for fusion based on average feature value, maximum

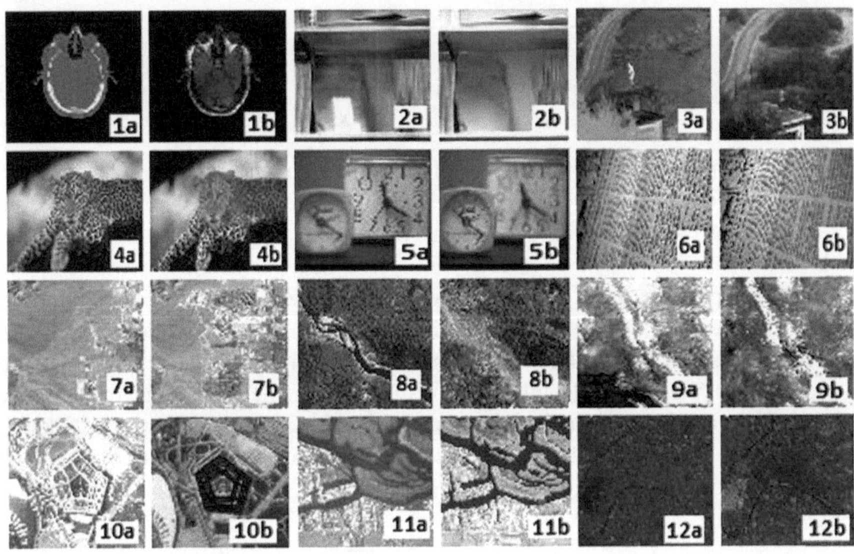

Fig. 1 *Row1* (*1–3*) multi sensor images, *row2* (*4–6*) multi focal images, *row3* (*7–9*) multi temporal image, *row4* (*10–12*) multi spectral images

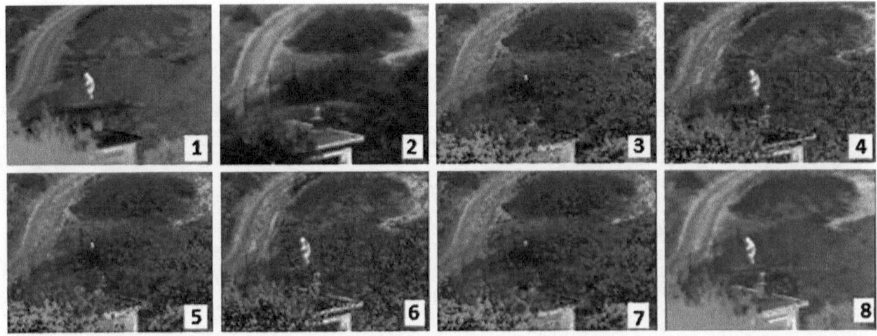

Fig. 2 Multi sensor images (484 × 363); *1* and *2* are input image; *3–8* are fused images using GLCM

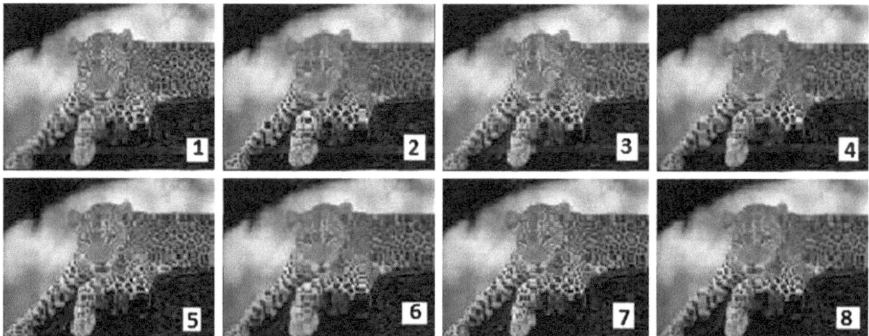

Fig. 3 Multi focal images (480 × 360); *1* and *2* are input image; *3–8* are fused images using GLCM

feature value and weighted average feature value based techniques. It has been observed that fusion using maximum feature value results is better since the output images have absorbed valid information from both the input images. A database of results is generated for the quality parameters which are analyzed to select the best feature for each method. A sample database of fused image using GLCM based fusion is shown in Figs. 2, 3, 4, 5 and Tables 1, 2, 3, 4, 5.

The results of image fusion are evaluated qualitatively by visual observation and also quantitatively by measurement of image quality metric. Many existing quantitative evaluation method requires an ideal fused image called a reference image. In our case, such an image was not available; hence quantitative evaluation was done which do not require a reference image.

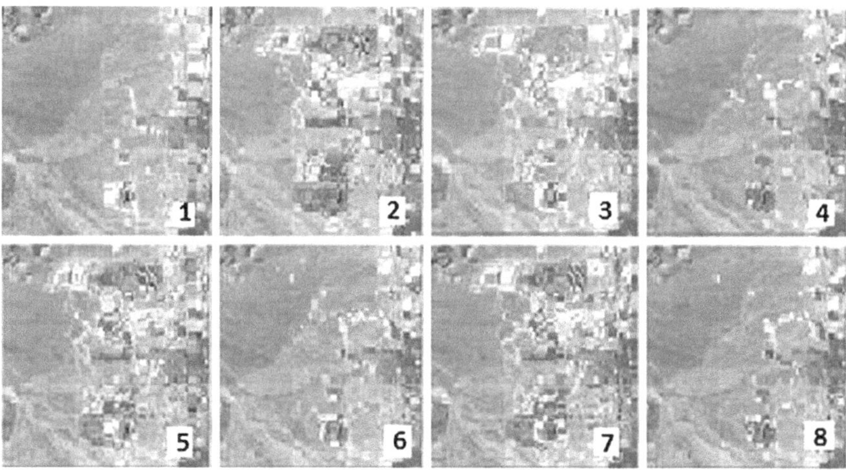

Fig. 4 Multi temporal images (200 × 200); *1* and *2* are input images; *3–8* are fused images using GLCM

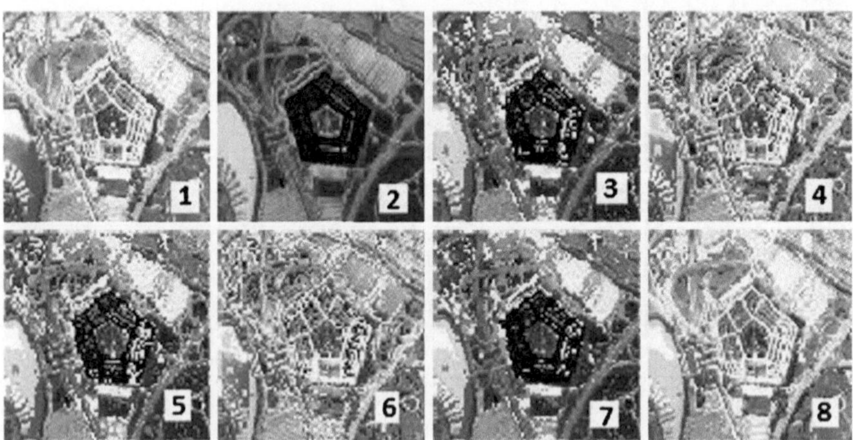

Fig. 5 Multispectral images of size 300 × 300; *1* and *2* are input images; *3–8* are fused images using GLCM

Experiment is conducted on a large set of images from different modalities and the results of image fusion using different image attributes are analyzed using quality assessment algorithms. This is to estimate how well information contained

Table 1 Quality metric parameters for multi sensor image fusion using GLCM

Quality metric parameters	Image type—multi sensor image fusion Texture feature extraction method—GLCM					
	Ang sec moment	Contrast	Max. probability	Entropy	Inv diff moment	Variance
Cross entropy	0.190	0.194	0.190	0.193	0.189	**1.180**
Wang and Bovik	0.227	0.334	0.223	0.328	0.227	**0.586**
G Piella variance	0.233	0.296	0.225	0.288	0.234	**0.592**
G Piella entropy	0.230	0.325	0.227	0.323	0.232	**0.461**
Piella spatial frequency	0.220	0.324	0.214	0.314	0.220	**0.604**

Table 2 Quality metric parameters for multi focal image fusion using GLCM

Quality metric parameters	Image type—multi focal image fusion Texture feature extraction method—GLCM					
	Ang sec moment	Contrast	Max. probability	Entropy	Inv diff moment	Variance
Cross entropy	0.008	**0.009**	0.007	0.008	0.008	0.007
Wang and Bovik	0.797	**0.864**	0.798	0.863	0.796	0.842
G Piella variance	0.821	**0.836**	0.820	0.836	0.821	0.834
G Piella entropy	0.797	**0.864**	0.798	0.863	0.796	0.842
Piella spatial frequency	0.748	**0.853**	0.751	0.852	0.747	0.815

Table 3 Quality metric parameters for multi spectral image fusion using GLCM

Quality metric parameters	Image type—multi temporal image fusion Texture feature extraction method—GLCM					
	Ang sec moment	Contrast	Max. probability	Entropy	Inv diff moment	Variance
Cross entropy	0.264	0.272	**0.246**	0.273	0.266	0.444
Wang and Bovik	0.549	0.728	0.551	0.717	0.544	**0.738**
G Piella Variance	0.587	0.668	0.575	0.659	0.583	**0.692**
G Piella entropy	0.552	0.72	0.55	0.708	0.549	**0.721**
Piella spatial frequency	0.549	0.72	0.551	0.709	0.547	**0.736**

Table 4 Quality metric parameters for multi temporal image fusion using GLCM

Quality metric parameters	Image type—multi spectral image fusion Texture feature extraction method—GLCM					
	Ang sec moment	Contrast	Max. probability	Entropy	Inv diff moment	Variance
Cross entropy	0.900	1.039	**0.819**	0.994	0.902	1.109
Wang and Bovik	0.355	0.540	0.337	0.505	0.360	**0.641**
G Piella variance	0.340	0.504	0.317	0.471	0.348	**0.641**
G Piella entropy	0.390	0.442	0.345	0.423	0.407	**0.481**
Piella spatial frequency	0.317	0.511	0.301	0.479	0.325	**0.620**

Table 5 Performance of different texture methods for fusion of multi modal imagery

Texture methods for fusion	Fusion of images from different modalities			
	Multi sensor	Multi focal	Multi temporal	Multi spectral
GLCM	Variance	Variance	Variance	Variance
Run length	Run percentage	Run percentage	Run percentage	Run percentage
Statistical	No suitable feature	No suitable feature	No suitable feature	No suitable feature
Tamura	Regularity	Regularity	Regularity	Regularity
Texture spectrum	Black_white_ symmetry	Black_white_ symmetry	Black_white_ symmetry	Black_white_ symmetry

in the source images are represented in the fused image. The best quality metric values are highlighted for each method. It is observed that statistical method of texture do not show any noticeable feature which could be considered for fusion. The results show that 'variance' feature from GLCM is best suited for fusion of all types of images considered in our present study. Similarly, analysis to find the best features for fusion are found for Run Length is 'Run Percentage'; for Tamura is 'Regularity'; for Texture Spectrum is 'Black and White Symmetry' and for

Statistical methods no unique feature found and the findings are tabulated in Table 5. Detail results are shown in the appendix.

4 Conclusion and Future Work

The feature extraction using texture methods is performed using GLCM, Run Length, Statistical, Tamura and Texture Spectrum methods. Extensive study and analysis of results show that for each of the texture feature extraction methods, there exists one feature which gives a comparatively fused image. Further it is seen that the statistical feature extraction method does not show any suitable feature for any set of input images. Quality parameters are used to assess the quality of the fused image.

Future research work in this direction could be to find out suitable weights for each of these best features found from each method. Instead of using one texture method for fusion, we could use a combination of different methods with appropriate weights for each feature. We could also use some of the more recent texture feature extraction methods to study their effects on fusion. The findings of this paper may be especially useful for fusion of medical images obtained from different modalities, where accuracy of fusion plays a major role in the diagnosis process. The findings may also be useful for surveillance applications, where fusion of multiple data from sensors such as Visible and IR could play an important role to ascertain the appearance of a target in a given scenario.

Acknowledgments The authors gratefully acknowledge VTU Research Fund, VTU Belgaum for sponsoring the Project at the Nitte Meenakshi Institute of Technology, Bangalore. The authors gratefully acknowledge Venkatesh G M, Research Associate of NMIT for his contribution during the course of this work. The authors acknowledge the constant support and encouragement provided by Dr. N.R. Shetty, Director NMIT and Dr. H C Nagaraj, Principal NMIT during the course of this work.

References

1. Sadjadi F (2005) Comparative image fusion analysis. In: IEEE computer society conference on computer vision and pattern recognition, CVPR'05, 2005
2. Wang Z, Ziou D, Armenakis C, Li D, Li Q (2005) A comparative analysis of image fusion methods. IEEE Trans Geosci Remote Sens 43(6):1391–1402
3. Liu Z, Blasch E, Xue Z, Zhao J, Laganiere R, Wu W (2012) Objective assessment of multiresolution image fusion algorithms for context enhancement in night vision: a comparative study. IEEE Trans Pattern Anal Mach Intell 34(1):94–109
4. Jian Yang Jingfeng Guo (2011) Image texture feature extraction method based on regional average binary gray level difference co-occurrence matrix. IEEE Int Conf Virtual Reality Vis 10(3):61–64

5. Gui Y, Chen M, Ma L, Chen Z (2011) Texel based regular and near-regular texture characterization. In: International conference on multimedia and signal processing (CMSP-2011), IEEE, 14–15 May 2011
6. Mumtaz A, Majid A, Mumtaz A (2008) Genetic algorithms and its application to image fusion. In: 2008 international conference on emerging technologies, IEEE-ICET, 18–19 October 2008
7. Sheng Z, Wen-Zhong S, Liu J, Zhu G-X, Tian J-W (2007) Multisource image fusion method using support value transform. IEEE Trans Image Process 16(17):1831–1839
8. Alam FI, Faruqui RU (2011) Optimized calculation of Haralick texture features. Eur J Sci Res IISN 1450-216X 50(4) 543–553
9. Bhiwani RJ, Khan MA, Agarwal SM (2010) Texture based pattern classification. Int J Comput Appl (0975-8887) 1(1):54
10. Borghys D et al (1997) Long range target detection in a cluttered environment using multi-sensor image sequences. In: Proceedings on signal processing, sensor fusion and target recognition IV—SPIE—USA (Orlando), 20–25 April 1997
11. Yi L, Yingle F, Jian X (2007) A new method based on fused features and fusion of multiple classifiers applied to texture segmentation. Ind Electron Appl, CIEA 2007
12. Haralick RM et al (1973) Textural features for image classification. IEEE Trans Syst Man Cybern SMC-3(6):610–621
13. Lohmann G (1995) Analysis and synthesis of textures: a co-occurrence-based approach. CVGIP 19(1):29–36
14. Verlinde P et al (1997) Data fusion for a long range target acquisition. In: Seventh symposium on multi sensor systems and data fusion for telecommunications, Agard, NATO, 1997
15. Galloway MM (1975) Texture analysis using gray level run lengths. CVGIP 4:172–179
16. Chen YQ, Mark SN, David WT (1995) Statistical geometrical features for texture classification. Pattern Recogn 28(4):537–552
17. Tamura H et al (1978) Textural features corresponding to visual perception. IEEE Trans Syst Man Cybern SMC-8(6)
18. He D-C, Wang L (1990) Texture unit, texture spectrum, and texture analysis. IEEE Trans Geosci Remote Sens 28(4):509–512
19. He D-C, Wang L (1990) Texture features based on texture spectrum. Pattern Recognit 24:905–910
20. Wang Z, Bovik AC (2002) A universal image quality index. IEEE Signal Process Lett 9(3):81–84
21. Piella G 2004 New quality measures for image fusion. In: Proceedings of international conference on information fusion, Sweden. 28 June–1 July 2004
22. Cvejic N, Seppänen T, Godsill SJ (2009) A nonreference image fusion metric based on the regional importance measure. IEEE J Sel Top Sign Proces 3(2):212–221

Performance Evaluation of Gabor Filter in Removing Rician Noise in MR Images

J. Papitha and D. Nedumaran

Abstract Magnetic Resonance images are contaminated by Rician distributed noise due to the presence of contrast-diminishing signal-dependent bias. In this paper, Gabor filter approach for bias removal in MR images is attempted. The filter was tested in four different brain MR images and the results obtained were compared qualitatively and quantitatively with the other four established filtering techniques. This study exhibits that the Gabor bias removal technique improved the contrast of the MR image that is found from the moderate increase in PSNR value and visual inspection by trained radiologist.

Keywords: Rician noise · Gabor filter · MRI · Denoising · Artifacts

1 Introduction

Magnetic resonance imaging (MRI) is a powerful imaging technique to visualize particularized internal structures of human body due to its ability to produce good contrast between various soft tissues of the body. In MRI machines, powerful magnetic field, radio frequency pulses and a computer are integrated as a system to produce detailed pictures of organs, soft tissues, bone and virtually all other internal body structures. MR images are suffered with serious artifacts, in particular bias due to the existence of a non-uniform brightness caused by the radio frequency inhomogeneity and intensity deviation of the device magnetic field.

J. Papitha · D. Nedumaran (✉)
Central Instrumentation and Service Laboratory, University of Madras, Maraimalai Campus, Guindy, Chennai 600 0025 Tamilnadu, INDIA
e-mail: dnmaran@gmail.com, dnmaran@yahoo.com

Mohan S. and S. S. Kumar (eds.), *Proceedings of the Fourth International Conference on Signal and Image Processing 2012 (ICSIP 2012)*, Lecture Notes in Electrical Engineering 221, DOI: 10.1007/978-81-322-0997-3_32,

In MR image processing, the estimation and removal of the bias artifact is an ascendant step to assess the quality and the consistency of the MR images.

Several techniques have been found in the literature for the removal of bias artifact. Gudbjartsson and Patz analyzed the magnitude and phase MRI images and found that the noise distribution is nearly Gaussian for the signal-to-noise ratio (SNR) value larger than two [1]. Nowak demonstrated a wavelet domain filter operated on the squared magnitude MR image using Chi square distribution to remove noise and bias effectively both in high as well as low SNR MR images, since Gaussian approximation and Rician distribution are no longer valid for low SNR images [2, 3]. Basu et al., presented a maximum a posteriori (MAP) estimation technique for filtering diffusion tensor magnetic resonance images, which operated directly on the diffusion weighted images and reduced the bias introduced by the Rician noise effectively [4]. Ardizzone et al., derived a bias correction model that was obtained with a particular configuration of the Gabor filter and was applied for MR image restoration purpose [5]. Aelterman et al., proposed combined bias removal and denoising using the squared and square root magnitude of the image, respectively in the wavelet domain, which showed a significant increase in the contrast of the image as well as drastic increase in the peak signal to noise ratio (PSNR) [6].

Daessle et al., attempted improved Non Local Means (NLMeans) filter for Rician noise removal in low SNR MR and diffused-weighted MR images and found that the proposed method outperformed the existing NLMeans technique in terms of PSNR value [7]. Aja-Fernández et al., derived a closed-form solution of the linear minimum mean square error (LMMSE) estimator for Rician noise removal using the local variance, the local mean, and the local mean square value, which reduced the noise dynamically through iteration and exhibited good performance in noise removal as well as feature preservation [8]. Coupe et al., proposed a robust Median Absolute Deviation (RMAD) estimator in the wavelet domain for Rician noise and tested on the real and synthetic images. The experimental results exhibited improvements in accuracy and robustness [9].

He and Greenshields proposed a nonlocal maximum likelihood (NLML) estimation method for Rician noise reduction, which optimally estimated the true signal from Rician noise in terms of SNR, contrast and error [10]. Dakua and Sahambi discussed the effect of Rician noise on image segmentation and found that the eigen values describing the contrast of the image were decreasing with increase in noise level [11]. Krissian and Aja-Fernández modified the speckle reducing anisotropic diffusion (SRAD) filter by accurately measuring the noise standard deviation using local variance of the image intensity. They also introduced a new matrix diffusion extension of scalar partial differential equation, which performed better smoothing of the boarders of the different tissues [12]. Rajeesh et al., proposed a wave atom shrinkage method to remove the Rician noise using a histogram based technique for threshold selection [13]. Nobi and Yousuf proposed a combined median and mean filter for removing the noise present in the ultrasound and MR images. The performance of the proposed algorithm was compared with the existing algorithms in terms of PSNR, SNR and RMSE values and found to be comparable with the other algorithms [14].

This paper presents a Gabor based approach to remove bias present in the MR images. The paper is organized as follows: In Sect. 2, we briefly reviewed the artifacts in MR images. In Sect. 3, the characteristics of Rician noise is discussed. Sections 4, 5 discusses the basics of Gabor filter and its functional parameters. Sections 6, 7, 8 discuss the testing and comparison of the proposed technique with the existing techniques. Finally, Sect. 9 summarizes the concluding remarks arrived out of this study.

2 Artifacts in MRI

MRI has various artifacts due to tangled MRI image formation and factors exotic to physiological requirements, which affect image formation and become visible on the resulting MRI. Brown and Semelka listed the different sources of artifacts that affect the MR images such as hardware, software, physiological phenomena and physics limitations [15]. Motion Artifacts in MRI are due to the blood flow direction relative to the slice orientation for arterial pulsations, swallowing, breathing, peristalsis, and physical movements, which tend to appear as ghosts along the phase-encoded direction. Aliasing is another artifact, which occurs when the field of view (FOV) is smaller than the body part due to under sampling in the phase or (rarely) frequency direction. Chemical shift artifact occurs, when the different resonant frequency of fat affects the frequency encoding direction. Further, truncation artifacts occur due to bright or dark lines that are seen parallel and next to borders of abrupt intensity change. Slice-overlap (cross-slice) artifacts exist due to the loss of signal seen in an image from a multi-angle, multi-slice acquisition. Magnetic susceptibility difference artifacts are variations in the magnetic field strength, which causes de-phasing of spins and frequency shifts of the surrounding tissue. Homogeneity artifacts cause signal intensity changes and image distortions due to hardware imperfections and susceptibility effects within the human body. Phase cancellation artifact and coherence artifacts are protocol related artifacts. In addition, external artifacts occur due to magnetic field distortions, measurement hardware and noise. Finally, digital imaging artifacts have a variety of appearances, which include phase wrap-around artifacts and problems arising from the digital Fourier transform.

3 Rician Noise

The bias artifact present in MR images are governed by Rician distribution, which are accurately approximated as Gaussian for SNR value larger than two and the MR magnitude image can be considered as additive white Gaussian noise with standard deviation [16]. This bias is signal-dependent and is large for low SNR (dark regions) and small for high SNR (bright regions). Further, bias reduces contrast between bright and dark areas. The Rician noise has two degrading effects such as random fluctuation of pixel values and the introduction of a signal-dependent bias to the data that reduces image contrast [17, 18].

Standard imaging techniques acquire complex k-space data, which is used to calculate the image as the magnitude of the Fourier transform of the k-space data. Since, the magnitude image obeys Rician distribution, the Rician PDF of the Rician distribution is given by,

$$p(x) = \frac{x}{\sigma^2} \exp\left(\frac{x^2 + A^2}{-2\sigma^2}\right) I_0\left(\frac{xA}{\sigma^2}\right) \tag{1}$$

where x is a noisy image intensity of Rician distribution, I_0 is the zero-order modified Bessel function of the first kind, σ is standard deviation and A is the true image intensity value.

4 Gabor Filter

Gabor filter has been employed as an efficient filtering technique for many image processing applications viz., texture segmentation, target detection, fractal dimension management, document analysis, edge detection, retina identification and image coding and image representation [19]. It can be conveniently applied to such applications by convolving the Fourier Transform of the respective harmonic function with the Fourier Transform of the Gaussian function. The filter output has a real and an imaginary component representing orthogonal directions. Gabor filters can also be viewed as a sinusoidal plane of particular frequency and orientation, modulated by a Gaussian envelope. Further, Gabor filter design technique based on autocorrelation measurements is used for estimating the PDF of the Gabor filter output employing a Rician model [20].

According to Kranauskas, the two dimensional Gabor filter $g(x, y)$ expressed as product of isotropic Gaussian and complex exponential plane wave [21]: Complex

$$g(x, y; \lambda, \theta, \psi, \sigma, \gamma) = \frac{\gamma}{2\pi\sigma^2} \exp\left(-\frac{x'^2 + \gamma^2 y'^2}{2\sigma^2}\right) \exp\left(i\left(2\pi\frac{x'}{\lambda} + \psi\right)\right) \tag{2}$$

Real

$$g(x, y; \lambda, \theta, \psi, \sigma, \gamma) = \frac{\gamma}{2\pi\sigma^2} \exp\left(-\frac{x'^2 + \gamma^2 y'^2}{2\sigma^2}\right) \cos\left(2\pi\frac{x'}{\lambda} + \psi\right) \tag{3}$$

Imaginary

$$g(x, y; \lambda, \theta, \psi, \sigma, \gamma) = \frac{\gamma}{2\pi\sigma^2} \exp\left(-\frac{x'^2 + \gamma^2 y'^2}{2\sigma^2}\right) \sin\left(2\pi \frac{x'}{\lambda} + \psi\right) \tag{4}$$

where

$$x' = x \cos\theta + y \sin\theta \tag{5}$$

and

$$y' = -x \sin\theta + y \cos\theta \tag{6}$$

In Eqs. (2, 3 and 4), λ represents the wavelength of the sinusoidal factor, θ represents the orientation of the normal to the parallel stripes of a Gabor function, ψ is the phase offset, σ is the standard deviation of the Gaussian envelope and γ is the spatial aspect ratio that specifies the ellipticity of the support of the Gabor function.

5 Gabor Filter Parameters

For efficient bias removal in MR images, the Gabor filter parameters θ, λ, ψ, γ and σ has been assigned with some optimum values. The orientation angle parameter θ can chose any angle between 0 to 2π value, which determines the preferred direction and the spatial orientation of the filter. In this work, we have fixed either 0 or 90° for minimizing the local entropy value of the filtered image. The parameter γ is the spatial aspect ratio. The phase offset is symmetric when $\psi = 0$ *or* π, whereas it is anti-symmetric when $\psi = -\pi/2$ *or* $\pi/2$. It has asymmetric mixtures for other values of ψ. The standard deviation σ of the Gaussian factor determines the size of the receptive field. The parameter λ is the wavelength and $1/\lambda$ is the spatial frequency of the cosine factor. The ratio σ/λ determines the spatial bandwidth and the number of excitatory and inhibitory stripe zones in the receptive field [22, 23]. We assigned the half-response spatial frequency bandwidth b (one octave) for the ratio σ/λ using the Eq. (7):

$$\frac{\sigma}{\lambda} = \frac{1}{\pi} \sqrt{\frac{\ln 2}{2} \frac{2^b + 1}{2^b - 1}}. \tag{7}$$

The parameters θ, λ, ψ, γ and σ are assigned with the following values for the calculation of the Gabor filter function.

1. sigma = standard deviation of this Gaussian factor, (0.25)
2. gamma = aspect ratio, (0.5)
3. psi = phase shift, [0 $\pi/2$]
4. lambda = wave length, (1)
5. theta = angle in rad, (0).

6 Comparison Methods

Several objective quality metrics have been reported in the literature. The most commonly used quality metrics is the peak signal to noise ratio (PSNR), which provide the image quality in terms of the squared error between the original and the processed images [24]. The PSNR is defined by the expression,

$$PSNR(dB) = 10^{*} \log \left[\frac{255^2}{MSE} \right] \tag{8}$$

$$MSE = \sum_{i=1}^{x} \sum_{j=1}^{y} \frac{(|A_{ij} - B_{ij}|)}{x^{*}y} \tag{9}$$

where *MSE* is the mean square error value calculated using Eq. (9). 255 is the number of gray scale resolution (8-bit) of the image, *x, y* are the number of pixels of the MR image, A_{ij} and B_{ij} are the threshold values of the original and the denoised MR images, respectively.

The similarity between the two images can be measured using the mean structural similarity (MSSIM) index represented by the Eq. (11).

$$SSIM(x, y) = \frac{(2\mu_x \mu_y + C_1)(2\sigma_{xy} + C_2)}{(\mu_x^2 + \mu_y^2 + C_1)(\sigma_x^2 + \sigma_y^2 + C_2)} \tag{10}$$

$$MSSIM(X, Y) = \frac{1}{M} \sum_{j=1}^{M} SSIM(x_i, y_j) \tag{11}$$

Where μ_x, μ_y is average,σ_x^2, σ_y^2 is variance and σ_{xy} is covariance of *x, y*.

Similarly, contrast-to-noise ratio (CNR) can be determined using Eq. (12) for the comparison of the image quality between the raw and filtered image.

$$CNR_{AB} = \frac{|S_A - S_B|}{\sigma_0} \tag{12}$$

where S_A and S_B are intensities for signal producing structures *A* and *B* in the region of interest and σ_o is the standard deviation of the pure image noise.

7 Algorithm Development and Image Details

The Gabor algorithm was developed in MATLAB 7.1 in the Window XP operating system running in the Pentium IV 3 GHz PC. The MR brain images used in this study were taken from Siemens Esaote ARTOSCAN C MRI Machine with a magnetic field intensity of 0.15 Tesla. The image data were stored in DICOM file

Table 1 Specification of the four different brain MR image used in this study

Image name/ format	SP H42.2	SP F29.2	SP H10.0	SP A5.4
Scan option	PFP/SAT1	PFP/SAT1	PFP	IR
MR acquisition type	3D	3D	3D	2D
Slice thickness	0.6999999880	0.6999999880	3	5
Reception time	24	24	49	8600
Echo time	7	7	40	86
Imaging frequency	63.677701	63.677701	63.6777696	63.677687
No.of phase encoding	131	131	148	210
Field of view	175 × 220	175 × 220	172 × 230	201 × 230
Row × column, size	512 × 408, 499 kB	512 × 408, 499 kB	320 × 240, 248 kB	320 × 280, 281 kB

format. The algorithm was tested in four different MR images of brain and the image details are given in Table 1.

8 Testing

For estimating the performance of the Gabor filter, a comparative study was performed with established denoising techniques such as Median filter, Discrete Topological Derivative (DTD) filter, Wiener filter and Rician noise reducing anisotropic diffusion (RNRAD) filter. All the five filters (Gabor filter and other four traditional filters) were tested on more than 200 MR images. From the 200 images, four different MR images (SP H42.2, SP F29.2, SP H10.0, SP A5.4) of the brain and their denoised images are given in Fig. 1 for estimating the image quality improvements achieved in each filtering technique. Visual inspection of these images by the trained radiologists opined that the Gabor filtered images have achieved more homogeneity than others. Additionally, the Gabor filtered image exhibits better contrast improvements over the original and other filtered images, which is due to the presence of false/unfiltered contrast information originated from the Rician noise.

Further, to estimate the performance of the filtering technique quantitatively, peak signal to noise ratio (PSNR) and mean square error (MSE) have been calculated for all the four images and are given in Table 2 and Table 3. The calculated PSNR values are also given as a histogram chart as shown in Fig. 2 for easy comparison of the results obtained. From the graph shown in Fig. 2 and the Tables 2 and 3, it is observed that the PSNR and MSE values are found to be very high and very low, respectively for the Gabor filter, which substantiate that the Gabor filter method is a better choice for Rician noise removal in MR images than the other filtering techniques. For qualitatively measuring the performance of the

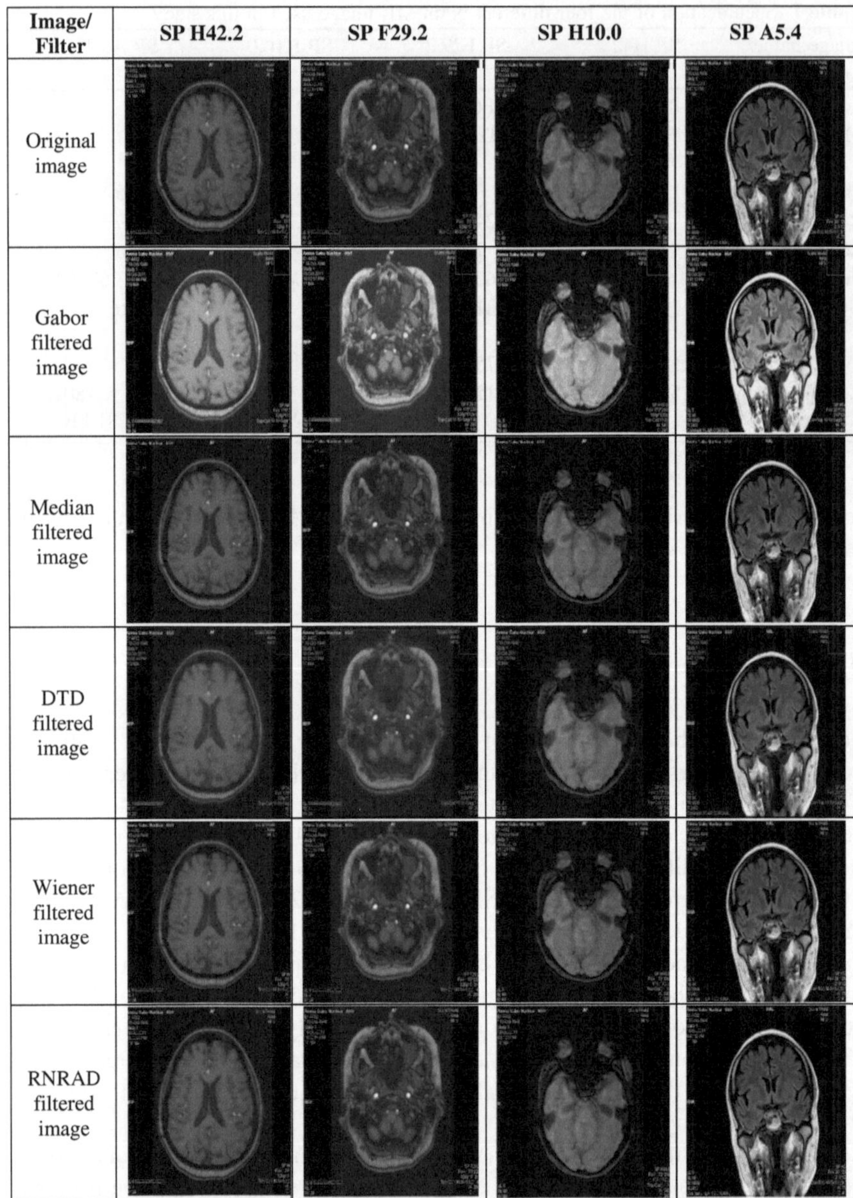

Fig. 1 Original and filtered MR images

filters, we have computed the mean structural similarity index matrix and contrast-to noise ratio values for all the four images and are given in the Tables 4 and 5, respectively. From the Tables 4 and 5, the values of MSSIM and CNR values are

Table 2 Calculated PSNR values of the denoising filters

PSNR values				
PSNR value/filter name	SP H42.2	SP F29.2	SP H10.0	SP A5.4
Gabor filter	65.8622	66.2068	67.1075	63.5461
RNRAD filter	43.7133	43.1582	43.7889	37.9522
Median filter	43.2673	43.1686	43.5315	41.7066
Wiener filter	42.1470	41.4953	42.2931	37.3915
DTD filter	40.6646	39.1612	40.3382	35.2579

Table 3 Calculated MSE values of the denoising filters

MSE value				
Image name/filter name	SP H42.2	SP F29.2	SP H10.0	SP A5.4
Gabor filter	0.0169	0.0156	0.0127	0.0287
RNRAD filter	2.7653	3.1424	2.7171	10.4199
Median filter	3.0644	3.1349	2.8836	4.3895
Wiener filter	3.9662	4.6084	3.835	11.8559
DTD filter	5.5799	7.888	6.0154	19.3773

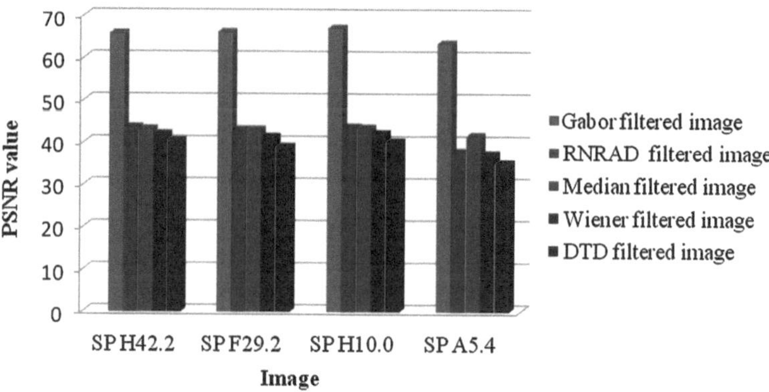

Fig. 2 Histogram plot of the calculated PSNR values of the four denoising filters

Table 4 Calculated MSSIM of the denoising filters

Mean structural similarity index matrix				
Image name/filter name	SP H42.2	SP F29.2	SP H10.0	SP A5.4
Gabor filter	0.9975	0.9977	0.9982	0.9964
RNRAD filter	0.9827	0.9791	0.9792	0.9663
Median filter	0.9381	0.9377	0.9442	0.9302
Wiener filter	0.9749	0.9692	0.9720	0.9527
DTD filter	0.8217	0.8122	0.7845	0.8702

Table 5 Calculated CNR of the denoising filters

Contrast to noise ratio				
Image name/filter name	SP H42.2	SP F29.2	SP H10.0	SP A5.4
Gabor filter	1.3997	1.5618	0.9628	0.5095
RNRAD filter	7.9047 e–005	8.7092 e–005	8.2741 e–005	3.4554 e–005
Median filter	6.6386 e–004	7.3317 e–004	5.9088 e–004	2.9247 e–004
Wiener filter	4.5403 e–006	4.1013 e–006	2.2233 e–006	1.8010 e–006
DTD filter	0.0017	0.8122	0.0016	5.5828 e–004

found to be high for the Gabor filter over the other traditional filters, which clearly indicated that the Gabor filter retains the original information of the image better than the other filters.

9 Conclusion

In this work, we implemented the Gabor filter techniques for the Rician noise removal in MR images. The performance of Gabor filtering method was estimated both qualitatively and quantitatively. From the PSNR and MSE values, Gabor filter is found to be better choice for Rician noise removal in MR images than the other common filtering methods. The experimental results showed that the Gabor filter removes the signal-dependent bias (Rician noise) effectively, which is evident from the performance metrics calculated and the improvements in the contrast of the MR images.

References

1. Gudbjartsson H, Patz S (1996) The Rician distribution of noisy MRI data. Magn Reson Med 36(2):332
2. Nowak RD (1999) Wavelet-based Rician noise removal for magnetic resonance imaging. IEEE Trans Image Process 8(10):1408–1419
3. Nowak RD, Gregg RL, Cooper TG, Siebert JE (2002) Removing Rician noise in MRI via wavelet-domain filtering. IEEE transactions on image processing
4. Bas S, Fletcher T, Whitaker R (2006) Rician noise removal in diffusion tensor MRI. Med Image Comput Comput-Assist Interv 9:117–125, Springer
5. Ardizzone E, Pirrone R, Mastrella M, Gambino O (2007) A Gabor-based technique for bias removal in MR Images. In: 29th annual international conference of the IEEE EMBS, France
6. Aelterman J, Goossens B, Pizurica A, Philips W (2008) Removal of correlated Rician noise in magnetic resonance imaging. In: 16th European signal processing conference (EUSIPCO 2008), Switzerland, pp 25–29
7. Daessle NW, Prima S, Coupe P, Morrissey SP, Barillot C (2008) Rician noise removal by non-local means filtering for low signal-to-noise ratio MRI: applications to DT-MRI. In: 11th international conference on medical image computing and computer-assisted intervention, New York

8. Fernandez SA, Lopez CA, Westin CF (2008) Noise and signal estimation in magnitude MRI and Rician distributed images. A LMMSE approach. IEEE Trans on Image Process 17(8):1383–1398
9. Coupe P, Manjon JV, Gedamu E, Arnold D, Robles M, Collins DL (2008) An object-based method for Rician noise estimation in MR images. Med Image Comput Computer-Assist Interv 5762:601–608
10. He L, Greenshields IR (2009) A nonlocal maximum likelihood estimation method for Rician noise reduction in MR image. IEEE Trans Med Imaging 28(2):165–172
11. Krissian K, Fernandez SA (2009) Noise-driven anisotropic diffusion filtering of MRI. IEEE Trans Image Process 18(10):2265–2274
12. Dakua SP, Sahambi JS (2010) Noise effect on LV image segmentation. In: Proceedings of the annual IEEE conference (INDICON), Ahmedabad, India, 233–236
13. Rajeesh J, Moni RS, Kumar SP, Krishnan TG (2010) Rician noise removal on MRI using wave atom transform with histogram based noise variance estimation. IEEE Trans Commun Control Comput Technol (ICCCT), 531–535
14. Nobi MN, Yousuf MA (2011) A new method to remove noise in magnetic resonance and ultrasound images. J Sci Res 3(1):81–89
15. Brown MA, Semelka RC (2003) MRI basic principles and applications, 3rd edn. Wiley-Liss, John Wiley and Sons. Inc, New Jersey
16. Sarode MV, Deshmukh PR (2010–11) Performance evaluation of Rician noise reduction algorithm in magnetic resonance images. J Emerg Trends Comput Inf Sci (CIS J) 2:39–44
17. Landini L, Positano V, Santarelli MF (2005) Advanced image processing in magnetic resonance imaging. CRC Press, Taylor & Francis Group. LLC
18. Gregg RL, Nowak RD (2002) Noise removal methods for high resolution MRI.In: Nuclear Science Symposium, IEEE, vol 2
19. Jain AK, Ratha NK, Lakshmanan S (1997).Object detection using Gabor filter. In: Pattern Recognition Society, Elsevier Science Ltd 30:295–309
20. Weldon TP, Higgins WE, Dunn DF (1996) Efficient Gabor filter design for texture segmentation. Pattern Rec Soc 29:2005–2015
21. Kranauskas J (2010) Accelerated calculation of Gabor features in spatial domain. Elektronika Ir Elektrotechnika 1(97):39–44
22. Petkov N, Subramanian E (2007) Motion detection, noise reduction texture suppression and contour enhancement by spatiotemporal Gabor filters with surround inhibition, Springer
23. Kruizinga P, Petkov N, Grigorescu SE (1999).Comparison of texture features based on Gabor filters. In: 10th international conference on image analysis and processing, Italy, pp 142–147
24. Turaga DS, Chen Y, Caviedes J (2004) No reference PSNR estimation for compressed pictures. Sig Process: Image Commun 19:173–184

Multi-Pass Unsymmetric Trimmed Median Filter for Salt-and-Pepper Noise Removal

Navaneeth K. Ramakrishnan, Pillai Praveen Thulasidharan, Arun D. Panicker and Madhu S. Nair

Abstract Restoration of original image corrupted with high density salt-and-pepper noise is still a challenging task. In this letter, we propose here a new method; Multi-Pass Unsymmetric Trimmed Median Filter (MPUTMF) [versions 'a' and 'b'], to restore an image affected with high density salt-and-pepper noise, with better edge preservation. The $MPUTMF_a$ can do the restoration within two passes over the noisy image using the preprocessed pixels obtained in the same pass, where as $MPUTMF_b$ can take up to six passes over the noisy image for restoration without using the preprocessed pixels. $MPUTMF_b$ is computationally efficient on single core processor systems where as $MPUTMF_b$ is well suited to be implemented on parallel processing systems or GPUs to achieve higher computational efficiency. The proposed methods are compared with conventional as well as advanced algorithms like Median Filter (MF), Adaptive Median Filter (AMF), Efficient Decision Based Algorithm (EDBA), Improved Decision Based Algorithm (IDBA) and Modified Decision Based Unsymmetric Trimmed Median Filter (MDBUTMF). The experimental analysis (visual and quantitative) shows that our method gives better results on images affected with high density salt-and-pepper

N. K. Ramakrishnan (✉) · P. P. Thulasidharan · M. S. Nair
Department of Computer Science, University of Kerala, Kariavattom, Thiruvananthapuram 695581, Kerala, India
e-mail: navaneeth.kr@gmail.com

P. P. Thulasidharan
e-mail: praveentpillai@gmail.com

M. S. Nair
e-mail: madhu_s_nair2001@yahoo.com

A. D. Panicker
Tata Elxsi, TechnoPark, Trivandrum, Kerala, India
e-mail: arundp@gmail.com

Mohan S. and S. S. Kumar (eds.), *Proceedings of the Fourth International Conference on Signal and Image Processing 2012 (ICSIP 2012)*, Lecture Notes in Electrical Engineering 221, DOI: 10.1007/978-81-322-0997-3_33, © Springer India 2013

noise. Peak Signal-to-Noise Ratio (PSNR) and Image Enhancement Factor (IEF) are used for quantitatively evaluating the results of proposed algorithm(s).

Keywords Image restoration · Salt-and-pepper noise · Median filter · Trimmed median filter · Impulse noise · Parallel processing

1 Introduction

Impulse Noise has always been of prime importance in all the noise model study and it is also the common noise which influences any form of the image due to the interference or any other un-favorable factors occurring during the image acquisition or transmission procedures. Many linear and non-linear filters have been proposed for removing the impulse noise especially salt-and-pepper noise. Linear filters like Mean Filter do not preserve the edge details due to the smoothening effect. Non-linear filters try to overcome this by either following an adaptive or decision based algorithm; but still these algorithm may prove inefficient with regard to high density noise and execution time.

The simplest non-linear filter like the Median Filter (MF) [1] is efficient with regards to low density noise but fails for high densities and also results in loss of edge details. AMF [2] consumes more execution time since the window size increases for high density. EDBA [3] produces line artifacts and does not preserve the edge details. IDBA [4] improves over the EDBA disadvantages but still proves to be inefficient in regards with preservation of edge details and execution time. In MDBUTMF [5], when the processing window contains all noisy pixels, the method suggests to replace the central pixel with the mean of the window pixel values. This resultant value may not be considered as a true value with respect to the image which results in the loss of information and edge details.

The proposed algorithms $MPUTMF_a$ and $MPUTMF_b$ provide satisfactory results both in terms of high density noise and execution time. This is achieved by following a multi-pass approach which attempts to find the nearest true value of a noisy pixel, by providing methods which can be adapted to execution optimization principles such as parallel processing.

2 Proposed Algorithm

Multi-Pass Unsymmetric Trimmed Median Filter (MPUTMF) method attempts to find the most adaptive non-noisy pixel within the given neighborhood that can replace a given noisy pixel. Though every algorithm functions with this regard, the problem arises when the entire pixel within a given neighborhood are noisy. For example, consider a 3×3 window where all the nine pixels within the window are either 255 or 0 or a combination of both. In such a situation, we cannot find an appropriate non-noisy value that can be used to replace the central pixel. The Modified Decision Based Unsymmetric Trimmed Median Filter (MDBUTMF) [5] handles this scenario by replacing the central pixel with the mean of the window pixel values. But the computed mean value may not be a good representation of the original pixel value and may result in loss of image details. MPUTMF overcomes the above problem by delaying the replacement of the central pixel until a non-noisy pixel is found within the neighborhood of the pixel using multi-pass.

The first pass of MPUTMF scans the entire image using a 3×3 sliding window and identifies all the noisy pixels which have at least one non-noisy pixel in their neighborhood. Such noisy pixels are replaced with median value computed using only the uncorrupted pixels in the neighborhood. The remaining noisy pixels in the image are labeled as "unprocessed". In the second pass, only the labeled pixels are taken into consideration and a 3×3 window around each pixel is selected in an attempt to find a non-noisy/preprocessed pixel within it. If at least one uncorrupted pixel is found in the selected window, then the noisy pixel is replaced with the median value computed from the uncorrupted pixels. Otherwise, the pixel is again labeled as "unprocessed" for further processing. This process continues until no labeled pixel remains in the image.

The first pass of $MPUTMF_a$ and $MPUTMF_b$ is similar but they differ in approach being used in subsequent passes. In each subsequent pass $MPUTMF_a$ utilizes the uncorrupted/restored pixels obtained from the previous pass as well as the processed pixels obtained from the previous iterations of the same pass for calculating the appropriate value to be restored. On the other hand, $MPUTMF_b$ utilizes only the uncorrupted/restored pixel values obtained from the previous pass, making it suitable for parallel processing implementation.

3 Detailed Algorithm

3.1 $MPUTMF_a$

3.1.1 Pass I

The input to Pass I is a noisy image and the output image consists of processed pixels and labeled pixels which represent a noisy pixel which no non-noisy pixel in its neighborhood after the Pass I. The processing steps are detailed below:

Step1: Select a 3×3 window of the noisy image. Assume that P_{ij} is the pixel being processed.

Step2: Do the following comparisons on the pixel being processed:

If $0 < P_{ij} < 255$ {The pixel is a non-noisy pixel and hence the value can be retained or left un-changed.}Else {The pixel is a noisy pixel i.e., its value is either 255 or 0. The following two cases are considered in this situation:

Case (1): The selected window contains some non-noisy pixels, i.e., pixels with intensity values other than 255 and 0. In this case an unsymmetric trimming operation is being performed for extracting uncorrupted pixels by eliminating the noisy pixels with values 0 and 255. The central noisy pixel is then restored with median value computed using the extracted uncorrupted pixel values.

Case (2): All the pixels in the window are noisy pixels i.e., the entire pixels in the window have value either 255 or 0. In this case the pixel is labeled as "unprocessed" indicating that it needs to be processed in Pass II.}

Step3: Repeat the above steps for all the pixels in the image.

3.1.2 Pass II

The input is the output image obtained from Pass I and the output image consists of all pixels with uncorrupted/restored pixel values. The processing steps of Pass II are detailed below:

Step1: Obtain a 3×3 window around the labeled pixel. Since it is a noisy pixel the following two cases are possible.

Case (1): The selected window might contain uncorrupted/restored pixels obtained from Pass I as well as the processed pixels obtained from previous iterations. In this case an unsymmetric trimming operation is being performed for extracting uncorrupted pixels by eliminating the noisy pixels with values 0 and 255. The central noisy pixel is then restored with median value computed using the extracted uncorrupted pixel values.

Case (2): All the pixels in the window are noisy pixels i.e., the entire pixels in the window have value either 255 or 0. In this case the pixel is again labeled as "unprocessed".}

Step2: Repeat the below Step 1 for every labeled pixel in the given input image.
Step3: Go for another iteration of the Pass II steps if at least one labeled pixel remains in the output image. Currently obtained output image is given as the input for the next Pass.

3.2 *MPUTMF$_b$*

3.2.1 Pass I

The Pass I of MPUTMF$_b$ is similar to Pass I of MPUTMF$_a$.

3.2.2 Pass II

The Pass II of $MPUTMF_b$ differs from $MPUTMF_a$ in that it utilizes only the uncorrupted/restored pixel values obtained from the previous pass, for calculating the appropriate value to be restored. This approach makes it suitable for parallel processing implementation.

4 Experimental Analysis

In this section a detailed simulation of different steps in $MPUTMF_a$ and $MPUTMF_b$ is discussed along with a comparative analysis of the proposed method with conventional/advanced algorithms.

The $MPUTMF_a$ algorithm contains mainly three types of intensity comparisons. First comparison checks whether the pixel being processed Pij is noisy or not. If Pij is not noisy ($0 < Pij < 255$), then the pixel is retained as it is. A sample 3×3 window which satisfies this condition is simulated in Fig. 1a.

In the window shown in Fig. 1a, Pij has an intensity value of 170, indicating that it is an uncorrupted pixel. Hence no changes are made on this pixel value. Second and third types of comparisons are made when the previous check fails. These situations are explained in the Case (1) and Case (2) steps in the Pass I algorithm. Case (1) step of the $MPUTMF_a$ algorithm handles the situation simulated in Fig. 1b. Here the 3×3 window contains both noisy and non-noisy pixels and the pixel being processed is a noisy one. Unsymmetric trimming operation is then performed to remove the noisy pixels from the window. The intensity values shown in the window are [34, 255, 140, 0, 255, 45, 128, 28, 255]. From these intensity values the noisy elements, 0 and 255 are trimmed to obtain a noise free set of values [34, 140, 45, 128, 28]. Median of these noise free values is computed to replace the noisy pixel Pij. In this case the replaced value will be 45.

In Case (2) the window contains noisy pixels only which is simulated in Fig. 1c. Taking mean of the window is not a good option because it may not be a good

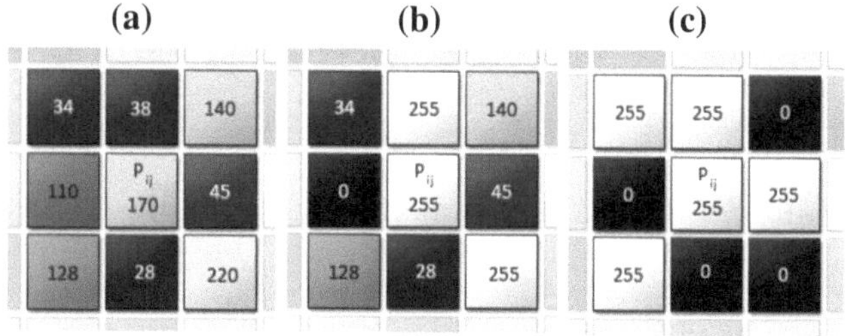

Fig. 1 Illustration of algorithm steps with 3×3 window **a** Non-noisy central pixel, **b** Noisy central pixel with at least one non-noisy pixel in the neighborhood and **c** Noisy central pixel within a noisy neighborhood

representation of nearby pixels. So the proposed algorithms $MPUTMF_a$ and $MPUTMF_b$ label the Pij pixel position as "unprocessed" to be considered for the next pass. In each subsequent pass $MPUTMF_a$ utilizes the uncorrupted/restored pixels obtained from the previous pass as well as the processed pixels obtained from the previous iterations of the same pass for calculating the appropriate value to be restored. On the other hand, $MPUTMF_b$ utilizes only the uncorrupted/restored pixel values obtained from the previous pass, making it suitable for parallel processing implementation.

The proposed algorithms are tested for performance both visually and quantitatively in terms of PSNR and IEF. Both algorithms are compared with conventional as well as advanced algorithms such as MF, AMF, EDBA, IDBA and MDBUTMF. Experimental analysis is carried out for all range of noise levels from 10 to 90 %.

PSNR and IEF value comparisons of $MPUTMF_a$ and $MPUTMF_b$ with other algorithms are shown in Figs. 2 and 3 respectively. From these figures it is evident that there is a significant increase in both PSNR and IEF values compared to MDBUTMF and other advanced algorithms. Comparison of the proposed method with other algorithms based on the computational time is shown in Fig. 4. A comparative performance analysis of different algorithms applied on corrupted images with different noise densities is shown Tables 1(on Lena image) and 2 (on Lifting body image). A good quality noise reduction of the proposed method is also evident from the visual analysis of the output images shown in Figs. 5(Lena image) and 6(Lifting body image). Good edge preserving capability of the proposed method compared with other algorithms has been proved in Fig. 7.

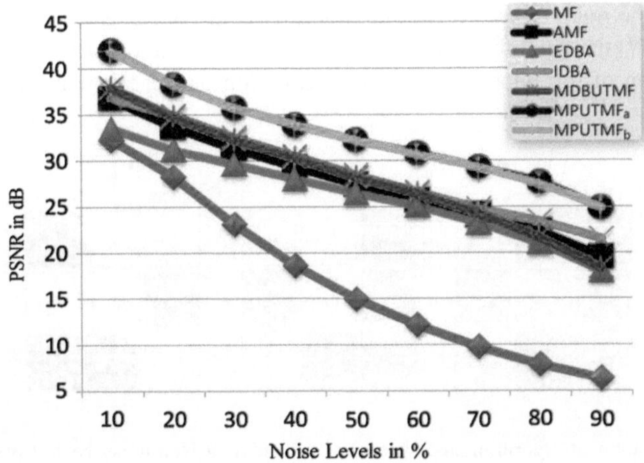

Fig. 2 PSNR against noise levels

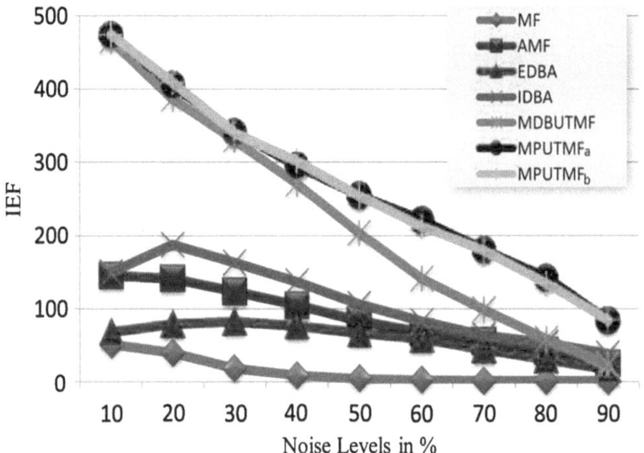

Fig. 3 IEF against noise levels

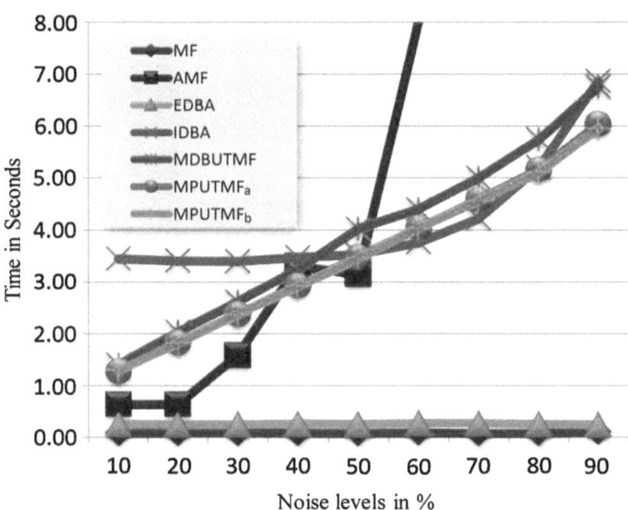

Fig. 4 Execution time against noise levels

Table 1 Comparative performance analysis of different algorithms applied on corrupted Lena image

Noise Density in %	MF			AMF			EDBA			IDBA			MDBUTMF			$MPUTMF_a$				$MPUTMF_b$			
	PSNR(dB)	IEF	Time(sec)	PSNR(dB)	IEF	Time(sec)	PSNR(dB)	IEF	Time(sec)	PSNR(dB)	IEF	Time(sec)	PSNR(dB)	IEF	Time(sec)	PSNR(dB)	IEF	Time(sec)	No.Passes	PSNR(dB)	IEF	Time(sec)	No.Passes
10	32.2	51.5	0.08	36.8	145.6	0.6	33.5	68.2	0.26	36.9	149.3	3.45	37.9	466.1	1.40	41.9	473.7	1.28	1	41.9	474.5	1.30	1
20	28.2	40.3	0.09	33.7	141.3	0.6	31.1	79.2	0.26	34.9	187.5	3.41	34.8	386.6	2.03	38.2	405.6	1.83	1	38.3	409.4	1.82	1
30	23.1	18.6	0.09	31.4	123.6	1.6	29.6	82.5	0.25	32.5	162.8	3.40	32.3	330.0	2.62	35.7	341.2	2.38	2	35.7	340.1	2.37	2
40	18.6	8.9	0.08	29.3	104.8	3.3	28.0	76.9	0.26	30.5	136.9	3.47	30.3	270.9	3.25	33.8	296.0	2.93	2	33.9	301.8	2.92	2
50	15.0	4.8	0.08	27.4	84.1	3.1	26.4	67.2	0.26	28.4	106.3	3.53	28.2	203.5	4.03	32.2	254.6	3.49	2	32.2	254.2	3.46	2
60	12.1	3.0	0.10	25.7	67.9	8	25.1	59.1	0.28	26.6	83.2	3.76	26.4	141.0	4.40	30.8	221.4	4.04	2	30.7	212.4	4.03	3
70	9.8	2.0	0.08	24.2	55.7	16	23.2	44.5	0.27	24.8	63.9	4.22	24.3	99.0	5.00	29.3	180.2	4.59	2	29.2	179.3	4.56	3
80	7.8	1.5	0.08	22.4	42.1	49	21.0	31.1	0.26	23.4	53.4	5.20	21.7	58.9	5.75	27.6	142.2	5.16	2	27.4	134.7	5.10	4
90	6.4	1.2	0.11	19.6	24.9	176	18.0	17.4	0.25	21.5	38.8	6.87	18.4	24.2	6.76	24.9	84.4	6.05	2	24.9	84.7	5.95	5

Table 2 Comparative performance analysis of different algorithms applied on corrupted lifting body image

Noise Density in %	MF			AMF			EDBA			IDBA			MDBUTMF			$MPUTMF_a$				$MPUTMF_b$			
	PSNR(dB)	IEF	Time(sec)	PSNR(dB)	IEF	Time(sec)	PSNR(dB)	IEF	Time(sec)	PSNR(dB)	IEF	Time(sec)	PSNR(dB)	IEF	Time(sec)	PSNR(dB)	IEF	Time(sec)	No.Passes	PSNR(dB)	IEF	Time(sec)	No.Passes
30	24.0	20.1	0.68	31.8	123.2	5.6	33.5	180.8	0.90	36.4	355.2	6.21	39.5	714.0	3.37	39.9	781.0	1.38	2	39.5	715.5	1.37	2
60	12.7	3.0	0.29	26.7	74.3	16	29.6	145.2	0.81	30.3	174.0	6.47	30.6	182.3	6.11	34.5	450.7	2.35	2	34.5	457.5	2.34	2
90	6.9	1.2	0.29	20.7	28.4	192	23.8	57.6	0.82	25.2	80.8	10.40	17.4	13.4	6.63	28.7	179.0	3.45	2	28.8	183.2	3.41	6

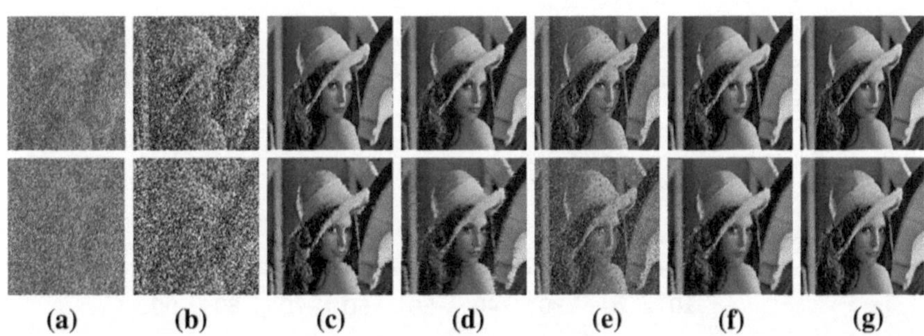

(a) **(b)** **(c)** **(d)** **(e)** **(f)** **(g)**

Fig. 5 *First row*: 80 % noise affected and restored Lena images, *Second row*: 90 % noise affected and restored Lena images. **a** Noise affected image **b** restored using MF **c** AMF **d** EDBA **e** MDBUTMF **f** IDBA **g** MPUTMF

Fig. 6 *First row*: 60 % noise affected and restored Lifting body images, *Second row*: 90 % noise affected and restored Lena images. **a** Noise affected image **b** restored using MF **c** AMF **d** EDBA **e** MDBUTMF **f** IDBA **g** MPUTMF

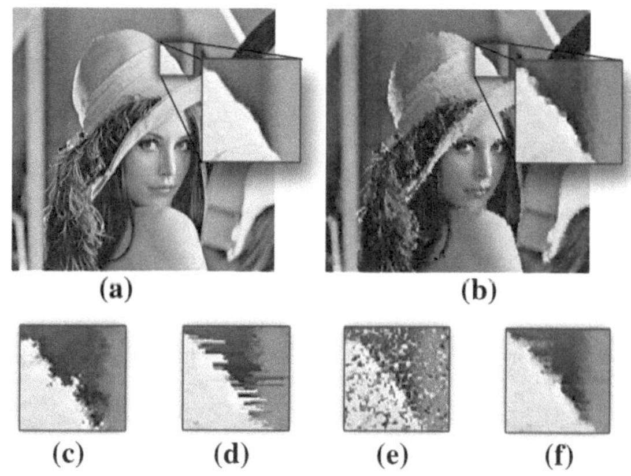

Fig. 7 Edge preservation example. **a** Original Lena image and a small area zoomed to view an edge. Restored images using **b** MPUTMF **c** AMF **d** EDBA **e** MDBUTMF **f** IDBA, from a 90 % noise affected Lena image

5 Conclusion

Our proposed algorithms $MPUTMF_a$ and $MPUTMF_b$ restore images, affected with different levels of Salt-and-pepper noise densities, with better visual quality through edge detail preservation. Experimental analysis shows that the proposed algorithm give better results than conventional/advanced algorithms like MF, AMF, EDBA, IDBA and MDBUTMF. Quantitative analysis using performance measures such as PSNR, IEF and computational time shows the superior performance of the proposed $MPUTMF_a$ and $MPUTMF_b$ methods.

References

1. Rafael C, Gonzalez, Woods RE (2009) Digital image processing, 3rd edn. Pearson Prentice Hall, New Jersey
2. Chan RH, Ho CW, Nikolova M (2005) Salt-and-pepper noise removal by median-type noise detectors and detail-preserving regularization. IEEE Trans Image Process 14(10):1479–1485
3. Srinivasan KS, Ebenezer D (2007) A new fast and efficient decision based algorithm for removal of high density impulse noise. IEEE Signal Process Lett 14(3):189–192
4. Nair MS, Revathy K Tatavarti R (2008) An improved decision- based algorithm for impulse noise removal, international congress on image and signal processing—CISP 2008 (China), vol 1. IEEE Computer Society Press, Vienna, pp 426–431
5. Esakkirajan S, Veerakumar T, Subramanyam AN, PremChand CH (2011) Removal of high density salt-and-pepper noise through modified decision based unsymmetric trimmed median filter. IEEE Signal Process Lett 18(5):287–290

Analysis and Evaluation of Keystroke Duration of User's Typing as a Distinctive Measure of Recognition

Dwijen Rudrapal and Smita Das

Abstract This paper illustrated one unique measure of keystroke patterns of user typing to prevent insider threats with combination of existing keystroke credentials. While insider threats may never be eliminated, it is worth noting that risk-mitigating counter measures exist. Every organization's security architecture must take insider threat into account. As reiterated trough out the security industry, defense in depth is the essential. Defense in depth requires security professional to take a layered approach to protecting information. In essence, we should never rely on just one security method or technology. Only user ID and passwords are not enough strong now a day to protect workstations from unauthorized access. Being remarkable advantages and to support multilevel authentication biometrics may be strong add-on. Recent research on behavioral biometrics i.e., keystroke dynamics proves its efficiency for stronger authentication. With numerous advantages keystroke pattern credential are employing in many organization to establish layered security mechanism. Work of last decade on keystroke dynamics also reveals some drawbacks. So research is going on to find more traits of user typing for more efficient authentication. Through this paper we proposed and experimentally proved one measure of keystroke pattern which is distinctive for each user and can make keystroke pattern credential stronger.

Keywords Identification · Biometrics · Keystroke dynamics · Degree of disorder · Digraph · Internal security

D. Rudrapal (✉) · S. Das
Department of Computer Science and Engineering, National Institute of Technology, Agartala, Tripura (w), India
e-mail: dwijen.rudrapal@gmail.com

S. Das
e-mail: smitadas.nita@gmail.com

Mohan S. and S. S. Kumar (eds.), *Proceedings of the Fourth International Conference on Signal and Image Processing 2012 (ICSIP 2012)*, Lecture Notes in Electrical Engineering 221, DOI: 10.1007/978-81-322-0997-3_34,

1 Introduction

Passwords have been the primary means of authenticating the users to computers since the beginning of access controls in computer systems. But passwords are struggling by several inadequacies. Regular password based Authentication policies triggered several issues on security and user convenience that, attacker or hacker can guess password and gain unauthorized access and claim as a real user. Few studies exposed that, users tend to choose passwords that can be broken by a thorough search of a relatively small subset of all possible passwords. In one study of 14,000 UNIX passwords, almost 25 % of the passwords were found by searching for words from a carefully formed "dictionary" of only 3×10^6 words [1]. This high success rate is not unusual despite the fact that there are roughly 2×10^{14} 8-character passwords consisting of digits and upper and lower case letters alone.

This breaches made weak conventional security mechanism and open multi-level authentication [2] methodology which is going to deploy many of the authentication system. Biometrics traits [3] are to add an extra layer over existing security setup. Researchers are concentrating on human characteristics which can't be stolen of lost. Observations of many recent research works reveled behavioral biometrics like Keystroke dynamics [4] could be advantageous in user identification, in addition of traditional authentication. Studies also indicate that, users, given proper data in a controlled environment could be accurately identified without the use of username and passwords too.

In this paper, we propose one measure for improving the keystroke dynamics credential for security of password-based applications by incorporating biometric information into the password. Specifically, our analysis is calculating some value which is almost unique for each user and will hardened password more. This hardened password can be tested for login purposes or used as a cryptographic key for encryption, virtual private network access, etc.

2 Overview of Authentication by Keystroke Dynamics

The act of confirming the truth of an attribute of a datum or entity is called Authentication. This might involve the ensuring that a product is what it's packaging and labeling or confirming claims to be identity of a person. The term identification refers to the first time a person is presented to a security device. Once the device registers the biometric parameters of the person every subsequent visit of that person is for the purpose of authentication.

Biometrics offers a valuable approach to extending current security technologies that make it far harder for fraud to take place by preventing ready impersonation of the authorized user.

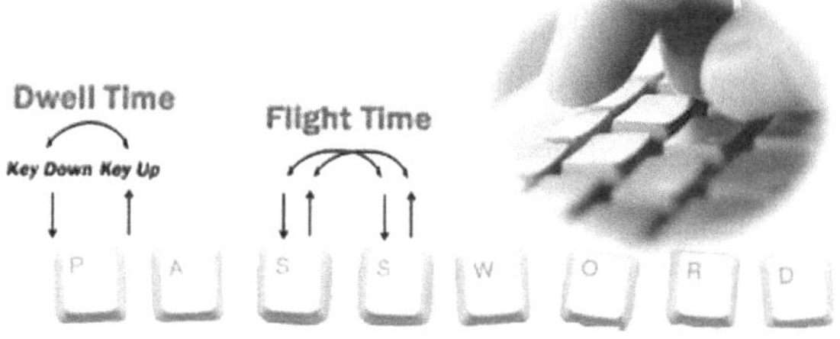

Fig. 1 Key events of user typing rhythm

A behavioral biometric is a measurable behavior trait of human being that is acquired over time and is used to identify or validate a person's identity. Keystroke dynamics falls within the category of behavioral biometrics. The idea behind keystroke dynamics is that people have different typing styles and by analyzing the timings of keystrokes, a person can be identified.

Keystroke dynamics dealings the series of key pressed and key released event timings while the user types a string. For example, if a user's password is 'password' then key pressed and key released events are captured for each character. These raw measurements can be recorded from almost any keyboard and can be recorded to determine Dwell time (the time between key down and key up) and Flight time (the time from "key down" to the next "key down" to the time between one "key up" and the next "key up") as represented in the figure below (Fig 1).

Once the keystroke timing data is captured, the same keystroke timing data is then processed through algorithm, which determines a primary pattern for future comparison.

As with any biometric technology applied to an authentication function, the technology is used for two major functions: enroll and verify user credentials.

Keystroke dynamics is the process of analyzing the way a user types on a keyboard and identify him based on his habitual typing rhythm. The features that are considered to be account on for authenticating purpose are:

- Keystrokes latencies (fight)
- Duration of a specific keystroke (dwell)
- Pressure (Force of keystrokes)
- Typing speed
- Frequency of error
- Overlapping of specific keys combination
- Method of error correction.

3 Motivation

Every user has a specific typing pattern that can be profiled and used for comparison to strengthen authentication. All the earlier experiments regarding keystroke dynamics have found a number of numerical strategies to be successful.

While experiments differ on what statistical tools were used, the essential idea from all of them is the creation of a reference profile for every user and a method of comparing a submitted sample with a profile.

- The Standard Deviation Method
- The Euclidean Distance Method
- The Ratio Method
- The Vector Disorder Method.

Research with neural networks as a method of classifying claimant vectors they were able to achieve FAR between 12.0 and 40.9 %.

The most recent patent granted for keystroke dynamics was in April 2007 and the authors of the patent [5] argue that global patterns do not exist and that any viable keystroke authentication system must have controlled typing situations. Meanwhile, new studies are now examining the feasibility of using keystroke dynamics on mobile phone devices. Initial results are promising with EER of around 12.8 %.

To strengthen this authentication credential of user by keystroke dynamics process we searched one new measure of keystroke patterns.

4 Proposed Methodology

Through this proposed study we tried to find out one additional quantify of keystroke dynamics by analyzing keystroke duration feature for distinguishing individuals. In various previous research works, researchers have shown that calculation of degree of disorder of digraphs and trigraphs are valuable features for validation of individual keystrokes.

This additional quantify of keystroke dynamics will make authentication credential more optimized.

4.1 Mining of Keystroke Patterns

When entering a character on a keyboard, a set of instructions is sent from the keyboard to the computer. In this thesis the parameters Key-pressed (also referred to as kp) and KeyWait (also referred to as digraph) are the ones used most frequently. Figure 2 shown below illustrates the time period of key events while user types something on a keyboard. The Key-pressed [4] is defined as the time from a key is pressed down, until the key is depressed (released). Every key can have different Key-pressed time for each user, depending on their general behaviour. Different keys and key combinations are likely to have a different Key-pressed time. This may be because of the angle the fingers have on each key, the familiarity with the text one is typing and the familiarity with a certain key combination.

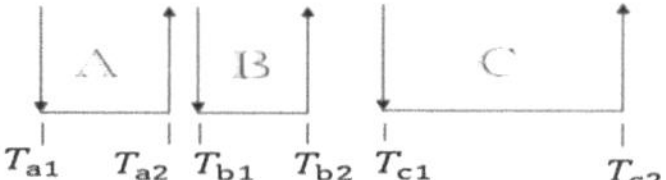

Fig. 2 Timing events of pressing keys by user

For example entering name or password, this has been entered frequently. There is no thinking when typing these two and becomes a habit. There is not much flexibility in how this parameter is used, and this study using time in keystroke, use this exact definition.

One important timing event of keystroke is duration [6] defined as the time that how long each key held down. In this study we have taken a fixed password and computed the time of every key pressed duration for further calculation.

Another measure called digraph, defined as two keys typed one after the other. In our study the duration of a digraph is not used.

Trigraph [8] is defined as three keys typed one after the other. The duration of trigraph is measured between pressing event of the first key and release of the third key.

Figure 2 drawn below shown how to measure the duration, digraph and trigraphs from the key events.

Each time a key is pressed down or released, an event is triggered. Each key press or key release event will have a timestamp which consists of the number of microseconds (say 120 ms) since 01.01.1900. The timestamp value is used for calculating the time from a key to be pressed down to it is released, and for calculating the time between the keys.

Here from the above image key pressed duration of a key "A" is time $T_{a1}-T_{a2}$. KeyWait time i.e., Down-Down time for diagraph (Here key "A" to key "B") is $T_{b1}-T_{a1}$. Considering trigraphs Down-Down time is $T_{c1}-T_{a1}$ (for key "A" to key "C").

4.2 Degree of Disorder Methodology

Method used in this thesis is the vector disorder method which is adapted from studies by Bergadano, Gunetti, and Picardi [7]. Instead of only absolute measure which depends on the comparison of absolute timing measurements, this approach opens relative measure vs. an absolute measure. Applying the vector disorder method, we sorted the claimant and reference profiles by the duration timings of fixed password keys.

A list of elements V of N elements, a simple measure of the degree of disorder [8] (or, simply, the disorder) of V with respect to its ordered complement V can be computed as the sum of the distances between the position of each element in V and the position of the same element in V.

The formula to calculate disorder is

$$\frac{|V|^2}{2} \quad \text{(if } |V| \text{ is even) Or } \frac{(|V|^2 - 1)}{2} \quad \text{(if } |V| \text{ is odd)}$$

Sorted array V will have a degree of disorder equal to 0 when the elements of the array V to which V is compared is in same order. We have the maximum disorder for an array V when its elements are in reverse order with compared to the elements in array V.

After normalization for any array V, its degree of disorder falls between 0 (if V is ordered) and 1 (if V is in reverse order).

4.3 Estimation of Disorder for Key Pressed Durations

Proposed work defines a measure of the distance for two samples. Experiments found in the literature (e.g., in Umphress and Williams [9], Joyce and Gupta [10], Brown and Rogers [11], and Obaidat and Sadoun [12]) normally use some combination of two basic measures:

- The duration of a key (how long a key is held down) and
- The latency between two consecutively typed keys (the elapsed time between the release of the first key and the depression of the second).

Through our experiments, we have taken the disorder value for keystroke duration to segregate user and validate with a suitable threshold value. This measure we calculated from a fixed text provided by the user to the system. For example, we calculated disorder on a fixed text "zigbag89" to depict the disorder of keystroke duration and shown in Fig. 3. Two sorted arrays have been shown as one is representing V and another array representing V.

While same letters are in the list, calculation of distance done like first occurred latter is calculating distance with the letter which is also first in the sorted list.

Maximum disorder given by $do_{max} = \frac{|V2|}{2}$, where |V| is the length of array V and it is even or $do_{max} = \frac{|V2|-1}{2}$if |V| odd. As in our example length of V is even i.e., 8, so, $do_{max} = 32$

For given array V total disorder is equal to

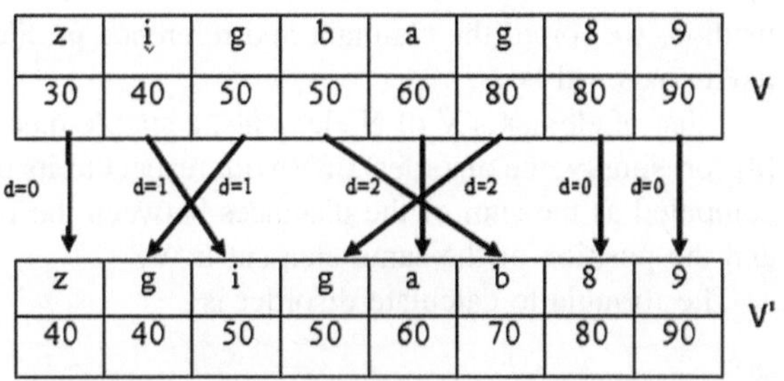

Fig. 3 Distance of the disordered element in sorted duration array

$$\text{do} = (0 + 1 + 1 + 2 + 0 + 2 + 0 + 0) = 6.$$

In order to normalize the disorder, we perform:

$$\text{do}_{\text{nor}} = \frac{\text{do}}{\text{do}_{\text{max}}} = \frac{06}{32} = 0.1875.$$

5 Simulation Results

Advantages with using existing resources, such as libraries and built in features, made C# the selected option for developing the login application. The .NET C# environment uses the System.Windows.Forms library with Control.KeyPress Events, Control.KeyDown Events and Control.KeyUp Events classes. These control classes are used to handle the input and output stream of data on the keyboard. Typically the input stream is used in these experiments as no commands are sent back to the keyboard. The application can run on any.NET 2.0 supporting OS, including Microsoft Windows 98 and newer.

Designing the layout and content of the login was done and tested carefully, making it very much standard without any confusing texts or extra features. Figure 4 shows a screenshot of the login window. It consists of one input field with the label UserID, one input field with the label password and three buttons. At the time of new registration user will enter necessary information along with multiple entry of password. Entry of password must be done carefully as the rhythm gets recorded for future authentication and identification. Same rhythm must be followed at the time of login.

Initially cursor is placed at UserID field. Tab button and backspace has not been allowed. Because each tab button as well as backspace is having key timing events as other keys used in password, which will make extracted keystroke features invalid. So, only mouse events are valid while traversing fields.

Figure 5 shown below the screen, where user registers self by providing the necessary information and typing password twice.

5.1 Result Analysis

Samples from some successful registration can be found in Table 1. This table has several values worth noticing. This sample provide the basic knowledge to how widely spread the data samples are and what can be considered as low or high values in the following sections. Another thing to recognize in these data is the repetitive behavior of some sample values.

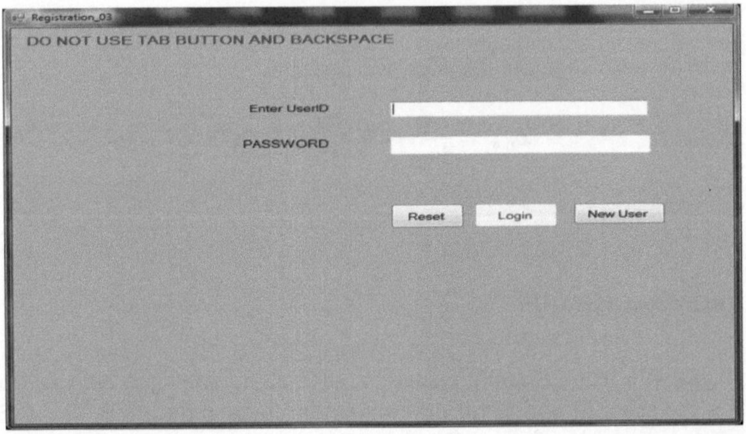

Fig. 4 Login screen of the application along with keystroke dynamics authentication

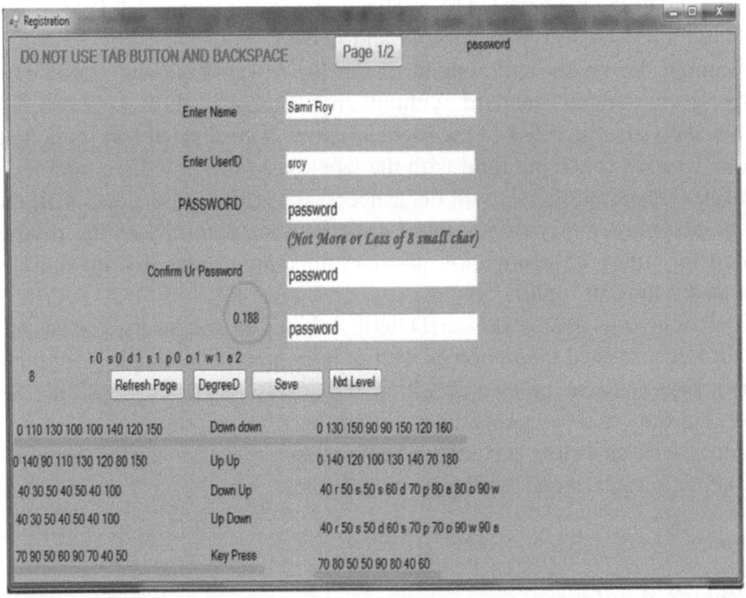

Fig. 5 Registration of user with calculation of key pressed disorder

In the above table it is observed that user has chosen passwords of 8 characters (Alphanumeric) each. It has observed that some passwords are repetitions of same character 8 times. For all users registered, everyone is given different disorder value of their key-pressed duration. In the experiment, in different sessions user has given data and found that the deviation of disorder value for keystroke duration

Table 1 Comparison key pressed disorder values of registered user templates

SN	U_ID	Password	Degree KP	Degree KP2	Deviatn
1	Drp	Password	0.188	0.188	0
2	Anish2	Password	0.313	0.313	0
3	Snehapriya	Password	0.5	0.375	0.125
4	Vishalika16	Vish1234	0.625	0.375	0.25
5	Csm	Ramadevi	0.438	0.188	0.25
6	Khan	Password	0.438	0.25	0.188
7	Skh	Shekhark	0.5	0.313	0.187
8	Kv	Nokia523	0.125	0.063	0.062
9	Kd	Shivkali	0.313	0.063	0.25
10	Abhinandan98	Abhinita	0.188	0	0.188
11	Moonmi	98545591	0.188	0.375	-0.187

Table 2 Comparison of mean deviation of key pressed disorder for fixed password

SSN	UserID	S-01	S-02	S-03	S-04	M-Deviatn
1	Drp	0.167	0.188	0.188	0.188	0.0105
2	Anish2	0.313	0.313	0.313	0.313	0
3	Snehapriya	0.476	0.5	0.5	0.5	0.012
4	Khan	0.475	0.47	0.438	0.47	0.0135
5	Sh	0.525	0.556	0.5	0.438	0.0155
6	Mradul	0.625	0.675	0.625	0.375	0.1

is identifying concern user. Table shows disorder values for each user in two different sessions here.

Analysis of the above table open one issue in FAR that, if any user setting password same character 8 times then it become easier for imposter to make false authorization. While in a password of 8 same characters flight time is negligible and can become closer for most of other user typing.

Table 2 depicts the mean deviation comparison of disorder values for few users. Disorder of keystroke duration captured in various sessions for a fixed password. This table enhance the uniqueness of disorder values user typing patterns.

5.2 Determination of Threshold Value

FAR and FRR values can be adjusted easily by changing the threshold [13] for accepting and rejecting users. Different situations require different emphasis on FAR or FRR. For this study one threshold value has defined. Table 1 show that the maximum difference of two extracted disorder value for key pressed duration is 0.25 for registered 21 users/subjects. So, based on the experiment result of this study, the best result of FAR and FRR will be for 0.25 threshold value.

6 Conclusion

This paper has provided an experimental illustration to enhance the keystroke dynamics credential as a biometric for authenticating access to workstations. Through this article we showed that calculation of degree of disorder on keystroke duration can be one important appraising of keystroke patterns. Observations of the illustrated methodology showed its efficiency and could be implemented to boost security in applications where logging-in is the necessity for the clients and also in multilevel authentication. In some observation we also encounter very nearer values for different user and this value with combination of existing keystroke credential [14] will significantly differentiate each user.

7 Limitation and Future Works

Keystroke dynamics is having a numerous challenges. Higher error rates and prospective problems with logical or physical access control convergence require a study of how keystroke dynamics fits into the overall organisations internal security strategy. Boundaries like Dynamic change in typing patterns, User's susceptibility to fatigue, Injury, skill of the user, Change of keyboard hardware, interruption while typing, mode while typing etc. In respect of these limitation data must be captured in a controlled environment. The proposed disorder of keystroke duration is also not distinguishable in all cases. This additional calculation of disorder can be combined with existing features and implemented in such application where a second layer of authentication can be employed for hardening the security or claimant validation procedure.

In this paper we selected a simple fixed password as well as their chosen password for all users. In future, same disorder can be compute among various user taking a strong password where characters are spreading over the keyboard, combination of special and shift keys. Same feature can be calculated also using different keyboard to compare uniqueness and adding this with the existing features of keystroke dynamics to determine FAR and FRR in biometric authentication.

References

1. Klein D (1990) Foiling the cracker: a survey of, and improvements to, password security. In: Proceedings of the 2nd USENIX security workshop, Aug 1990
2. Ahmed AAE, Traore I (2005) Anomaly intrusion detection based on biometrics," Information assurance workshop, IAW '05. Proceedings from the sixth annual IEEE SMC, pp 452–453, 15–17 June 2005
3. Jain AK, Ross A, Prabhakar S (2004) An introduction to biometric recognition. IEEE Trans Circuits Syst Video Technol, Spec Issue on Image Video-Based Biometrics 14(1)

4. Monrose F, Rubin A (1997) Authentication via keystroke dynamics. 4th ACM conference on computer and communications security, pp 48–56
5. Postley HJ, Bender SS (2007) Key sequence rhythm recognition system and method. US patent 7,206,938
6. Revett K (2009) A bioinformatics based approach to user authentication via keystroke dynamics. Int J Control Autom Syst 7(1):7–15
7. Bergadano F, Gunetti D, Picardi C (2002) User authentication through keystroke dynamics. ACM Trans Inf Syst Secur 5(4):367–397
8. Bergadano F, Gunetti D, Picardi C (2002) User authentication through keystroke dynamics ACM ACM Trans Inf Syst Secur 5(4):367–397
9. Umphress D, Williams G (1985) Identity verification through key-board characteristics. Int J Man Mach Stud 23(3):263–274
10. Joyce R, Gupta G (1990) Identity authorization based on keystroke latencies. Commun ACM 33(2):168–176
11. Brown M, Rogers SJ (1994) A practical approach to user authentication. Computer security applications conference, 1994. Proceedings of 10th annual, pp 108–116, 5-9 Dec 1994
12. Obaidat MS, Sadoun B (1997) Verification of Computer Users Using Keystroke Dynamics. IEEE Trans Syst Man Cybern-Part B 24(2):261–269
13. Martı J et al (2007) Recognizing individual typing patterns IbPRIA 2007, Part II, LNCS 4478, 323–330
14. Rudrapal D et al (2012) A study and analysis of keystroke dynamics and its enhancement for proficient user authentication ICECT-2012, Apr 6–8th

Analysis of Color Moment as a Low Level Feature in Improvement of Content Based Image Retrieval

Smita Das and Dwijen Rudrapal

Abstract In the recent past, the rapid intensification of the Internet has significantly enhanced the quantity of image collections accessible owing to the simplicity with which images are being formed or stored. CBIR (Content Based Image Retrieval) phenomenon is therefore highly encouraged by this requirement of unbeaten and proficient exploration of the large image databases. Consequently, the low level feature extraction of the visual contents of an image and their analysis is very significant in terms of CBIR. These low level features can be colour, texture and shape features. As Colour based image retrieval procedure is the trendiest of all these feature extraction algorithms, hence in this paper the color moments of the IIue, Saturation, and Value (IISV) component images in HSV color space are used as feature extraction algorithm. After the successful calculation of features for extraction, similarity computation is done using Euclidean Distance in between the test image and object images and finally the image retrieval is done. Analysis of this paper shows that the training time required for individual image, as well as, all the images in the database is very small which provides instantaneous retrieval. The estimation of the proposed approach is conceded out using the standard precision, recall and f-score measures, and the experimental results demonstrate that the proposed method has higher accuracy and retrieval rate than the conventional methods.

Keywords Image retrieval · Feature extraction algorithm · HSV color space · Color moments · Euclidean distance · Precision · Recall · F-score

---

S. Das (✉) · D. Rudrapal
Department of Computer Science and Engineering, National Institute of Technology, Agartala, Tripura (w), India
e-mail: smitadas.nita@gmail.com

D. Rudrapal
e-mail: dwijen.rudrapal@gmail.com

Mohan S. and S. S. Kumar (eds.), *Proceedings of the Fourth International Conference on Signal and Image Processing 2012 (ICSIP 2012)*, Lecture Notes in Electrical Engineering 221, DOI: 10.1007/978-81-322-0997-3_35,

1 Introduction

Started in the late 1970s, content based image retrieval (CBIR) [1] gains huge momentum in the early 90s and has delivered quite a few foremost breakthroughs. Early generations of image retrieval was on the basis of texts or keyword based search which was an expected addition of text based image retrieval. In this loom, first the manual annotation of the images are done by keywords, and then image retrieval is accomplished by their annotations. However, this process has several drawbacks. For e.g., the manual effort required to annotate the image database is cumbersome. As the volume of database will increase, effort will also increase. To conquer these drawbacks, a substitute method of content based image retrieval (CBIR) was proposed in the early 1990s. CBIR at the beginning used the low-level [2] image features as an alternative of the keyword features to retrieve images. Its benefit over keyword based image retrieval [3] is that the content of the image always stays consistent and feature extraction can be done automatically which reduces manual effort. As CBIR has its imperative applications [4] in the fields of art collections, digital libraries, retail catalogs, medical diagnosis, advertising, crime detection, entertainment and many more therefore, the desired distinctiveness of a CBIR system is high retrieval efficiency and less computational complexity. As color is the most important low level feature, hence in our paper we have chosen color as a low level feature for image retrieval from the image database. More specifically, we have chosen the colour moment descriptor in HSV space [5] for feature extraction.

2 Motivation and Objectives

The main purpose of this paper is to relate the CBIR system which takes into account the low level features of image or more predominantly, the Color feature [6, 7] which is more widespread when compared to high level features. The chief objective of this paper is to show the effectiveness of CBIR using color moment feature as compared with other feature extraction methods. After the accumulation of feature extraction, similarity computation in between test image and database images is done using Euclidian Distance. On the basis of distance, image retrieval is completed. After the completion of image retrieval, an analysis on colour moment feature extraction is done for finding the retrieval effectiveness. This is calculated on the basis of matrices like precision, recall and f-score which shows the efficiency of the retrieval method.

3 Recent CBIR Development

CBIR consists of the following steps in system recognition:

1. Collection of Data: By extracting significant features, each image in the image data base has to be represented competently.
2. Building feature database: Using similarity measurement between test image and object images of database relevant images has to be retrieved.
3. Searching: Search for a particular image in the database is done on the basis of feature descriptors.
4. Retrieval: Finally, similarity computation is done and on the basis of minimum distance, all relevant images are retrieved.

While building up the feature database for images, Color feature plays an important role as it is the most perceptive and understandable feature of the image. Various research papers usually adopt histograms to explain it. Depending on the signature mining approach used existing color-based general-purpose image retrieval systems approximately can be categorized into three sections: histogram, color layout, and region-based search. Also, histogram-based [8] exploration methods are further investigated in two different color spaces namely: RGB and HSV spaces.

Countless methods can be used to explain colour feature, some of them are colour histogram [9], colour structure descriptor (CSD) and scalable colour descriptor (SCD) [10], colour correlograms [11], colour moment [12] etc., (Fig. 1).

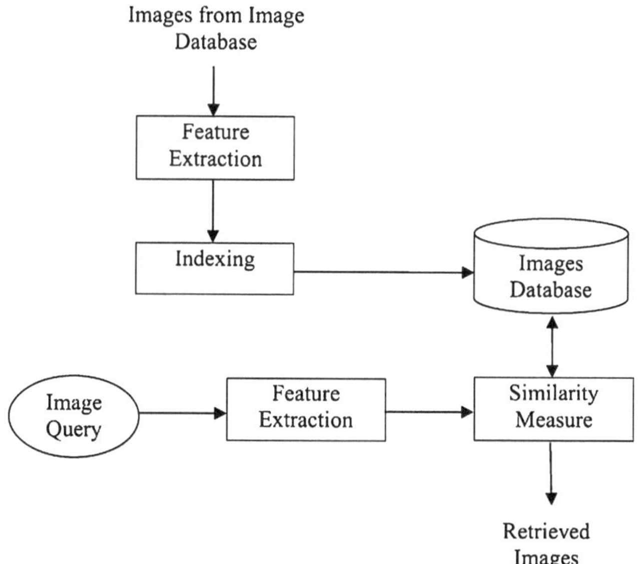

Fig. 1 Proposed CBIR architecture

4 Proposed Scheme

As Color is one of the most crucial features of the image, hence, to further decrease the feature vector dimension, we have used the color moment feature. Color moment feature not only reduces the feature vector dimension but also accelerates retrieval speed by taking very nominal time in train the images in database.

4.1 Feature Extraction Using Color Moment

Colour space is a very important issue while considering the feature extraction using colour moment. From current research it has been uncovered that RGB model diverge a lot with human colour sensitivity system, on the other hand, HSV model connects very well with human visualization perception. Hence, we preferred HSV colour space model in this paper. Also, HSV colour space is axis reliant of each other in visualization and space distance is steady with human vision features. With a very easy nonlinear transformation the adaptation from RGB to HSV can be achieved.

In various learning, it has been depicted that there are many low order color moments but most of the color distribution information is prohibited by mainly three low order moments [13] as follows: firstly, Mean color is captured by the first order moment (μ_c), secondly the standard deviation by second order moment (σ_c) and skewness by third order moment (θ_c). The below mentioned mathematical formulas are used in calculation of these three low order color moments for each of the three color panes in HSV color space:

$$\mu_c = \frac{1}{MN}\sum_{i=1}^{M}\sum_{j=1}^{N} P_{ij}^{c},$$

$$\sigma_c = \left[\frac{1}{MN}\sum_{i=1}^{M}\sum_{j=1}^{N}\left(P_{ij}^{c} - \mu_c\right)^2\right]^{\frac{1}{2}}$$

$$\theta_c = \left[\frac{1}{MN}\sum_{i=1}^{M}\sum_{j=1}^{N}\left(P_{ij}^{c} - \mu_c\right)^3\right]^{\frac{1}{3}}$$

where P_{ij}^{C} is the value of the cth color component of the color pixel in the ith row and jth column of the image.

For calculation of these moments first the image has to be converted from RGB to HSV color plane. Next, the image in HSV space is converted to the three components i.e., H (hue), S (Saturation) and V (Value). Followed by this step, contrast adjustment using the image's histogram i.e., histogram equalization of each of the three components is done. Finally, the calculation of the three moments (mean, SD and skewness) is done using the above mentioned formulae for each of the three components' histogram respectively. Therefore, per color plane three moments i.e., in totality only nine parameters are evaluated to depict each color image which is a very small vector dimension (Fig. 2).

4.2 Calculation of Precision and Recall

To establish the Image Retrieval effectiveness two metrics are extensively used named as: Precision and Recall. Precision is the portion of retrieved instances that are significant, whereas Recall is the division of appropriate instances that are retrieved. Yet in simpler stipulations, high recall means that the process or algorithm returns most of the relevant outcome. On the other hand, high precision denotes that an algorithm has returned more relevant results than irrelevant. In general, we can have the following formula:

$$\text{Precision} = \text{TP} / (\text{TP} + \text{FP}); \text{Recall} = \text{TP} / (\text{TP} + \text{FN})$$

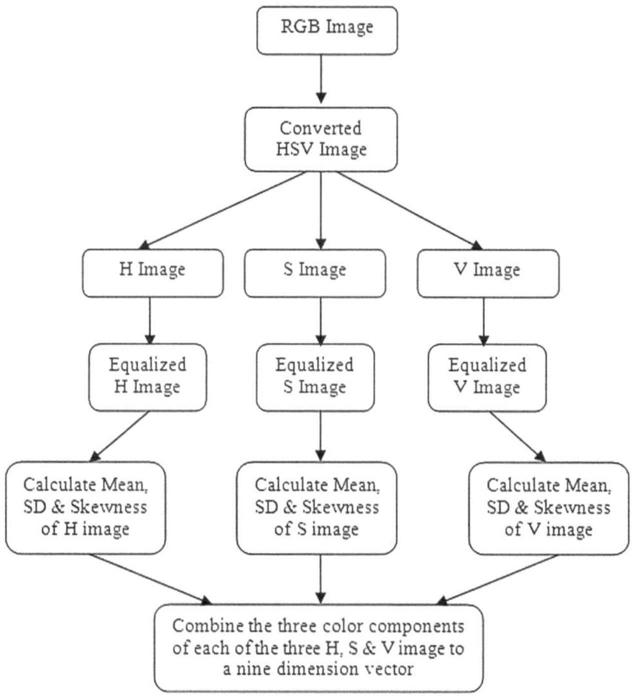

Fig. 2 Flowchart for feature extraction using color moment

4.3 Calculation of F-Score

In some calculations Precision is used separately or along with recall. But sometimes these two measures are required to be used together which derives the F Score or f-measure to provide a single dimension for a system. In this paper also we make an average over multiple retrieval queries. A technique that combines precision and recall together is the harmonic mean of them and is known as the traditional F-measure or balanced F-score given by the following formula:

$$\text{F} = 2 * \text{ precision } * \text{ recall } / (\text{precision } + \text{ recall})$$

In this formula, we can find that recall and precision are evenly weighted.

4.4 Comparison of Training Time

In this paper, analysis is done solitarily on Colour Moment feature extraction technique. Therefore, we can comment that as there is only a single feature extraction method used, so training time required for the images of the database is very nominal. The following figure tabulates the different instances of total execution time for all 400 images in the database: (Table 1).

Here, instances means in different hardware conditions, i.e., in different computers or OS. Each single image in database is trained using colour moment feature extraction algorithm. For the experiments we have taken 10 categories of images each having 40 images and total 400 images in.jpg format. The following figure tabulates the average (Mean) time and their Standard Deviation (SD) Values for each image category in database (Table 2).

5 Image Retrieval Using Similarity Measurement

The different similarity measurement techniques for image retrieval can fall into two categories: Distance Measurement and Correlation Measurement. Here, we

Table 1 Total training time of 400 images in database

* Instance	Total training time (in seconds)
1	297.7382
2	309.6783
3	307.6944
4	302.7582
5	298.8357
Mean time	303.3410

Table 2 Total training time of individual category of images in the database

Image category	Mean (average) time (s)	SD	Image category	Mean (average) time (s)	SD
Animal	0.4516	0.1162	Green lake	0.5500	0.1377
Arbogreens	0.5151	0.1454	Sky	0.6227	0.1614
Beer	0.4700	0.1184	Stadium	0.5363	0.1176
Cambridge	0.5385	0.1346	Swiss mountain	0.4861	0.1255
Cherries	0.4967	0.1395	Tigers	0.4963	0.1493

have used the Distance Measurement method which measures a definite distance between database image and query image by using the Euclidean distance [14] from the following formula:

$$D(q,s) = \left\{ \sum_{i=0}^{L-1} (q_i - s_i)^2 \right\}^{1/2}$$

where, $q = (q_0, q_1, \ldots q_{L-1})$ is the feature vector of query image, and $s = (s_0, s_1, \ldots s_{L-1})$ is the feature vector of the database images and L is the dimension of image feature.

6 Analysis of the Experimental Result Using Simulation

The database used in the simulation work for this paper work is available at www.cs.cmu.edu/afs/cs/ In this database, 11 sets of color images for testing the algorithms for content-based retrieval is available out of which we have chosen 10 sets of images. The Format of the image is.jpeg. From the database around 650 images are trained using the Color Moment features for testing purpose. From the database 10 classes of image categories are selected namely: animal, arborgreens, cherries, skies, mountains, animals etc. The experimental results are shown below.

In Fig. 3, image retrieval using color moment feature for a test image has been shown.

6.1 Determining Threshold Value

After image retrieval, to show its effectiveness, we need to set some threshold value upon which only retrieved images will be established. For experimental reason, we have chosen six values unto which retrieved images are shown. From the retrieval of images, precision and recall are calculated and we have checked in which threshold highest precision is achieved. For instance, we have chosen the values as following:

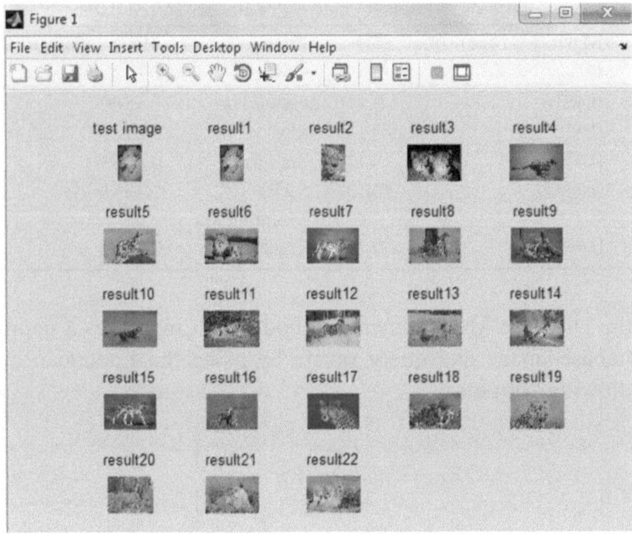

Fig. 3 Image retrieval using color moment feature for test image 1

Th1 = 0.5; Th2 = 0.8; Th3 = 1; Th4 = 1.2; Th5 = 1.5; Th6 = 2.0;

From these values we have got the maximum highest precision on the value Th3 = 1. Hence, we have set this particular value as threshold. For setting threshold, we have taken 20 test images arbitrarily chosen from the image database in a separate folder and named it as "test" which consists of the test images in.jpg format. In the below figure, the precision (highest), mean precision (of 6 values) and the standard deviation of the precision is shown for the 20 test images after retrieval as per threshold value 1 (Table 3).

Table 3 Precision, mean and SD of test images

Image name	Precision	Mean	SD	Image name	Precision	Mean	SD
A1.jpg	0.5000	0.4243	0.0438	Gl1.jpg	0.7778	0.5120	0.1444
A2.jpg	0.7778	0.4562	0.1881	Gl2.jpg	0.6667	0.5077	0.0954
Arb1.jpg	0.2429	0.1820	0.0501	Sky1.jpg	0.5000	0.3133	0.1004
Arb2.jpg	0.5714	0.4343	0.0789	Sky2.jpg	0.5833	0.4472	0.1039
Br1.jpg	0.9130	0.7798	0.0922	Std1.jpg	0.9524	0.8109	0.1301
Br2.jpg	0.9412	0.8193	0.1094	Std2.jpg	0.9239	0.7707	0.1574
Cam1.jpg	0.5000	0.3168	0.1267	Sw1.jpg	0.5000	0.2417	0.1350
Cam2.jpg	0.5000	0.3046	0.1361	Sw2.jpg	0.8897	0.5097	0.3593
Ch1.jpg	0.8750	0.5662	0.2344	T1.jpg	0.4118	0.3237	0.0961
Ch2.jpg	0.8750	0.4937	0.2399	T2.jpg	0.7500	0.5046	0.1571

6.2 *Efficiency of Retrieval*

The efficiency of image retrieval can be understood by the following graphs of precision, recall and f-score values:

After the observation of the graph of Fig. 4, anybody can easily understand that in threshold 1, highest significant images are retrieved.

In Fig. 5, the probability/fraction of significant retrieval in different threshold values are shown. From the graph it can be commented that when the threshold value is increasing, the chances of returning the significant image is getting higher.

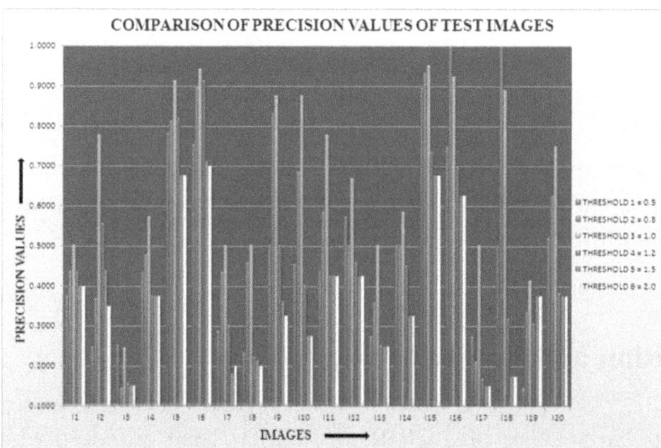

Fig. 4 Comparison of precision values at different threshold values

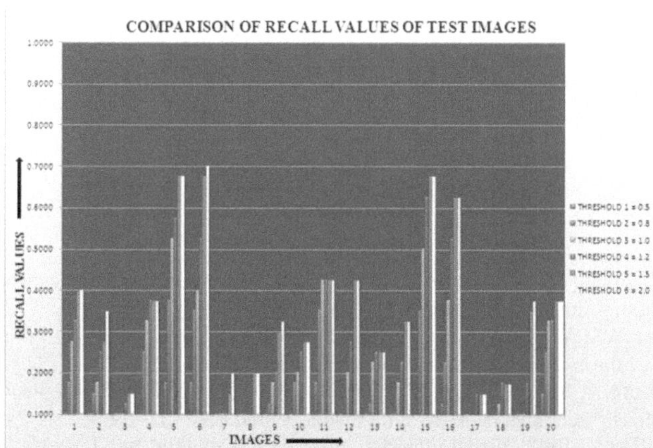

Fig. 5 Comparison of recall values at different thresholds

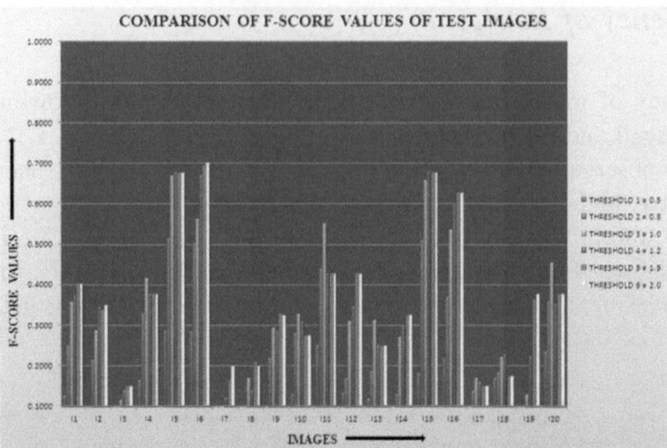

Fig. 6 Comparison of F-score values at different thresholds

In Fig. 6, the single matrix combining recall and precision named as f-score is obtained which shows the harmonic mean of the above two matrixes in different threshold values.

7 Conclusion and Future Work

In this paper, an analysis of a CBIR method has been proposed which uses the color moment feature in HSV space as a feature extraction method. Investigational results for ten categories of images depicted higher retrieval accuracy in terms of precision values and lower computation speed in terms of training time. This method has faster retrieval speed than other methods due to its simple feature extraction method bearing low feature vector dimension. In future work, we can try to add other low level features along with color moment feature and try to find the retrieval effectiveness.

References

1. Gevers T, Smeulders AWM (1999) Content-based image retrieval by viewpoint-invariant image indexing. Image Vis Comput 17(7):475–488
2. Smeulders AWM, Worring M, Satini S, Gupta A, Jain R (2000) Content—based image retrieval at the end of the early years. IEEE Trans Pattern Anal Mach Intell 22(12):1349–1380
3. Zhu L, Zhang A, Rao A, Srihari R (2000) Keyblock: an approach for content-based image retrieval. In: Proceedings of ACM multimedia conference, 157–166
4. Gupta A, Jain R (1997) Visual information retrieval. Commun ACM 40(5):71–79

5. Ma JQ (2009) Content Based Image Retrieval with HSV color space and texture feature. International conference on web information systems and mining (WISM '09), 7–8 November 2009, Shanghai, China
6. Tamura H, Yokoya N (1984) Image database systems: a survey. Pattern Recogn 17(1):29–43
7. Datta R, Joshi D, Jiali, Wang JZ (2008) Image retrieval: ideas, influences and trends of the new age. ACM Trans Comput Surv 40(2)
8. Jeong S (2001) Histogram based colour image retrieval Psycs221/EE362 Project report, Mar 2001
9. Wang JZ (2001) Integrated region-based image retrieval. Kluwer Academic Publishers, Boston
10. Brunelli R, Mich O (2001) Histograms analysis for image retrieval. Pattern Recogn 34(8):1625–1637
11. Del Bimbo A, Mugnaini M, Pala P, Turco Picasso F (1997) Visual querying by color perceptive regions. In: Proceedings of the 2nd international conference on visual information systems, San Diego, pp 125–131
12. Huang J, Kumar SR, Mitra M, Zhu WJ, Zabih R (1997) Image indexing using color correlograms. IEEE International Conference Computer Vision and Pattern Recognition, San Juan, Puerto Rico, pp 762–768
13. Huang Z et al (2010) CBIR using color moment and gabor texture feature. In: Proceedings of 9th international conference on machine learning and cybernetics in Qingdao, 11–14
14. Shih FT, Mitchell OR (1992) A mathematical morphology approach to Euclidean distance transformation. IEEE Trans Image Process 1(2):197–204

Enhancing Security in Medical Image Communication with JPEG2000 Compression and Lossless Watermarking

A. Umamageswari and G. R. Suresh

Abstract Medical image content authentication is very important, since medical image contents are more and more widely distributed. Reversible watermarking becomes a promising technique to embed the information into medical images. In this paper, we define the Region of Interest (ROI) in an image and trying to embed data in Region of Non Interest. When medical image shared through network, for the compression purpose we proposed the JPEG2000 algorithm and to improve the integrity and Authenticity hash value of the image is found by using MD5 and encrypted using RSA to form the DS (Digital Signature). DS and patient information is embedded into DICOM images. Strict authentication can be achieved by providing high security in accessing the secured medical images by medical experts which are available in the websites using Kerberos technique.

Keywords Reversible watermarking · Medical image security · Medical image compression · Authenticity and integrity · RSA · HD5 hash function · JPEG2000 compression · Kerberos

---

A. Umamageswari (✉)
Sathyabama University, Chennai, India
e-mail: r.umaramesh@gmail.com

G. R. Suresh
Easwari Engineering College, Chennai, India
e-mail: sureshgr@rediffmail.com

Mohan S. and S. S. Kumar (eds.), *Proceedings of the Fourth International Conference on Signal and Image Processing 2012 (ICSIP 2012)*, Lecture Notes in Electrical Engineering 221, DOI: 10.1007/978-81-322-0997-3_36,

1 Introduction

Medical image communication is used in a variety of application like telesurgery and telediagnosis [1, 2] with the advances internet technology, Especially in healthcare, images can be cross-exchange in correct time allowing new medical practice[3]. Image compression is useful to reduce the size of an image during communication, so the bandwidth can be effectively utilized. JPEG2000 offers numerous advantages over the JPEG standard. It also offers both lossy and lossless compression. When high quality is a concern, JPEG2000 process promises a higher quality final image, even when using lossy compression and also it offers higher compression ratios. The JPEG2000 image compression system has a rate distortion advantage over the original JPEG [4, 5]. Data encryption techniques and Digital Signature algorithms are important on protecting confidential information [6]. To generate the Digital Signature, hash value of the medical image (Covering image) is calculated using MD5 Algorithm. The algorithm is an iterative, one way hash function that can process image to produce a condensed representation called a message Digest. The algorithm enables the integrity of a message to be determined and any change to the message will, with a very high probability, result in a different message Digest [7, 8]. The Rivest-Shamir-Adleman (RSA) scheme has since the time reigned supreme as the most widely accepted and implemented general-purpose approach to public-key encryption. RSA makes use of an expression with exponential. Four possible approaches to attacking the RSA Algorithm are Brute force attack, Mathematical attack, timing and chosen cipher text attack [9, 10, 11].

Medical image knowledge digest consists of patient information like patient name, patient-ID, disease description, procedures with doctor's information [12].Combination of medical image knowledge digest and digital signature of the medical image will be the watermark. This watermark is embedded into the image which has to be shared by using lossless watermarking technique. The data hiding scheme should have a large embedding capacity to carry more general information. The goals of the reversible watermarking are to protect the copyrights and can recover the original image. Reversible watermarking provides robustness, imperceptibility, high embedding capacity and readily retrieving capacity [13]. A reversible data hiding scheme and a reversible image authentication scheme can also be defined as the schemes which can recover the original image from the embedded image [14, 15]. Security can be defined in the term of integrity, authenticity, confidentiality and availability. This paper discusses mainly on authenticity, i.e., providing knowledge digest belongs to the correct patient information. An unimportant area of an image (RONI) is watermarked. In this approach we leave the information of interest (ROI) for the diagnosis purpose [16]. After embedding the watermark into an image, image quality can be calculated by peak signal to noise ratio, or PSNR, and the root mean square error (RMSE) and compression ratio (CR).Compression ratio should be minimum and PSNR should be maximum for better quality image. Compression ratio can be calculated by the ratio between the size of the image before compression and size of the image after compression.

$$Compression\,Ratio = (Original\,Image\,Size|Compressed\,Image\,Size)$$

.

Here we make use of PSNR to quantify the distortion between the original image I and watermarked image Iw [17, 18].

$$PSNR(I, Iw) = 10log_{10}(\left(\left(2^p - 1\right)^2 |MSE\right)\right)$$

$$MSE = \frac{1}{MN}[\sum_{i=0}^{M}\sum_{j=0}^{N}\left[\tilde{f}(m,n) - f(m,n)\right]^2\right]$$

The watermarked images are shared through the web sites. The medical experts who are accessing the images should be registered with the website with their user id and password. The strict authentication can be provided to those medical experts by using Kerberos. Kerberos introduces intermediate server which has the database all the medical experts should register their user id and passwords with this database [19]. The intermediate authentication server produces ticket to access the medical images which are available in the websites, so the doctors registered properly with the websites through this Kerberos only can able to access the message.

2 Methodology Used

2.1 JPEG2000 Image Compression

The JPEG 2000 image compression consists of four basic steps in the algorithm-pre-process, transformation. In our work we implemented JPEG2000 compression without quantization because medical images contains sensitive information, these information should not get lost during compression. JPEG2000 utilizes a new coding method called Embedded Block Coding with Optimized Truncation (EBCOT).

Step 1: Pre-processing: Pre-processing step will center the gray scale intensity values. We subtracted 127 from each intensity value in the image matrix.

Step 2: Transformation: JPEG2000 uses discrete wavelet Transformation (DWT). For lossless compression, we use the DWT in conjunction with the LeGall53 and perform the computation using lifting method. We compute 2–3 iterations of the DWT.

Step 3: Quantization: Above-mentioned two steps are enough for lossless compression.

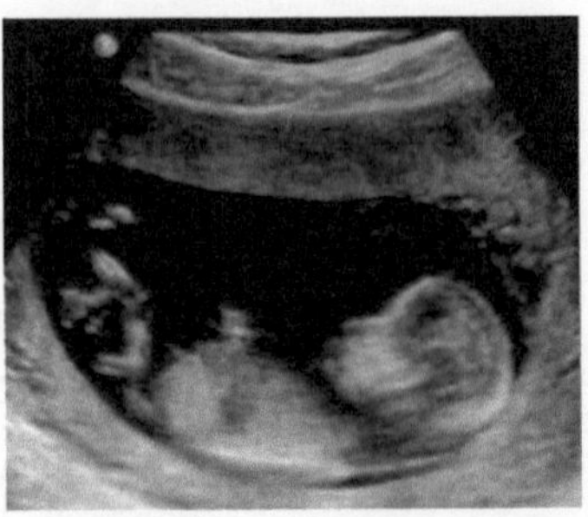

Fig. 1 Image before compression

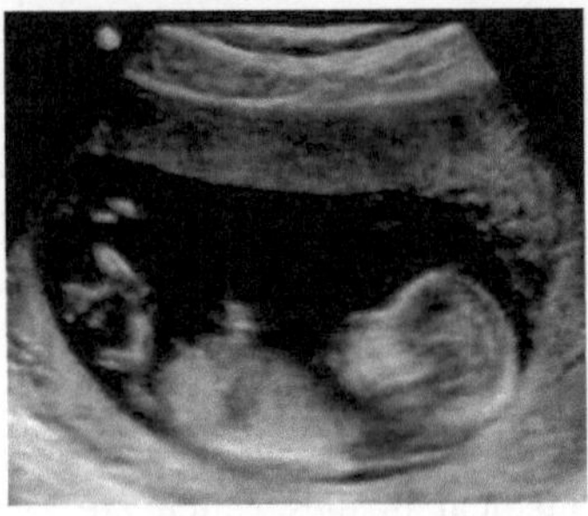

Fig. 2 Image after compression

Step 4: We simply used EBCOT to code the elements of the wavelet Transformation constructed with the LeGall filter. We can store the image using 215,544 bits. The original image, in raw format requires 307,200 bits of storage so the lossless method represents a saving of about 30 %. The compression rate is 5.6 bpp. Figure 1 shows the input US image of size 246◇205 before compression and Fig. 2 shows the same image after applying JPEG2000 compression

From the Figs. 1 and 2 we can ensure that image size has been decreased from 41 to 4.64 KB, so compression ratio will be 0.113. When comparing the compression ratio of JPEG with JPEG2000, JPEG2000 giving better compression ratio with lossless compression.

2.2 Digital Signature Using RSA Approach

Authentication is maintained through the Digital Signature (DS). This DS is computed over the input medical image. We use this signature to verify the reliability of the information. The difference between the signature and the reconstructed will indicate the information has been corrupted during transmission. We used RSA approach to generate the Digital Signature (DS). This signature can be produced with the Ticket created by the following Kerberos algorithm. This ticket is acts as the web server key to access the image from the Websites. Hash value of

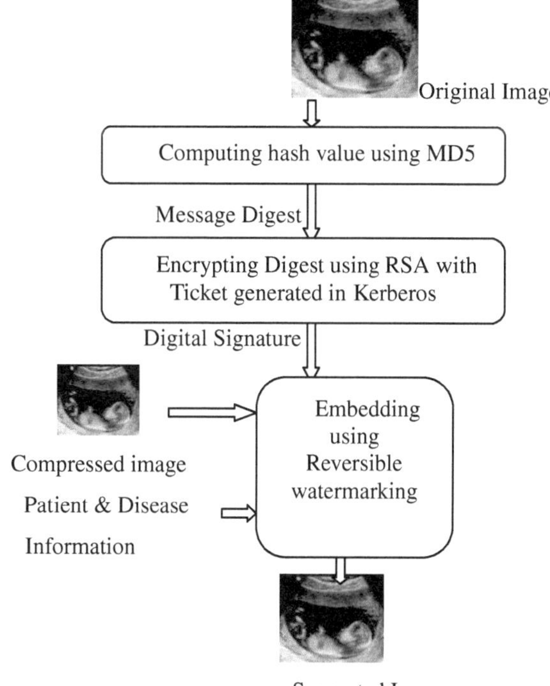

Fig. 3 Embedding procedure and authentication procedure

the input image is computed by using MD5 algorithm. MD5 accepts the image values and produces the 128 bit constant output as the hash value. This hash value will be encrypted using RSA approach.

The combination of Patient information, Disease information and DS is called as Watermark. This watermark is embedded inside the image using reversible watermarking in the sender side. In the receiver side the signature and patient and disease information is extracted from the suspected image and hash value of the original image is also computed in the receiver side because we used reversible watermarking, the hash value is encrypted using RSA approach to find the digital signature then this DS is compared with the Signature extracted from the suspected image, If these two signatures are same we can say that no alteration in the suspected image during transmission (Fig. 3).

2.3 Reversible/Lossless Watermarking

In reversible watermarking, we embed a watermark in a digital image I, and obtain the watermarked image Iw. The authenticator can remove the watermark from Iw to restore the original image and also the watermark we have embedded. The extracted image is same as the original image, because medical images having sensitive information these images should not be altered during embedding process, for this purpose only we proposed reversible watermarking. If the amount of information need to embed is larger than the embedding area, most of the techniques rely on lossless compression on the original values in the embedding area, and the space saved from compression will be used for embedding the watermark. We are using difference expansion method for reversible watermarking. This scheme usually generates some small values to represent the features of the original image. Then we expand the generated values to embed the bits of watermark information. The watermark information is embedded in the LSB parts of the expanded values. Then the watermarked image is reconstructed by using the modified values. In our method we will embed the watermark in the difference of the pixel values. For a pair of pixel values(x, y) in a grey scale image, $0 \leq x$, $y \leq 255$, define their (integer) average l and difference h as

$$l = (x + y)/2$$

$$h = x - y$$

where x and y are two adjacent pixels (Fig. 4).

2.4 Algorithm for Kerberos

The Kerberos authentication model relies on a secret key symmetric encryption scheme and the concept of dual encryption to provide secure authentication across a possibly insecure network. Authentication tickets are delivered to Kerberos medical experts encrypted in two keys.

Step 1: The medical expert wishing access to an authenticated target service provides his/her username and password to the system he/she is using. The system used by the medical expert has no record of the user's username and password.

Step 2: The user system sends a request to the Kerberos initial ticketing service requesting a ticket-granting ticket for the user whose user name it has been given. This request is totally unauthenticated.

Step 3: The initial ticketing service creates a unique session key (Ksession) and sends back to the user a dual-encrypted ticket-granting ticket and session key in the form

$$\{\{\text{Ttgs}, \text{ksession}\}\text{Ktgs}, \text{Ksession}\}\text{Kuser}$$

The user attempts to decrypt the TGT using his/her password as a key. If the decryption succeeds, the user can be certain that the user is authentic.

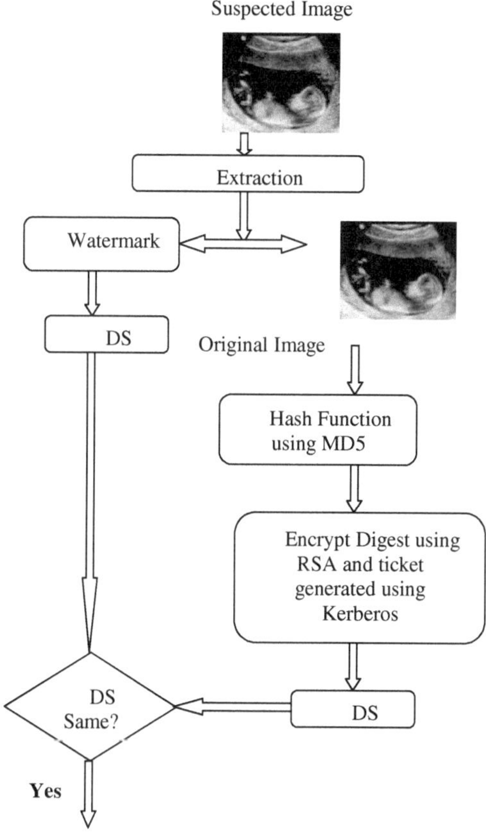

Fig. 4 Watermark extraction and authentication verification

Step 4: When the medical expert attempts to use a particular target service, the user sends a service ticket request to the Kerberos ticket granting service.

$$\{\text{TGT}, \{\text{request}, \text{User ID}, \text{Time}\}\text{Ksession}\}$$

where $\text{TGT} = \{\text{Ttgs}, \text{ksession}\}\text{Ktgs}$

Step 5: The Kerberos ticket granting service uses its own secret key (Ktgs) to decrypt the TGT in the request it has received, then uses the session key (Ksession) in that TGT to decrypt the rest of the request.

Step 6: The user decrypts the service ticket it has received using the session key provided to yield the service session key and an encrypted service ticket.

$$(\{\text{Tservice}, \ \text{kservice} - \text{session}\}\text{Kservice})$$

The medical experts can access the watermarked medical images available in the websites through this Ticket produced by the ticket granting ticket. These tickets are reusable.

3 Results and Discussions

The proposed methodology has been simulated in C# .Net using around 100 digital Ultrasonic images (US) images. These images were taken from public databases like community and public health and NIH. The images in the databases were in different formats. We brought it to the various sizes of medical US images, 8 bits per pixel and represented in PNG format. We have taken only five images for discussions. Table 1 shows the PSNR and Compression ratio (CR) of those five images when JPEG is used for compression and Reversible watermarking with RSA approach and Kerberos is used for authentication, reliability and integrity maintenance Table 2 shows the PSNR and Compression ratio (CR) of the proposed algorithms mentioned in methodologies used. When we are comparing the compression ratio of the existing and proposed algorithms our proposed method only gives better CR. When CR of the first US image taken into consideration from Tables 1 and 2 it is 3.03 in previous algorithms but it is 4.11 in proposed algorithm. So our proposed method giving better results for compression. This JPEG2000 is lossless compression only so sensitive information in the medical image will not get lost. If our medical image is compressed a lot then we can insert more amount of information into an image. So obviously Capacity Ratio will be increased. For this paper we didn't take capacity ratio for Discussions.

Table 1 PSNR and CR of Existing algorithm (digital signature with JPEG compression) in [3]

Sample images	PSNR Value in dB	CR value
I1	52.63	3.03
I2	53.19	2.90
I3	48.51	2.97
I4	48.23	2.88
I5	50.45	2.89

Table 2 PSNR and CR of proposed algorithm

Sample images	PSNR value	CR value
I1	60.72	4.11
I2	59.28	3.57
I3	54.58	4.49
I4	58.52	3.92
I5	54.45	3.78

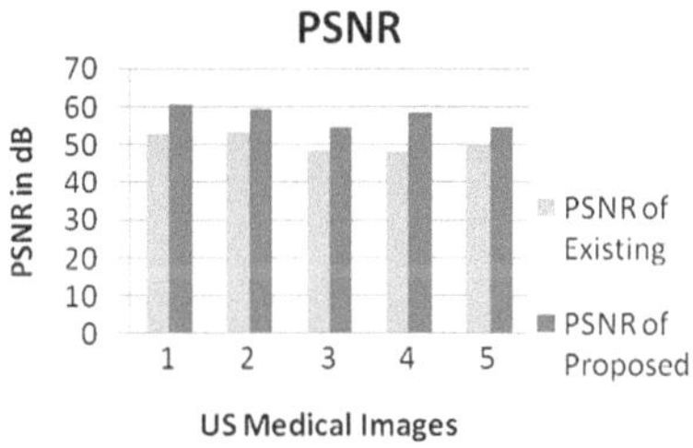

Fig. 5 Comparative results of PSNR for existing and proposed

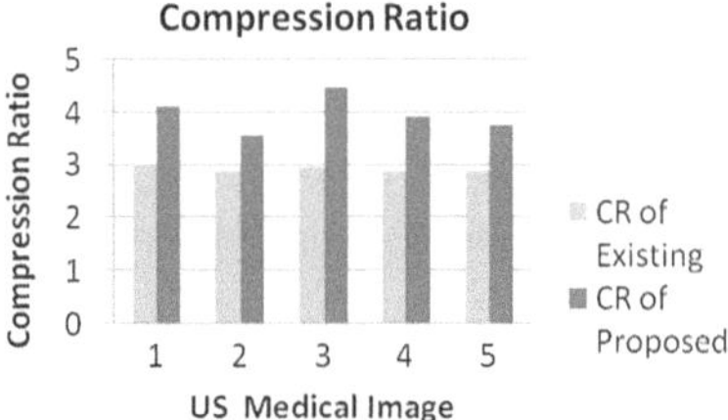

Fig. 6 Comparative results of compression ratio (*CR*) for existing and proposed

When we take the second parameter PSNR, PSNR also best in our proposed methodology. PSNR value in our existing method is only 52.63 dB but in our proposed methodology it is 60.72 dB. It is applicable for all the messages used for our discussions. From the Tables 1 and 2 we can say that the PSNR values of the existing algorithm is better in proposed algorithm.

The following graphs shown in Figs. 5 and 6 also shows the comparison of PSNR value of the existing and proposed methodology and Compression Ratio (CR) of the existing and proposed methodology respectively. From these two figures we can definitely conclude that our proposed algorithm giving better PSNR value and Compression Ratio (CR) value.

Figure 5 shows the PSNR comparison of previous algorithm and proposed algorithm. From the figure we can say that the PSNR value is more in proposed algorithm only. Figure 6 shows the Compression ratio of previous algorithm and proposed algorithm. From the figure we can say that the CR is more in proposed algorithm only.

Above Figs. 7 and 8 shows the implemented results of watermark embedding and extraction process respectively.

4 Conclusion

A medical image security system based on lossless watermarking to achieve authentication, reliability and integrity was designed and implemented in this paper. A strict authentication was achieved through Kerberos. The digital signature (DS), Patient information and Disease information also embedded inside the image. It has completely solves the problem of integrity, reliability and authentication of

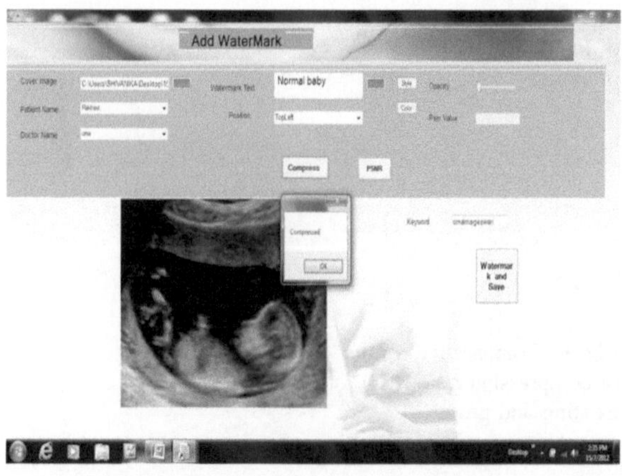

Fig. 7 Implementation of watermark embedding

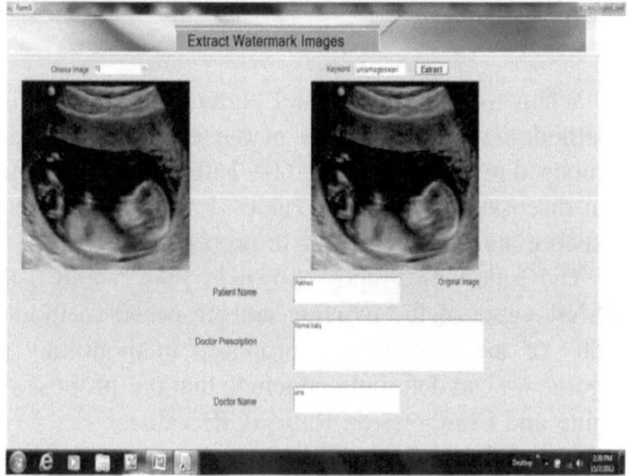

Fig. 8 Implementation of watermark extraction

medical image and also we can embed large amount of data inside the medical image without any distortion in an image. Since it requires secret key for both embedding process and extraction process it gives more authentication to our medical images. Moreover, these keys are tickets generated by the authenticated web server by using Kerberos algorithm, so it has high security.

5 Future Enhancement

In future we can achieve higher compression ratio (CR) by introducing JPEG-LS algorithm for compression, so we can embed more information inside an image. Implementation of RSA approach is difficult so we can introduce other security algorithm.

References

1. Coatrieux G, Guillou C le, Cauvin J, Roux CH (2009) Reversible watermarking for knowledge digest embedding and reliability control in medical images. IEEE Trans Info Technol Biomed 13(2)
2. Coatrieux G, Lamard M, Daccache W, Puentes J, Roux C (2005) A low distortion and reversible watermark application in angiographic images of the retina. In: Proceedings IEEE-EMBC conference, Shanghai pp 2224–2227
3. Pan W, Coatrieux G, Cuppens-Boulahia N, Cuppens F, Roux CH (2010) Medical image integrity control combining digital signature and lossless watermarking, published in 2nd SSETOP international workshop on autonomous and spontaneous security, Saint malo, Version 1–14
4. Marcellin MW, Garmish MJ, Bilgin A, Bolick MP (2000) An overview of JPEG-2000. In: proceedings of IEEE data compression conference, pp 523–541
5. ISO (2000) JPEG2000 image codingsystem, ISO/IEC FCD 15444-1,JPEG2000 part I Final committee draft, Version 1.0
6. Kuang LQ, Zhang Y, Xie H (2009) A medical image authentication system based on reversible digital watermarking. In: IEEE, 1st international conference on information science and engineering (ICISE 2009), pp 1047–1050
7. Zain JM (2012) Strict authentication watermarking with JPEG compression (SAW-JPEG) for medical images. Euro J Sci Res 42(2):232–241. ISSN 450-216X
8. Zhaol GC, Yang X, Zhoul B Wei W RSA-based digital image encryption algorithm in wireless sensor networks. In: proceedings second international conference on signal processing systems, Version 2, pp 640–643
9. Rivest R, Shamir A, Adleman l (1978) A method for obtaining digital signatures and public-key cryptosystems. Commun ACM 21(2):120–126
10. Kuo WC, Chen MY (2005) A modified(t,n) threshold proxy signature scheme based on the RSA cryptosystem. In: Information technology and applications, ICITA vol 2, pp 576–579
11. Harn L, Ren J (2008) Efficient identity-based RSA multisignatures. Comput Secur 27:12–15
12. Coatrieux G, Guillou C le, Cauvin J, Locarnu L, Roux CH (2006) Enhancing shared medical image functionalities with image knowledge digest and watermarking, presented in the IEEE EMBC conference international technology applied biomedicine (ITAB 2006), Joannina
13. Chang CC, Lin IC (20047) Remarks in fingerprint-based remote user authentication scheme using smart cards. ACM Operating Syst Rev 38(3):91–100
14. Chang CC, Tai WL, Lin MH (2005) A reversible data hiding scheme with modified side match vector quantization. In: proceedings of the international conference on advanced information networking and applications, vol 1, Tai-Wan, pp 947–952 Mar 2005
15. Keyvanpour MR, Merrikh-Bayat F (2010) A new encryption method for secure embedding in image watermarking. In: procedings of third international conference on advanced computer theory and engineering, vol 2, pp 402–407
16. Coatrieux G, Quantin C, Montagner J, Francois MF, Allert A, Roux CH (2008) Watermarking medical images with anonymous patient identification to verify authenticity, in ehealth beyond the horizon, vol 136, pp 667–672
17. Umamageswari A, FerniUkrit M, Suresh GR (2011) A survey on security in medical image communication. Int J Comput Appl 30(3):0975–8887
18. FerniUkrit M, Umamageswari A, Suresh GR (2011) A survey on lossless compression techniques for dynamic images. Int J Comput Appl 31(8):0975–8887
19. William S (2010) Cryptography and network security

Proposal of a Search for Rotation Based Independent Component Analysis (SRICA) Algorithm

Bhaveshkumar C. Dharmani

Abstract Existing Independent Component Analysis (ICA) algorithms are using varying independence measures derived through varying independence definitions and approximations. It will be interesting to study the effect of these variation on ICA solution and applications. This study require an ICA algorithm allowing use of varying independence measures as optimization criteria, assuring global solution and being truly blind. The article derives and verifies experimentally for the need, the Search for Rotation based ICA (SRICA) algorithm. It uses the fact that the independent components can be found by rotation of the whiten components. It uses Genetic Algorithm (GA), as a global search technique, to find the optimal angle of rotation. Also, the required study through SRICA finds minimization of sum of marginal entropies with kernel method for density estimation as the best independence measure, in terms of source matching, compare to the used four other independence measures.

Keywords Independent Component Analysis (ICA) · Blind Source Separation (BSS) · Genetic Algorithm (GA) · Search for Rotation based ICA (SRICA) · Overlearning

B. C. Dharmani (✉)
Dhirubhai Ambani Institute of Information and Communication Technology (DA-IICT), Gandhinagar 382007 Gujarat, India
e-mail: dharmanibc@gmail.com

Mohan S. and S. S. Kumar (eds.), *Proceedings of the Fourth International Conference on Signal and Image Processing 2012 (ICSIP 2012)*, Lecture Notes in Electrical Engineering 221, DOI: 10.1007/978-81-322-0997-3_37, © Springer India 2013

1 Introduction

Independent Component Analysis (ICA) is an established tool for both Component Analysis (CA) and Blind Source Separation (BSS) [3, 6, 7].

There are existing many algebraic and neural net techniques [4, 6, 9] for linear ICA. The methods and their performances vary based on the selected independence definition and measure as an optimization criteria and the used optimization technique. The ideal definition and measure of independence requires probability density function (pdf) into consideration. And then, all gradient based optimization techniques need to take derivative of that pdf based measure with respect to the unmixing or mixing matrix being estimated. So, in practice, approximations to that independence measures and (or) weakening of independence definition are must [4, Sect. 2.6, 5, Sect. 2.4] and so is, for all the existing methods, whether algebraic or neural net. The weakening of independence definition is provided by using either non-Gaussianity based or pairwise independence based or cumulant based definitions of independence. The approximations are derived using truncated version of either a Gram-Charlier expansion or an Edgeworth expansion of pdf or entropy. In the midst of these varying definitions, measures and approximations of independence, it will be interesting to study how the ICA solution is affected. The study requires to use an optimization technique which permits use of all these measures.

Also, all gradient based optimization methods used have poor global convergence. In case of a neural net techniques, the nonlinearity used for learning has to be a function of pdf of the components to be estimated. In the absence of this knowledge, family of densities e.g. superGaussian or subGaussian; is used as an approximation to select the nonlinearity. This requires to have some prior knowledge of densities to be estimated and so violates the blind assumption.

Overall, an ICA algorithm—allowing use of varying independence measures, assuring global solution and being truly blind—is required and can not be obtained through gradient based optimization technique. This leads to have an ICA algorithm based on global search techniques.

Towards the study on the effects of varying independence measures and achieving the required ICA algorithm, this article has following contributions:

1. It proposes a novel Search for Rotation based ICA (SRICA) algorithm. The SRICA exploits a fact that the independent components (ICs) can be obtained through orthogonal transformation, more specifically by rotation, of the whiten components. Use of global search methods in ICA is not new. All previous such proposals [13, 14] try to estimate all entries of an orthogonal unmixing matrix directly with just a constraint that the norm is unity. Compare to such unbounded search, the SRICA searches for an optimal angle of rotation from the set of all possible angle of rotations.
2. The study of ICA solution using varying independence measures finds minimization of the sum of marginal entropies, with kernel based pdf estimation, as the best independence measure on an average, compare to other four used independence measures.

The suggested SRICA algorithm using required basics has been derived in Sect. 2. SRICA characteristics and implementation issues have been discussed in Sect. 3. Section 4 comes with the experiments for SRICA verification and required study with varying independence measures. Based on them, Sect. 5 derives some results and discussion. Then, Sect. 6 concludes the article.

2 Linear ICA Problem and SRICA Solution

The Independent Component Analysis (ICA) model explains generation of an observed random vector **x**, as a linear transformation to another latent (hidden) random vector **s**. Mathematically,

$$\mathbf{x} = \mathbf{A}\mathbf{s} \tag{1}$$

where, $\mathbf{x} = [x_1; x_2; \ldots; x_m]$; $\mathbf{s} = [s_1; s_2; \ldots; s_n]$; x_i, s_i are random variables with values in $\mathcal{R}$; $m >\ = 2$ and **A** is full rank. The problem of ICA is to estimate both the unknowns **A** and **s**, with the only assumption of s_i being mutually *most independent possible* (*m.i.p.*)[1] random variables with respect to a given independence measure. If **W** is the estimated inverse of the mixing matrix **A** then for the estimated source random vector **y**

$$\mathbf{y} = \mathbf{A}^{-1}\mathbf{x} = \mathbf{W}\mathbf{x} = \mathbf{W}\mathbf{A}\mathbf{s} \tag{2}$$

As, $\mathbf{x} = \mathbf{A}\mathbf{s} = (\mathbf{A}\Lambda^{-1}\mathbf{P}^{-1})(\mathbf{P}\Lambda\mathbf{s})$, for any permutation matrix **P** and a scaling matrix Λ, there are going to be scaling and permutation ambiguities in the estimated components. The conditions to estimate both the sources and the mixing matrix uniquely, have been discussed in [5, 8].

2.1 The SRICA Solution

The method can be summarized in two steps.

1. Find zero mean whiten components using PCA (Principle Component Analysis) through EVD (Eigen Value Decomposition) or SVD (Singular Value Decomposition), as a compulsory step. Let a zero mean observed mixture data matrix **x**, be linearly transformed through a whitening matrix **V**, to give a zero mean, univariant, whiten data matrix **z**. $\mathbf{z} = \mathbf{V}\mathbf{x} = \mathbf{V}\mathbf{A}\mathbf{s}$.
2. Search for the optimal n-dimensional angle of rotation, through a global search technique, to transform the whiten components into independent components.

[1] Whether mention specifically or not, *m.i.p.* implies *m.i.p.* with respect to the given independence measure only.

The optimality can be defined through either maximization of independence or minimization of dependency. Let **R** be the rotation orthogonal matrix. Then,

$$\mathbf{y} = \mathbf{Rz} = \mathbf{RVAs} = \mathbf{WAs} \tag{3}$$

where, $\mathbf{W} = \mathbf{RV}$ is the estimated unmixing matrix.

As the algorithm is searching for a perfect angle of rotation, let it be called the *Search for Rotation based ICA* (*SRICA*) algorithm. The current article uses GA as the global search method.

How to rotate in a higher dimensional space? A two dimensional rotation can be achieved through Given's rotation. To convert a 3-dimensional rotation matrix Q into an identity matrix I, we need three, 2-dimensional rotations.

$$R_{yz}R_{xz}R_{xy}Q = \begin{bmatrix} 1 & 0 & 0 \\ 0 & 1 & 0 \\ 0 & 0 & 1 \end{bmatrix}$$

$\Rightarrow R = R_{yz}R_{xz}R_{xy} = Q'$ or $Q = R_{xy}^{-1}R_{xz}^{-1}R_{yz}^{-1} = R'$ An $n \times n$ rotation matrix will have $M = (n-1) + (n-2) + \ldots + 1 = n(n-1)/2$ entries below the diagonal to zero. So, it can be concluded that rotation in n dimension can be performed through $n(n-1)/2$ two dimensional rotations sequentially performed.

The Genetic Algorithm GA tries to imitate the evolution process, where natural selection is the guiding principle. Nature selects those, which are fit to survive. The fitness is measured based on the optimization function called the fitness criteria.

GA represents the solution as a binary string or a float variable. GA, randomly selects a set of possible solutions, called initial population, from the given solution space (total population). Assigns fitness to all. Then assuming the current population as the parent, generates a new child population using selection, crossover and mutation operators. The operators are some definite rules, using random number generation and fitness. The child becomes parents for the next generations. Thus, iteratively it achieves the optimal solution. The GA, with elitist model, keeps track of the best obtained so far. The stopping criteria can be based on number of iterations or optimality.

The optimization criteria or the fitness functions As per the definition of independence an asymmetric distance measure Kullback-Leibler Divergence (KLD) between the two distributions, the joint distribution $p(\mathbf{y})$ and the product of the marginal distributions $\prod_{i=1}^{n} p(y_i)$, is a natural measure of independence.

$$KLD(\mathbf{y}, y_i) = \int p(\mathbf{y}) \log \frac{p(\mathbf{y})}{\prod_{i=1}^{n} p(y_i)} d\mathbf{y} \tag{4}$$

where, $d\mathbf{y} = dy_1 dy_2 ... dy_n$

Equation 4 also interprets KLD, as a measure of mutual information $I(\cdot)$ among the random variables.

$$mikld(\mathbf{y}, y_i) = KLD(\mathbf{y}, y_i) = I(y_1, y_2, ..., y_n) \tag{5}$$

The measure is zero when both the distributions are same or the y_is are mutually independent. Accordingly, minimization of KLD or the mutual information is a good contrast function for ICA [5]. The other independence measures used for the experiments in Sect. 4 are defined as under.

$$jskld(\mathbf{y}, y_i) = KLD(\mathbf{y}, \mathbf{M}) + KLD(y_i, \mathbf{M}) \tag{6}$$

$$minhyi(y_i) = \sum_{i=1}^{n} Hy_i \tag{7}$$

$$normk4(y_i) = \sum_{i=1}^{n} |kurt(y_i) - 3| \tag{8}$$

where, $M = \frac{1}{2}(p(\mathbf{y}) + \prod_{i=1}^{n} p(y_i))$; $H(\cdot)$ implies Entropy; $kurt(\cdot)$ implies kurtosis and $|\cdot|$ implies the absolute value.

3 SRICA Characteristics and Algorithm Complexity

For data whitening, the first component of SRICA, numerically stable versions are available. GA, the second component, is a technique, applicable to optimize an objective function which is constrained or unconstrained or non differentiable or discontinuous or probabilistic or complex. So, the optimization criteria for SRICA, can be any given measure of independence. It could be completely blind. But, if available either a partial or full prior information, regarding the possible distribution of few or all sources or the mixing matrix, then that could be easily incorporated in the fitness criteria. Again, GA has a very good characteristic to converge to a global optima and thus finds if not optimal, near optimal solution. Off course, all these advantages come at the cost of more computations. There can be derived many variants of the SRICA by varying the global search technique, by varying the optimization criteria and by varying the pdf estimation technique. Overall, SRICA facilitates to tailor made or customize the algorithm, as on the performance or application requirement. This makes it suitable for the current experiment, as well, for the general applications in ICA, BSS and semi-BSS.

For an n-dimensional ICA problem, the GA has to search for optimality of $m = n(n-1)/2$ variables. If these m variables are independently affecting the fitness function f i.e. if the fitness function can be represented as $f(\theta_1, \theta_2, ..., \theta_m) = \sum_{i=1}^{m} f(\theta_i)$ then $f(\cdot)$ is called a decomposable (separable) function. In case of a decomposable fitness function, the optimal value of parameters can be found by just running the GA for each of the variable one by one, irrespective of their order. So, the algorithm complexity for continuous or float type of m variable, elitist GA with decomposable fitness function, would have been just $O(km)$ or maximally $O(km \ln m)$ [15], where assumed number of function evaluation for a single variable is k and number of computations for a function evaluation is 1. But, for an ICA problem, there is no independence measure, to be used as a fitness function, which can be decomposed into m independent rotation angles. In that case of non-decomposable (nonseparable) fitness function, the algorithmic complexity may go up to km^m or $kexp(m \ln m)$ theoretically. In the Evolutionary Algorithm (EA) literature, these type of nonseparable optimization functions with many variables are considered still as a problem and face the problem of 'curse of dimensionality'. So, none of the optimization algorithm, if used for ICA is free from this problem. In this article, GA has been used as a representative of the EA community. Also, towards solving the problem of nonseparability and reducing the algorithm complexity, there had been tried many variants of GA and other EAs—may be either successful partially or failed by the author. But, the scope of the article limits representing them.

4 Simulations

There have been designed two experiments with the aim to check the validity of the SRICA algorithm. To study the performances of different independence measure on the source separation SRICA is used with five different variants, four of them as defined in Sect. 2.1 and the fifth, the *radH* as per entropy definition in [12]. For the pdf estimation, the kernel method with Gaussian kernel had been used. For all the experiments following GA parameters were tuned to balance exploration and exploitation abilities. The selection of the individuals was done through normalized geometric distribution based on ranking through fitness, with the probability to select the best being 0.08. The float GA was using—arithmetic crossover (linear combination of the parents) and heuristic crossover (linear extrapolation, with the child generated near the more fit parent)—both with equal probability and total probability of crossover being 0.90. The mutation probability was kept to be 0.95 for the initial 10 % of the total generations. Then, for the remaining 90 % generations, it was reduced to 0.05. The initial high mutation probability takes care about the exploration and then the lower mutation probabilities allow for exploitation. In the experiments the population size was taken to be 10 and the maximum number of generations were kept to be 15. The termination criteria for all the experiments was based on maximum generations. For all the simulations in the article, the results of SRICA, are compared with that of FastICA (deflation mode and *tanh* nonlinearity) [10], EFICA [11], NPICA [2] and RADICAL (without augmentation) [12] with their standard parameters.

As mentioned, **y** is a permuted and scaled to univariant version of **s**. If **u** is zero mean, univariant source matrix, then $\mathbf{u} = \mathbf{D}^{-1}(\mathbf{s} - \bar{\mathbf{s}})$ where, **D** is a diagonal matrix, with inverse of the standard deviations of all s_i as the diagonal. So, from Eq. 3

$$\mathbf{y} = \mathbf{PWADu} = \mathbf{Gu} \tag{9}$$

where P is a permutation matrix and $\mathbf{G} = \mathbf{PWAD}$ is the gain matrix. Ideally the so called performance or gain matrix **G** should be an identity matrix. Based on this criteria, the *Amari Performance Index* (*API*) measures the deviation from diagonalization, of the gain matrix.

$$API(G) = \sum_{i=1}^{n}\left(\sum_{j=1}^{n}\frac{|g_{ij}|}{max_k|g_{ik}|} - 1\right) + \sum_{j=1}^{n}\left(\sum_{i=1}^{n}\frac{|g_{ij}|}{max_k|g_{kj}|} - 1\right) \tag{10}$$

So, better the performance, the measure should be more nearer to zero. The following experiments are done.

Experiment#1: To verify the performance of SRICA and varying independence measures against varying distributions There have been generated mixtures of two independent and identically distributed sources, with varying distributions and sample size (N) of 1000. The results were obtained using 50 simulation trials, for 21 types of distributions. The first 18 types (a to r) of distributions are suggested by [1] and three more types of distributions were added to test the performance of the ICA algorithms against skewed sources. The s type is a GGD with skewness $s = -0.25$ (left skewed) and kurtosis $k = 3.75$ and the t type is a GGD with skewness $s = 0.75$ (right skewed) and kurtosis $k = 0$. Both the distributions are generated using Power Method with parameters $b = 0.75031534111078$, $c = -0.02734119591845$, $d = 0.07699282409939$ for s type and $b = 1.11251460048528$, $c = 0.17363001955694$ and $d = -0.05033444870926$ for t type. The u type is a Rayleigh distribution with $\beta = 1$ and so the corresponding $s = 0.631$ and $k = 0.245$.

Experiment#2: To verify the performance of SRICA for more than two sources This experiment was to separate a mixture of 4 independent sources with Generalized Gaussian distributions (GGD(2.2)) and sample size of N = 4000. For SRICA, number of bins per random variable are taken as 5. The genetic algorithm used for SRICA, is of float type, six search variables (for four sources) and with fixed population size of 60 and 90 generations as the stopping criteria.

5 Results and Observations

The results of the Experiment#1 are tabulated in the Table 1. The SRICA algorithm using all the five independence measures gave results comparable with the other algorithms. Also, the SRICA variants using all the five measures gave very few cases with $API > 0.10$ (shown in boldface italics). This indicates that SRICA

Table 1 Performances of varying independence measures as fitness measures in SRICA; comparison with other ICA algorithms; using API as the performance measure

pdf	**srica** *jskld* $(\mathbf{y}, y_i)$	**srica** *mikld* $(\mathbf{y}, y_i)$	**srica** *minhyi* $(\sum y_i)$	**srica** *normk4*	**srica** *radH* $(\sum y_i)$	**fastica**	**efica**	**npica**	**radical**
a	0.0203	0.0213	0.0194	0.0283	0.0285	0.0237	0.0240	**0.0190**	0.0225
b	0.0296	0.0306	0.0227	0.0374	0.0282	0.0371	**0.0200**	0.0230	0.0288
c	0.0188	0.0185	0.0147	0.0157	0.0180	0.0260	**0.0107**	0.0120	0.0204
d	0.0342	**0.0302**	0.0359	0.0396	0.0492	0.0431	0.0545	0.0359	0.0424
e	0.0199	0.0196	0.0137	0.0312	0.0126	0.0393	0.0310	**0.0098**	0.0131
f	0.0139	0.0142	0.0140	0.0244	0.0161	0.0215	0.0308	0.0164	**0.0137**
g	**0.0116**	0.0118	0.0118	0.0130	0.0153	0.0166	0.0205	***0.9564***	0.0135
h	0.0329	0.0332	0.0357	0.0291	0.0537	0.0380	**0.0279**	0.0379	0.0451
i	0.0505	0.0505	0.0564	0.0554	0.1561	0.0677	**0.0497**	0.0588	***0.1781***
j	0.0127	**0.0127**	0.0133	0.0348	0.0148	***0.8306***	***0.9030***	***0.4772***	0.0148
k	0.0194	0.0185	0.0177	0.0471	0.0439	0.0496	0.0465	**0.0163**	0.0281
l	**0.0308**	0.0322	0.0390	0.0593	0.0696	0.0699	0.0470	0.0394	0.0654
m	0.0129	0.0128	0.0124	0.0226	0.0141	0.0478	0.0218	**0.0096**	0.0118
n	0.0164	0.0166	0.0164	0.0298	0.0211	***0.1584***	0.0386	**0.0121**	0.0207
o	0.0233	0.0234	0.0274	0.0231	0.0503	0.0476	**0.0217**	0.0261	0.0572
p	0.0185	0.0187	0.0137	0.0252	0.0194	0.0576	0.0209	**0.0122**	0.0178
q	0.0229	0.0237	0.0198	***0.1280***	0.0310	0.4149	0.3466	**0.0186**	0.0276
r	0.0246	0.0249	**0.0233**	0.0409	0.0503	0.0469	0.0395	0.0237	0.0503
s	0.0361	0.0356	**0.0336**	0.0374	0.0477	0.0349	0.0386	0.0348	0.0413
t	0.0251	0.0247	0.0135	***0.4143***	0.0108	***0.1325***	***0.1382***	0.0124	**0.0121**
u	**0.3388**	***0.3388***	***0.3273***	***0.3330***	***0.2657***	***0.5721***	***0.3865***	***0.4968***	***0.4254***
mean	0.0387	0.0387	**0.0372**	0.0700	0.0484	***0.1322***	***0.1104***	***0.1118***	0.0548
rand	0.0220	0.0227	**0.0190**	0.0393	0.0232	0.0578	0.0388	0.0191	0.0198

Number of sources = 2 with varying source distributions; GA parameters: float version; population size = 10; generations = 15; Kernel density estimation method with Gaussian kernel; Each entry is a median of the 50 trials

is the best at being uniformly good for any type of distributions. This validates the use of SRICA algorithm for varying distributions. Also, the results show that as an independence measure $minHyi(y_i)$ serves as the best on an average. Though not intended, it was expected that the SRICA searching for the best from the complete solution space would have found the best always. Instead, the results indicate either the components found by GA were not actually '*m.i.p.*' and the GA did not converge properly or the actual sources were not '*m.i.p.*'. To verify further, the canonical independence measure $mikld(\mathbf{y}, y_i)$ of the estimated sources by all the algorithms were compared, as in Table 2. Also, similar results for Experiment#2 are tabulated in Table 3.

It could be observed that the estimated sources by SRICA with $mikld(\mathbf{y}, y_i)$ measure were the most independent with respect to that measure, most of the times, compare to the estimated components by other algorithms. It proves that GA converged most of the times to the optimal or near optimal. But, best with respect

Table 2 Comparison of independence measure $mikld(\mathbf{y}, y_i)$ of the estimated sources through ICA algorithms in Table 1; Number of sources = 2 with varying source distributions; Kernel density estimation method with Gaussian kernel; Each entry is a median of the 50 trials

pdf	**srica** *jskld* $(\mathbf{y}, y_i)$	**srica** *mikld* $(\mathbf{y}, y_i)$	**srica** *minhyi* $(\sum y_i)$	**srica** *normk4*	**srica** *radH* $(\sum y_i)$	**fastica**	**efica**	**npica**	**radical**	**Sources**
a	0.3479	0.3474	0.3430	0.3538	0.3435	**0.3408**	0.3469	0.3465	0.3476	0.3469
b	0.8340	0.8281	0.8236	0.8304	0.8391	**0.8087**	0.8266	0.8348	0.8344	0.8351
c	0.0057	**0.0057**	0.0058	0.0057	0.0059	0.0060	0.0060	0.0059	0.0059	0.0060
d	0.0356	**0.0354**	0.0355	0.0359	0.0363	0.0364	0.0363	0.0357	0.0359	0.0586
e	0.0229	**0.0229**	0.0230	0.0247	0.0234	0.0264	0.0248	0.0228	0.0233	0.0234
f	0.0133	**0.0133**	0.0133	0.0148	0.0140	0.0145	0.0166	0.0167	0.0136	0.0176
g	**0.0044**	0.0045	0.0045	0.0046	0.0050	0.0057	0.0079	0.6093	0.0046	**0.0006**
h	0.0133	**0.0133**	0.0134	0.0134	0.0147	0.0136	0.0135	0.0136	0.0146	**0.0044**
i	0.0153	**0.0153**	0.0154	0.0154	0.0183	0.0157	0.0154	0.0155	0.0179	**0.0070**
j	0.0058	**0.0058**	0.0059	0.0146	0.0065	0.6701	0.6783	0.3246	0.0060	**0.0007**
k	0.0122	0.0122	0.0123	0.0130	0.0137	0.0135	0.0128	**0.0121**	0.0126	**0.0045**
l	0.0140	0.0139	0.0139	0.0152	0.0159	0.0154	0.0148	**0.0139**	0.0150	**0.0074**
m	0.0089	**0.0089**	0.0091	0.0094	0.0089	0.0118	0.0094	0.0096	0.0089	**0.0031**
n	0.0102	**0.0102**	0.0102	0.0114	0.0105	0.0275	0.0130	0.0105	0.0103	**0.0037**
o	0.0106	0.0106	0.0106	0.0106	0.0116	0.0112	0.0106	**0.0106**	0.0120	**0.0016**
p	**0.0081**	0.0083	0.0083	0.0086	0.0084	0.0116	0.0086	0.0082	0.0084	**0.0039**
q	0.0133	**0.0133**	0.0133	0.0221	0.0143	0.0474	0.0531	0.0136	0.0137	**0.0033**
r	0.0113	**0.0113**	0.0114	0.0123	0.0130	0.0127	0.0122	0.0114	0.0129	**0.0016**
s	0.0350	**0.0347**	0.0348	0.0357	0.0356	0.0359	0.0353	0.0350	0.0351	0.0358
t	0.0134	**0.0132**	0.0133	0.1087	0.0137	0.0260	0.0259	0.0137	0.0135	0.0141
u	0.0182	**0.0180**	0.0183	0.0195	0.0196	0.0207	0.0201	0.0196	0.0196	0.0206
mean	0.0692	0.0689	**0.0685**	0.0752	0.0701	0.1034	0.1042	0.1135	0.0698	**0.0667**
rand	0.0131	**0.0129**	0.0129	0.0152	0.0138	0.0169	0.0156	0.0136	0.0133	**0.0055**

to the independence measure do not necessarily imply best API performance measure. Vis-a-verse, best API performance measure does not necessarily imply best independence measure. Overall, the SRICA algorithm had been successful in finding the *m.i.p.* components but while doing that it had not been able to follow the actual sources.

There could be two reasons for this. One of the reasons from the observations is, many times, the estimated sources were more independent than the actual sources. This indicates that the sources were neither independent (*mikld* measure 0) nor *m.i.p.*. That means, the sources were not following the ICA independence assumption. The other reason is that the actual sources were not exactly uncorrelated. The SRICA searches for the solution, which is exactly decorrelated. So, from the given solution space SRICA found *m.i.p.* but the actual solution was not at all in the solution space.

Generally, the phenomena of getting estimated components more independent than the sources is not new in ICA algorithms and is explained as overlearning [16]. But, the observations raises many queries. The laboratory generated signals were also very frequently neither *m.i.p.* with respect to the used independence

Table 3 Comparison of performances of SRICA algorithm, using measures API and ($mikld(\mathbf{y}, y_i)$) with pdf estimation through histogram technique, with that of FastICA, EFICA and NPICA; to separate the mixture of four ggd(2.2) randomly distributed signals, described in Experiment#2

Sources pdf	srica ($mikld(\mathbf{y}, y_i), Hist$)	fastica	efica	npica
4 - ggd(2.2)	0.3625	0.3354	**0.3097**	0.4078
($mikld(\mathbf{y}, y_i), Hist$) measure for the actual and estimated sources				
1.3569	**0.3787**	0.4273	0.4556	0.4456

measure nor even uncorrelated. The real world signals, as well the increased number of sources would further enhance this violation of ICA independence assumption. So, the observations motivate to study further the violation of independence assumption by the sources in ICA applications and its' consequences, specifically in BSS applications. How much valid the solution will be, through a blindly selected ICA algorithm for BSS? Could there be defined a '*source following ability*' of an independence measure that would be a measure of the ability of an independence measure to follow the sources i.e. maximizing that independence measure will maximize the source matching? Answers to these questions and the further study are beyond the scope of the current article and left for the future work.

6 Conclusion

There has been proposed novel SRICA algorithm that has been best at being consistently good with respect to the source matching criteria for varying distributions. The algorithm has facilitated the study of the effects of various independence definitions and measures on ICA solution. The study has resulted declaring minimization of sum of marginal entropies, with kernel based pdf estimation method as the mean wise best independence measure, in terms of source matching, among the used other four independence measure. The article also motivates to work in future towards at least two directions. One is, solving the nonseparable function optimization problem and applying that for the ICA solution and the another is, to study the violation of independence assumption in ICA applications and consequences.

References

1. Bach F, Jordan M (2003) Kernel independent component analysis. J Mach Learn Res 3:1–48
2. Boscolo R, Pan H, Roychowdhury V (2004) Independent component analysis based on nonparametric density estimation. IEEE Trans Neural Netw 15(1):55–65
3. Cardoso JF (1998) Blind signal separation: statistical principles. Proc IEEE 9(10):2009–2025

4. Cardoso JF (1999) High-order contrasts for independent component analysis. Neural Comput 11(1):157–192
5. Comon P (1994) Independent component analysis—a new concept? Signal Process 36:287–314
6. Comon P, Jutten C (2010) Handbook of blind source separation: independent component analysis and applications, 1st edn. Academic Press, Burlington
7. Duda RO, Hart PE, Stork DG (2000) Pattern classification, 2nd edn. Wiley-Interscience, New York
8. Eriksson J, Koivunen V (2004) Identifiability, separability, and uniqueness of linear ICA models. IEEE Signal Process Lett 11(7):601–604
9. Hyvärinen A, Karhunen J, Oja E (2001) Independent component analysis. Wiley, New York, 481+xxii pp. http://www.cis.hut.fi/projects/ica/book/
10. Hyvärinen A, Oja E (1997) A fast fixed-point algorithm for independent component analysis. Neural Comput 9(7):1483–1492
11. Koldovskỳ Z, Tichavskỳ P, Oja E (2006) Efficient variant of algorithm fastica for independent component analysis attaining the Cramer-Rao lower bound. IEEE Trans. Neural Netw 17:1265–1277
12. Learned-Miller E, John W III (2003) ICA using spacings estimates of entropy. J Mach Learn Res 4:1271–1295
13. Palaniappan R, Navin C (2006) Genetic algorithm based independent component analysis to separate noise from electrocardiogram signals. In: 2006 IEEE international conference on engineering of intelligent systems, pp 1–5
14. Rojas F, Rojas I, Clemente R, Puntonet C (2001) Nonlinear blind source separation using genetic algorithms. In: Proceedings of international conference on independent component analysis and signal separation (ICA2001), Citeseer, pp 400–405
15. Salomon R (1996) Re-evaluating genetic algorithm performance under coordinate rotation of benchmark functions. A survey of some theoretical and practical aspects of genetic algorithms. BioSystems 39(3):263–278
16. Sarela J, Vigario R (2003) Overlearning in marginal distribution-based ICA: analysis and solutions. J Mach Learn Res 4:1447–1469

Local Fractal Dimension-Based Color Quantization for Error Diffusion Techniques

Mohammed Hassan and Chakravarthy Bhagvati

Abstract Reconstruction of an image with a limited number of colors (color palette) leads to highly visible degradations in image quality known as false contours. A way to overcome this problem is to perform dithering techniques. In this paper we propose a color quantization method for use with color dithering techniques. It makes use of local fractal dimensions to allocate larger weight to pixels in low activity regions where false contours in an image are most likely to occur. The results show that our method significantly removes false contours and color impulses as well as preserves textures that are commonly lost in high activity regions when applying dithering techniques to color quantized images.

Keywords Color quantization · Fractal dimensions · Error diffusion · Combined quantization · Error diffusion.

1 Introduction

Color quantization is the process of reducing the number of colors presented in a digital color image while maintaining the appearance of the original as much as possible. Originally, color quantization has been used to satisfy the display hardware constraints that allow a limited number of colors to be displayed simultaneously. Today the original motivation of color quantization has changed due to availability of inexpensive full color displays. However, color quantization is still

M. Hassan (✉) · C. Bhagvati
Department of Computer and Information Sciences, University of Hyderabad, Hyderabad 500046, India
e-mail: priester@icmpe.cnrs.fr

Mohan S. and S. S. Kumar (eds.), *Proceedings of the Fourth International Conference on Signal and Image Processing 2012 (ICSIP 2012)*, Lecture Notes in Electrical Engineering 221, DOI: 10.1007/978-81-322-0997-3_38,

an important problem in the fields of image processing and computer graphics; it can be used in mobile and hand-held devices where memory is usually small [1], for low-cost display and printing devices where only a small number of colors can be displayed or printed simultaneously [2], it also can be used in lossy compression techniques [3]. Another aspect of importance of color quantization is that the human visual system can't perceive more than 20,000 different colors at any given time [4] while a full color image may contain up to 16 million different colors; this large number of colors makes it difficult to handle a variety of color-based tasks such as computing histograms or other useful statistics. The most commonly used algorithms for color quantization are: median-cut [5], center-cut [6], octree [7], k-means [8], Wu's quantizer [9], fuzzy C-means [10], and SOM [11].

Color quantization process is considered in two stages: the selection of an optimal color palette and the mapping of each pixel of the image to a color from the color palette. Since the color palette is limited, some disturbing degradations such as false contours are visible on delivered color quantized image. A way to overcome this problem is to perform dithering techniques that make use of the averaging property of the human eye to the colors in a neighborhood of the point of interest and create the illusion of more colors. A dithering technique called error diffusion achieves this effect by distributing the error encountered in quantizing a pixel to neighboring pixels. This results in an alternation of palette colors between neighbor pixels and is perceived as a new color by the eye. Some well-known error diffusion filters are Floyd-Steinberg [12], Stucki [13], and Jarvice-Judice-Ninke [14].

False contours in an image are most likely to occur in fat or low activity regions. As the human eyes are less sensitive to quantization errors in high activity regions than in low activity regions, introducing more colors to regions with low activity will improve the removal of false contours in that regions after dithering. Based on that fact, we proposed a color quantization algorithm for the use with color dithering techniques. We used local fractal dimensions (LFDs) to give more importance (weight) to colors in low activity regions than those in high activity regions in a way that better results will be obtained after dithering.

There is some previous work for jointly quantizing and dithering color images in the literature. Orchard [15] used binary tree splitting (BTS) to create the color palette that minimizes the total squared difference between the actual and the quantized images and then modified dithering techniques are combined with quantization process. A modified BTS quantizer of [15] is proposed in [16] in such a way that at leaves the pair of cluster centers are given equal displacements from each other following the split in order to span a larger volume of color space. This creates an illusion of more colors after dithering. In competitive learning and error diffusion algorithm [17], the quantization error is diffused to neighboring pixels during learning process. In [18] three new fuzzy methods to jointly quantize and dither the color images are presented; in two of them a fraction of the quantization error is distributed to neighboring pixels during creating color palettes and then perform quantization and dithering as usual with the resultant palette, while the third method is to enlarge the convex hull of the quantization colors to obtain a color palette which is more suitable for dithering (Fig. 1).

Fig. 1 Testing images. **a** Tree (512 × 345). **b** Woman (512 × 345). **c** Flowers (345 × 345). **d** House (345 × 345)

The paper is organized as follows: Section 2 explains the local fractal dimensions. In Sect. 3, the color palette construction is described. Section 4 presents the color quantization process with error diffusion. Results of the study are discussed in Sect. 5 and are followed by the conclusions.

2 Local Fractal Dimensions

Fractal dimensions of imaged 3D surfaces can be used as a measure to perceptually distinguish between smooth and rough textured regions [19, 20]. It has been widely applied to many fields of digital image processing, such as texture classification and segmentation[21–24], image compression [25]. Sarkar and Chaudhuri [26, 27] proposed a method to find the fractal dimension of images known as differential box counting method. For an image of $M \times M$ pixels scaled down to a size $s \times s$ where $M/2 \geq s > 1$ and s is an integer, we can cover the entire image with boxes of sides $s \times s \times s'$ in the vertical direction. If the total number of gray levels is G then $\lfloor G/s' \rfloor = \lfloor M/s \rfloor$. The contribution of the $(i,j)th$ grid is computed as

$$n_r = l - k + 1 \tag{1}$$

where the minimum and maximum gray level of the image in the $(i,j)th$ grid fall in box number k and l, respectively. Taking the contribution of all grids, $N_r = \sum_{i,j} n_r$. N_r is counted for different values of r, i.e., different values of s. Then using the following equation:

$$1 = N_r \times r^D \text{ or} D = \log(N_r)/\log(1/r) \tag{2}$$

The fractal dimension D can be estimated from the least square linear fit of $log(N_r)$ against $log(1/r)$.

Figure 2 shows the original images(left) and their local fractal dimension (LFD) maps (right) where every position (i,j) in the LFD map represents the LFD estimation centered at position (i,j) in the original image. The LFD values are transformed from the range 2.0–3.0 into the rage 0–255. It is obvious from Fig. 2 that the smooth regions have small LFD values (dark) while the busy ones have large LFD values (bright), so it would be easy to distinguish between such regions and give more importance (weight) to the image pixels in the smooth regions than those in the busy ones during the color palette construction process.

3 Color Palette Construction

In this section, we present the proposed algorithm for color image quantization using local fractal dimension (LFD) maps. To describe the proposed algorithm, we first explain the k-means [8] algorithm which is the basis of our proposed algorithm. Generally the k-means algorithm seeks to minimize an objective function which is defined as

$$j = \sum_{i=1}^{k} \sum_{C \in S_i} W(C) \| C - \bar{C}_{S_i} \|^2 \tag{3}$$

where $\bar{C}_{S_i}$ is the center of the cluster S_i, k is the number of clusters, and $W(C)$ is the weight factor of the pixel C.

Initially the k-means algorithm starts with a set of k centers that are randomly selected. At each iterative step, scan through all the pixels of the original image

Fig. 2 LFD map of an image: (*left*) the original images and (*right*) their LFD maps

and assign each pixel to the nearest center in the sense of minimizing (3). Then the new centers are calculated as

$$\bar{C}_{S_i} = \frac{\sum_{C \in S_i} W(C) \times C}{\sum_{C \in S_i} W(C)} \quad i = 1, 2, .., k \tag{4}$$

The steps are repeated until the algorithm converges or the number of iterations reaches a specified value. Usually in the conventional k-means algorithm the weight factor $W(C) = 1$ for all pixels, therefore all pixels are treated equally regardless of their spatial distribution.

Our proposed color quantization algorithm follows the same steps as the k-means algorithm except that the weight factor $W(C)$ of pixel C at the $(i,j)th$ position in the original image is equal to the inverse of its corresponding LFD value at the $(i,j)^{th}$ position in the LFD map. In this way pixels are not equally treated where pixels in low activity regions are given larger weight (importance) than those in high activity regions. As a result more levels of colors will be

Fig. 3 Quantization results of 'Tree' image. **a** The original image. **b** LFD map of **a**. **c** k-means 16 colors. **d** k-means 32 colors, **e** Proposed algorithm 16 colors **f** Proposed algorithm 32 colors

presented in low activity regions and this will improve the perceived quality in such regions after dithering techniques are applied as shown in the next section. Figure 3 shows the color quantization outputs for the conventional k-means and the proposed algorithm for 16 and 32 colors. It is clear that the proposed algorithm allocates more levels of colors to the gradient sky region than the conventional k-means does to the same region for both 16 and 32 colors.

4 Color Quantization and Error Diffusion

Once a color palette has been designed, the quantized image is produced by mapping each pixel to the closest color from the color palette. Regardless of the color quantization algorithm used false contours are among the most disturbing degradations appeared in the resultant color quantized image. One way to overcome this problem is to perform error diffusion techniques. In conventional systems for error diffusion the pixels are chosen in raster ordering from left to right and top to bottom. Each pixel is quantized and the quantization error is propagated

Fig. 4 Dithered quantization results of 'Tree' image. **a** k-means 16 colors. **b** LFD map of original image 'Tree' image. **c** Proposed algorithm 16 colors. **d** flat-region map of original 'Tree' image (high activity regions in white color and low activity regions in black color)

Table 1 Average of the S-CIELab difference (ΔE) values of the dithered color quantized testing images

Image	#Colors	Orchard [15]	k-means+ED	Proposed
Tree	32	8.3087	8.3087	4.5739
	64	5.5770	4.3125	3.9637
	128	4.7970	3.9157	3.5735
	256	3.9583	3.3197	2.4642
Woman	32	0.1553	0.1553	0.1553
	64	0.1553	0.1553	0.1553
	128	0.1553	0.0770	0.0082
	256	0.1553	0.0728	0.0081
Flowers	32	10.0690	10.2834	9.2855
	64	9.1078	9.1150	8.2021
	128	9.3083	7.9035	7.5109
	256	8.0565	7.3997	7.1625
House	32	0.3273	0.3273	0.1975
	64	0.1900	0.2389	0.0213
	128	0.0649	0.1031	0.0153
	256	0.0251	0.0363	0.0069

forward to the neighboring pixels that have not yet been quantized with a casual filter. One problem associated with this approach that the textures in high activity regions may be distorted due to the accumulated quantization error (see Fig. 4a).

To resolve this probelm we threshold the LFD map using the Otsu's global image threshold method [28] to create a flat-region map to be used as a guide to distinguish between low and high activity regions. The Floyd-Steinberg error diffusion filter [12] is then applied only to pixels in low activity regions where false contours in an image are most likely to occur. In this way false contours in low activity regions will be removed while preserving the textures in high activity regions. Figure 4b and d show the LFD map of the 'Tree' image and its corresponding flat-region map respectively.

5 Results and Discussions

In Fig. 4(a) the full color image 'Tree' has been quantized to 16 colors using the conventional k-means where the Floyd-Steinberg error diffusion [12] has been incorporated within the pixel mapping process. It is observed that the application of the Floyd-Steinberg error diffusion causes the creation of color impulses and some false contours remain visible in the sky region as well as distorting textures in grass region. Those color impulses and false contours are the result of insufficient color levels presented in the gradient sky region, while the distorted textures are the result of unnecessarily dithering high activity regions where false contours in an image are invisible in such regions. On the other hand Fig. 4c shows the proposed color quantization outputs of 16 colors. It is clear that the false contours are completely removed in the sky region compared to the conventional k-means

Table 2 PSNR metric comparison of the dithered color quantized testing images

Image	#Colors	Orchard [15]	k-means+ED	Proposed
Tree	32	30.179	29.854	31.666
	64	32.305	32.014	35.214
	128	34.972	34.910	37.509
	256	37.335	36.924	39.129
Woman	32	30.143	29.039	30.724
	64	32.369	32.682	34.297
	128	35.002	34.986	36.682
	256	37.646	37.128	38.880
Flowers	32	28.833	28.756	30.017
	64	30.333	31.605	32.124
	128	33.039	33.527	35.405
	256	35.517	35.483	37.446
House	32	28.867	29.575	32.310
	64	31.159	31.753	34.518
	128	33.965	34.162	35.949
	256	36.043	36.287	38.059

Table 3 SSIM metric comparison of the dithered color quantized testing images

Image	#Colors	Orchard [15]	k-means+ED	Proposed
Tree	32	0.887	0.883	0.891
	64	0.914	0.924	0.934
	128	0.951	0.954	0.955
	256	0.969	0.968	0.975
Woman	32	0.776	0.778	0.808
	64	0.831	0.858	0.872
	128	0.892	0.905	0.921
	256	0.934	0.933	0.950
Flowers	32	0.750	0.776	0.794
	64	0.813	0.833	0.874
	128	0.853	0.902	0.911
	256	0.897	0.927	0.937
House	32	0.795	0.829	0.873
	64	0.873	0.881	0.916
	128	0.916	0.924	0.934
	256	0.938	0.947	0.957

and the result is free from color impulses. This is due to introducing more color levels in gradient sky region than the conventional k-means (see Fig. 3c and e). Therefore dithering low activity regions needs small diffused error to jump into another close palette color. This results a high alternation of palette colors between neighboring pixels to give full illusion of smooth gradation. It is also noted that textures in the grass region are preserved due to not performing dithering in high activity regions.

For the objective evaluation, three image quality metrics have been used: the widely used peak signal-to-noise ratio metric (PSNR) where the higher PSNR value means that the distorted image is the closer to the original, the structural similarity measure (SSIM) [29] to measure the structure of the resultant dithered quantized images, and the S-CIELab color reproduction metric (ΔE) [30] which is commonly used for measuring the quality of halftoned images. Tables 1 and 2 show the performance comparison of Orchard's method [15], k-means with error diffusion, and the proposed method in terms of the average ΔE and PSNR metrics for the testing images in Fig. 1. The results show that the proposed method is superior to the others. It is also clear from Table 3 that the proposed method preserves the structure of the resultant dithered quantized images more than the other methods.

6 Conclusion

In this paper, a color quantization algorithm to be used with error diffusion techniques is proposed. It makes use of local fractal dimensions to allocate larger weight to pixels in low activity regions where false contours in an image are most likely to occur. Simulation results showed that the proposed algorithm gives a remarkable improvement in the quality of dithered color quantized images both subjectively and objectively in terms of S-CIELAB, SSIM, and PSNR image quality metrics as compared with other algorithms.

References

1. Rui X, Chang C, Srikanthan T (2002) On the initialization and training methods for Kohonen self-organizing feature maps in color image quantization. In: First IEEE International Workshop on Electronic Design, Test and Applications, 321–325 2002
2. Scheunders P (1997) A genetic C-means clustering algorithm applied to color image quantization. Pattern Recogn 30:859–866
3. Velho L, Gomes J, Sobreiro M (1997) Color image quantization by pairwise clustering. In: 10th brazilian symposium on computer graphics and image processing, IEEE Computer Society, 203–207 1997
4. Sharma G (1996) Digital color imaging, CRC Press (1996)
5. Heckbert P (1982) Color image quantization for frame buffer display. ACM Trans Comput Graphics (SIGGRAPH) 16:297–307
6. Joy G, Xiang Z (1993) Center-cut for color-image quantization. Visual Computation 10:62–66
7. Gervautz M, Purgathofer W (1988) A simple method for color quantization: octree quantization. In: New trends in computer graphics, Springer, Verlag, pp 219–231
8. Lloyd SP (1982) Least squares quantization in PCM. IEEE Trans Inf Theory 28:129–137
9. Wu X (1991) Efficient statistical computations for optimal color quantization. Graphics Gems, 11, J. Arvo, Ed. New York: Academic, 126–133

10. Lim YW, Lee SU (1990) On the color image segmentation algorithm based on the thresholding and the fuzzy C-means. techniques. Pattern Recogn 23:935–952
11. Dekker A H (1994) Kohonen neural networks for optimal colour quantization. Netw Computat Neural Syst 5:351–367
12. Floyd RW, Steinberg L (1976) An adaptive algorithm for spatial grey scale. In: Proceedings of the society of information display 17:75–77 1976
13. Stucki P (1981) Mecca-a multiple-error correcting computation algorithm for bilevel image hardcopy reproduction. Research Report RZ1060, IBM Research Laboratory
14. Jarvis JF, Judice CN, Ninke WH (1976) A survey of techniques for the display of continuous tone pictures on bilevel displays. Comput Graphics Image Process 5:13–40
15. Orchard MT, Bouman CA (1991) Color quantization of images. IEEE Trans Signal Process 39:2677–2690
16. Akarun L, Ozdemir D, Yalcin O (1996) A modified quantization algorithm for dithering of color images. Electron Lett 32:1185–1186
17. Scheunders P, De Backer S (1997) Joint quantization and error diffusion of color images using competitive learning. In: International conference on image processing, 811–814 1997
18. Ozdemir D, Akarun L (2001) Fuzzy algorithms for combined quantization and dithering. IEEE Trans Image Process 10:923–931
19. Pentland A (1984) Fractal-based description of natural scenes. IEEE Trans Pattern Anal Mach Intell 6:661–674
20. Pentland A (1986) Shading into texture. Artif Intell 29:147–170
21. Liu Y, Li Y (1997) Image feature extraction and segmentation using fractal dimension. In: Proceedings of international conference on information, communications and, signal processing, 975–979 1997
22. Potlapalli H, Luo RC (1998) Fractal-based classification of natural textures. IEEE Trans Industr Electron 45:142–150
23. Lance EE, Kaplan M (1999) Extended fractal analysis for texture classification and segmentation. IEEE Trans Image Process 8:1572–1585
24. Novianto S, Suzuki Y, Maeda J (2003) Near optimum estimation of local fractal dimension for image segmentation. Pattern Recogn Lett 24:365–374
25. Matsumoto H, Sasazaki K (2008) Color image compression with vector quantization. In: IEEE conference on soft computing in industrial applications, 84–88 2008
26. Sarkar N, Chaudhuri BB (1992) An efficient approach to estimate fractal dimension of textural images. Pattern Recogn 25:1035–1041
27. Sarkar N, Chaudhuri BB (1994) An efficient differential box-counting approach to compute fractal dimension of image. IEEE Trans Syst Man Cybern B Cybern 24:115–120
28. Otsu N (1979) A threshold selection method from gray-level histograms. IEEE Trans Syst Man Cybern B Cybern 9:62–66
29. Wang Z, Bovik C, Sheikh H, Simoncelli P (2004) Image quality assessment: from error measurement to structural similarity. IEEE Trans Image Process 13:600–612
30. Zhang X, Wandell BA (1996) A spatial extension of CIELAB for digital color image reproduction. In: The SID symposium technical digest, 27, 731–734 1996

Enhanced Dynamic Error Concealment Algorithm for Multiview Coding Based on Lost MBs Sizes and Adaptively Selected Candidates MBs

Mohamed Ebian, Mohamed El-Sharkawy and Salwa El-Ramly

Abstract Multi-view video sequences are compressed using advanced high compression ratio tools called H.264/AVC multi-view coding (MVC). MVC is very sensitive to errors due to its structure as a result Error Concealment (EC) algorithms are developed at the decoder to improve the quality of decoded MVC frames. In this paper efficient multi-view EC algorithms are proposed for intra and inter MBs. For Intra MBs, Spatial-Inter View (SIV) algorithm is proposed. For inter MBs, the proposed algorithm is adaptively changed according to the lost MB size. In addition, the algorithm dynamically candidates MBs according to the current used view. Then, the inter 16×16 MBs are enhanced by a new method called Weighted Block Motion and Disparity Concealment (WBMDC). The other inter MBs types are enhanced using a proposed method called the Overlapped Block Motion Compensation (OBMC). It is shown that the proposed algorithm leads to better subjective quality in addition to higher PSNR values. By applying the proposed algorithms, the gain is improved by up to 12.84 and 1.1 dB comparing to no applying any EC algorithms and applying normal EC algorithm respectively at 22 % Loss Rate (LR).

Keywords Multi-view coding · Error concealment · Boundary matching algorithm · Weighted pixel average

M. Ebian (✉) · S. El-Ramly
Electronics and Communications Department, Faculty of Engineering, Ain Shams University, Al Waili, Egypt
e-mail: m3bian@yahoo.com

S. El-Ramly
e-mail: sramlye@netscape.net

M. El-Sharkawy
Electronics and Communications Engineering Department, Egypt–Japan University of Science and Technology (E-JUST), Borg El-Arab, Alexandria, Egypt
e-mail: dspsys@aol.com

Mohan S. and S. S. Kumar (eds.), *Proceedings of the Fourth International Conference on Signal and Image Processing 2012 (ICSIP 2012)*, Lecture Notes in Electrical Engineering 221, DOI: 10.1007/978-81-322-0997-3_39,

1 Introduction

Multi View Video (MVV) is multiple video streams shot by several cameras around a single scene simultaneously. Compression is certainly needed since the increase in the amount of data is quit substantial. To reduce the data transition and storage requirements, new compression techniques exploit not only the temporal correlation in the single video but also the inter view correlation between the adjacent videos. Those techniques are being extensively studied in H.264/AVC Multi-View Coding (MVC). New applications such as 3DTV and Free-Viewpoint Television (FFV) services are now available depending on MVC. MVC utilizes the considerable amount of inter-view redundancies between adjacent views for further compression [1]. In MVC, Motion Vectors (MV) are generated from motion compensation between frames in the same view and Disparity Vectors (DV) are generated from disparity compensation between frames of adjacent views as shown in the MVC prediction structure in Fig. 1.

In this paper, a proposed algorithm exploits the inter-view and intra-view spatio-temporal correlation to conceal lost MBs to get higher PSNR and higher subjective quality. The proposed algorithm initially conceals lost inter MBs. It dynamically changes its behavior according to lost MBs size. In addition, the algorithm adaptively candidates MBs according to the current used view in EC process. During the proposed algorithm execution, new candidate MBs are generated according to motion direction to get more matched MBs instead of using fixed candidates MBs as in [2].

After initial concealment two proposed enhancement methods are used depending on lost MB size. For inter 16×16 MBs, WBMDC is applied using best MB in each reference frame of other views. Those best MBs are added with different factors producing best matched MB. For other inter MBs types, OBMC is applied which uses pre defined weighting matrices [3]. The rest of this paper is organized as follows: Section 2 presents basic EC algorithms, Sect. 3 presents the

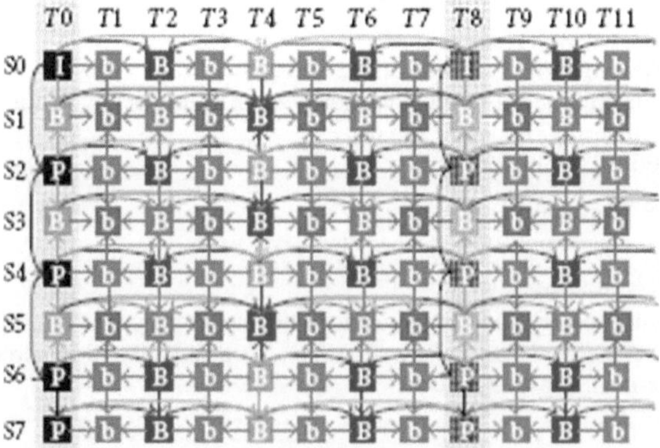

Fig. 1 Prediction structure of the multi-view video coding [1]

proposed EC algorithms, Sect. 4 presents simulation results and Sect. 5 concludes the paper.

2 Bases of Error Concealment Algorithms

The Boundary Matching Algorithm (BMA) is considered as the basic motion compensation EC technique recommended in H.264/AVC standard for temporal concealment. With MVC, BMA utilizes the disparity vectors of inter-view as well as the motion vectors [4]. The selection between the motion/disparity compensated MB is based on the smallest values of Sum of Absolute Differences (SAD). Another EC technique is called the Outer BMA (OBMA). The OBMA is generated from the BMA but it gives the differences between the two pixels wide outer boundary of the replacing MB and the same external boundary of the lost corrupted. This offers significantly better concealment performance than BMA with the same complexity of calculations [3].

The algorithm presented in [5] enhances the results of BMA and OBMA to obtain better results in stereoscopic video coding. In our proposed algorithm, the algorithm in [5] is applied to MVC but with some modifications and more calculations complexity to enhance initially concealed MBs. Due to the prediction structure of MVC any MBs loss results in errors propagations in other frames of adjacent views [4]. When a lost MB shall be concealed in MVC, reference MBs are searched in the temporally neighboring frames in the current view and in the neighboring frames of other camera views to the left and to the right of the considered view. The best reference or a weighted average between two references is then selected to be copied into the missing area.

3 Proposed Error Concealment Algorithm

The block diagram of the proposed MVC concealment technique is shown in Fig. 2.

In the proposed algorithm, the damaged intra MB is concealed and enhanced using (SIV) algorithm. In SIV algorithm, the spatial EC method called Weighted Pixel Average (WPA) is applied first. The WPA values are obtained inversely proportional to the distance between the reference pixel and the interpolated pixel. After applying WPA, the best DVs are obtained using pixels inside the lost MB and pixels surrounding lost MB position in other views to enhance the initially concealed Intra MB.

If zthe damaged MB is inter MB, the algorithm first determines the reference frames in all views. Those reference frames may be located in the same view or in other views. Then the proposed search engine is applied over all reference frames.

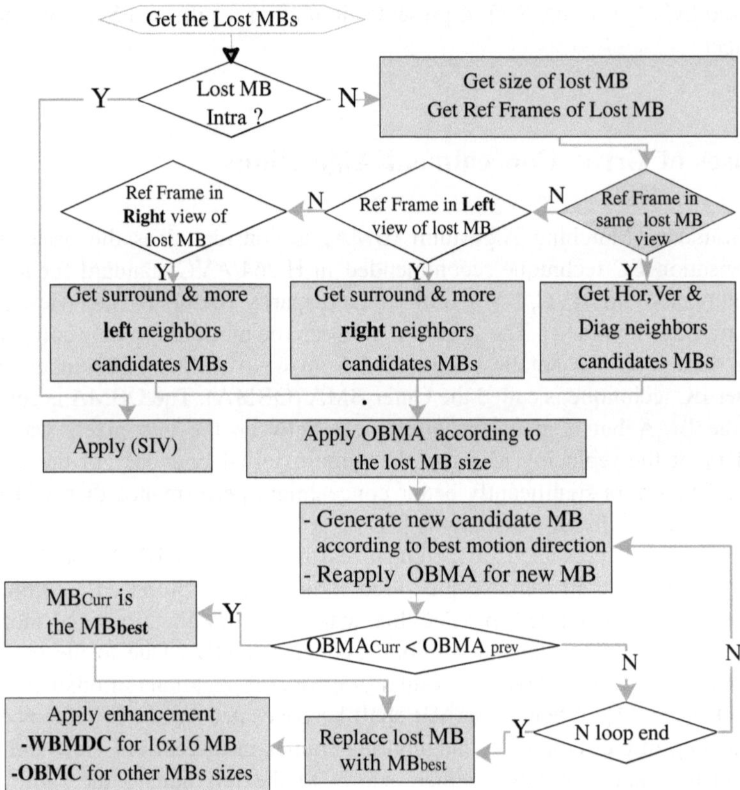

Fig. 2 The proposed algorithm

The candidate MBs in the proposed algorithm are adaptively selected according to the view contains current reference frame being processed with next rules:

- If the view of the current reference frame is the same view of lost MB so the candidate MBs are selected from the horizontal, vertical and diagonal neighborhood MBs.
- If the view of the current reference frame is the left view of lost MB view so the candidate MBs are surrounding neighboring MBs (4 neighboring, 4 corners) plus more right neighboring MBs.
- If the view of the current reference frame is the right view of lost MB view then the candidate MBs are surrounding neighboring MBs (4 neighboring, 4 corners) plus more left neighboring MBs.

The selection for more left candidate MBs at right view and right candidate MBs at left view is explained in Fig. 3.

Lost MB is assumed to be placed in intermediate view (current view). For the left view, the location of best MB is properly placed at the right side of lost MB position. As a result, there is more probability to find more matched MBs at the

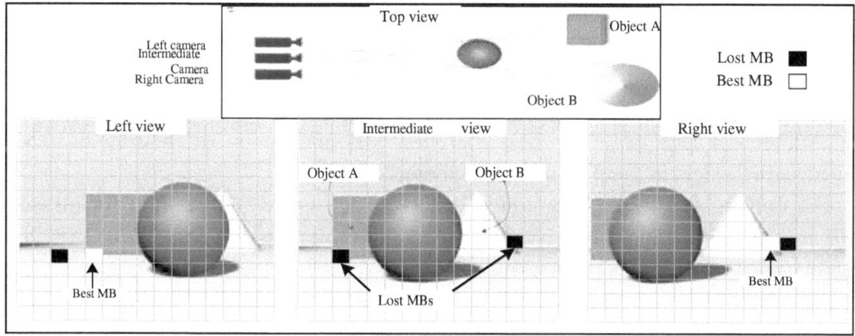

Fig. 3 Location of lost MB and best matched MB with respect to adjacent left and right views

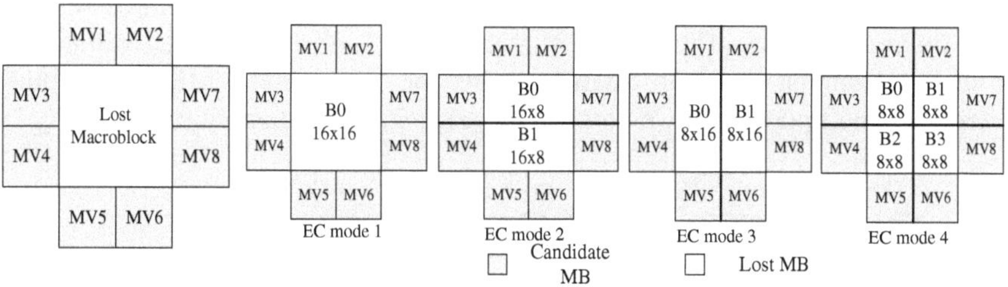

Fig. 4 MV candidate for lost MB and four EC modes

right side of lost MB. So more right MBs are candidates for proposed search algorithm.

For the right view, the location of best MB is properly placed at the left side of lost MB position. As a result, there is more probability to find more matched MBs at the left side of lost MB. So more left MBs are candidates for the proposed search algorithm. Then, the MB size is determined using the size of surrounding neighboring MBs [3]. The determined MB modes will be one of the following four modes as shown in Fig. 4.

- **Mode 1** (**16** × **16**): The set of concealed MVs candidates for block 0 is {V1, V2,…, V8} where V1, V2,…,V8 are referred to all MVs located around lost MB.
- **Mode 2** (**16** × **8**): The sets of concealed MV candidates for blocks 0 and 1 are {MV1, MV2, MV3, MV7} and {MV4, MV5, MV6, MV8}, respectively.
- **Mode 3** (**8** × **16**): The sets of MV candidates for blocks 0 and 1 are {MV1, MV3, MV4, MV5} and {MV2, MV6, MV7, MV8}, respectively.
- **Mode 4** (**8** × **8**): The sets of MV candidates for blocks 0, 1, 2 and 3 are {MV1, MV3}, {MV2, MV7}, {MV4, MV5} and {MV6, MV8}, respectively.

After selecting the most suitable partition type for the lost MB, each partition of the lost MB is concealed by applying the proposed method using the first level candidate set of MBs. For better MBs replacement, a self generated candidate MB algorithm is proposed. This algorithm uses the best selected MB direction as a step to get more candidate MBs in the same motion direction. With each step, if the OBMA value of new candidate MB is lower than the previous one then this new candidate MB is selected as best matched MB and so on till getting best MB through the selected loop number N where N is 5.

For enhancing the initially concealed MBs, the OBMC is applied for all inter MB modes *expect* 16 × 16 MB. The OBMC exploits the division of lost MB. It is mainly used to avoid the deblocking effects after initial concealment of lost MB. In OBMC the initial concealed MB is split into four 8 × 8 blocks, and each of these blocks is processed individually by predefined weighting matrices [3] and neighboring MBs pixels to be better matched.

For 16 × 16 inter MBs enhancement, WBMDC is applied. In WBMDC, OBMA is first applied to get the best MB $\mathbf{P_{best}}$ which is similar to the lost MB in all reference frames. During applying OBMA, we get the most similar MBs in each reference frames of the lost MB in all views. The best MB $\mathbf{P_{best}}$ is improved to be more similar to the lost MB by using the other similar MBs in the other references multiplied with weights values ω1, ω2, ω3 and apply the formula (1):

$$P_{lost}(i,j) = \frac{\omega_1 * P_{best}(i,j) + \omega_2 * P^1_{replac}(i,j) + \sum \omega_3 * P^{2,3}_{replac}(i,j)}{\omega_1 + \omega_2 + \sum \omega_3} \quad (1)$$

where

P^1_{replac} Best pixels in other temporal reference frames if MB P_{best} in same view or other disparity reference frame if MB P_{best} in other views.

$P^{2,3}_{replac}$ Best pixels in other disparity two reference frames if MB $\mathbf{P_{best}}$ in the same view or other temporary reference frame if MB $\mathbf{P_{best}}$ in other views.

The weights $w1$, $w2$, and $w3$ are set to 5, 4, and 3, respectively, for better results.

4 Simulation and Results

The proposed algorithm is applied to Joint MVC reference software [6] for 50 frames of size 640 × 480 of ballroom sequences with frame rate 30 Hz. An error mask is applied for multiview stream to get Loss Rate (LR) of about 22 % for all MBs as shown in Fig. 6a. Then the error mask ratio is decreased to 15, 11 and 5 % to test the proposed algorithm assuming lost MBs locations are known.

Table 1 shows the PSNR results of the concealed Ballroom sequence with no error occur, error occur without applying EC, applying normal OBMA, applying the proposed algorithm without adaptive candidate MBs and applying full proposed algorithm (with adaptive candidate MBs).

Table 1 PSNR (dB) results for ballroom test sequence

Applied EC algorithm	Loss rate (LR)			
	5 %	11 %	15 %	22 %
No error occur	33.41	33.41	33.41	33.41
Error occur without applying EC	20.75	17.82	16.48	14.75
Normal OBMA	30.39	28.8	27.94	26.49
Proposed OBMA without adaptive candidate MBs	30.50	29.10	28.50	27.14
Full algorithm	30.86	29.46	28.95	27.59

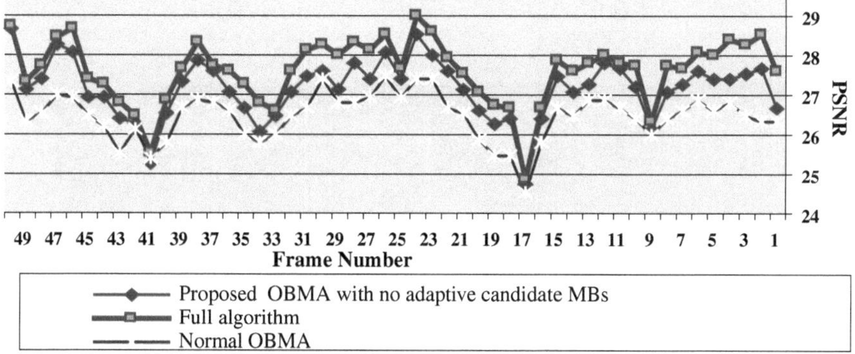

Fig. 5 PSNR of ballroom frames

By applying the full proposed algorithm, the PSNR results for *ballroom* sequences is improved up to 1.1 dB comparing with applying normal OBMA at LR equals to 22 %. As seen from Table 1 the improvements of PSNR differences are increased with higher LR.

Using adaptive candidate MBs according to view of the current reference frame in the proposed algorithm saves processing time due to using specific candidates MBs for each reference frame than using large collection of candidate neighborhood MBs for all reference frames [2].

In Fig. 5, PSNR for all concealed frames are higher than others concealed by normal OBMA and are slightly higher than the OBMA with fixed candidate MBs. The obtained PSNR of concealed frames by the full proposed algorithm are higher than normal OBMA and higher than proposed algorithm with fixed candidate MBs however the visual quality by full algorithm is better as shown in Fig. 6. The yellow dashed arrows in Fig. 6d indicate visual details improvements of the proposed algorithm compared to normal OBMA in Fig. 6b. The red arrows in Fig. 6d indicate visual improvements using adaptive candidate MBs with each reference frame than proposed OBMA without adaptive candidate MBs as in Fig. 6c.

Fig. 6 Subjective quality comparison for frame 26 of ballroom sequence. **a** Corrupted frame. **b** Concealed by normal OBMA. **c** Concealed by proposed OBMA but without adaptive candidate MBs. **d** Concealed by full algorithm

Those improvements are obtained due to making sure that the lost MB is replaced with the same MB size and using better candidate MBs with higher probability to be matched to lost MBs.

5 Conclusion

In this paper a proposed EC algorithm is suggested to conceal lost MBs in MVC. This algorithm is dynamically operates according to the lost MB size and adaptively selects the candidate MBs according to view of current reference frame being processed. Two enhancement algorithms are tan proposed according to lost MBs size to increase total PSNR. The proposed algorithm provides considerable better gain in the objective and subjective quality with lower processing time comparing with normal OBMA method if we use all possible candidate MBs are used for all reference frames.

References

1. Chen Y, Wang YK, Ugur K, Hannuksela MM, Lainema J, Gabbouj M (2009) The emerging MVC standard for 3D video services. EURASIP J Adv Signal Process 2009:13, Article ID 786015
2. Micallef BW, Debono CJ, Farrugia RA (2011) Performance of enhanced error concealment techniques in multi-view video coding systems. In: IEEE international conference on systems, signals & image processing, Sarajevo, Bosnia
3. Chapaneri SV (2011) Content-adaptive refined error concealment schemes for H.264/AVC video coding. Int J Comput Appl 27(7):36–43
4. Micallef BW, Debono CJ (2010) Error concealment techniques for multi-view video. IFIP wireless days conference, Venice, Italy
5. Xiang X, Zhao D, Wang Q, Ji X, Gao W (2007) A novel error concealment method for stereoscopic video coding. IEEE ICIP 2007, Texas, USA
6. ISO/IEC MPEG & ITU-T VCEG (2009) Reference software—joint multiview video coding model (JMVC 5.0.5). JVT-AE207, Joint multiview video coding software (JMVC) 8.0

References

Multi-Resolution 3D Human Brain Visualization Through Streaming

V. Vani, R. Pradeep Kumar and S. Mohan

Abstract The well established techniques for video and audio streaming have become the primary inspiration to focus our research towards 3D Streaming. In 3D streaming, the model is buffered and rendered instead of complete download. The central idea of this paper is to visualize the typical 3D human brain model at different resolutions based on the client's requirement at a rapid rate with necessary interactions. This is achieved by initially streaming the simplified version of 3D human brain and then the refinements are streamed based on the client's requirement. The experiment results and analysis affirms the fact that this type of multi resolution streaming reduces the initial time to present the rendered 3D model to the requested clients. Experiments are further conducted to compare the quality of the multi resolution models that are being rendered with respect to the original model.

Keywords Multi-resolution model · 3D streaming · 3D rendering · 3D visualization

V. Vani (✉)
Department of Information Technology, Dr. N. G. P. IT, Affiliated to Anna University, Coimbatore, India
e-mail: vasudevan.vani@gmail.com

R. Pradeep Kumar
Department of Computer Science Engineering, Adithya IT, Affiliated to Anna University, Coimbatore, India
e-mail: rpk.ind@gmail.com

Mohan S
Department of Computer Science Engineering, Dr. N. G. P. IT, Affiliated to Anna University, Coimbatore, India
e-mail: s.mohan77@gmail.com

Mohan S. and S. S. Kumar (eds.), *Proceedings of the Fourth International Conference on Signal and Image Processing 2012 (ICSIP 2012)*, Lecture Notes in Electrical Engineering 221, DOI: 10.1007/978-81-322-0997-3_40,

1 Introduction

In recent years, 3D real time computer graphics [1] over the web have given rise to the development of 3D Virtual Environments (VE) [2]. The 3D virtual environment which comprises of set of 3D scenes and the scene in turn consists of set of 3D complex meshes is available on the web and accessed by multiple users at the same time. Therefore, an efficient mode of transmitting the 3D complex meshes across the web has become the need of the hour. In order to cater to the need, the streaming [3] approach which is promising to deliver the content as per the client request has evolved and few attempts are already made by the researchers. However, in this paper, an attempt is made specifically to stream and deliver the 3D human brain with the required quality requested by the client who could be a medical student/practitioner who wants to visualize the 3D human brain and study in detail about its anatomy by interacting with it. The system he uses to stream and render the 3D human brain may have limited buffering capacity from few KBs to MBs.

Therefore, based on the buffering limits of the client, the corresponding LODs can be streamed from server to the client. Also, in order to ensure the quality of the rendered 3D mesh compared with the original mesh a detailed experiment is conducted and results are analyzed. In the process of developing a 3D visualization of human brain with required resolution through steaming firstly, Geometric Level Of Details (GLOD) [4] library which was developed on top of OpenGL is used at the server end. The GLOD, which focuses on the geometry details of the mesh, has included polygon reduction technique [5] to reduce the level of details from the original mesh and gives different levels of meshes by reduction process. This process is carried out and required levels of multi resolution meshes are identified and stored in the server for a specific 3D Human brain scene which has got only one complex 3D mesh. Secondly at the server end, the quality of the derived LODs with respect to the original mesh (LOD14) is also estimated. Finally the real time streaming is implemented with serialization and deserialization technique and results are analyzed. In streaming, if the requested scene is the 3D human brain then the initial LOD (Level 1) with very less geometrical information is streamed by computing the total space required and the time spent for streaming. Then, if the total space estimated is available, the streamed data is buffered and rendered at the client end with the rendering time estimated. On the client end, the client can interact (move, rotate, zoom in/out) with the rendered LOD with the help of keyboard and pointing device. Also, either based on the special key press (+/−) or based on the *Euclidean distance* between updated camera position and a random point taken from the screen space next/previous LOD would be streamed and rendered if the space criteria is satisfied. This process is repeated by ensuring that the levels remain between 1 and 14.

The rest of the paper is organized as follows. In Sect. 2, we present the overview of Multi resolution 3D model. Section 3 describes the distance measures used to check the quality of the original and reduced multi-resolution mesh. In Sect. 4, with the 3D human brain model being considered as a specific case, we discussed the proposed work which includes the streaming and 3D visualization process. In Sect. 5, we discussed the experiment results and analysis. Section 6 concludes by reflecting on the insights we gained from our model and its implications.

2 Multi-Resolution 3D Model: An Overview

Highly detailed geometric models are necessary to satisfy the demands of user in 3D virtual world for realism in computer graphics [5]. But, due to the increasing complexity of the models, it makes the virtual world developers feel expensive to store, transmit, and render requested 3D data by the client. In order to store, transmit and render the complex models at the rapid rate, the multi-resolution models [6] are developed whenever, a representation at low resolution is adequate to meet the needs of the application. Multi-resolution models offer the possibility to manipulate representations of objects at different levels of detail (LOD) [7–9] and accuracy, depending on the needs of each specific application. In addition to the benefits of speeding up visualization, multi-resolution techniques also support the interactive modeling of detailed objects.

2.1 Types of Hierarchies

The level of detail pipeline consists of three basic stages [4]: geometric *simplification*, *adaptation*, and *rendering*. The simplification process takes in "*flat*" geometry and produces a multi-resolution *hierarchy* [7, 9] from it. There are several types of multi-resolution hierarchies, the classification of hierarchies as (1) *Discrete* (2) *Continuous* (3) *View-dependent* or (4) *Hierarchical* is based on the multi resolution granularity as shown in Fig. 1.

(1) *Discrete* hierarchies (*Discrete Level of Detail* (*DLOD*)): refers to the creation of several *static levels of detail* [7, 9] which are swapped out directly for each other and it is analogous to *mip-maps* for texturing. Discrete hierarchies are extremely computation efficient as far as rendering is concerned.
(2) The DLOD is the simplest and most common form of simplification hierarchy.

It encodes multiple LODs at very coarse granularity. In this, each successive LOD would be half the complexity of the predecessors and the choice of granularity doubles the storage requirements with respect to higher resolution model. Each LOD may also contain the error value indicating the fidelity/quality of it. This kind of simplification is unaware of viewing directions since, the LODs are estimated offline during preprocessing. Therefore, it uniformly reduces object detail. Thus, Discrete LODs are also referred to as isotropic or view-independent LODs.

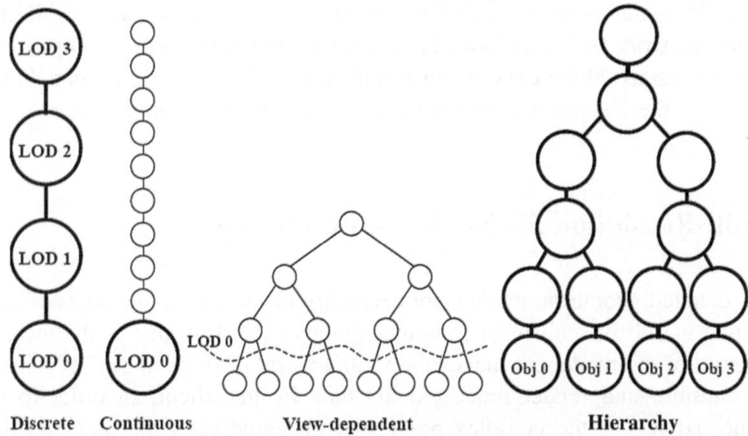

Fig. 1 Types of hierarchies [7], continuous LOD is large no. of LODs encoded by representing small changes from one LOD to next. LOD in the view-dependent hierarchy is represented as a cut across the tree. HLOD is similar to view-dependent, but it is much more coarse-grained, allowing faster management and more efficient rendering. A cut across an HLOD hierarchy essentially selects a set of discrete LODs to represent the scene

The *benefits* of DLOD are described as below:

- LODs are easily compiled into an optimized form for efficient rendering with the help of processes such as
  - Triangular strip reconstruction
  - Re-ordering of vertices based on vertex cache size
  - Indexed vertex arrays.
- The management of one or more discrete hierarchies is not too computationally expensive within the interactive application (i.e.,) with the given circumstances for each hierarchy appropriate DLOD can be chosen.
- DLODs are most useful when the data comprises of one or more objects that are relatively small in spatial extent. In this case, single choice of resolution is appropriate for the entire model.
- DLOD is used by a class of models ranging from 3D scanned objects to that of virtual environments.

(3) *Continuous* hierarchies (*Continuous Level of Detail* (*CLOD*)): creates a *progressive data structure* [7, 9] from which it can extract a continuous spectrum of detail at run time and it is also referred as *dynamic levels of detail*. Though it is more computationally complex during rendering, it addresses many of the limitations of DLOD.

The *benefits* of CLOD are described as below:

- It allows more exact choice of number of primitives to use for an object.
- It enables more subtle transition between LOD.

- It provides convenient representation for progressive transmission of data.
- Since it allows only linear progression and not selective refinement, it is used for same class models as that of DLODs.

(4) *View dependent LOD* extends CLOD by creating a *hierarchical data structure* [7, 9] from which it extracts a mesh tailored to the given viewpoint. The *benefit* of the view-dependent hierarchy is overwhelming though there is a run time overhead in adjusting the detail across the model on the fly.

(5) *Hierarchical Level of Detail* (*HLOD*) [7, 10, 11] works at scene level with multiple objects and has combined discrete hierarchy with view dependent hierarchy. This LOD hierarchy is more coarse-grained as compared to a typical view-dependent hierarchy and roughly corresponds to the notion of hierarchical scene structure. This coarse-grained representation is beneficial to visualize complex scenes with large number of objects, reducing per object overhead for given rendering frame rate both in LOD selection process as well as the number of primitives to be rendered.

3 Distance Measures

One of the efficient geometric error metrics [12–15] used to calculate the quality of the rendered 3D LOD is the distance. It gives the distance between the simplified vertices and the planes of the original surface. *Root Mean Square* (RMS) difference is estimated between two LODs and it is treated to be simplest possible estimation. Also, the *Hausdorff distance* is considered to be the most popular distance measure which compares a pair of LODs with different connectivity and calculates the similarity of two point sets by computing one sided distance ***Dist(LOD₁, LOD₂)***, where ***Dist*** is the *Euclidean distance* between two points $p(x, y, z)$ and $p'(x', y', z')$.

$$\boldsymbol{Dist}(\boldsymbol{p}, \boldsymbol{LOD_2}) = \min_{\mathbf{p'} \in \mathbf{LOD_2}} \left(\sqrt{(x - x')^2 + (y - y')^2 + (z - z')^2} \right)$$

$$\boldsymbol{Dist}(\boldsymbol{LOD_1}, \boldsymbol{LOD_2}) = \max_{\mathbf{p} \in \mathbf{LOD_1}} \boldsymbol{Dist}(\boldsymbol{p}, \boldsymbol{LOD_2})$$

The ***Dist*** in general is not symmetrical i.e.

$$\boldsymbol{Dist}(\boldsymbol{LOD_1}, \boldsymbol{LOD_2}) \neq \boldsymbol{Dist}(\boldsymbol{LOD_2}, \boldsymbol{LOD_1})$$

where ***Dist*****(*****LOD*****$_1$, *****LOD*****$_2$)** is referred as *forward distance* and ***Dist*****(*****LOD*****$_2$, *****LOD*****$_1$)** is referred as *backward distance*.

The two sided (symmetric) distance which provides a more accurate measurement can be estimated by taking the

$$\boldsymbol{Max}(\boldsymbol{Dist}(\boldsymbol{LOD}_1, \boldsymbol{LOD}_2)\boldsymbol{Dist}(\boldsymbol{LOD}_2, \boldsymbol{LOD}_1)).$$

Therefore, the *Hausdorff distance* has been used to find the geometric error between a pair of 3-D mesh models which is shortly referred as LODs.

4 Proposed Work

The proposed work attempts to visualize the 3D human brain streamed from the server and rendered at the client. Figure 2 gives the overview of the 3D Streaming and Rendering system which has got two major modules.

(1) Server Module
(2) Client Module.

4.1 Server Module

In this module, required levels of LODs are estimated and its quality with respect to the original well refined mesh is measured. This measure is would give the client an insight on the quality of the rendered mesh based on his request. Based on the client's request, the initial LOD is streamed by serializing the 3D data. Serializing is the process of packing the 3D data which comprises of set of vertices and faces for the level of LOD that is selected for streaming. Further, the server module streams the LODs based on the interactions made by the client.

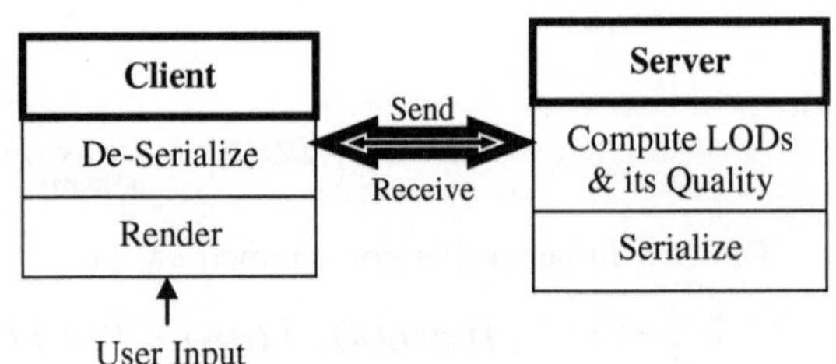

Fig. 2 Overview of 3D streaming and rendering system

4.2 Client Module

In this module, client establishes a connection with the server. Once the connection is successfully established, it requests for a scene from the server and if the scene is present then the initial LOD is streamed from the server. In client end, the streamed data is de-serialized in order to get the vertices and faces of the corresponding LOD and then it is rendered. Also the time taken for rendering process is also estimated. Figure 3 illustrates overall workflow of the system with the client and server interactions.

5 Experiment Results and Its Implications

Experiments are conducted to achieve 3D visualization of human brain through streaming of multi resolution meshes based on the client request and further interactions. The experimental setup was run on an Intel Core2 Duo CPU P8600 @ 2.4 GHz with 4 GB RAM and ATI Radeon 1 GB graphics card system for both server and client. For experimental purpose, we have considered 14 levels of LOD from coarse mesh (LOD1) to refined (LOD14) mesh. Table 1 gives multi-resolution 3D human brain models.

Table 2 describes the attributes of the LODs considered which includes number of vertices and faces considered in each LOD with its corresponding % of vertices and % of faces with respect to LOD14 (original mesh). Figure 4 shows the % of vertices and % of faces represents the corresponding LODs. Also, in order to understand the quality of the mesh (LOD) being rendered, we have considered *hausdorff distance* measure which would estimate the quality of the LOD being rendered with respect to LOD14 (original mesh). Figure 5, highlights the quality of the LODs streamed and rendered for 3D visualization using one sided and two sided hausdorff distance.

As specified in Sect. 3, hausdorff distance is a generic distance measure which defines the distance between two empty sets and it is useful in measuring the error between surfaces. We have considered both one sided and two sided hausdorff distance measure as some times one sided distance would end up in less accurate measure than that of the two sided distance as depicted in Fig. 6. However, according to the results highlighted for our LODs, it clearly shows that $\boldsymbol{Dist(LOD_2, LOD_1)} \gg \boldsymbol{Dist(LOD_1, LOD_2)}$ in all comparisons. Where $\boldsymbol{LOD_2}$ is considered to be the original mesh ($\boldsymbol{LOD_{14}}$) and $\boldsymbol{LOD_1}$ is substituted with $\boldsymbol{LOD_1}$ to $\boldsymbol{LOD_{13}}$. Table 3 and Fig. 7 highlight the turnaround time required to stream each LODs. The results shown in Tables 1–3 and Figs. 4, 5 and 7 clearly indicates that by exploiting the

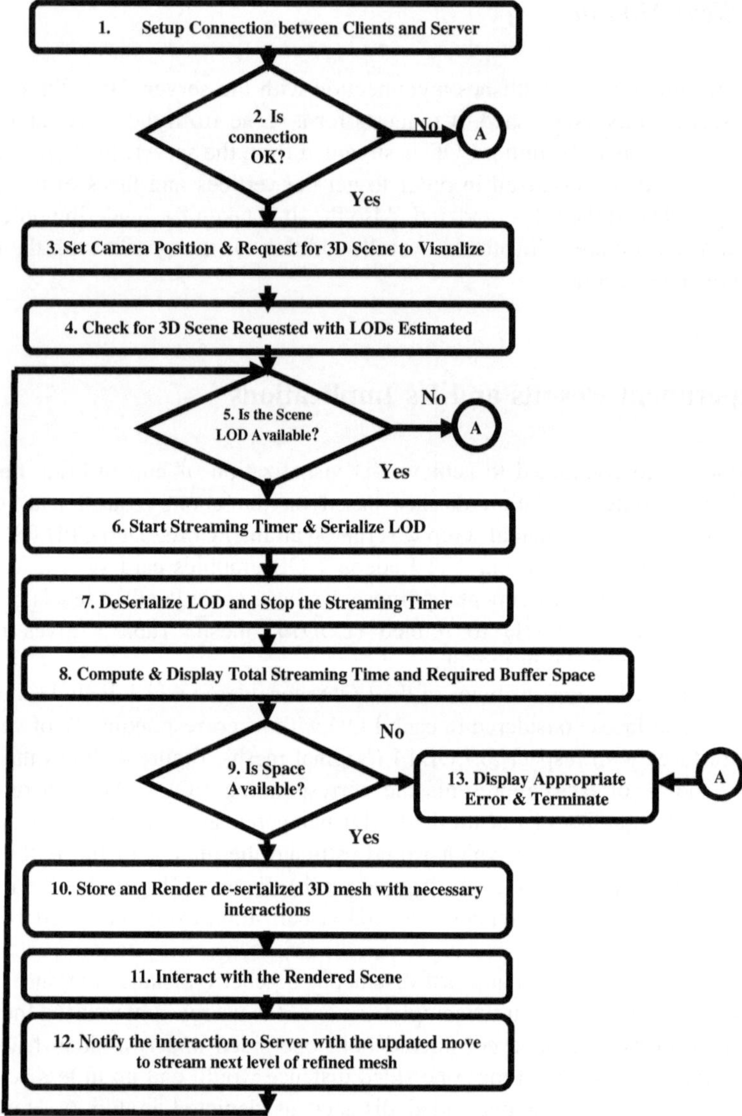

Fig. 3 Work flow of the streaming system

human perception characteristics, it is enough if we stream and render LOD_9 which could present all the parts of the brain with approximately 3 % of the vertices and 3 % of the faces and at the error rate of .09 distance from LOD_9 and

Table 1 Multi-resolution 3D human brain models (coarse LOD to refined LOD)

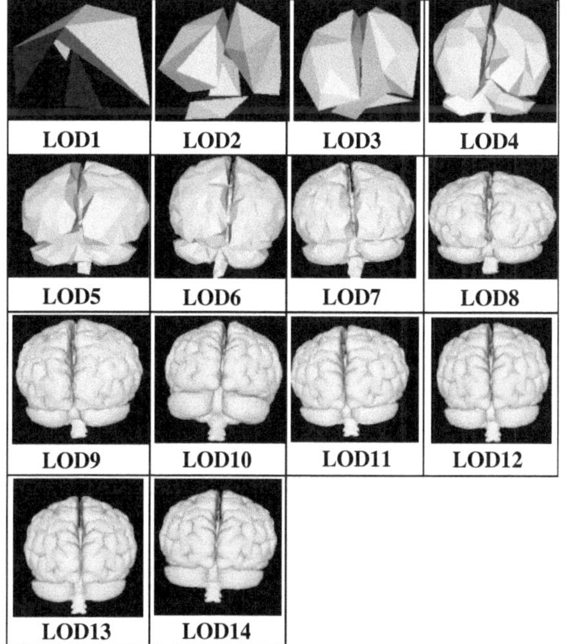

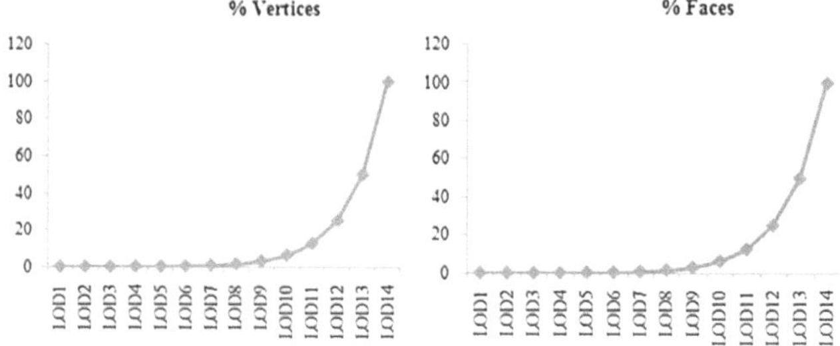

Fig. 4 Percentage of vertices and faces rendered w.r.t. LODs

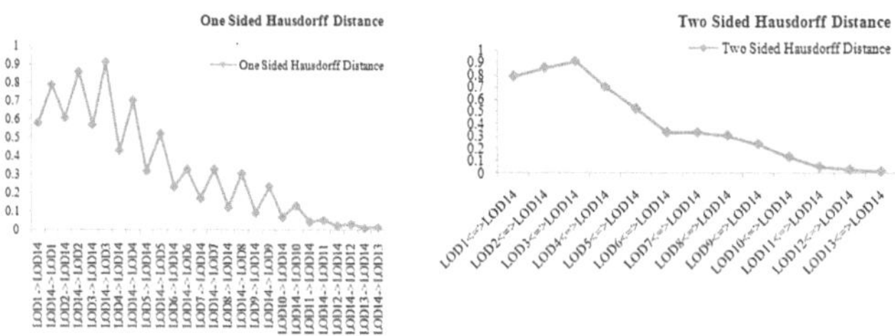

Fig. 5 Quality of the LOD estimated using **a** one sided and **b** two sided Hausdorff distances

Table 2 Attributes of 3D human brain LODs

LOD	No. of vertices	No. of faces	% vertices	% faces
LOD1	12	16	0.0041	0.0027
LOD2	39	71	0.0133	0.0121
LOD3	74	142	0.0252	0.0241
LOD4	289	574	0.0983	0.0976
LOD5	578	1148	0.1966	0.1952
LOD6	1152	2296	0.3918	0.3905
LOD7	2299	4594	0.7819	0.7813
LOD8	4593	9188	1.5622	1.5625
LOD9	9186	18376	3.1244	3.125
LOD10	18375	36752	6.2497	6.25
LOD11	36748	73504	12.4988	12.5
LOD12	73500	147008	24.999	25
LOD13	147004	294016	49.9993	50
LOD14	294012	588032	100	100

Table 3 Streaming and rendering time of discrete LODs

LOD	Streaming time (s)	Rendering time (s)	Turnaround time (s)
LOD1	0.024	1.51	1.534
LOD2	0.029	1.526	1.555
LOD3	0.064	1.458	1.522
LOD4	0.092	1.554	1.646
LOD5	0.07	1.742	1.812
LOD6	0.108	1.8	1.908
LOD7	0.146	1.96	2.106
LOD8	0.184	2.39	2.574
LOD9	0.19	3.593	3.783
LOD10	0.208	4.008	4.216
LOD11	0.293	6.196	6.489
LOD12	0.333	8.1	8.433
LOD13	0.666	9.24	9.906
LOD14	1.275	11.034	12.309

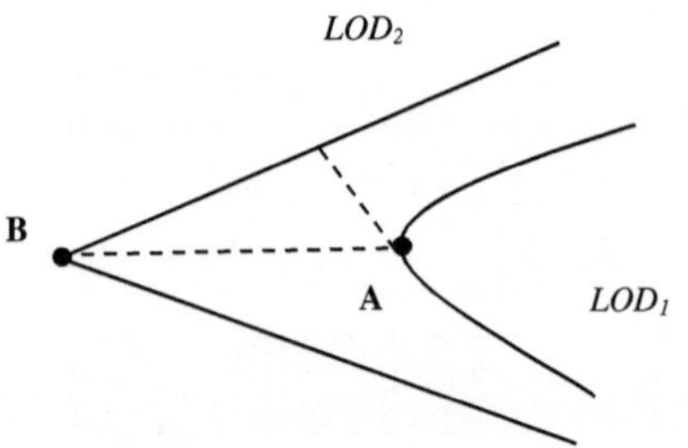

Fig. 6 In this case $Dist(LOD_1, LOD_2)$ will remain smaller than $Dist(LOD_2, LOD_1)$ since here $Dist(A, LOD_2) \ll Dist(B, LOD_1)$

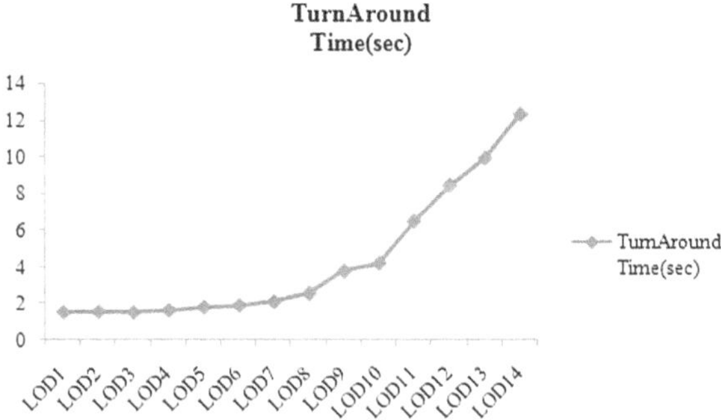

Fig. 7 Turnaround time of discrete LODs

$\boldsymbol{LOD_{14}}$ and .23 distance from $\boldsymbol{LOD_{14}}$ and $\boldsymbol{LOD_{9}}$. Also, the buffering capacity required is only 576 KB which is only 2.68 % of the total capacity required to transmit LOD_{14}. From Table 1 and Fig. 7, we could infer that if it is sufficient to stream and render $\boldsymbol{LOD_{9}}$ then the turnaround time required is only 3.78 s where we can save 8.5 s.

6 Conclusion

In this paper an attempt has been made to build a 3D human brain visualization system through streaming by exploiting human perception and multi-resolution capabilities. It is demonstrated with experimental results that the rendering latency and streaming time are reduced to a greater extent by streaming and rendering appropriate multi resolution model instead of transmitting the original mesh itself. 3D visualization is made possible with the predominant data being streamed and rendered with necessary interactions for further refinement if needed. The results achieved is overwhelming and would have certainly a greater impact when we have to stream and render much more complex scenes with time and space constraints enforced on it.

References

1. Brutzman D, Macedonia M, Rhyne T-M, McCanne S (1997) Internetworked 3D computer graphics: beyond the bottlenecks and roadblocks. ACM SIGCOMM '97
2. Blanc AL, Bunt J, Petch J, Kwok Y (2005) The virtual learning space: an interactive 3D environment. In: Proceedings of the tenth international conference on 3D web technology (Bangor, United Kingdom). Web3D '05. ACM, New York, pp 93–102. doi:http://doi.acm.org/10.1145/1050491.1050505, 29 March–01 April 2005
3. Deb S, Narayanan PJ (2004) Design of a geometry streaming system. In: Proceedings of ICVGIP, pp 296–301
4. Cohen JD, Luebke D, Duca N, Schubert B, Niski C (2004) GLOD: a minimal interface for geometric level of detail, p 10. Johns Hopkins University Graphics Lab Technical Report

5. Hoppe H (1996) Progressive meshes. In: Proceedings of the 23rd annual conference on computer graphics and interactive techniques (SIGGRAPH '96). ACM, NY, USA, pp 99–108. doi:10.1145/237170.237216
6. Shaffer E, Garland M (2005) A multi-resolution representation for massive meshes. IEEE Trans Vis Comput Graph 11(2)
7. Cohen JD, Manocha D (2005) Model simplification for interactive visualization. In: Johnson C, Hansen C (eds) Visualization handbook, p 13, Chapter 20. Elsevier Butterworth-Heinemann, pp 393–410
8. Ma H, Huang T, Wang Y (2010) Multi-resolution recognition of 3D objects based on visual resolution limits. Pattern Recognit Lett 31(3):259–266. ISSN 0167-8655. 10.1016/j.patrec.2009.08.015
9. Xin L (2011) Research on LOD technology in virtual reality. Energy Procedia 13:5144–5149. ISSN 1876-6102. 10.1016/j.egypro.2011.12.142
10. Erikson C, Manocha D, Baxter WV III (2001) HLODs for faster display of large static and dynamic environments. In: Proceedings of the symposium on interactive 3D graphics (I3D '01). ACM, NY, USA, pp 111–120. doi:10.1145/364338.364376
11. Charalambos JP (2010) Hoq-driven HLOD refinement: using hardware occlusion queries information to drive hierarchical level-of-detail refinement. VDM, Saarbrücken
12. Aspert N, Santa-Cruz D, Ebrahimi T (2002) MESH : measuring errors between surfaces using the Hausdorff distance. In: Proceedings of the IEEE international conference in multimedia and expo (ICME), vol 1, pp 705–708, Lausanne, Switzerland, 26–29 Aug 2002
13. Bulbul A, Capin T, Lavoué G, Preda M (2011) Assessing visual quality of 3D polygonal models. IEEE Signal Process Mag 28(6):80–90
14. Cignoni P, Montani C, Scopigno R (1998) A comparison of mesh simplification algorithms. Comput Graph 22(1):37–54
15. Cignoni P, Rocchini C, Scopigno R (1998) Metro: measuring error on simplified surfaces. Comput Graph Forum 17(2):167–174

Profiling User Interactions of 3D Complex Meshes for Predictive Streaming and Rendering

V. Vani, R. Pradeep Kumar and Mohan S

Abstract Inspired by the *cache model*, a *predictive agent* is analytically constructed to determine the user navigation based on the patterns derived out of user profiles. The user profiling is derived based on the user interactions made by the diversified set of users over different 3D models. An attempt has been made to analyze how efficiently the prediction works to stream a 3D model based on the pre determined *transition path* generated out of the user profiles. The transition paths for various models are generated by exploiting the properties of *Markov Chain model*. The analytics collected from the transition paths affirm that the predictive agent lessens the *rendering latency* significantly. The rendering latency is lessened by streaming the required data well before it is requested from the server to the client. The streaming and rendering process with user interactions from client would stream and render only the visible portion of the 3D models while ensuring that there is no compromise on the visual quality of the objects. This paper mainly focuses on profiling the user interactions during the navigation of 3D meshes and analyses various outcome of it.

Keywords User profiling · Web 3D · 3D streaming · Predictive agent · 3D modeling and rendering · 3D virtual environment · Transition path

---

V. Vani (✉)
Department of Information Technology, Dr. N. G. P. IT, Affiliated to Anna University, Chennai, India
e-mail: vasudevan.vani@gmail.com

R. Pradeep Kumar
Department of Computer Science Engineering, Adithya IT, Affiliated to Anna University, Chennai, India
e-mail: rpk.ind@gmail.com

Mohan S
Department of Computer Science Engineering, Dr. N. G. P. IT, Affiliated to Anna University, Chennai, India
e-mail: s.mohan77@gmail.com

Mohan S. and S. S. Kumar (eds.), *Proceedings of the Fourth International Conference on Signal and Image Processing 2012 (ICSIP 2012)*, Lecture Notes in Electrical Engineering 221, DOI: 10.1007/978-81-322-0997-3_41,

1 Introduction

3D modeling and rendering over the network has been the advancement in the recent research as the application of 3D over web is seamless. While creating a photo realistic virtual environment, the major challenge is to stream the 3D models within the available network bandwidth. At the same time, the visual quality and the delay in response to user navigation may considerably get affected. Hence a system which can promise a better virtual 3D environment over the existing network without placing any constraint on the user navigation is the need of the hour. In this paper, an attempt is made towards reducing the waiting time of the user/client during the interaction further by predicting the operation that would be performed by the users.

2 3D Streaming: An Overview

3D streaming [1–3] is the process of delivering 3D content in real-time for the users over a network. 3D streaming is carried in such a way that the interactivity and visual qualities of content may match as closely as if they were stored locally. The main resource bottleneck here is usually assumed to be the bandwidth and not rendering or processing power of the clients. To achieve this goal, simplification of the model and transmission of the content based on the user's view are two dominant strategies adopted. The model simplification [4] and transmission strategies exploit the resolution of the model with respect to the user's view. When the user's view point is far off from the 3D model's screen space then only the coarse low (resolution) mesh is brought to the client and in case when the user's view point is closer to the 3D model's screen space then the refined (high resolution) mesh is brought to the client. Therefore, the Multi-resolution models [5, 6] offer the possibility to manipulate representations of 3D objects at different levels of detail (LOD). It is possible to adapt to different hierarchies of LODs [7–9] based on the application requirement and mostly view dependent LODs which would incrementally bring in the required quality of the 3D mesh is highly used in the Virtual Environments by effectively utilizing the available bandwidth.

However, apart from the bandwidth, we also need to consider the rendering latency. Rendering latency could be reduced up to some extent by considering the multi resolution 3D model based on the user's view point. An attempt is made to reduce the rendering latency further by predictive the user's next move.

As most of the 3D streaming and rendering systems deals only with what the user is viewing (in the frustum) and not the actual way the user interacts, the main objective of the proposed work is to analyze the user interactions and determine the relationship between the interaction elements and the streaming (and rendered) elements of that 3D model.

This would determine the amount of data that ought to be sent to the client well before it is demanded by the client by prediction. This approach would definitely result in the reduction of rendering latency.

3 Proposed Work

3.1 Predictive Model (P_rm)

The proposed predictive model is based on understanding the user navigation in the virtual world. Based on the current position of user navigation, only visible vertices and faces of the selected triangular meshes are brought to the client machine during visualization. At the same time, based on the previous history collated from various user inputs, the next set of predicted vertices and faces are also pushed to the client with the help of the Predictive Agent (PA).

A ***Predictive Agent*** (***PA***) is built after successful offline analysis that was carried out on user profiles collected from 55 different users (aged 18–22, from engineering institutions with good visual and computer senses. The reason is that, this age group spends more time on gaming and relatively having a better understanding of the interfaces to navigate). As part of user analysis, the speed of the key press, total session time spent by every user, visual coverage of the model and pattern of the keys/buttons pressed are taken across complex 3D models/meshes. For experimentation purpose, complex 3D models of sizes 26 MB(Armadillo) and 45 MB (Brain) are considered with different shapes. Various shapes where the basic building block is a triangular mesh are considered in order to profile the moves if the 3D shape is oriented either horizontally/vertically. Based on the shapes even the user movement would vary if the user wishes to check the visual appearance of the entire 3D shapes.

The PA contains conventional transition probabilities [10, 11] of users when they move from one state to another state. During the transition, the maximum probability from a given step to current step is chosen for prediction and further towards predictive streaming. This algorithm uses greedy approach by considering the maximum probability (compared to all states transition probability) to move from a given state to the next. These transition probability paths generated for various models are used to predict the user interactions at every state. This would help further to optimize the 3D streaming and rendering over the network by reducing the time delay between user request and response.

3.1.1 Analytical Model

The main objective of the proposed work is to develop an analytical model based on the user interaction while viewing the 3D models over the network. The central idea is to predict the user navigation and construct an analytical model for every 3D object (3D triangular meshes) using the ***PA***. This predictive model hence would be useful in bringing the necessary surfaces during streaming. This prediction is useful to reduce the rendering & response time. To construct the predictive model (***Predictive Agent***: ***PA***), the following notations have been used:

Let $\mathbf{S_v}$ be a set of mesh vertices in the server and $\mathbf{S_f}$ be a set of corresponding mesh faces in the server for the selected 3D mesh. Let $\mathbf{C_v}$ be the set of mesh vertices in the client
where $\mathbf{C_v} \subseteq \mathbf{S_v}$ and $\mathbf{C_f}$ be the set of corresponding mesh faces in the client where $\mathbf{C_f} \subseteq \mathbf{S_f}$.

On an Operation $\mathbf{O_i}$, which can be an arbitrary rotation($\boldsymbol{\theta x}$, $\boldsymbol{\theta y}$, $\boldsymbol{\theta z}$) or Zoom in/Zoom out (Z_{in}, Z_{out}) or Translate (T_x, T_y, Y_z), $\mathbf{C_v}$ and $\mathbf{C_f}$ can undergo a change if it's a rotation operation in terms of Vertices and Faces as follows: $\pm\Delta$ **{V**$_\mathbf{i}$**} and**

$\pm\Delta$ **{F**$_\mathbf{i}$**}.**

For $\pm\Delta \{\mathbf{V_i}\}$: $+\Delta\{\mathbf{V_i}\} \subseteq \mathbf{S_v}$, $-\Delta\{\mathbf{V_i}\} \subseteq \mathbf{Cv}$,

where $+\Delta\{\mathbf{Vi}\}$ is the set of vertices chosen from **Sv** and $-\Delta\{\mathbf{V_i}\}$ is the set of vertices chosen out from **Cv**.

For $\pm\Delta \{\mathbf{F_i}\}$: $+\Delta\{\mathbf{F_i}\} \subseteq \mathbf{S_f}$, $-\Delta\{\mathbf{F_i}\} \subseteq \mathbf{C_f}$,

where $+\Delta\{\mathbf{F_i}\}$ is the set of faces chosen from $\mathbf{S_f}$ and $-\Delta\{\mathbf{F_i}\}$ is the set of faces chosen out from $\mathbf{C_f}$.

Table 1 summarizes the notations used in our model.

Table 1 Notations

$\mathbf{S_v}$	Server vertex set
$\mathbf{S_f}$	Server face set
$\mathbf{C_v}$	Client vertex set
$\mathbf{C_f}$	Client face set
$\mathbf{O_i}$	ith operation
$\Delta \{\mathbf{V_i}\}$	Vertices changes
$\Delta \{\mathbf{F_i}\}$	Faces changes
$\mathbf{R_\theta}$	Rotation
$\boldsymbol{\theta_x}$	Rotation about x axis
$\boldsymbol{\theta_y}$	Rotation about y axis
$\boldsymbol{\theta_z}$	Rotation about z axis
$\mathbf{T_x}$	Translate about x axis
$\mathbf{T_y}$	Translate about y axis
$\mathbf{T_z}$	Translate about z axis
$\mathbf{Z_{in}}$	Zoom in
$\mathbf{Z_{out}}$	Zoom out

3.1.2 Operation Profiling

To profile the interaction performed by the user, basically the Rotation operation $\mathbf{R}_\theta$ in any one of the directions: $+\theta_x/-\theta_x$, $+\theta_y/-\theta_y$, $+\theta_z/-\theta_z$ and also Translation/Scaling with a fixed Translation/scale factors are considered.

For every key press/mouse move during the rotation, a fixed angle of rotation is applied on the 3D object and outcome of the rotation generates updated eye position and eye orientation (*eye refers to the camera position, which is the view point of the user in 3D world*). Based on this operation, the speed of rotation is estimated based on the number of key pressed per second. The key presses would determine the amount of angle being rotated per second.

Based on the rotation output, the amount of change in the vertices and faces ($+\Delta$ $\{V_i\}$ **and** $+\Delta$ $\{F_i\}$) that ought to be transmitted to the client is predicted. The predicted faces and vertices only are transmitted to the client. The prediction, hence, would reduce the rendering latency based on the client input.

3.1.3 User Profiling

To construct the predictive agent, an offline analysis has been carried out by considering 55 user profiles taken from a range of novice to professionals interacting with 3D virtual world. The user profiles include, rate at which the key is pressed/mouse button clicked with a drag and the actual key/mouse/scroll button that is pressed per user session on various complex 3D meshes are considered for analysis. Using the collated user profiles, operation patterns are determined by considering the transition probabilities. At each step of constructing the transition path, transition from maximum probable state to all other possible states based on the user profiles are extracted by exploiting the markov chain model and greedy approach. As per the greedy approach, at every transition step locally optimal choice (state with maximum probability) is considered and a complete transition path is constructed. Also, as per the markov chain process, summation of the transition probabilities from one state to all possible states should be 1. Based on the constructed transition path with maximum probabilities considered at every step, the predictive agent for a specific model is built. This process is considered to be a training session for the users before they actually navigate the virtual world. Once trained, the users would be able to get the rendered 3D models with a better response time across the network while interacting with the 3D web. Table 2 describes various keys that can be used by the user and the corresponding operations performed in response to the key press are specified. Also, we have carefully studied the existing systems and identified all possible operations that the user is intended to perform in a 3D environment and consolidated it.

Similar to Table 2, the Table 3 describes various mouse moves the user can perform during his/her interactions in the 3D environment. Here also, most common practice adopted in the 3D virtual environment is followed. However, it is

Table 2 User key press

S.No	Key pressed	Operation performed	Key value
1	LEFT	Rotate −X	1
2	RIGHT	Rotate +X	2
3	Up	Rotate −Y	3
4	Down	Rotate +Y	4
5	PgUp	Rotate +Z	5
6	PgDown	Rotate −Z	6
7	Tab	Move right	7
8	Backspace	Move left	8
9	X	Move up	<
10	Y	Move down	>
11	O	Zoom out	0
12	Z	Zoom In	9

Table 3 User mouse movement

S.No	Mouse move	Operation performed	Key value
1	Left + drag	Rotate	LG
2	Right + drag	Zoom in/out	RG
3	Left + right + drag	Move	LRG
4	Scroll	Zoom in/out	U/D

a common user practice that most of the time the interactions are made using the keys rather than mouse. Study reveals that this is due to the linear degree of movement of key press and non-linear movement of the mouse operation. Evidences are that the commercial graphic hardware consoles use joystick movements which make a linear movement while playing the games. However, we have conducted the detailed study and analyzed extensively using keys as well as mouse interactions.

4 Result and Analysis

We have used two different 3D models (Table 4) which contain only meshes to experiment the user interactions and build a rigid predictive agent. Meshes are considered so that the number of culled faces can be easily computed. 3D meshes with its number of vertices, faces, triangle strips and file size considered is given in Table 5. The 3D meshes are carefully chosen so that it differs in shape and complexity (like total no. of vertices and faces). The experimental setup was run on an Intel Core2 Duo CPU P8600 @ 2.4 GHz with 4 GB RAM and ATI Radeon 1 GB graphics card system. The system has been used to simulate a server environment where the server was streaming the 3D meshes as requested by user

Table 4 Thumbnails of actual and rotated/zoomed 3D meshes using key press/ mouse moves (#http:// www1.cs.columbia.edu/ ~cs4162/models)

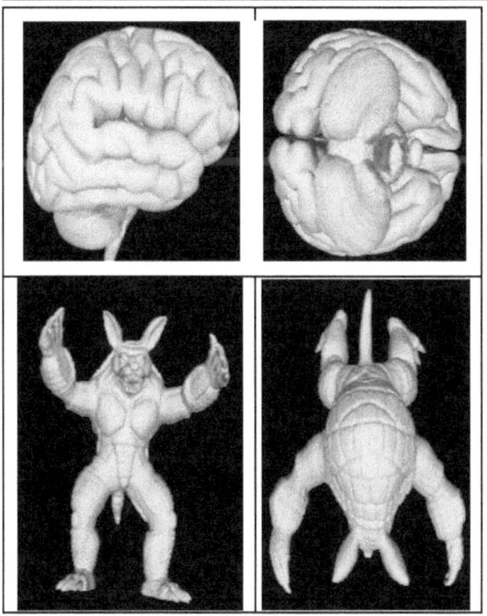

from a client machine based on the navigation. The client module was run on the machines with the following configuration which does not have exclusive graphics card. The configuration of 55 client machines are Intel Core 2 Duo CPU P6550 @ 2.3 GHz with 1 GB RAM. The profiling collected from 55 client machines helps in rendering the 3D meshes more efficiently.

4.1 Analysis of User Profiling

We have collected the profiling of 55 different users (aged 18–22, from engineering institutions with good visual and computer senses). For each 3D model, the users/clients are asked to navigate through the object with various key strokes/ mouse moves as defined earlier and user manual is also circulated among the users to get a fair idea about key/mouse press and the corresponding operation being performed. In addition to it, all the users are instructed properly on how to use the key strokes/mouse moves to navigate and visualize the 3D meshes. Number of times a key/mouse button pressed is counted. Later, the probability of pressing each key stroke/mouse button and the transition probability of moving from one key/mouse button to other key/mouse button is calculated as in Fig. 4a and b.

Figure 1 describes the overall visual coverage of 3D mesh models considered and could infer that brain model is covered by maximum number of users (29 users, 53 %) than that of armadillo model (13 users, 24 %) out of 55 users. This visual coverage is estimated based on the interactions of the model and the operation performed and the total number of vertices and faces covered by them.

Table 5 3D objects and its attributes

S.No	Model	No. of vertices	No. of faces	No. of triangle strips	File size (MB)
1.	Armadillo	1,72,974	3,45,944	3,37,958	26
2.	Brain	2,94,012	5,88,032	5,83,877	44

After applying visibility culling algorithm, the study reveals that 40 % of the meshes would be saved without rendering in a client machine for a complex mesh. It implies that only 60 % of meshes would be rendered and viewed by maximum number of users who covers the entire model.

Figure 2a and b highlights time spent on the model (session time) by each of the 55 users. From the result obtained, we could see, some of the users spent more 560 s that is around 10 min on the brain model with 1446 key/mouse interactions (Fig. 3a). Also, on the other hand 353 s are spent on the brain model with 4030 key presses. The results imply that the interactions performed by the user also influence his psychological aspects such how much he/she is interested on the model he is viewing, how long he takes to press the next key (think time) etc. can also be analyzed. Similarly, if we consider the Armadillo model, maximum 266 s is spent on a model with 1988 key/mouse interactions (Fig. 3b) and on the other hand maximum of 3129 key/mouse interactions is done for the duration of 211 s. Therefore, the session time and think time can be exploited to push the predicted faces and vertices well before it is requested.

Figure 4a shows the constructed transition path based on the transition probability. Based on the 55 users interactions with brain model consolidated, the transition path is constructed. From the start state **S**, the next probable state from the available 15 states is calculated by identifying the state with maximum probability. Accordingly, the state 3 from S is the termed to be the most probable state determined based on the user profiles. From state 3, next probable state is determined once again and the process is repeated till we reach the finish state from only two states with equal probability .5 and number of user involved in that move remains to be 1. This transition probability which estimated by considering only the unique moves by removing the duplicate states such as (3–3, 4–4) can be used as a look up table in the predictive agent to perform predictive streaming and rendering. Also, the consolidated user interactions reveals the fact that the probability of moving from state X_i to X_i is high compared to transition from state X_i to X_j where $i \neq j$. Therefore, the predictive agent is built to get the utmost efficiency by considering both the constructed transition path with unique user interactions

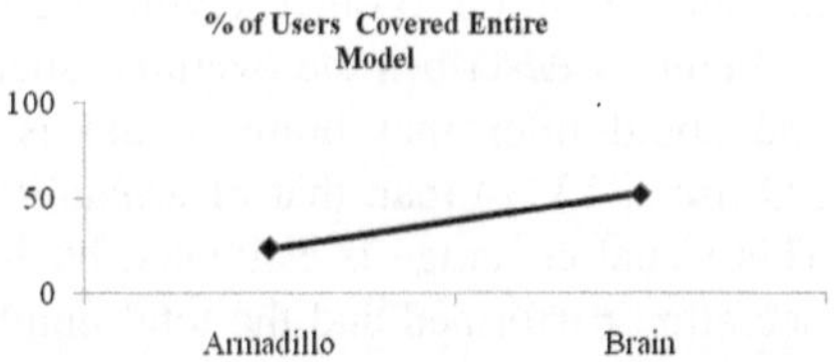

Fig. 1 Percentage of users covered entire model

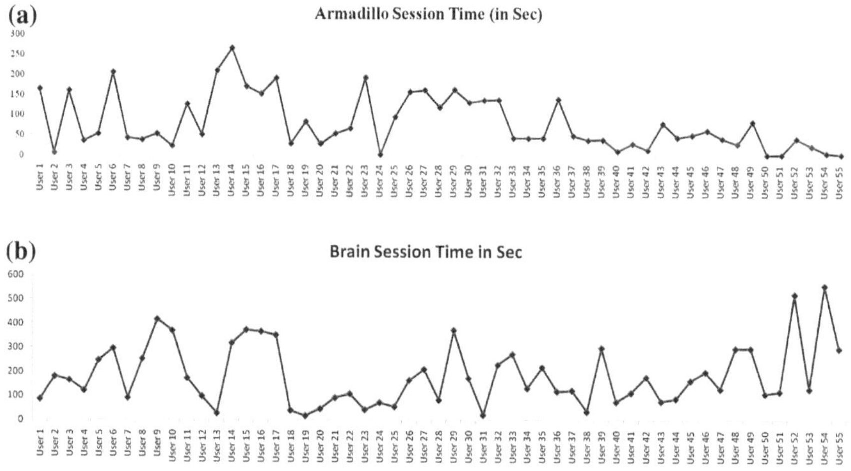

Fig. 2 **a** Users' session time (in sec) for Armadillo 3D mesh. **b** Users' session time (in sec) for Brain 3D mesh

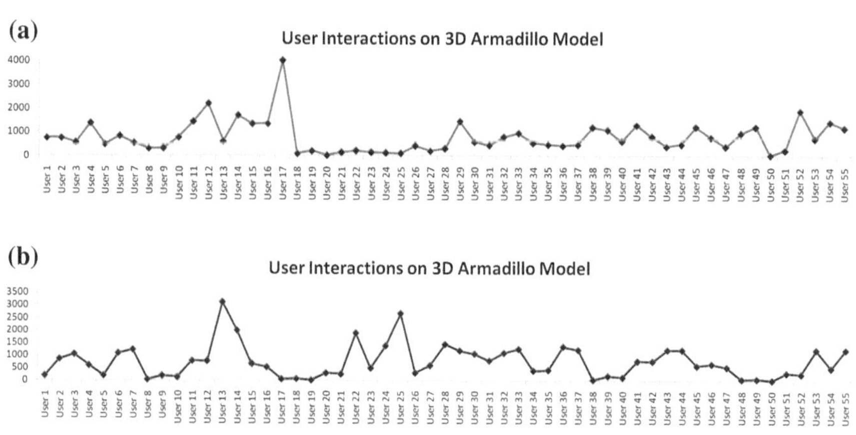

Fig. 3 **a** Users' interactions on Armadillo 3D mesh. **b** Users' interactions on brain 3D mesh

while moving from state X_{n-1}(previous) to X_n (current) as well as the calculated cumulative probability of X_i to X_i. Figure 4b shows the transition path generated for Armadillo model. Figure 4a and b gives a conclusion that based on the shape of the model the interactions of the users varies and this is also considered while implementing a predictive based 3D streaming and rendering to achieve swift response from the server for every interaction.

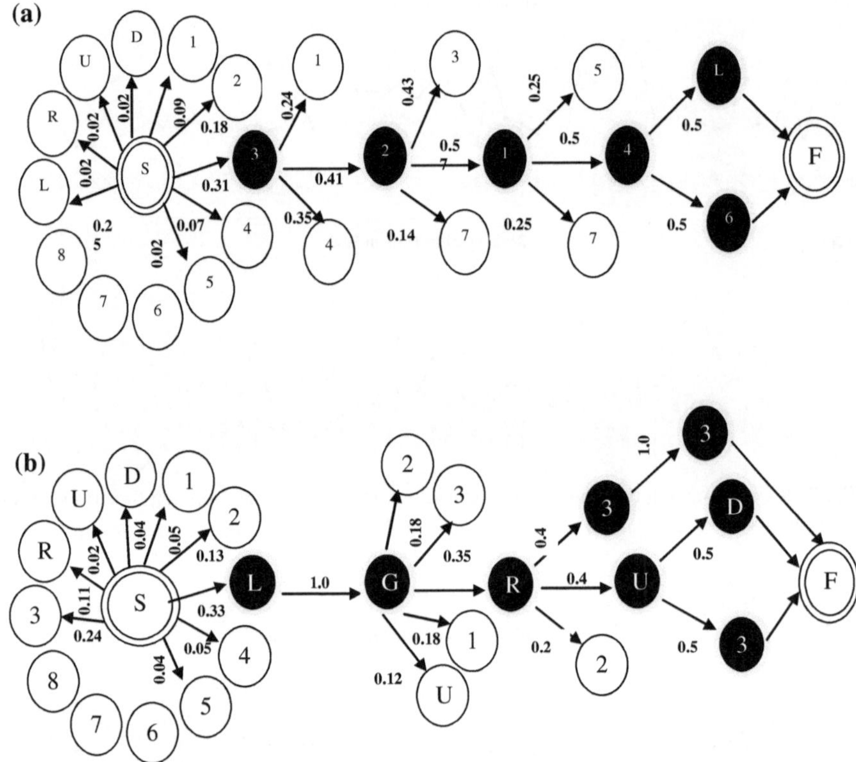

Fig. 4 **a** Transition path for brain model. **b** Transition path for Armadillo model

5 Conclusion

This paper explored the possibility of implementing a predictive based 3D streaming and rendering by exploiting the user interactions performed by various users across two complex models differ in size and shape. With the predicted user profiling, it is affirmed that the 3D meshes can be rendered with minimum latency. The transition paths generated and the analysis performed based on the think time and session time shows that the predictive model for 3D streaming has a significant impact in bringing the 3D models to the client system without compromising on the visual quality. Future extension of the proposed system would be introducing multiple objects in a more dynamic 3D environment.

References

1. Deb S, Narayanan PJ (2004) Design of a geometry streaming system. In: Proceedings of ICVGIP, pp 296–230
2. Lin N-S, Huang T-H, Chen B-Y (2007) 3D model streaming based on JPEG 2000. IEEE Trans Consumer Electron (TCE) 53(1)
3. Schroeder WJ, Zarge JA, Lorensen WE (1992) Decimation of triangle meshes. SIGGRAPH Comput Graph 26:65–70, 2 July 1992. doi:10.1145/142920.134010
4. Hoppe H (1996) Progressive meshes. In: Proceedings of SIGGRAPH, pp 99–108
5. Cheng W (2008) Streaming of 3D progressive meshes. In: Proceedings of the 16th ACM international conference on multimedia (MM '08). ACM, New York, USA, pp 1047–1050. doi:10.1145/1459359.1459570
6. Cheng W, Ooi WT, Mondet S, Grigoras R, Morin G (2011) Modeling progressive mesh streaming: does data dependency matter? ACM Trans Multimedia Comput Commun Appl 7, 2:24, Article 10 (March 2011). doi:10.1145/1925101.1925105
7. Cohen JD, Manocha D (2005) Model simplification for interactive visualization. In: Johnson C, Hansen C (eds) Visualization handbook, p 13. Elsevier Butterworth-Heinemann. Chapter 20, pp 393–410
8. Ma H, Huang T, Wang Y (2010) Multi-resolution recognition of 3D objects based on visual resolution limits. Pattern Recognit Lett 31(3–1):259–266. ISSN 0167-8655. 10.1016/j.patrec.2009.08.015
9. Xin L (2011) Research on LOD technology in virtual reality. Energy Procedia 13:5144–5149. ISSN 1876-6102. 10.1016/j.egypro.2011.12.142
10. Benoit G (2005) Application of Markov chains in an interactive information retrieval system. Inf Process Manage 41:843–857, 4 July 2005. doi:10.1016/j.ipm.2004.06.005
11. Jeong DH, Ji S-Y, Ribarsky W, Chang R (2011) A state transition approach to understanding users' interactions. IEEE VAST 2011:285-286

Multi-Table Reinforcement Learning for Visual Object Recognition

Monica Piñol, Angel D. Sappa and Ricardo Toledo

Abstract This paper presents a bag of feature based method for visual object recognition. Our contribution is focussed on the selection of the best feature descriptor. It is implemented by using a novel multi-table reinforcement learning method that selects among five of classical descriptors (i.e., Spin, SIFT, SURF, C-SIFT and PHOW) the one that best describes each image. Experimental results and comparisons are provided showing the improvements achieved with the proposed approach.

Keywords Object recognition · Artificial intelligence · Reinforcement learning

1 Introduction

Bag of features (BoF) has become one of the most widely used approaches for visual object recognition (e.g., [1–4]). It consists of four steps. Firstly, it finds the interest points (detectors) and describes them (descriptors) in order to characterize

M. Piñol (✉) · A. D. Sappa · R. Toledo
Computer Vision Center and Computer Science Department, Universitat Autònoma de Barcelona, 08193 Bellaterra, Barcelona, Spain
e-mail: mpinyol@cvc.uab.es
URL: www.cvc.uab.es

A. D. Sappa
e-mail: asappa@cvc.uab.es
URL: www.cvc.uab.es

R. Toledo
e-mail: ricard@cvc.uab.es
URL: www.cvc.uab.es

Mohan S. and S. S. Kumar (eds.), *Proceedings of the Fourth International Conference on Signal and Image Processing 2012 (ICSIP 2012)*, Lecture Notes in Electrical Engineering 221, DOI: 10.1007/978-81-322-0997-3_42,

the given object at a higher abstraction level. Secondly, the extracted feature points, from all the images in the training set, are structured in a kind of dictionary of words. This dictionary of words is obtained through a learning process and will be used during in the next step. Thirdly, each of the images from the training set is represented by means of a histogram that count the number of times a given word appears. Finally, the obtained histograms are used to train a support vector machine (SVM), which classifies the given images. The BoF is a flexible architecture and the fourth steps mentioned above can be implemented through the use of different algorithms. Hence, the final result will depend on the selection of the right algorithm for each step.

From the four steps mentioned above, a particular attention should be given to the first one, since it represents the most sensible and results are highly dependent on the right descriptor selection. Since the use of different descriptors implies different results, one could think that the best option for this first step is to concatenate as much descriptors as possible. Hence, a given image will be represented by all possible different descriptors. Unfortunately, such a kind of *brute-force* strategy in most of the cases is not feasible due to the fact that it introduces noise. Recently, in [5], the authors present a method to select the best descriptor for each image using *reinforcement learning* (RL). RL is a simple method that allows to learn the best action under a *trail-and-error* framework based on a set of user defined *states*. Although interesting results have been obtained in [5], and most of the time the approach converges to the best descriptor, the problem now is how to define a reliable state to be used during the RL learning process. The current work tackles this problem by proposing a strategy that helps to select the state that maximize the result.

RL has been largely used in the robotics community during the last two decades. Recently, it has attracted the attention in the computer vision field to address problems such as image segmentation or object recognition, just to mention a few. For instance, in the segmentation domain the RL method is used to select the appropriate threshold (e.g., [6, 7]). In [8], the authors propose a RL based face recognition technique that is able to learn the best feature from each image. Similarly, in the object recognition field, [9] presents a RL technique using first order logic. Finally, RL has also been used for learning interest points [10, 11] or for selecting methods for classification [12].

In this paper we present a BoF based approach for the object recognition. As mentioned above the current work is focussed on the first step by using RL. More precisely, the current work contributes with a novel scheme for selecting the best state in the RL method. This scheme results in a multi-table formulation. Regarding the rest of steps of BoF, in the current implementation we use a kd-tree in the second step and a support vector machine in the fourth step. The reminder of the paper is organized as follow. Section 2 presents a brief summary of the RL. Then, Sect. 3 details the proposed method. Experimental results are provided in Sect. 4 and conclusions and future work are given in Sect. 5.

2 Reinforcement Learning

As mentioned above the current work proposes the use of a multi-table RL strategy for finding the best descriptor that characterizes a given image. In this section a brief description of RL is presented just to introduce the notations and the definition of the used elements (see [13] for more details).

The RL is a learning method used in those cases where the agent does not have a prior knowledge about which is the correct action to take. The RL is a Markov decision process intended to learn how an *agent* ought to take an *action* in a given *environment* so that a *reward* is maximized. These concepts are defined with the following tuple $\langle S, A, \delta, \tau \rangle$, where: S is a set of environment states; A is a set of actions; δ is a transition function $\delta{:}S \times A \rightarrow S$ and τ is a reward/punishment function, $\tau{:}S \times A \rightarrow \Re$.

By using the definitions presented above the RL methods works as follow: for a given state s_z, the agent selects the action a_h that maximize the expected reward r based on the τ function. In other words, by applying the action a_h in the state s_z a new state s_{z+1} and a reward r are obtained. Figure 1 illustrates the interaction between the agent and the environment.

The RL can be solved using dynamic programming, Monte Carlo method and temporal difference learning. The temporal difference learning is used in the current work because it does not require a model and it is fully incremental [13]. More concretely, the used framework is based on the *Q-learning* algorithm [14]. In our work, the current state s_z is only affected by the previous visits but not for the future since the Markov decision problem is of first order [8]. The δ and τ functions are nondeterministic; hence each element of the Q-table, for an iteration n, is computed as follow:

$$Q_n(s_z, a_h) \longleftarrow (1 - \alpha_n) Q_{n-1}(s_z, a_h) + \alpha_n [r + \gamma \max_{a'} Q_{n-1}(s_z, a')], \tag{1}$$

$$\alpha_n = \frac{1}{1 + \mathbf{visits}_n(s_z, a_h)}, \tag{2}$$

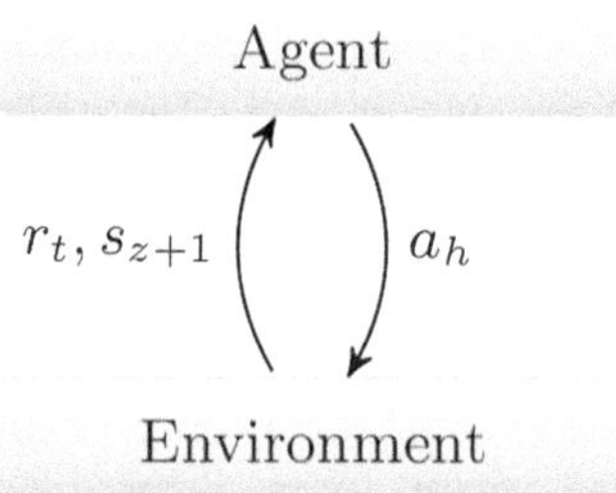

Fig. 1 Illustration of interaction between agent and the environment

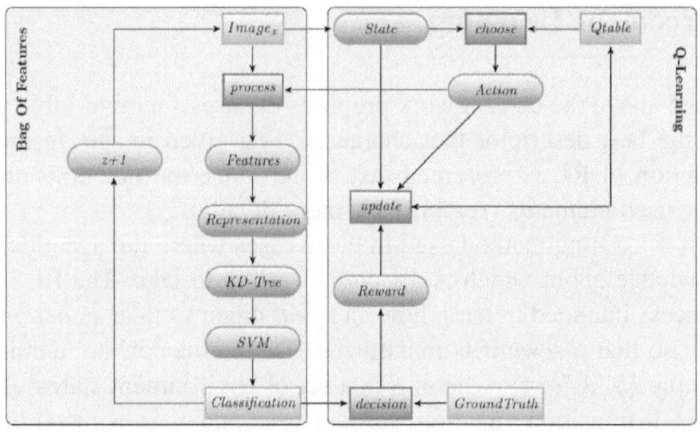

Fig. 2 Illustration of learning the best descriptor for each image using Q-learning

where γ is a discount factor for future reinforcements and is defined as $0 \leq \gamma < 1$. The Eq. 2 is the value of α_n for a nondeterministic world and **visits** is the number of iterations visiting the Q-table at the tuple (s_z, a_h) [15].

3 Proposed Method

The proposed approach, as mentioned in Sect. 1, uses the classical BoF ([1, 2]) for object recognition; our work is particularly focussed on the first step of BoF. In other words, we propose a multi-table reinforcement learning based strategy to select the best descriptor for each image from a set that contains the most widely used in the literature. This section presents the definition of the main elements of RL as well as the proposed strategy to combine the Q-table. Figure 2 illustrates the BoF (see *left-side*) with the proposed RL for the first step (see *right-side*). The three remainder steps of BoF are implemented following the state of the art, hence they are not detailed in this section (see [4] for more details).

3.1 Tuple Definition

This section aims at describing the different elements used in the RL formulation of the current work. The tuple $\langle S, A, \delta, \tau \rangle$ is defined as follows:

State definition: A state is defined as a set of characteristics from the given image. In order to tackle the challenge of defining a single representative state [5] that could be used with different databases, in the current work we propose the use four different state definitions. In all the cases the states are defined by a set of clusters obtained from the extracted vectors of characteristics—by using k-means. The vector for a given image is defined by extracting the information in a

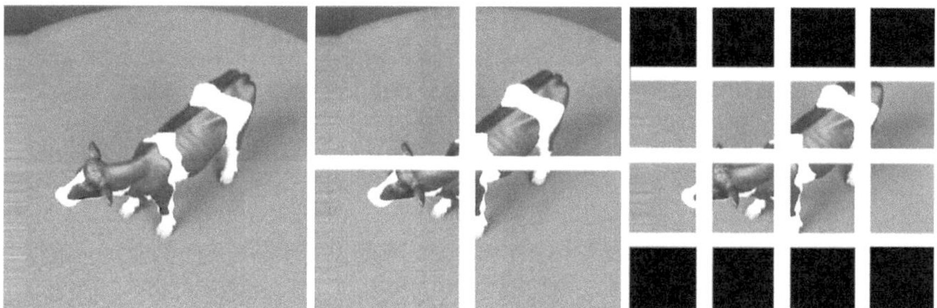

Fig. 3 (*left*) Image from ETH database [16]. (*middle*) Image splitted up into four squares. (*right*) Image splitted up into sixteen squares, but only eight of them are used

structured way. For instance, in the case of Fig. 3 the vector of characteristics is defined with 13 elements. The first element of that vector is obtained extracting information from Fig. 3 (*left*); then, Fig. 3 (*middle*) contributes with the next four elements of the vector; finally, the last eight elements are obtained from Fig. 3 (*right*). Once the vectors from all the given training images have been obtained a k-means clustering is used to compute the states that will be used for the Q-learning.

(1) $L^*a^*b^*$ ***state definition***: This state uses the $L^*a^*b^*$ color space. This color space is obtained by converting the RGB to XYZ and then, XYZ to $L^*a^*b^*$ (see [17, 18] for more details). The L^* represents the luminance of the image, a^* represents the difference between the red and green colors, and the b^* is the difference between the yellow and blue colors. As mentioned above the given image is split up into 13 squares (see Fig. 3, for each one of them the median value of $L^*a^*b^*$ is computed. Note that since $L^*a^*b^*$ has three components this state definition results in a vector of 39 elements.
(2) ***Gradient state definition***: This state definition uses the gradient in x and y directions. The gradient provides edges, but in this case, the state is defined by extracting the median of values in x and y. The partition shown in Fig. 3 is also used here. Hence, this state is defined with a vector of 26 elements.
(3) ***Entropy state definition***: The entropy measures the uncertainty of the information. In this case, the information is computed using the same partition shown in Fig. 3, but from the corresponding gray-scale image instead of the RGB color one. For each element of the partition the entropy is computed as follow:

$$E = -\sum_{i=1}^{N} p_i log_2(p_i), \qquad (3)$$

where, $p(x)$ is the histogram of image data. In this case the state is defined with a vector of 13 elements.
(4) ***Histogram of interest point state definition***: This particular state is defined using all the descriptors of the set of actions. It works as follow; for each

image from the training set, it extracts all the interested points and describe them accordingly. After that, similarly to the process of BoF, it constructs a dictionary and find a representation of the image interested points. Finally, a vector with 50 elements (10 elements per descriptor) is extracted to represent the state.

Actions: In the current work, the actions are a set of descriptors. In this case, the RL learns the best descriptor for each image. Note that there is a large number of descriptors in the literature [19] the five most representative descriptors are selected for this work: SIFT [20], PHOW [4], C-SIFT [21], SURF [3] and Spin [22].

δ function: Usually, in the RL, the δ functions is defined as $\delta : s_z \times a_h \rightarrow s_{z+1}$. But, in this work, the δ function does not give a new state. In the current work, after applying an action a_h to the state s_z it generates a new representation of the image (features). The features obtained in this stage are used in the BoF for classifying the object. After that, the process continues through a new image. Summarizing, given a state s_z and applying an action a_h we obtain a new image from the training set and this new image does not have any similitude with the previous image.

τ function:

Theτ functions is defined by $\tau : s_z \times a_h \rightarrow \Re$, when the classification step gives the same label than the given object, the τ function gives a reward, and when the label does not match with the ground truth, the τ provides a punishment.

3.2 Combination of Q-tables

Like in [5], the joining process of BoF and RL is used to train the Q-table. Figure 2 shows this training process, which works as follow. For a given image, the agent extracts the state and applies a descriptor selected from the Q-table using the exploration/exploitation trade off. In the current work the ε-greedy algorithm is used as a strategy for the exploration/exploitation. After applying the descriptor, the agent follow the BoF scheme using the kdtree algorithm and the support vector machine. Once it finishes, the agent obtains a label of the classified image. The agent compares the obtained label with the ground-truth and obtains a reward/ punishment. This information is used to update the Q-table according to Eq. 1, using the reward obtained before, the state and the applied action. Finally, after completing a whole iteration the agent extracts a new image from the training set and starts the training process again. In the current work this process is applied four times, one time per state definition. As a result we obtain four Q-tables. Now the question is to define which one should be used for a given image.

In the current work we propose a simple strategy for combining the four Q-tables computed as mentioned above. Actually, the information is not combined; the strategy consists in selecting the action from that Q-table where the

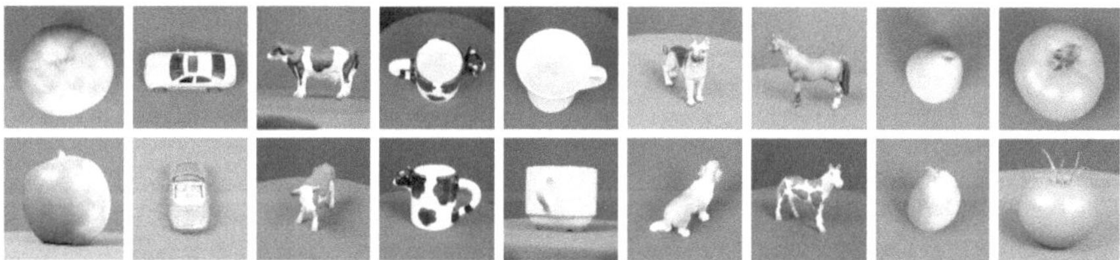

Fig. 4 Some of the objects contained in the nine classes of ETH database

reward is maximized. As will be presented in next section this simple strategy allows to improve results with respect to state-of-the-art, which only work with a single state definition.

4 Experimental Results

The proposed approach has been evaluated using two different databases. Additionally, it has been compared with a recent approach [5] as well as with others two BoF implementations where the first step consists of: (*i*) just a single descriptor; and (ii) a RL based approach with different states definition.

The first experiment is using the ETH database. Figure 4 shows the nine classes we have selected (i.e., apple, car, cow, cowcup, cup, dog, horse, pear and tomato) for testing and comparing the proposed approach. Each of these classes contain 45 images, which were randomly selected. These 45 images are split up into three sets: 15 images for training the BoF; 15 images for training the Q-table and 15 images for testing. The process of training starts creating for each definition of the states the corresponding Q-table. The process is repeated 60.000 times and the values of $\gamma = 0.9$ and $\varepsilon = 0.2$ are used.

In order to have a first comparison of the results obtained with the proposed approach, the performance of BoF for each descriptor is computed. Table 1 shows that the best performance is obtained when BoF uses the PHOW descriptor (74.81 % of recognition ratio). In Table 2 the performance of using the BoF with the RL method is presented. The first four rows depict the performance independently obtained for every state definition. In this case (BoF with RL) the best performance is reached with the $L^*a^*b^*$ state definition (82.4 % of recognition

Table 1 Performance of BoF using a single descriptor

Descriptor	Performance (%)
Spin	60.00
SIFT	61.48
SURF	62.96
C-SIFT	68.15
PHOW	74.81

Table 2 Performance of BoF with RL

State Definition	Performance (%)
$L^*a^*b^*$	82.2
Gradient	77.8
Entropy	78.5
Histogram of words	77.1
App. presented in [5]	81.4
Proposed approach (multi-table RL)	**83.4**

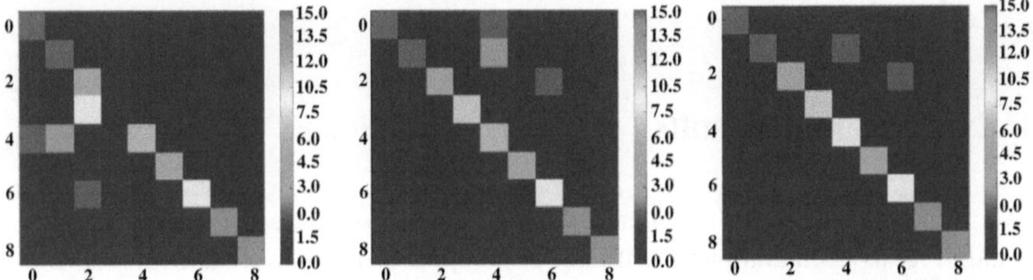

Fig. 5 Different confusion matrices for ETH database. (*left*) Using only the PHOW descriptor (74.81 %). (*middle*) Using $L^*a^*b^*$ state definition (82.2 %). (*right*) Using the proposed method (83.7 %)

ratio). Additionally, Table 2 shows the result obtained using the proposal presented in [5] (see fifth row). Finally, the performance obtained with the proposed approach (multi-table RL) is depicted in the last row. It can be appreciated that the best performance is obtained by using the strategy proposed in the current work (83.4 % of recognition ratio). Some of these results are presented in Fig. 5 by means of the corresponding confusions matrices. Figure 5 (*left*) shows the confusion matrix resulting when BoF with a single descriptor is used, in this case the PHOW descriptor. The confusion matrix presented in Fig. 5 (*middle*) corresponds to the BoF using RL and the $L^*a^*b^*$ state definition. Finally, Fig. 5 (*right*) depicts the confusion matrix resulting from the proposed approach.

A similar comparison to the one presented above has been performed with another database to validate the proposed approach. In this case, the COIL database [23], which contains 100 classes has been selected (Fig. 6 shows some of the objects contained in the COIL database). Each of these classes contain 45 images, which are split up into three sets: 15 images for training the BoF, 15 images for training RL and 15 for testing. In this case the process is repeated 600.000 times and the values of γ and ε are the same as those used in the first experiment ($\gamma = 0.9$ and $\varepsilon = 0.2$).

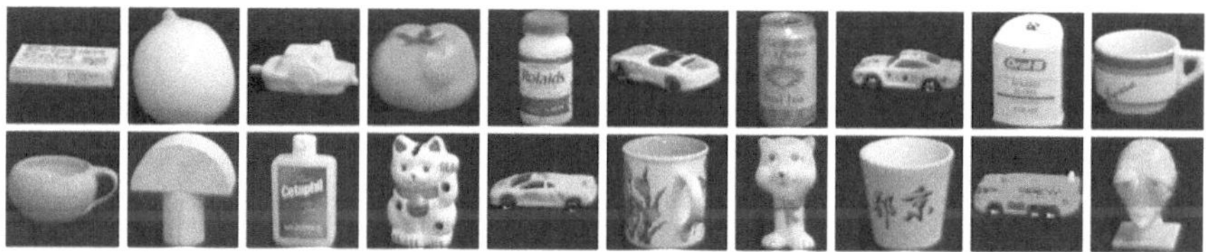

Fig. 6 Some of the objects contained in the COIL database (the whole database contains 100 classes, each class contains 45 images)

Table 3 Performance of BoF using a single descriptor

Descriptor	Performance (%)
Spin	83
SIFT	92.2
SURF	82.27
C-SIFT	94.47
PHOW	98.3

Table 4 Performance of BoF with RL

State Definition	Performance (%)
$L^*a^*b^*$	98.53
Gradient	98.8
Entropy	98.6
App. presented in[5]	98.3
Proposed approach (multi-table RL)	**99.0**

The performance of BoF for each descriptor is computed and presented Table 3; it can be appreciated that the best performance is again achieved using the PHOW descriptor (98.3 % of recognition ratio). In Table 4 the performance obtained when the BoF is used with the RL method is presented. The first four rows present the performance obtained for each of the state definitions introduced in Sect. 3.1. In this case the best performance corresponds to the BoF with RL and with the gradient state definition (98.8 % of recognition ratio). In order to compare the results with [5], the Table 4 presents in the fifth row the recognition ratio. Finally, in the last row of the Table 4, the performance of the proposed strategy is shown; note that in this case it reaches 99.0 % of recognition ratio. The confusion matrices corresponding to three of the examples presented in Tables 3 and 4 are presented in Fig. 7. In the left side the confusion matrix using BoF with a single descriptor (PHOW descriptor) is presented; the middle illustration corresponds to BoF with RL when the best state definition is used (gradient state definition). Finally, the right side illustration depicts the results obtained with the proposed approach. Note that in this case since the is a larger number of objects in the database and the recognition ratios are about 99 %, the confusion matrices are almost a diagonal line.

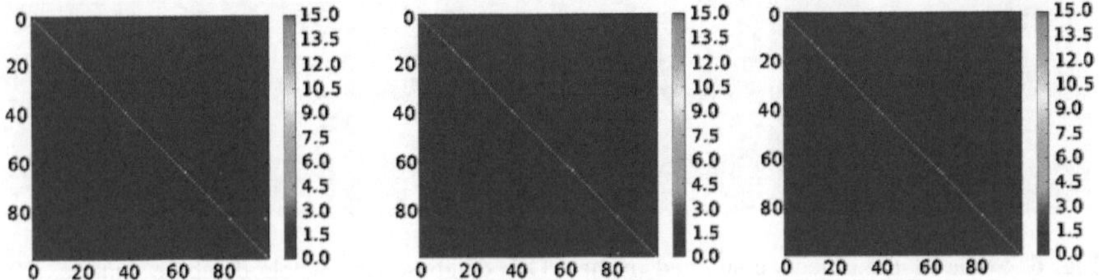

Fig. 7 Different confusion matrices for COIL database. (*left*) Using only the PHOW descriptor (98.31 %). (*middle*) Using gradient state definition (98.8 %). (*right*) Using the proposed method (99.0 %)

5 Conclusions and Future Work

This paper presents a BoF based approach for visual object recognition. We propose to improve classical BoF by means of the use of a RL strategy for selecting the best descriptor for each image. Our contribution lies on a novel method that allows the use of a multi-table strategy in the RL. This multi-table strategy allows to pick up the best state definition for each image. Experimental results are obtained using the BoF with the RL with two databases: ETH and COIL. In the first database, PHOW is the best descriptor and results in a 74.81 % of recognition ratio, the recognition ratios reaches 83.4 % using the proposed method. In the second database, the best single descriptor is also PHOW and in this case a recognition ratio of 98.3 % is reached, however, with the proposed method it could be also improved up to the 99.0 % of recognition ratio. Future work will be focused on the combination of descriptors and new state definitions in order to further improve the performance.

Acknowledgments This work was partially supported by the Spanish Government under Research Program Consolider Ingenio 2010: MIPRCV (CSD2007-00018) and Project TIN2011-25606. Monica Piñol was supported by Universitat Autònoma de Barcelona grant PIF 471-01-8/09.

References

1. Csurka G, Dance CR, Fan L, Willamowski J, Bray C (2004) Visual categorization with bags of keypoints. In: Workshop on statistical learning in computer vision, Proceedings of the European conference on computer vision (2004), pp 1–22
2. Fei-Fei L, Perona P (2005) A bayesian hierarchical model for learning natural scene categories. In: Proceedings of IEEE conference on computer vision and pattern recognition, pp 524–531
3. Bay H, Tuytelaars T, Gool LV (2006) Surf: speeded up robust features. In: Proceedings of the European conference on computer vision, pp 404–417
4. Bosch A, Zisserman A, Muñoz X (2007) Image classification using random forests and ferns. In: Proceedings of international conference on computer vision

5. Piñol M, Sappa AD, López A, Toledo R (2012) Feature selection based on reinforcement learning for object recognition. In: adaptive learning agent workshop, pp 4–8
6. Shokri M, Tizhoosh HR (2008) A reinforcement agent for threshold fusion. Appl Soft Comput 8:174–181
7. Sahba F, Tizhoosh HR, Salama M (2007) Application of opposition-based reinforcement learning in image segmentation. In: IEEE Symposium on Computational Intelligence in Image and Signal Processing, Honolulu, HI, pp 246–251
8. Harandi MT, Ahmadabadi, MN, Araabi, BN (2004) Face recognition using reinforcement learning. Proc IEEE Int conf image process 4:2709–2712
9. Häming K, Peters G (2010) Learning scan paths for object recognition with relational reinforcement learning. In: Proceedings of the 7th IASTED international conference on signal processing, pattern recognition and applications, vol 678. Innsbruck, Austria, p 253
10. Jodogne S (2005) Reinforcement learning of perceptual classes using q learning updates. In: Proceedings of the 23rd IASTED international multi-conference on artificial intelligence and applications, pp 445–450
11. Jodogne S, Piater JH (2004) Interactive selection of visual features through reinforcement learning. In: Proceedings of 24th SGAI international conference on innovative techniques and applications of artificial intelligence, pp 285–298
12. Bianchi R, Ramisa A, de Mántaras R (2010) Automatic selection of object recognition methods using reinforcement learning. Adv Mach Learn 1:421–439
13. Sutton R, Barto A (1998) Reinforcement learning: an introduction Cambridge Univ Press, MA (1998)
14. Watkins CJCH (1989) Learning from delayed rewards. Ph.D. thesis, King's College, Cambridge
15. Mitchell TM (1997) Machine learning. McGraw-Hill Science/Engineering/Math, New York
16. Leibe B, Schiele B (2003) Analyzing appearance and contour based methods for object categorization. Proc IEEE Conf Comput Vis Pattern Recogn 2:409
17. Ruzon MA, Tomasi C (2001) Edge, junction, and corner detection using color distributions. IEEE Trans Pattern Anal Mach Intell 23:1281–1295
18. Martin DR, Fowlkes CC, Malik J (2004) Learning to detect natural image boundaries using local brightness, color, and texture cues. IEEE Trans Pattern Anal Mach Intell 26:530–549
19. Mikolajczyk K, Schmid C (2005) A performance evaluation of local descriptors. IEEE Trans on Pattern Anal Mach Intell 27(10):1615–1630
20. Lowe D (2004) Distinctive image features from scale invariant keypoints. Int J Comput Vision 2:91–110
21. van de Sande KEA, Gevers T, Snoek CGM (2010) Evaluating color descriptors for object and scene recognition. IEEE Trans Pattern Anal Mach Intell 32:1582–1596
22. Lazebnik S, Schmid C, Ponce J (2005) A sparse texture representation using local affine regions. IEEE Trans Pattern Anal Mach Intell 27(8):1265–1278
23. Nene SA, Nayar SK, Murase H (1996) Columbia object image library (COIL-100). Technical report (Feb 1996)

DFT-Based Feature Extraction and Intensity Mapped Contrast Enhancement for Enhanced Iris Recognition

S. M. Rakesh, G. S. P. Sandeep, K. Manikantan and S. Ramachandran

Abstract Iris Recognition (IR) under varying contrast conditions with low gradience is challenging, and exacting contrast invariant features is an effective approach to solve this problem. In this paper, we propose two novel techniques viz., *Intensity Mapped Contrast Enhancement (IMCE)* and *Double symmetric rectangular hyperbolic based DFT (DsrhDFT) extraction.* IMCE is a preprocessing technique used to increase the gradience between brighter and darker pixels of an image, thereby obtaining the salient iris features. DsrhDFT is used to extract prominent shift-invariant features, and a Binary Particle Swarm Optimization (BPSO) based feature selection algorithm is used to search the feature space for optimal feature subset. Individual stages of the IR system are examined and an attempt is made to improve each stage. Experimental results obtained by applying the proposed algorithm on Phoenix, MMU and IITD iris databases, show the promising performance of the IMCE+DsrhDFT for iris recognition. A significant increase in the recognition rate and a substantial reduction in the number of features is observed.

Keywords Iris recognition · Feature extraction · Feature selection · Discrete fourier transform · Binary particle swarm optimization

S. M. Rakesh · G. S. P. Sandeep · K. Manikantan (✉)
M S Ramaiah Institute of Technology, Bangalore, Karnataka, India
e-mail: kmanikantan2005@yahoo.co.in

S. M. Rakesh
e-mail: rakeshsm91@gmail.com

G. S. P. Sandeep
e-mail: sandeepgsp17@gmail.com

S. Ramachandran
S J B Institute of Technology, Bangalore, Karnataka, India
e-mail: ramachandr@gmail.com

Mohan S. and S. S. Kumar (eds.), *Proceedings of the Fourth International Conference on Signal and Image Processing 2012 (ICSIP 2012)*, Lecture Notes in Electrical Engineering 221, DOI: 10.1007/978-81-322-0997-3_43,

1 Introduction

Biometrics [1] is giving a significant contribution in automated person identification as it aims in identifying each individual using various physiological characteristics such as fingerprints, face, iris, retina etc. Human recognition using Iris has been the most significant discovery as each individual has different iris patterns compared to others and the main advantage is that these patterns are invariant to ageing factors. In pattern recognition the key issue is the relation between interclass and intraclass variability. Identification can be reliably classified only if the variability among the different instances of a given class is less than the variability between different classes which is also satisfied by the properties of the Human Iris. Reference [2] explains Iris Recognition based on 2D-Gabor filters including many other normalizing, enhancement and invariant properties which are covered in Ref. [3]. Iris Recognition using Circular Symmetric Filters to enhance the capture of texture information of iris and a Nearest Feature Line (NFL) based iris matching approach was proposed in Ref. [4]. Extracting the tissue edge features from the iris and their matching using Identity verification, Texture analysis and Image segmentation are introduced in Ref. [5]. Reference [6] proposes Iris Recognition based on boundary localization using DWT as the feature extractor with SVM classifier. A novel iris coding method based on differences of discrete cosine transform coefficients of overlapped angular patches from normalized iris images is introduced in Ref. [7]. A comparative study of Iris recognition using Log-Gabor, Haar wavelet, DCT and DFT based features and related research works are discussed in Ref. [8].

Figure 1 represents the general flow of an Iris recognition system. The test iris image is acquired and preprocessed. The key features are obtained using an extractor. The matching is done comparing these features with the stored features. In this paper, *Log transformation* is used as an image preprocessing step to enhance the intensity level of the image. Two dimensional DFT is used as a feature extractor. *Binary Particle Swarm Optimization (BPSO)* [9] is adopted to select the optimized features from the extracted feature set, thereby further reducing the

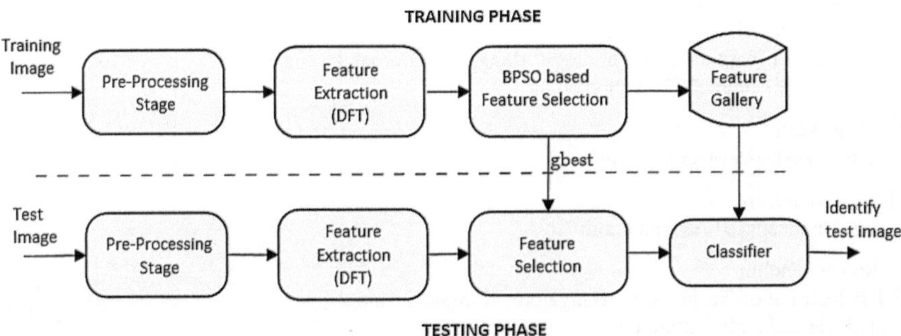

Fig. 1 General Iris recognition system

computational time. The obtained features are pixels which represent the properties of the whole image such as illumination, edges etc. proportional to their respective amplitudes. *Euclidean Classifier* performs Iris matching.

In this paper, we propose two novel approaches to enhance the Iris Recognition based on Two-Dimensional DFT as a feature extractor, which are shown below:

i. *Intensity Mapped Contrast Enhancement (IMCE)*:
Iris images in the databases used suffer from light insufficiency, lack of focus etc. during their capture. The proposed IMCE technique brings out a good contrast by increasing the gradience between brighter and darker pixels, resulting in enhanced iris features.
ii. *Double symmetric rectangular hyperbolic based DFT (DsrhDFT) extraction*:
The efficiency of an extractor depends on the extraction of required key features through precise geometrical shapes based on the spectrum obtained. The resultant shifted distribution of features after DFT is not only concentrated at the center, but also extended along perpendicular bisectors to edges of the spectrum, which is efficiently extracted by the proposed Double symmetric rectangular hyperbola.

The rest of the paper is organized as follows: Sect. 2 deals with image preprocessing based on the illumination of the iris image. Section 3 introduces the proposed contrast enhancement technique to obtain key features of the iris. The application of 2D-DFT as a feature extractor and the proposed *Dsrh* shape are discussed in Sect. 4. Section 5 introduces BPSO which selects optimized features saving the computational time along with Euclidean classifier for eye matching. Section 6 illustrates the proposed IR system and the experimental results obtained with three databases namely Phoenix, IITD, MMU. Section 7 summarizes the results obtained comparing with the proposed techniques.

2 Fundamental Image Preprocessing

Image preprocessing is a basic step done to improve the performance of subsequent procedures so that relevant information loss is reduced [10]. The Iris image obtained after converting to grayscale is not suitable for explicit feature extraction since the tissue edges are not distinct which may lead to erroneous recognition.

2.1 Logarithmic Transform

The original Iris image is converted to logarithmic domain thereby enhancing the image with respect to illumination. The above technique is performed by processing the image according to Eq. 1.

$$I(enhanced) = c \times \log(1 + I) \tag{1}$$

The resultant illumination varies proportionally with c. The value of the constant c is chosen based on intensity level of iris image in respective database.

3 Proposed Intensity Mapped Contrast Enhancement Technique (IMCE)

Contrast enhancement is based on the emphasis of the difference between the brightness levels in an image to improve its perceptual quality [11]. Various contrast enhancement techniques [12] in spatial domain have been proposed using histogram modifications [13] and several Unsharp Masking techniques [14].

The original iris image (Fig. 2a) and the image obtained after logarithmic transform have smooth variations thereby not reflecting the detailed pattern of the iris as shown in Fig. 2b. To overcome this problem, a high contrast iris image containing detailed tissue edge features is obtained by IMCE as shown in Fig. 3. The above process is implemented by our proposed technique indicated by Eqs. 2–4, where $f(x, y)$ is the Iris image to be preprocessed and $F(x, y)$ is the resultant intensity mapped contrast enhanced image and n depends on the bit map of the respective Iris image. The images used here are 8 bit mapped, hence the maximum intensity level that can be obtained is 255.

Block diagram in Fig. 3 shows the flow of the proposed IMCE technique.

i. The original iris image (Fig. 3i) of Phoenix database is complemented through Eq. 4 to obtain Fig. 3ii which results in brighter pixels.
ii. The original image is now subtracted from the complemented image to obtain an intermediate image shown in Fig. 3iii. During this process, brighter pixels (higher value) of the complemented image are subtracted from the darker pixels (lower value) of the original image preserving the brightness and the darker pixels are subtracted from brighter pixels that results in negative amplitudes, which are considered as 0 value or absolute black. This brings out appreciable difference between the darker and brighter parts of the image.

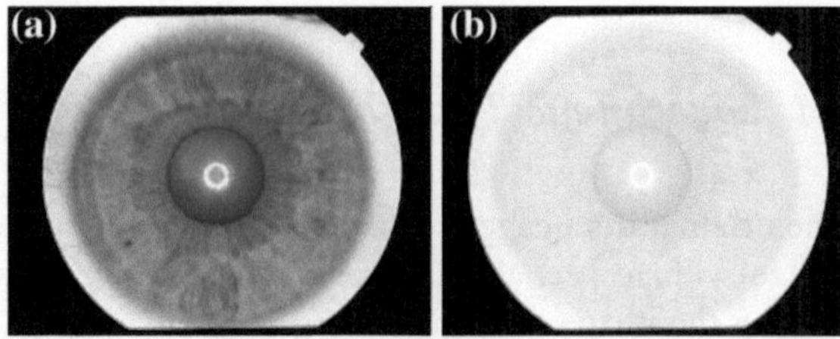

Fig. 2 Sample phoenix image **a** RGB to gray **b** applying log transform

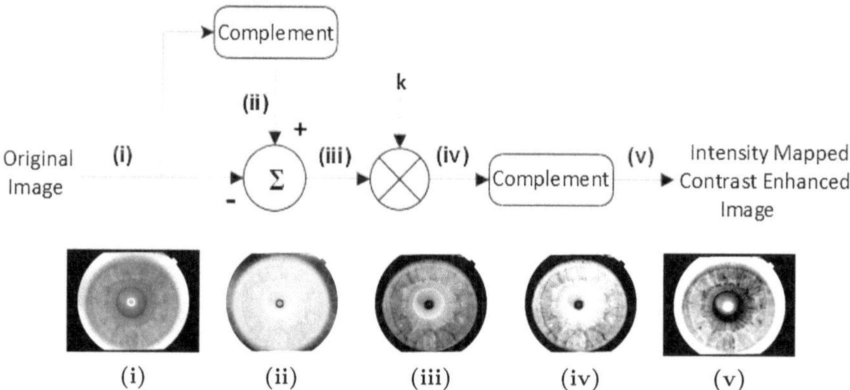

Fig. 3 Various blocks in IMCE (*upper*), and the images obtained at the output of each block (*lower*)

iii. The obtained image is now scaled by a factor k (greater than 1). The scaling factor further increases the mid-range intensity levels whereas it has no effect on the negative amplitudes as it is still treated as 0. This process further increases the gradient resulting in a high contrast image (Fig. 3iv).
iv. The resultant image is still in the complemented form as the original image is subtracted from the complemented image. Hence the image is again complemented resulting in IMCE image as shown in Fig. 3v.

$$\overline{f(x,y)} = complement(f(x,y)) \quad (2)$$

$$F(x,y) = \overline{(\overline{f(x,y)} - f(x,y)) \times k} \quad (3)$$

$$complement = [2^n - 1] - f(x,y) \quad (4)$$

The value of k is always chosen greater than 1 as a value below 1 will decrease the contrast of the iris image. An optimum k value is selected in the range of 1–1.5 for the databases used, to obtain the required contrast.

4 Proposed DsrhDFT Based Feature Extraction

DFT is a complex transform which converts the image from spatial domain to frequency domain whose equation is described in Eq. 5. The frequency components obtained are complex in nature having both real and imaginary parts. Since

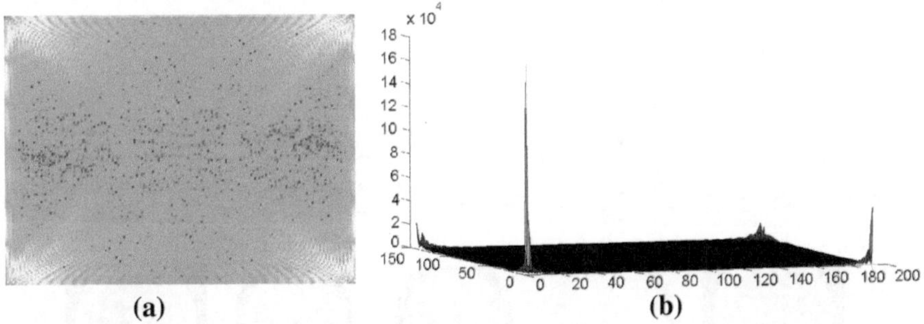

Fig. 4 DFT of Fig. 3v. **a** Frequency spectrum **b** surf diagram

the frequency components are in complex, for further processing only the magnitude of each component is calculated and processed as discussed in Ref. [15].

$$F(u,v) = \frac{1}{MN}\sum_{x=0}^{M-1}\sum_{y=0}^{N-1} f(x,y)e^{-j2\pi\left(\frac{ux}{M}+\frac{vy}{N}\right)} \tag{5}$$

Two-Dimensional DFT is applied to the preprocessed image to obtain the frequency spectrum as shown in the Fig. 4a. The features of an image are the low frequency components having higher amplitudes which are adequate for representing the key patterns of the respective image. The features are distributed at the 4 corners of the obtained spectrum. The high amplitudes of the important features are clearly shown in Fig. 4b which have peaks at the corners. Since the extraction of these features is complicated, the spectrum is shifted to the center in a symmetric manner, exploiting the shift invariance property of DFT. The application of Logarithmic Transform as a preprocessing step results in a feature distribution which are not only concentrated at the center rather few extend along the perpendicular bisectors to the edges of the spectrum as shown in Fig. 5a and in the surf diagram illustrated in Fig. 5b. Considering the above distribution of the features a novel idea of *Double symmetric rectangular hyperbola* is used for extracting the features as shown in Fig. 6.

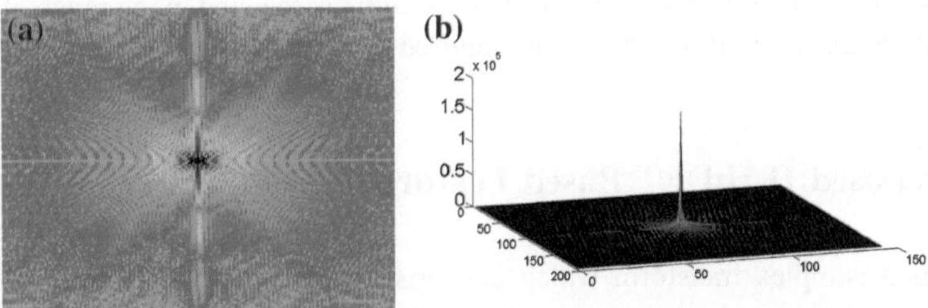

Fig. 5 Centered spectrum using fftshift [21] **a** frequency spectrum **b** surf diagram

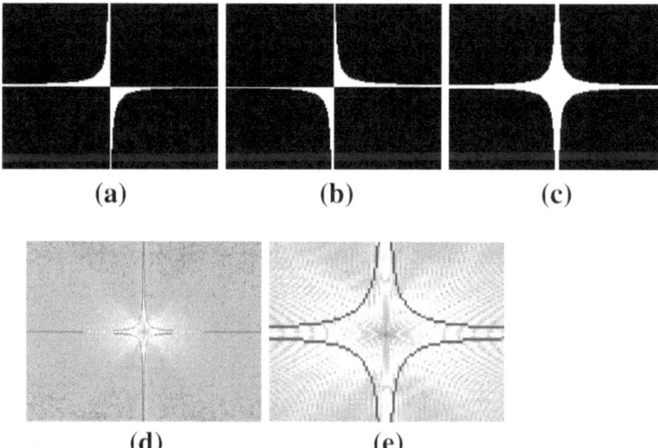

Fig. 6 Generation of Double symmetric rectangular hyperbola mask. **a** $xy = r^2$ **b** $xy = -r^2$ **c** $xy = \pm r^2$ **d** DSRH on centered DFT spectrum **e** magnified centered spectrum

This optimized Double symmetric rectangular hyperbolic based extractor mask is generated by the combination of two rectangular hyperbolas as shown in Fig. 6a, b whose equation is given by $xy = \pm r^2$ (Fig. 6c). From the property of Hyperbolas, a circle can be drawn with its center coinciding with the center of the spectrum touching the 4 parts of the Dsrh whose radius is $\sqrt{2}$ times the value of r. The value r is chosen such that the features at the center are completely enclosed in this circle which also ensures extraction of features along the asymptotic region, which is found to be in the range of 1–3. The size of the iris image also affects the value r, a higher value of r is chosen for larger iris image. The extraction of the features using the above Dsrh mask is illustrated in Fig. 6d, e shows the extraction of the center features ensuring enhanced extraction.

5 Feature Selection Using Binary Particle Swarm Optimization

Particle Swarm Optimization (PSO) was introduced by Eberhart and Kennedy in 1995 based on the idea of collaborative behavior and swarming in biological populations inspired by the social behavior of bird flocking or fish schooling [16]. PSO uses a set of swarms associated with two variables namely the position and the velocity. PSO keeps updating the velocity of the particle relative to its previous velocity until an optimized convergence is achieved. The above methodology is obtained using Eq. 10 in which $v_i{}^t$ and $x_i{}^t$ represent the velocity and position of the particle respectively with c_1 (*cognitive factor*) and c_2 (*social factor*) decide the rate of convergence and ω is the inertial weight. The values of c_1, c_2 and ω are 2, 2, and 0.6 respectively with a swarm size of 30. The velocity of particle depends on $pbest_i$ which is the best position previously visited by the particle and *gbest* which is the overall best position visited by any particle. The fitness function evaluates the quality of evolved particles in terms of their ability to maximize the class. The pbest and gbest are assigned based on the highest Fitness Function value which is described in Eq. 11. Let $N_1, N_2, N_3 \ldots N_L$ denote the images within the classes, Let $M_1, M_2, M_3 \ldots M_L$ denote the class means of the respective classes which are calculated by Eq. 6 and M_0 is the grand mean in the feature space calculated by Eq. 7. Accordingly the fitness function is described in Eq. 11.

$$M_i = \frac{1}{N_i} \sum_{j=1}^{L} W_j^{(i)} \tag{6}$$

$$M_0 = \frac{1}{N} \sum_{i=1}^{L} N_i M_i \tag{7}$$

$$f(x) = \frac{1}{1 + e^{v_i^{t+1}}} \tag{8}$$

$$x_{id} = \begin{cases} 1 & \text{if } rand_3 < f(x), \\ 0 & \text{otherwise} \end{cases} \tag{9}$$

where $W_j^{(i)}$, $j = 1, 2, 3 \ldots N_i$ represents the sample images of the corresponding class w_i.

$$v_i^{t+1} = \omega \times v_i^t + c_1 \times rand_1 \times (pbest_i - x_i^t) + c_2 \times rand_2 \times (gbest - x_i^t) \tag{10}$$

Binary Particle Swarm Optimization (BPSO) [9] is a modified binary version of the PSO in which the continuous velocities of the swarms are represented by binary 1's and 0's. There are various feature selective algorithms like BPSO, GA (Genetic algorithm) etc. In GA, chromosomes share information with each other continuously resulting in the movement of whole group together towards the optimum area. In BPSO the global best particle shares information among the remaining particles resulting in a one way information sharing. BPSO is thereby preferred in recognition [17] as it reduces the time to converge to the optimum area reducing the computational time. The particle velocity v_i^{t+1} is restricted to (0,1) using the sigmoid transformation given by Eq. 8. The position of each particle is updated depending on the condition Eq. 9. Thus the extracted features are selectively mapped to 1's which are the essential features of the iris image discarding

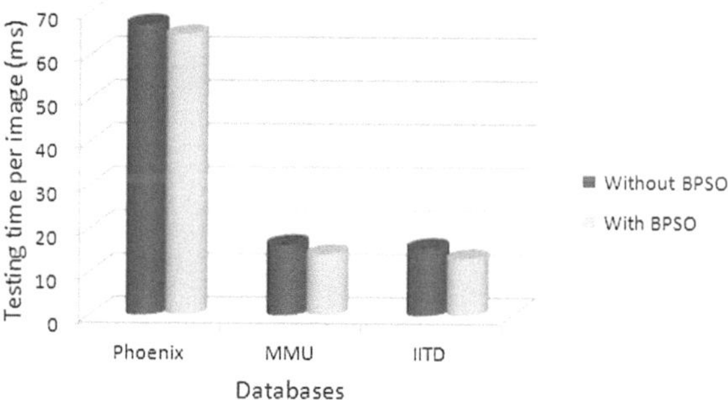

Fig. 7 Illustration of reduced testing time with BPSO

the 0's, resulting in a reduced feature set avoiding complexity and less computational time. Figure 7 shows that the addition of BPSO reduces the testing time of the recognition process in all the iris databases used.

$$F = \sqrt{\sum_{i=1}^{L} (M_i - M_0)^t (M_i - M_0)} \tag{11}$$

$$E = \sqrt{\sum_{i=1}^{N} (p_i - q_i)^2} \tag{12}$$

where $(rand_1, rand_2, rand_3) \in (0,1)$. The fitness function used for feature selection computes a feature vector which minimizes intraclass variance, while maximizing interclass variance ensuring reliable classification. Euclidean classifier calculates the distance between two corresponding points. This technique is used to measure the similarity between the features of the test image with the feature gallery obtained during the training process. The N-dimensional distance between them is calculated using Eq. 12, where p_i is one of the feature vector, q_i is the

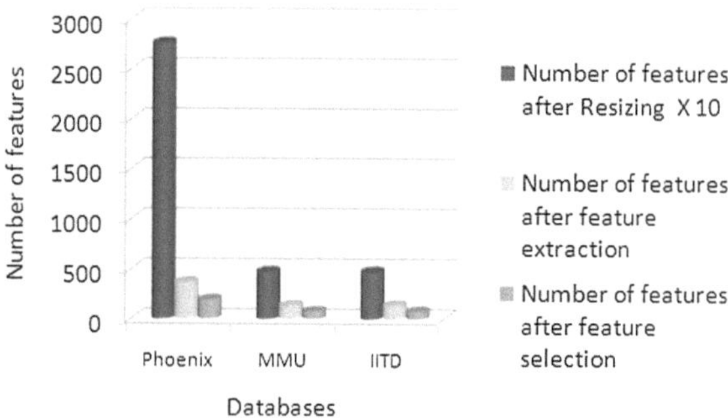

Fig. 8 Dimensionality reduction at various stages for three databases

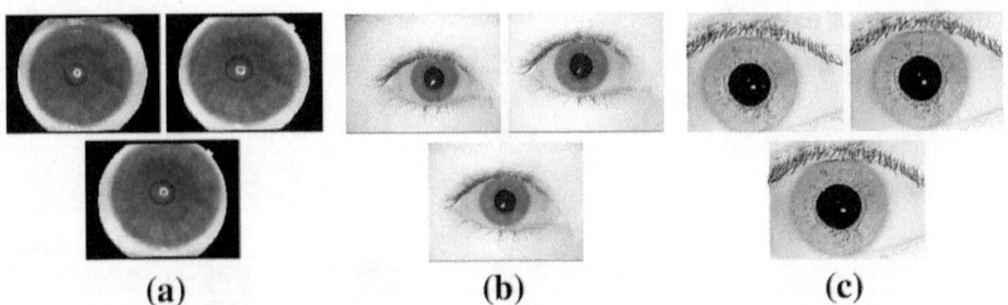

Fig. 9 Sample images of **a** Phoenix database **b** MMU database **c** IITD database

feature vector of the test image and N is the number of features extracted. The feature vector which gives the least distance is the matched image (Figs. 8, 9).

6 Discussion of Proposed IR System and Experimental Results

The block diagram of the proposed IR system is shown in Fig. 10. The experiments following the proposed method are carried out on three different databases namely Phoenix, Multi Media University (MMU) and Indian Institute of Technology Delhi (IITD). These databases have been selected to test the proposed IR system under various conditions, Phoenix database contains high resolution images of only the iris whereas the other two databases contain images of eye with noise from eye lashes. Figure 10 is applicable for Phoenix and MMU databases, Log transformation is discarded for IITD database as it is sufficiently illuminated, resulting in an other experiment.

The constant k of proposed Intensity Mapped Contrast Enhancement technique in Sect. 3 is database dependent. The variation of recognition rate for different values of constant k is shown in Fig. 11d. The optimum values of k are chosen to be 1.1, 1.2 and 1.3 for IITD, MMU and Phoenix databases respectively.

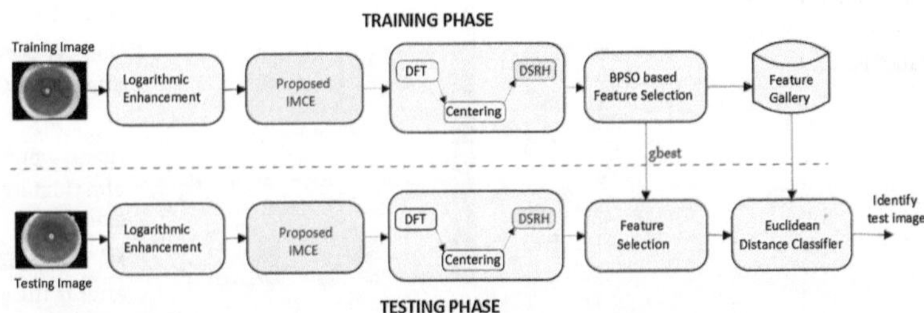

Fig. 10 Block diagram of the proposed Iris recognition system

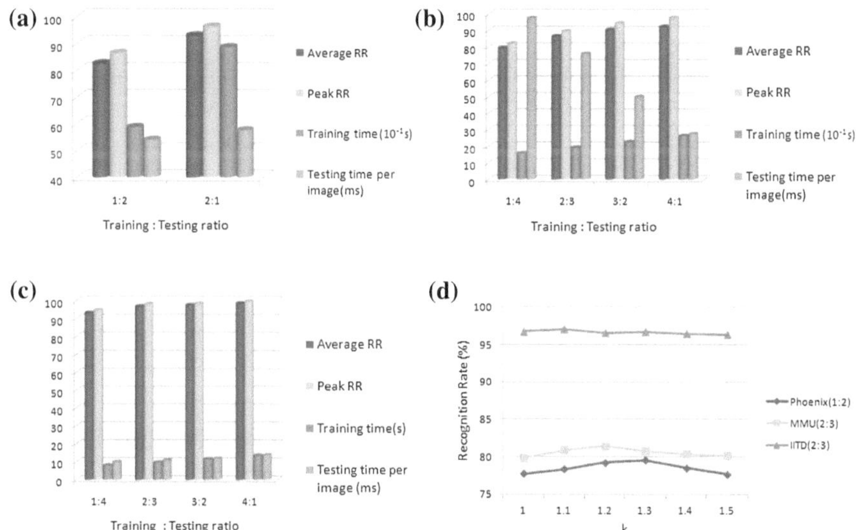

Fig. 11 Iris recognition results using the proposed IMCE+DsrhDFT method for **a** Phoenix database **b** MMU database **c** IITD database **d** Recognition rate versus k

6.1 Experiment 1 : Phoenix and MMU Database

The experiments were performed on Phoenix database [18] and MMU database [19] whose sample images are shown in Fig. 9a, b respectively. Phoenix database consists of 64 subjects with 3 images of each eye. The size of each image is 768×576 pixels which is scaled down to 192×144 pixels. The optimized features are obtained from these original pixels through various stages as illustrated in Fig. 8. The first 3 experiments tracks the importance of the image preprocessing steps and are conducted for all training to testing set ratios. The proposed IMCE technique with DFT has significantly improved the recognition rate by 13 % and further application of Log transformation lays foundation for better contrast enhancement resulting in an increase of 5 %, thereby enhancing Iris Recognition as shown in Table 1a. Further experiments covers the contribution of the geometry on Extractor, among which the proposed *Dsrh* provides better recognition rate using less number of features relative to circular and rectangular shapes as seen in Table 1d. The experimental results of proposed method for all Training:Testing ratios is shown in Fig. 11a.

MMU database contains 45 subjects with 5 images for each eye. The size of each image is 320×240 which is scaled down to 80×60 pixels. A 2 % increase is observed with application of IMCE technique and Dsrh proves to be a better extractor shape compared to the rectangular and circular based feature extractor which is indicated in Table 1b. The experimental results of proposed method for all Training to Testing set ratios is as shown in Fig. 11b.

Table 1 Recognition results highlighting the advantage of the proposed method

(a)

Method Adopted	Recognition rate (%) vs TTR[a] 1:2	2:1
Only DsrhDFT	64.95	76.41
IMCE + DsrhDFT	77.53	87.08
LOG + IMCE + RDFT	79.06	91.25
LOG + IMCE + CDFT	80.36	91.56
LOG + IMCE + DsrhDFT	**82.64**	**93.04**

[a] Training Testing ratios

(b)

Method Adopted	Recognition rate (%) vs TTR 1:4	2:3	3:2	4:1
Only DsrhDFT	67.48	78.56	83.55	85.48
IMCE + DsrhDFT	67.61	80.34	85	87.4
LOG + IMCE + RDFT	72.25	82.81	86.9	89.18
LOG + IMCE + CDFT	73.77	84.04	87.48	90.48
LOG + IMCE + DsrhDFT	**79.59**	**86.56**	**90.51**	**92.44**

(c)

Method Adopted	Recognition rate (%) vs TTR 1:4	2:3	3:2	4:1
Only DsrhDFT	92.06	96.00	97.28	98.31
IMCE + RDFT	92.56	96.29	97.36	98.21
IMCE + CDFT	93.10	96.42	97.67	98.66
LOG + IMCE + DsrhDFT	92.99	95.57	96.99	97.41
IMCE + DsrhDFT	**93.28**	**96.86**	**97.67**	**98.60**

(d)

Extraction Method	Recognition Rate (%)	Number of features Selected
Square	86.96	87
Circle	87.48	84
Dsrh	**90.51**	**82**

(a) Phoenix database (b) MMU database (c) IITD database (d) Optimized results using Dsrh compared to conventional square and circular DFT extractions

6.2 Experiment 2 : IITD Database

The experiments were performed on IIT Delhi database [20]. The sample images are shown in Fig. 9c. The database contains 224 subjects with 5 images for each eye. The size of each image is 320 × 240 pixels which is scaled down to 80 × 60 pixels. The database is sufficiently illuminated for efficient working of IMCE technique discarding the need of Log transformation. The removal of Log transformation results in the accumulation of all the features at the center. The application of Dsrh results in the extraction of redundant features, as a result, the change in the Recognition Rate with respect to CDFT is negligible as shown in Table 1c. The experimental results of proposed method for all Training to Testing set ratios is shown in Fig. 11c.

7 Conclusions

A novel approach for a flexible Iris Recognition (IR) system is proposed which uses the combination of Intensity Mapped Contrast Enhancement (IMCE) for enhancing the finer details of the iris edges, Double symmetric rectangular hyperbolic DFT (DsrhDFT) for feature extraction, and a BPSO-based feature selection. $IMCE + DsrhDFT$ have played a key role and have been the main contributors for the high recognition rates (RR) being obtained. Due to the application of DFT and BPSO, a substantial reduction in the number of features has been observed. A successful attempt has been made to equally handle all image variations (low contrast, illumination, shift-variance). The proposed method exhibits extremely good performance under low contrast (Phoenix database). The experimental results indicate that the proposed method has performed well under severe illumination conditions with top RR having reached 97.7 % for MMU database. It is also successful in tackling the most challenging task of shift variance in IR with average RR of 96.8 % for IITD database with a Training to Testing ratio of 2:3. Using the technique of IMCE, top RR of 96.31 % was obtained for Phoenix database with a Training to Testing ratio of 2:1. On a PC with Intel(R) Core 2 Duo, 2.4 GHz CPU and 3 GB RAM, $IMCE + DsrhDFT$ costs an average testing time of 57.63 ms per image (Phoenix database) using MATLAB(R) [21]. This may still be a limitation of IMCE for real time applications. Hence, a future research issue could be to develop fast computation methods for IMCE.

This paper uses a simple Euclidean classifier. By using other classifiers such as SVM, Random Forest etc, and using suitable noise removal techniques, the performance of the IR system is expected to improve substantially.

References

1. Jain A, Bolle R, Pankanti S (2000) Biometrics: the future of identification. IEEE comput soc 33(2):46–49
2. Daugman J (2004) How Iris recognition works. IEEE Trans Circuits Syst Video Technol 14(1):21–30

3. Daugman J (2007) New methods in Iris recognition. IEEE Trans Syst Man Cybern 37(5):1167–1175
4. Li M, Yunhong W, Tan T (2002) Iris recognition using circular symmetric filters. In: Proceedings of 16th international conference on pattern recognition, vol 2, pp 414–417
5. Kevin B, Karen H, Flynn, P (2008) Image understanding for Iris biometrics: A survey. Comput Vis Image Underst 110:281–307
6. Sung H, Lim J, Park J-H, Lee Y (2004) Iris recognition using collarette boundary localization. In: 17th international conference on pattern recognition, vol 4, pp 857–860
7. Monro D, Rakshit S, Zhang D (2007) DCT-based Iris recognition. IEEE Trans Pattern Anal Mach Intell 29(4):586–595
8. Kumar A, Passi A (2010) Comparison and combination of Iris matchers for reliable personal authentication. Pattern Recogn 43(3):1016–1026
9. Kennedy J, Eberhart R (1997) A discrete binary version of the particle swarm algorithm. IEEE Int Conf Syst Man Cybern 5:4104–4108
10. Duda R, Hart P, Stork D (2000) Pattern classification, 2nd edn. Wiley-Interscience, New York
11. Gonzalez R, Woods R (2008) Digital image processing, 3rd edn. Prentice Hall, Upper Saddle River
12. Wang D, Vagnucci AH (1983) Digital image enhancement. Comput Vis Graph Image Process 24(3):363–381
13. Sun C, Ruan SJ, Shie MC, Pai TW (2005) Dynamic contrast enhancement based on histogram specification. IEEE Trans Consum Electron 51(4):1300–1305
14. Badamchizadeh MA, Aghagolzadeh A (2004) Comparative study of unsharp masking methods for image enhancement. In: 3rd international conference on image and graphics, pp 27–30
15. Imtiaz H (2011) A spectral domain local feature extraction algorithm for face recognition. Int J Secur 5(2):62–73
16. Kennedy J, Eberhart RC (1995) Particle swarm optimization. In: IEEE international conference on neural network, vol 4, issue 4, pp 1942–1948
17. Ramadan RM, Abdel-Kader RF (2009) Face recognition using particle swarm optimization-based selected features. Int J Signal Process Image Process Pattern Recognit 2(2):57–59
18. Phoenix Database, http://www.inf.upol.cz/iris/
19. MMU Database, http://pesona.mmu.edu.my/~ccteo/
20. IIT Delhi database, http://web.iitd.ac.in/biometrics/DatabaseIris.htm
21. Matlab, http://www.mathworks.in/

Programmable Gaussian Noise Generator to Test and Characterize Onboard Subcarrier Systems in Satellite Communication

K. K. Raghunandana, P. N. Ravichandran, Sunil Kulkarni, H. S. Vasudeva Murthy and M. Vanitha

Abstract The paper presents design, simulation and implementation of programmable Gaussian noise generator in hardware. The communication system performance is evaluated for additive Gaussian noise, this necessitates the design of highly accurate, programmable noise generator which is dealt in this paper. The Box-Muller method is used for Gaussian noise generator and the programmable variance is implemented in hardware through multiplying DAC. The Box-Muller method requires two uniform random generators inputs and mathematical operations for these inputs which are implemented in hardware using CORDIC algorithms. The Uniform noise generators and Gaussian noise generator are simulated in MATLAB Simulink®, finite word length effect is analyzed with bit true simulation and the same were implemented in hardware. The whole design is accommodated in a Xilinx xc4vsx35-10ff668 FPGA, the multiplying DAC circuit is realized as a separate circuit, with control inputs for noise variance control. The design is tested in real time for different subcarrier frequency and data modulation and receiver performance is evaluated in terms of BER characterized by E_b/N_0.

K. K. Raghunandana (✉) · P. N. Ravichandran · S. Kulkarni · H. S. Vasudeva Murthy · M. Vanitha
Digital Systems Group, ISRO Satellite Centre, Bangalore 560017, India
e-mail: kkraghu@isac.gov.in

P. N. Ravichandran
e-mail: pnravi@isac.gov.in

S. Kulkarni
e-mail: kulkarni@isac.gov.in

H. S. Vasudeva Murthy
e-mail: hmurthy@isac.gov.in

M. Vanitha
e-mail: vani@isac.gov.in

Mohan S. and S. S. Kumar (eds.), *Proceedings of the Fourth International Conference on Signal and Image Processing 2012 (ICSIP 2012)*, Lecture Notes in Electrical Engineering 221, DOI: 10.1007/978-81-322-0997-3_44,

Keywords Gaussian · Noise · CORDIC · Box-muller · DAC · Variance

1 Introduction

The quality of the Gaussian noise is very crucial in evaluating the performance of the communication systems. The non availability or expensive off the shelf Gaussian noise generators to test and characterize the on board receivers necessitated the development of simple, low cost, high performance and accurate noise generator.

The Gaussian noise can be generated in both analog and digital domain. The analog methods can produce truly random numbers but are sensitive to operational environment, hence digital methods for generating Gaussian random variables are preferred over analog methods. The random numbers generated by digital methods are pseudo random, but the period can be made sufficiently large such that the sequences never repeat themselves even in the largest practical situations.

The method chosen for the Gaussian noise generator is based on transformation of the uniform random variables, which can be implemented effectively in hardware with area, speed efficient and has programmable noise variance.

The programmable Gaussian noise generator will be used to test and characterize the performance of the on board subcarrier demodulator system. The subcarrier signal is BPSK/QPSK modulated and the demodulator performance is evaluated in terms of Bit Error Rate (BER) characterized by E_b/N_0. Onboard systems are qualified for BER of 10^{-6} using implemented noise generator.

The generation of Gaussian noise in digital domain and converting to analog signal with DAC gives limited range of programmability for noise variance. To overcome this limitation a new method is adopted which is dealt in detail.

The Gaussian noise generator is implemented using Box-Muller method. This method requires two independent uniform random generators, Mersenne twister and Tausworthe generators are used. The Box-Muller method requires trigonometric, logarithmic and arithmetic functions which are implemented using Coordinate Rotational Digital Computer (CORDIC) [1]. The mathematical operations are implemented in hardware in a fixed point format.

The Multiplying DAC provides the control signals to vary the amplitude of the reference signal through the control inputs these control inputs are used to provide programmable noise variance.

The paper is organized as follows: Sect. 2 deals with Gaussian noise generator Architecture, Sect. 3 deals with the selection of uniform random generators and hardware efficient algorithms Sect. 4 deals with the simulation and implementation. Section 5 deals with results and discussion. Section 6 deals conclusion at the end.

2 Programmable Gaussian Noise Generator Architecture

The Box-Muller method of Gaussian noise generator use the following equations [2]

$$G_1 = \cos(2\pi U_2)\sqrt{-2\sigma^2 \ln(U_1)} \quad (1)$$

$$G_2 = \mathrm{Sin}(2\pi U_2)\sqrt{-2\sigma^2 \ln(U_1)} \quad (2)$$

where U_1 and U_2 are uniform random generators, σ is the Gaussian noise variance. Gaussian noise is generated in Digital domain in FPGA, it is converted to analog through DAC, the DAC output is amplified and given as the reference voltage to the multiplying DAC. The multiplying DAC control inputs are used for noise variance control. The Fig. 1 shows the overall block diagram of programmable noise generator.

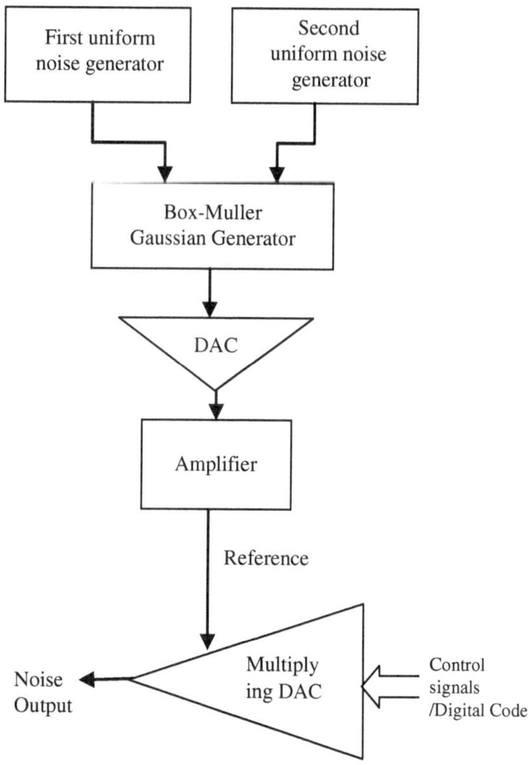

Fig. 1 Block diagram of programmable noise generator

3 Selection of Uniform Random Generators and Hardware Efficient Algorithms

The Box-Muller method of Gaussian noise generator requires the two independent uniform noise generators. The criterion for selecting the random generators are they should pass all randomness testes [3], the repetition rate should be more and they can be implemented in hardware with minimum resources. The Mersenne twister (MT) [4] and Tausworthe generator [5] meet these criterion hence they are used for uniform noise generators.

MT19937, with 32-bit word length is used in development. MT has period of 2^{19937}–1. The Tausworthe with 32 bit word length is used which will generate pseudo numbers with period 2^{88}.

The equation $G_1 = \cos(2\pi U_2)\sqrt{-2\sigma^2 \ln(U_1)}$ is used for noise generation, since $\sqrt{\sigma^2} = \sigma$, the variance of the noise can be controlled by a multiplying the remaining terms in the equation $(\cos(2\pi U_2)\sqrt{-2\ln(U_1)})$ by a programmable σ. For VHDL implementation the unit variance is chosen.

To implement the trigonometric and arithmetic functions the Coordinate Rotation Digital computer (CORDIC) algorithms are chosen, the CORDIC uses simple shift and add methods along with the look up table to implement the functions in a pipeline architecture which is more hardware and speed efficient.

CORDIC in circular coordinate system has two modes of operation

- Rotation Mode (RM)
- Vector mode (VM).

In rotation mode (RM) a given vector with initial value x_0, y_0 is rotated by an angle z_0 by iterative means with final rotated vector having values $x_n, y_n, z_n = 0$. In vector mode (VM) a given vector x_0, y_0 with $z_0 = 0$ rotates the input vector through whatever angle necessary to align the result vector with x axis computing the magnitude and angle of a given vector.

In Rotation mode (RM) unified CORDIC equations are given by

$$x_{i+1} = x_i - y_i.d_i.2^{-i} \tag{3}$$

$$y_{i+1} = y_i + x_i.d_i.2^{-i} \tag{4}$$

$$z_{i+1} = z_i - d_i.\tan^{-1}(2^{-i}) \tag{5}$$

where $d_i = -1$ if $z_i < 0$, $+1$ *otherwise*

This provides the following result

$$x_n = A_n[x_0 \cos z_0 - y_0 \sin z_0] \tag{6}$$

$$y_n = A_n[y_0 \cos z_0 + x_0 \sin z_0] \tag{7}$$

$$z_n = 0 \tag{8}$$

$$A_n = \prod_n \sqrt{1 + 2^{-2i}} \tag{9}$$

$$K_i = 1 \Big/ \sqrt{1 + 2^{-2i}} \tag{10}$$

For implementing the CORDIC system it important to choose required number format, word length appropriately, since the computational accuracy and precision is limited by number of rotation angles chosen and word length used for three input variables x, y, z. Numerical accuracy of the CORDIC algorithm can be calculated by the examination of truncation and approximation errors. Truncation errors are due to the finite word length and approximation errors are due to the finite number of iterations. For n iterations and n bits accuracy, data paths should be of size $(n + (logn + 2))$ bits which gives an upper bound error for CORDIC.

The 32 bit random numbers generated from MT and Tausworthe are uniform random numbers in the range (0, 1), the following trigonometric functions in Box-Muller method are implemented using CORDIC algorithms [6, 7], the final output format is chosen to cover DAC full scale output value.

3.1 Natural Log

The 32 bit random number from MT is used as input to the Natural Logarithm function, here the CORDIC algorithm is used in the Hyperbolic, vectoring mode.

The natural logarithm of number ‘a’ is calculated as
if $X_0 = a + 1$, $Y_0 = a - 1$, $Z_0 = 0$ then after ‘n’ iterations
if $Y_n \approx 0$ then
$X_n = 2.Sqrt(a)$, $Y_n = 0$, $Z_n = 0.5 \ln(a)$.

3.2 Square Root

The output of natural log function block is 0.5ln (a), the actual square root required is for the function −2.ln (a), the natural log output is left shifted by 2 bits to multiply it by 4 and negated to get the function −2.ln (a), the square root is also obtained by using CORDIC algorithm in Hyperbolic Vectoring mode, the square root is calculated as

if $X_0 = a + 1$, $Y_0 = a - 1$, $Z_0 = 0$
then after ‘n’ iteration if $Y_n \approx 0$ then
$X_n = 2.Sqrt(a)$, $Y_n = 0$, $Z_n = 0.5 \ln(a)$
The X_n has square root of ‘a’ multiplied by 2, so output of X_n is right shifted by 1 bit to divide by 2.

3.3 Multiplier

The multiplier is implemented in CORDIC as Linear rotation mode, if a and b are two number to be rotated then
$X_0 = a$, $Y_0 = 0$, $Z_0 = b$
then after 'n' iteration $Z_n \approx 0$ then
X_n a, $Y_n = a.b$, $Z_n = 0$.

3.4 Sine/Cosine Function

The sine/Cosine function is implemented in CORDIC as Circular rotation mode. If U_1 is the number to be rotated then
$X_0 = 1/K_m$, $Y_0 = 0$ and $Z_0 = a$
then after 'n' iterations $Z_n \approx 0$, (K_m is the CORDIC gain.)
$X_n = \cos(a)$, $Y_n = \sin(a)$ and $Z_n = 0$.

3.5 Calculation of Noise Variance

For a given E_b/N_o [2] the noise variance is calculated as follows, during each symbol interval, the energy in the signal time is given by

$E = f(A,T)$

where 'A' signal amplitude and 'T' is the symbol duration,

$E_b = f(E, symbol\ rate)$

The sampling rate is symbol rate multiplied by the number of samples per symbol.

The variance of a random noise process equals the process auto correlation at lag zero $\sigma^2 = R(0)$, for white noise R(0) is infinite, but we are concerned with noise power contained within some particular bandwidth of interest, for a given white noise density of N_0, the noise power in a bandwidth B is

$$P = \sigma^2 = N_0 B \quad (11)$$

To satisfy the uniform sampling theorem and prevent aliasing the sampling rate must be chosen such that

$R_s \geq 2B$ for real sampling

$$B \text{ for complex signal} \quad (12)$$

Thus

$\sigma^2 = N_0 B \leq N_0 R_s/2$ for real sampling

$$N_0 R_s \quad \text{for complex sampling.} \quad (13)$$

4 Simulations and Implementation

The uniform random generators and unit variance Gaussian noise generator using Box-Muller method is simulated in MATLAB Simulink®, for the normal data. The bit true simulation is done for the fixed point data format, the same were implemented in VHDL and simulation is done in MODELSIM. The Fig. 2 show the overall bit true implementation of the Box-Muller Algorithm.

The Fig. 3 shows the MATLAB bit true simulation result of Box-Muller method, the Fig. 4 show the VHDL MODELSIM simulation results. The Fig. 5 shows the overall setup used for the Receiver evaluation.

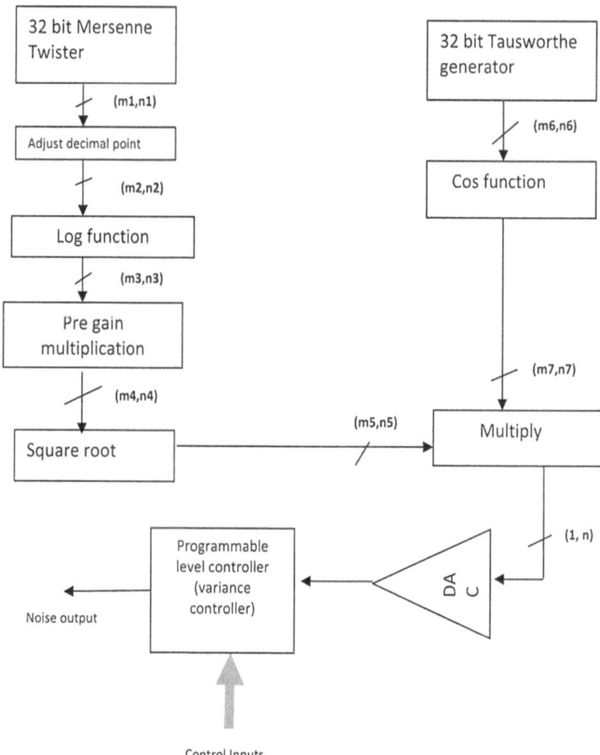

Fig. 2 Bit true implementation of box-muller method

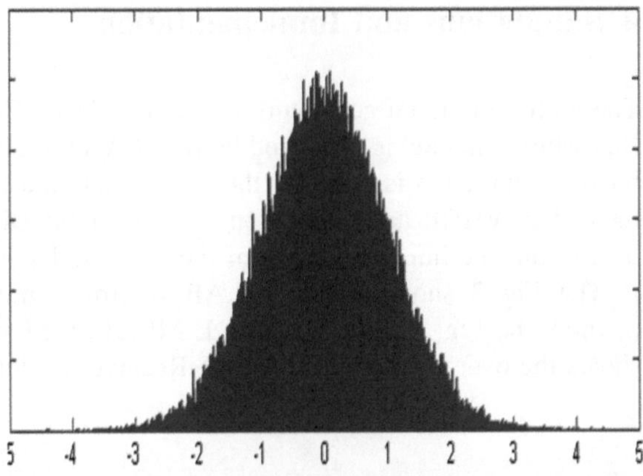

Fig. 3 MATLAB bit true simulation

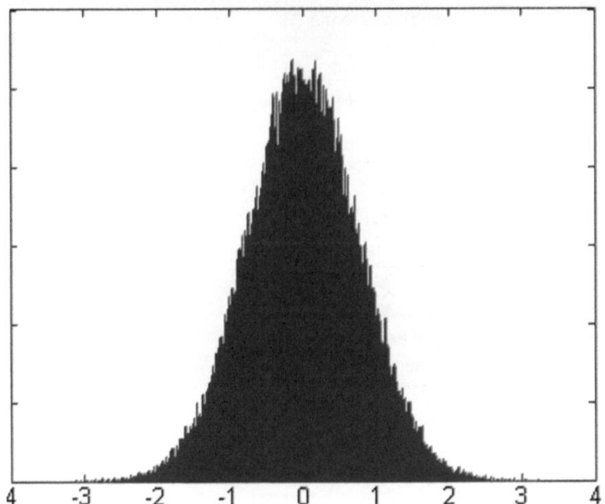

Fig. 4 MODELSIM VHDL simulated data

5 Results and Discussion

The uniform random number generators using Tausworthe method and Mersenne twister are simulated and analyzed in MATLAB Simulink® and same were implemented in VHDL. The results are matching.

The Box-Muller method implemented in MATLAB using Bit true simulation which is verified with the VHDL implemented results at each stage results are matching.

The VHDL simulated data is imported to MATLAB and analyzed for the statistical properties and distribution, results are as per theoretical calculation.

The Hardware programmable noise variance controller circuit is developed with the multiplying DAC and real time noise variance value against the programmed value is verified and results are validated.

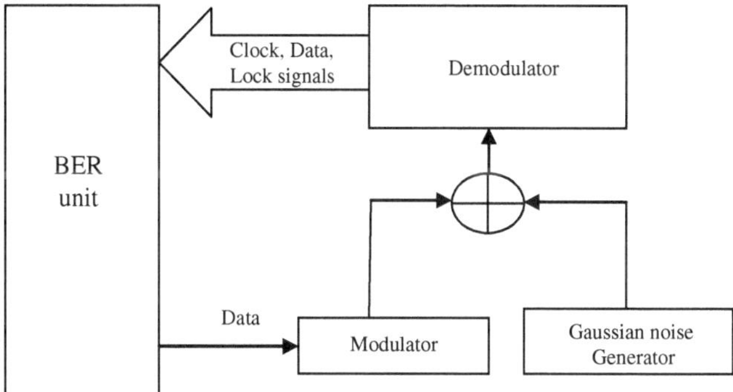

Fig. 5 Block diagram of setup to evaluate receiver performance

The E_b/N_0 v/s BER of BPSK demodulator is verified with the different noise variance values, a performance deviation of 0.5 dB is observed with standard noise generator it can be related to finite word length implementation effect.

6 Conclusion

The programmable variance Gaussian generator is used to characterize performance metric of the subcarrier receiver system in terms of E_b/N_0 and S/N. The noise generated is uncorrelated over long duration of time hence noise generator can be used for the performance evaluation of LDPC, Turbo codes ... etc.

Acknowledgments Authors would like to thank and acknowledge the support and encouragement received from E. Vasantha Deputy Director-DCA, Dr. Shivakumar S. K, Director, ISRO Satellite Centre, Bangalore.

References

1. Volder JE (1959) The CORDIC trigonometric computing technique. IRE Trans Electron Comput 8(3):330:334
2. Rorabaugh CB (1997) Digital filter designers handbook. MacGraw Hill, INC
3. L'Ecuyer P (1998) Uniform random number generators. In: Proceedings of the 1998 winter simulation conference
4. Matsumoto M, T Nishimura (1998) Meresenne twister: a 623-dimensionally equidistributed uniform pseudo-random number generator. ACM Trans Model Comput Simul 8(1):3–30
5. L'Ecuyer P (1996) Maximally equidistributed combined tausworthe generators. Math Comput 65:213, 203–213
6. Lee D-U, Luk W, Villasenor JD, Cheung PYK (2004) A gaussian noise generator for hardware-based simulations. IEEE Trans Comput 53(12)
7. Andraka R (1998) A survey of CORDIC algorithms for FPGA based computers. In: Proceedings of ACM/SIGDA Sixth International Symposium on FPGAs, Feb 1998. Monterrey, CA, pp 191–200

A Novel Block De-noising Algorithm for Sounds of Indian Musical Instruments with Modified Threshold in Wavelet Domain

Raghavendra Sharma and V. Prem Pyara

Abstract The problem of estimating the musical instrument sound signal, corrupted by additive white Gaussian noise has been of interest to many researchers for practical as well as theoretical reasons. The removal of white Gaussian noise is difficult as it persists at all the frequencies in the signal. Many of the methods, especially those based on wavelet technique have become popular, due to the number of advantages over the traditional methods. It has been shown that wavelet based thresholding is simple and optimal solution, also guarantees better rate of convergence. In this paper, a novel DWT based algorithm using block de-noising along with modified threshold is proposed. For experimental purpose, the sound signals of shehnai, dafli and flute are taken. The signal is first divided into the multiple blocks of samples and then both hard and soft thresholding methods are used on each block. All the blocks obtained after individual block de-noising are concatenated to get the final de-noised signal. When the sound signal corrupted with variable percentage of Gaussian noise, passed through this algorithm; significant improvement in PSNR is observed over normal wavelet thresholding method. The quality of sound signal obtained through this algorithm is perceptually close to original signal.

Keywords Block denoising · Gaussian noise · Wavelet thresholding · Wavelet coefficients

R. Sharma (✉) · V. P. Pyara
Department of Electrical Engineering, Dayalbagh Educational Institute, Agra 282010, India
e-mail: raghsharma2000@yahoo.com

Mohan S. and S. S. Kumar (eds.), *Proceedings of the Fourth International Conference on Signal and Image Processing 2012 (ICSIP 2012)*, Lecture Notes in Electrical Engineering 221, DOI: 10.1007/978-81-322-0997-3_45,

1 Introduction

Signal de-noising has emerged as one standard testbed application for wavelet based representations. In [1, 2], the discrete wavelet transform and filter bank theory is discussed. In [3], Donoho shows that for various functional spaces, the wavelet basis is an optimal basis for data compression and de-noising. Moreover, wavelet based de-noising can be effectively carried out by simple thresholding schemes. The papers by Donoho et al. [4, 5] analyze a series of de-noising algorithms based on thresholding in the wavelet transform domain. The optimality property discussed in [3] was connected to the ability of the wavelet basis to capture most of the signal energy in a small number of coefficients. The standard wavelet basis was shown to be optimal in this regard for representing objects that have local singularities.

Estimating a non stationary signal, corrupted by additive white Gaussian noise has been of interest of many researchers for practical as well as theoretical reasons. The problem of recovering the signal from noise corrupted signal as close as possible to original signal is main cause of concern. Traditional de-noising schemes are based on linear methods, where most common choice is filters. In recent past nonlinear methods, especially based on wavelets have become popular. In [6–8], wavelet based de-noising methods for speech enhancements are discussed. In first method traditional spatially selective noise filtration technique is proposed and second method is based on un-decimated discrete wavelet transform. These methods can be used for edge detection satisfactorily. Wavelet methods have been used for 1D or 2D signal analysis, producing and analyzing irregular signals. The fundamental idea behind wavelets is to analyze according to scale. The signal is decomposed into several scales, which represent different frequency bands at each scale. At each scale the wavelet coefficients are compared to the threshold value and the coefficients with larger value than threshold are retained. In [9] modified threshold function is given which enhances the PSNR value and the comparative analysis of different thresholding techniques is discussed in [10]. The other techniques based on soft and hard thresholding are given in [11, 12] for speech signal, where soft thresholding perform better than hard thresholding method, since hard thresholding is a keep or kill procedure and sometimes pure noise coefficients may pass the hard threshold and appear as annoying blips in the output. This paper is organized as follows; Sect. 2 gives theory of wavelet de-noising along with the block de-noising scheme. The novel algorithm for de-noising the musical instrument sound signal is elaborated in Sect. 3. In Sect. 4, experimental results based on the algorithm in Sect. 3 are discussed. Finally concluding remarks are given in Sect. 5.

2 De-Noising

Noise is the undesired information that contaminates any signal. The information regarding the type of noise plays significant role in signal de-noising process. The removal of Gaussian noise is very significant as it is evenly distributed over the entire signal. Here de-noising of the signal is briefly explained in the following sub-sections.

2.1 Wavelet Thresholding

Donoho and Johnstone pioneered the work of filtering the additive white Gaussian noise using wavelet thresholding. From the properties and behavior of the wavelets, it plays major role in signal de-noising. Wavelet thresholding is explained as decomposition of signal into wavelet coefficients, comparing the detail coefficients with a given threshold value, and shrinking these coefficients close to zero to take away the effect of noise in the signal. The signal is reconstructed with these modified coefficients and the process is known as Inverse Discrete Wavelet Transform (IDWT). During thresholding, a wavelet coefficient is compared with a given threshold and is set to zero if its value is less than the threshold; otherwise, it is retained or modified depending upon the threshold rule. Thresholding distinguishes between the coefficients due to the additive noise and ones consisting the important information.

Let us take a musical instrument sound signal $\mathrm{s_i}$, and corrupt it by additive white Gaussian noise $\mathrm{z_i} \sim \mathrm{N}(0, \sigma^2)$, the resultant noisy signal is given by:

$$\mathrm{y_i} = \mathrm{s_i} + \mathrm{z_i} \quad (\mathrm{i} = 0, 1, 2, \ldots, \mathrm{N} - 1) \tag{1}$$

From this noisy signal $\mathrm{y_i}$, we want to find an approximation $\tilde{s}_i$ to the original signal s_i, that minimizes the mean square error,

$$\left\| s - \tilde{s}^2 \right\| = \frac{1}{N} \sum_{i=0}^{N-1} |s_i - \tilde{s}_i|^2 \tag{2}$$

where $s = [s_{0,} s_1 \ldots . s_{N-1}]^T$ and $\tilde{s} = [\tilde{s}_0, \tilde{s}_1 \ldots \tilde{s}_{N-1}]^T$. Let W is orthogonal wavelet transformation, then Eq. (1) is written as:

$$d_j = c_j + \in_j \tag{3}$$

where $d = Wy$, $\mathrm{c} = Ws$ and $\in = Wz$. Since W is an orthogonal transform, $\in_j$ are also i.i.d Gaussian random variables with $\in_j \sim N(0, \sigma^2)$. Now, let T(.) be a wavelet thresholding function, then the wavelet thresholding based denoising scheme can be expressed as:

$$\tilde{s} = w^{-1}(T(Wy)) \tag{4}$$

First of all take the wavelet transform of the noisy signal and pass it through thresholding function T(.). Take the inverse wavelet transform of the output and $\tilde{s}$ is obtained.

The thresholding function T(.), which is also known as wavelet shrinkage function is categorized as hard thresholding and soft thresholding function. The hard thresholding function retain the wavelet coefficients which are greater than the threshold λ and sets all other to zero. The hard thresholding is defined as;

$$\mathrm{f_h(x)} = \begin{cases} x, & if|\mathrm{x}| \geq \lambda \\ 0, & otherwise \end{cases} \tag{5}$$

The threshold λ is chosen according to the signal energy and the standard deviation σ of the noise. If the wavelet coefficient is greater than λ, then it is assumed that it is significant and contributes to the original signal. Otherwise it is due to the noise and discarded. The soft thresholding function shrinks the wavelet coefficients by λ towards zero. Hence this function is also called as shrinkage function. The soft thresholding function is defined as;

$$\mathrm{f_s(x)} = \begin{cases} \mathrm{x} - \lambda, & \mathrm{if}|\mathrm{x}| \geq \lambda \\ 0 & \mathrm{if}|\mathrm{x}| < \lambda \\ \mathrm{x} + \lambda, & \mathrm{if}|\mathrm{x}| \leq \lambda \end{cases} \tag{6}$$

It is shown in Fig. 1 that hard thresholding function is discontinuous at $|\mathrm{x}| = \lambda$, due to this discontinuity hard thresholding function results in abrupt artifacts in de-noised signal, especially when the noise level is significant. In [13], we see that the soft thresholding gives lesser amount of mean square error. Due to this reason soft thresholding is preferred over hard thresholding, but in case of some signals, we could see that hard thresholding results in lesser amount of mean square error.

2.2 Block De-Noising

Most of the musical instrument sound signals are far too long to be processed in their entirety; for example a 10 s sarangi sound signal sampled at 44.1 K Hz will contain 441,000 samples. Thus, as with spectral methods of noise reduction, it is necessary to divide the time domain signal in multiple blocks and process the each block individually. The important task is to choose the block length. Berger at el.

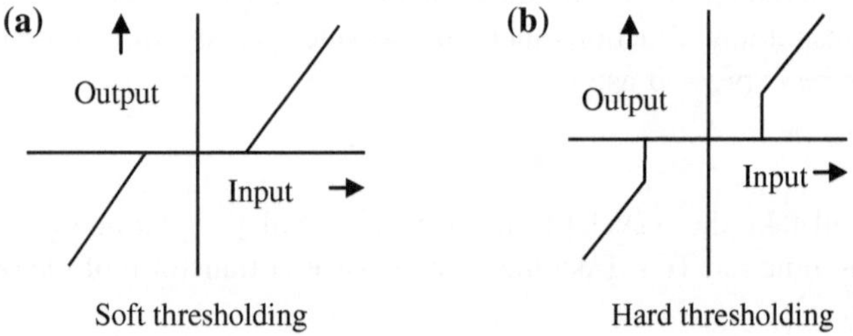

Fig. 1 **a** Soft thresholding. **b** Hard thresholding

[14] shows that, blocks which are too shorts fail to pick important time structures of the signal. Conversely, blocks which are too long miss cause the algorithm to miss the important transient details in the musical instrument sound signal. Due to the binary splitting nature of the tree bases in wavelet analysis to decompose the signal, it is better to choose the length of each block with a number of samples to a power of two Fig. 2.

Donoho and Johnstone derived a general optimal universal threshold for the Gaussian white noise under a mean square error (MSE) criterion described in (2). However this threshold is not ideal for musical instrument sound signals due to poor correlation between the MSE and subjective quality and the more realistic presence of correlated noise. Here we use a new time frequency dependent threshold estimation method. In this method first of all the standard deviation of the noise, σ is calculated for each block. For given σ, we calculate the threshold for each block.

Noise component removal by thresholding the wavelet coefficients is based on the observation that in musical instrument sound signal, energy is mostly concentrated in small number of wavelet dimensions. The coefficients of these dimensions are relatively very large compared to other dimensions or to any other signal like noise that has its energy spread over a large number of coefficients. Hence by setting smaller coefficients to be zero, we can optimally eliminate noise while preserving important information of the signal. In wavelet domain noise is characterized by smaller coefficients, while signal energy is concentrated in larger coefficients. This feature is useful for eliminating noise from signal by choosing the appropriate threshold. Generally the selected threshold is multiplied by the median value of the detail coefficients at some specified level which is called threshold processing.

At each level of decomposition, the standard deviation of the noisy signal is calculated. The standard deviation is calculated by Eq. (7);

$$\sigma_j = \frac{median|c_j|)}{0.6745} \tag{7}$$

where c_j are high frequency wavelet coefficients at jth level of decomposition, which are used to identify the noise components and σ_j is Median Absolute Deviation (MAD) at this level. This standard deviation can be further used to set the threshold value based on the noise energy at that level. The modified threshold value [15] can be obtained by the Eq. (8);

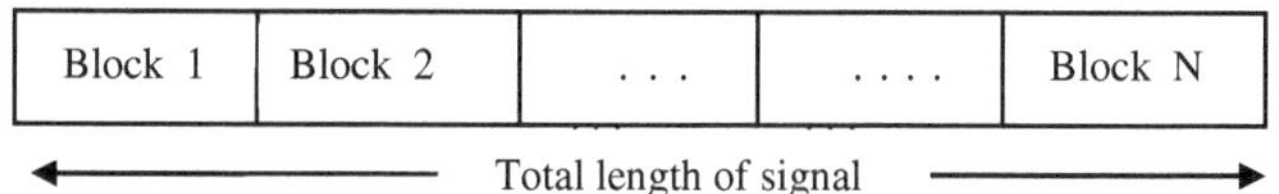

Fig. 2 Block formation of signal

$$T_h = \sigma_j \sqrt{2\log(L_j log_2 L_j)} \quad (8)$$

where T_h is threshold value and L_j is the length of each block of noisy signal. Here in this paper the threshold is dependent upon detail coefficients at each level.

3 De-Noising Algorithm

The proposed wavelet based block denoising algorithm for reduction of white Gaussian noise is explained in the following steps;

Step 1: Take a musical instrument sound signal of suitable length.
Step 2: Add White Gaussian Noise to the original signal depending upon the standard deviation σ.
Step 3: Divide the noisy signal into blocks of different length depending upon the length of the signal in time domain, and the number of samples should be to a power of two.
Step 4: Compute the Discrete wavelet transform of one block of the noisy signal at level 1.
Step 5: Estimate the standard deviation of the noise using (7) and determine the threshold value using (8), then apply the different thresholding techniques for time and level dependent wavelet coefficients using (5) and (6).
Step 6: Take Inverse Discrete wavelet Transform (IDWT) of the coefficients obtained through step 5, which has reduced noise.
Step 7: Calculate mean square error (MSE), peak signal to noise ratio (PSNR) for de-noised signal.
Step 8: Repeat steps 4 to step 7 for other level of decomposition 2–4.
Step 9: Concatenate all the blocks of the de-noised signals obtained through step 8 and do averaging operation for MSE and PSNR of the musical instrument sound signal.

The complete de-noising algorithm is shown graphically in Fig. 3.

4 Experimental Results

The de-noising algorithm developed in the previous section is applied to the sound samples of the various Indian musical instruments sampled at 44.1 K samples per second. For experimental purpose the sounds of three musical instruments shehnai, dafli and flute are taken. For comparing the performance of the various wavelets for musical instrument sound signals, six wavelets haar, db10, sym3, coif5, dmey and bior 2.2 are taken. Besides observing the performance of the wavelets, the effect of decomposition is also discussed. For comparing the performance and

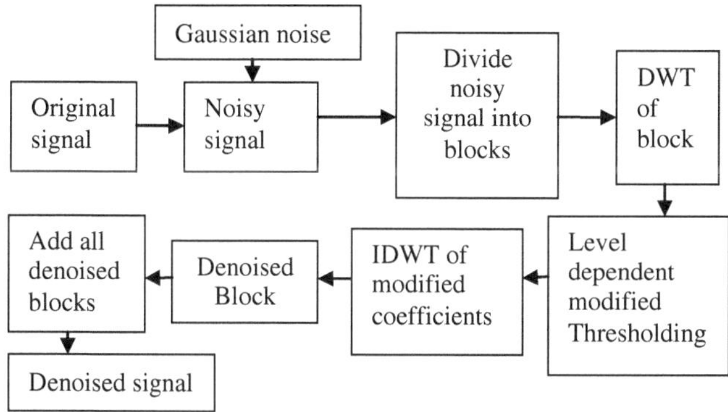

Fig. 3 Block de-noising algorithm with modified threshold

measurement of quality of de-noising, the peak signal to noise ratio (PSNR) is determined between the original signal S_i and the signal de-noised S_d, by our algorithm. We have also compared our algorithm with wavelet thresholding method developed previously.

$$\mathrm{PSNR} = 10\log_{10}\left(\frac{S_{max}^2}{\mathrm{MSE}}\right) \tag{9}$$

where S_{max} is the maximum value of the signal and is given by,

$$S_{max} = \max(\max(S_i), \max\ (S_d)) \tag{10}$$

And MSE is Mean Square error given by,

$$\mathrm{MSE} = \frac{1}{N}\sum_{l=1}^{N}[S_d(l) - S_i(l)]^2 \tag{11}$$

where N is the length of the signal. The PSNR values obtained for different wavelets applied on shehnai, dafli and flute signals at second level of decomposition are shown in Tables 1, 2 and 3, where WM = wavelet method and PM = proposed method.

It is observed from Tables 1, 2 and 3 that as we increase the percentage of noise, PSNR values are decreased and the PSNR values obtained through the proposed algorithm are more than the normal wavelet thresholding methods. Hence shehnai sound is denoised with sym 3 wavelet at level 2 and $\sigma = 0.04$, dafli sound with dmey wavelet and flute sound with sym3 wavelet respectively due to maximum PSNR values. The different signals of the musical instrument sound are shown in Figs. 4, 5 and 6.

Table 1 PSNR values after decomposition of shehnai sound at level 2

Wavelet	$\sigma = 0.01$				$\sigma = 0.02$			
	T_h = soft		T_h = hard		T_h = soft		T_h = hard	
	WM	PM	WM	PM	WM	PM	WM	PM
haar	24.75	25.05	23.81	24.33	24.65	24.83	23.78	27.28
db10	35.32	35.63	35.05	36.58	37.75	39.77	38.42	43.72
sym3	31.23	31.55	30.39	31.27	31.26	31.6	34.91	35.13
coif 5	35.55	36.69	35.17	36.08	37.58	37.97	38.73	41.60
dmey	38.29	39.04	38.45	40.79	42.03	42.40	42.43	43.29
bior2.2	27.75	28.31	26.65	27.67	27.07	27.35	28.50	29.43
Wavelet	$\sigma = 0.03$				$\sigma = 0.04$			
	T_h = soft		T_h = hard		T_h = soft		T_h = hard	
	WM	PM	WM	PM	WM	PM	WM	PM
haar	24.36	24.55	27.60	27.61	24.59	25.12	26.54	26.65
db10	40.18	41.05	36.28	36.57	35.16	35.38	35.38	35.39
sym3	32.30	33.55	33.65	37.45	34.78	39.60	35.37	37.26
coif 5	42.20	43.42	42.06	42.11	35.58	35.16	34.95	35.18
dmey	37.70	37.90	37.54	37.68	33.26	34.26	33.34	34.37
bior2.2	27.31	27.52	28.11	28.59	29.38	29.53	28.66	29.44

Table 2 PSNR values after decomposition of dafli sound at level 2

Wavelet	$\sigma = 0.01$				$\sigma = 0.02$			
	T_h = soft		T_h = hard		T_h = soft		T_h = hard	
	WM	PM	WM	PM	WM	PM	WM	PM
haar	47.98	55.55	49.94	50.41	40.41	41.66	40.65	42.16
db10	49.32	50.37	45.44	45.90	38.36	38.56	39.59	40.03
sym 3	44.70	45.54	45.5	45.50	41.63	40.76	38.66	38.96
coif 5	45.25	46.84	46.09	47.01	41.21	42.07	39.97	40.30
dmey	47.65	47.39	45.26	46.24	39.99	40.19	39.38	40.53
bior2.2	55.87	52.29	47.47	47.58	38.77	38.72	39.68	43.52
Wavelet	$\sigma = 0.03$				$\sigma = 0.04$			
	T_h = soft		T_h = hard		T_h = soft		T_h = hard	
	WM	PM	WM	PM	WM	PM	WM	PM
haar	36.70	36.82	35.89	36.20	34.20	34.47	33.16	33.22
db10	36.00	36.26	36.58	36.67	32.96	33.10	33.22	33.64
sym 3	35.49	35.89	36.51	36.72	34.10	34.61	33.29	33.58
coif 5	36.42	36.58	37.38	37.70	33.02	33.69	33.57	33.93
dmey	35.54	35.61	36.25	38.13	33.48	34.29	33.72	34.95
bior2.2	36.20	36.14	35.89	35.74	33.22	34.18	33.23	34.40

Table 3 PSNR values after decomposition of flute sound at level 2

Wavelet	$\sigma = 0.01$				$\sigma = 0.02$			
	T_h = soft		T_h = hard		T_h = soft		T_h = hard	
	WM	PM	WM	PM	WM	PM	WM	PM
haar	24.11	24.87	34.10	32.70	23.54	24.59	34.11	35.83
db10	49.41	49.89	52.17	54.45	40.25	40.78	51.90	53.03
sym 3	38.41	44.14	39.92	42.99	41.07	42.03	41.02	43.19
coif 5	50.05	50.38	53.87	53.40	51.43	52.86	47.86	47.16
dmey	43.67	44.10	45.25	45.47	48.89	49.36	47.58	47.48
bior2.2	28.96	29.56	31.13	31.86	31.16	32.52	31.09	32.67
Wavelet	$\sigma = 0.03$				$\sigma = 0.04$			
	T_h = soft		T_h = hard		T_h = soft		T_h = hard	
	WM	PM	WM	PM	WM	PM	WM	PM
haar	24.11	24.04	22.75	22.34	22.31	22.63	22.31	22.39
db10	37.05	38.21	36.52	37.89	35.45	36.16	34.65	37.17
sym 3	46.99	47.15	46.18	46.37	37.93	37.99	35.99	40.27
coif 5	36.61	36.89	36.31	41.07	34.24	34.82	35.44	35.48
dmey	36.82	37.24	36.97	36.01	34.17	34.53	34.50	35.22
bior2.2	29.21	33.57	28.92	29.61	30.02	31.24	29.73	30.43

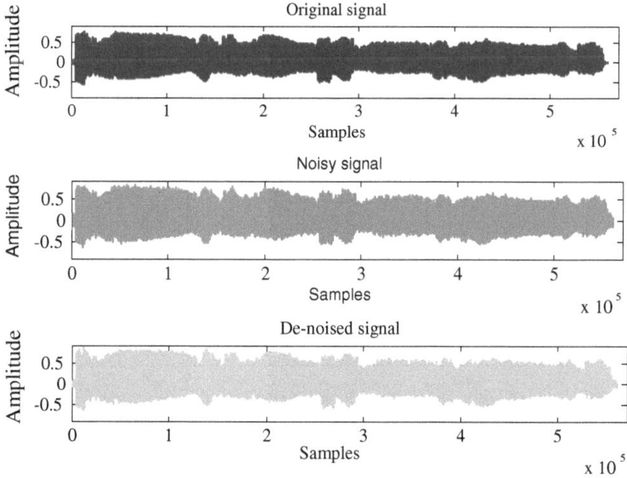

Fig. 4 Original, noisy and de-noised shehnai signal with wavelet sym3 at level 2

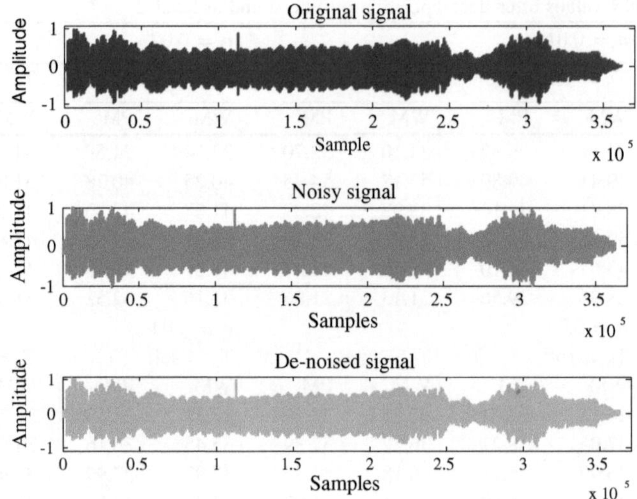

Fig. 5 Original, noisy and de-noised flute signal with wavelet sym3 at level 2

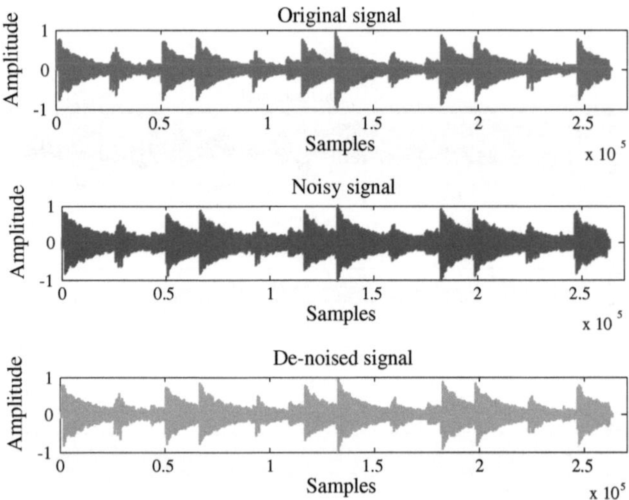

Fig. 6 Original, noisy and de-noised dafli signal with wavelet dmey at level 2

5 Conclusions

In this paper, wavelet transform is used for de-noising shehnai, dafli and flute sound signal corrupted with white Gaussian noise. Firstly, sound signal is divided into multiple blocks then de-noising of each block is performed in wavelet domain by thresholding the wavelet coefficients. When each block is de-noised, then all the blocks are concatenated to form the final de-noised signal. It is observed that when modified threshold is used, the PSNR values obtained are more than the normal wavelet thresholding method. The signal is decomposed till the second level and the de-noised signal gives better hearing perception. During the analysis it is also observed that the hard thresholding is better than the soft thresholding sometimes because the PSNR values obtained are more in case of hard thresholding. When we divide the signal into blocks then the wavelet coefficients contributing to the noise effect are heavily reduced due to which the hearing perception of the de-noised signal is improved. Higher thresholds remove the noise well but some part of the original signal is also removed because it is not possible to remove the noise without affecting the original signal.

References

1. Sharma R, Pyara VP (2012) A novel approach to synthesize sounds of some Indian musical instruments using DWT. Int J Comput Appl 45(13):19–22
2. Sharma R, Pyara VP (2012) A comparative analysis of mean square error adaptive filter algorithms for generation of modified scaling and wavelet function. Int J Eng Sci Technol 4(4):1396–1401
3. Donoho DL (1993) Unconditional bases are optimal bases for data and for statistical estimation. Appl Comput Harmonic Anal 1:100–115
4. Donoho DL, Johnstone IM (1995) Adapting to unknown smoothness via wavelet shrinkage. J Am Stat Assoc 90:1200–1224
5. Donoho DL, Johnstone IM, Kerkyacharian G, Picard D (1995) Wavelet shrinkage: asymptopia. J Roy Stat Soc 57(2):301–369
6. Cho N, Kuo CJ (2010) Sparse representation of musical signals using source-specific dictionaries. IEEE Trans, 3rd edn. Academic Press
7. Serrano EP (2002) Fabio.: application of the wavelet transform to acoustic emission signals processing. IEEE Sig Process Lett 17(11):913–916
8. Pan P, Zhang L, Dai G, Zhang H (1999) Two denoising methods by wavelet transform. IEEE Trans Sig Process 47(12):3401–3406
9. Aggarwal R et al (2011) Noise reduction of speech signal using wavelet transform with modified universal threshold. IJCA 20(5):14–19
10. Balakrishnan V, Borges N, Parchment L (2008) Wavelet denoising and speech enhancement. IEEE Conf Image Process 1–5
11. Gultekin O, Erer I, Kaplan M (2009) Empirical mode decomposition based de-noising for high resolution direction of arrival estimation. EUSIPCO 1983–1986
12. Chen Y (2009) Realization of wavelet soft threshold de-noising technology based on visual instrument. IEEE Int Jt Conf Artif Intell 849–852
13. Chang SG, Yu B, Vetterli M (2000) Adaptive wavelet thresholding for image denoising and compression. IEEE Trans Image Process 9:1532–1546
14. Berger J, Coifman RR, Maxim JG (1994) Removing noise from music using local trigonometric bases and wavelet packets. J Audio Eng Soc 42(10):808–818
15. Johnson MT, Yuan X, Ren Y (2007) Speech signal enhancement through adaptive wavelet thresholding. Speech Commun 49(10):123–133

Nonlocal Total Variation Based Speckle Denoising Model

Arundhati Bagchi Misra and Hyeona Lim

Abstract A large range of methods covering various fields of mathematics are available for denoising an image. The initial denoising models are derived from energy minimization using nonlinear partial differential equations (PDEs). The filtering models based on smoothing operators have also been used for denoising. Among them the recently developed nonlocal means method proposed by Buades, Coll and Morel in 2005 is quite successful. Though the method is very accurate, it is very slow and hence quite impractical. In 2008, Gilboa and Osher extended some known PDE and variational techniques in image processing to the nonlocal framework and proposed the nonlocal total variation method for Gaussian noise. We used this idea to develop a nonlocal model for speckle noise. Here we have extended the speckle model introduced by Krissian et al. in 2005 to the nonlocal framework. The Split Bregman scheme is used solve this new model.

Keywords Image denoising · Speckle denoising models · Nonlocal means · Nonlocal PDE · Nonlocal TV

A. B. Misra (✉)
Department of Mathematical Sciences, Saginaw Valley State University, University Center, MI 48710, USA
e-mail: abmisra@svsu.edu

H. Lim
Department of Mathematics and Statistics, Mississippi State University, Mississippi, MS 39762, USA
e-mail: hlim@math.msstate.edu

Mohan S. and S. S. Kumar (eds.), *Proceedings of the Fourth International Conference on Signal and Image Processing 2012 (ICSIP 2012)*, Lecture Notes in Electrical Engineering 221, DOI: 10.1007/978-81-322-0997-3_46,

1 Introduction

Image restoration is a very important process and is often necessary as a pre-processing for other imaging techniques such as segmentation and compression. In general, an observed image f is represented by the equation

$$f = u + n, \tag{1}$$

where u is the original noise free image, f is the observed noisy image and n is the Gaussian noise. Here $u,f : \Omega \subset \mathrm{IR}^2 \rightarrow \mathrm{IR}$. For any denoising model, the main objective is to reconstruct u from an observed image f.

In 1992 Rudin et al. [8] proposed the total variation (TV) denoising model as the minimization problem:

$$\min_u \int_\Omega |\nabla u| d\mathbf{x} \tag{2}$$

subject to the constraints,

$$\int_\Omega f d\,\mathbf{x} = \int_\Omega u\, d\mathbf{x} \tag{3}$$

$$\text{and} \int_\Omega \frac{1}{2}(f - u)^2 d\mathbf{x} = \sigma^2, \tag{4}$$

where σ is the standard deviation of the noise n. These constraints ensure that the resulting image and the observed images are quite close to each other.

Combining the above constraints the TV functional is obtained by

$$F(u) = \int_\Omega |\nabla u| d\mathbf{x} + \frac{\lambda}{2} \int_\Omega (f - u)^2 d\mathbf{x}. \tag{5}$$

Here λ is a constraint parameter. The equivalent Euler–Lagrange equation gives the TV denoising model as

$$\frac{\partial u}{\partial t} - \nabla \cdot \left(\frac{\nabla u}{|\nabla u|} \right) = \lambda(f - u). \tag{6}$$

To avoid singularities, it was regularized by using $|\nabla u| \approx |\nabla^{\varepsilon} u| = (u_x^2 + u_y^2 + \varepsilon^2)^{1/2}$.

2 Preliminaries

Here we present the initial speckle denoising model and the filtering based nonlocal means method.

2.1 Speckle Denoising Model

In 2005 Krissian et al. [6] considered the speckle noise model by

$$f = u + \sqrt{u}n, \tag{7}$$

where u is the desired image to find, n is Gaussian noise and f denotes the observed image. Thus, we have $n = \frac{f-u}{\sqrt{u}}$. Then replacing $f - u$ in (5) with the new expression for n, the minimization functional for speckle denoising is given by

$$F(u) = \int_{\Omega} \left[|\nabla u| + \frac{\lambda}{2} \left(\frac{f-u}{\sqrt{u}} \right)^2 \right] d\mathbf{x}. \tag{8}$$

From energy minimization of this functional, the TV-based speckle denoising model can be derived as:

$$\frac{\partial u}{\partial t} - \frac{u^2}{f+u} |\nabla^{\varepsilon} u| \nabla \cdot \left(\frac{\nabla u}{|\nabla^{\varepsilon} u|} \right) = \lambda |\nabla^{\varepsilon} u| (f - u). \tag{9}$$

2.2 Nonlocal Means Method

In 2005 Buades et al. proposed the new state of art image denoising algorithm known as nonlocal means algorithm [2]. The algorithm is given by the formula

$$NL[u](\mathbf{x}) = \frac{1}{C(\mathbf{x})} \int_{\Omega} e^{-\frac{(G_a \star |u(\mathbf{x}+\cdot) - u(\mathbf{y}+\cdot)|^2)(0)}{h^2}} u(\mathbf{y}) dy, \tag{10}$$

where $C(\mathbf{x}) = \int_{\Omega} e^{-\frac{(G_a \star |u(\mathbf{x}+\cdot) - u(\mathbf{z}+\cdot)|^2)(0)}{h^2}} dz$. Here G_a is the Gaussian kernel with standard deviation a, h is a filtering parameter and $u(\mathbf{x} + \cdot)$ denotes the neighborhood of the pixel $\mathbf{x}$. Here each pixel value is denoised using the weighted average of all the pixels in the image. Thus for a given discrete noisy image $u = \{u(i) : i \in I\}$, the estimated value $NL[u](i)$, for a pixel i, is computed as

$$NL[u](i) = \sum_{j \in I} w(i,j)u(j), \tag{11}$$

where the weight $w(i,j)$ depends on the similarity of the i and j pixels and satisfies the conditions $0 \leq w(i,j) \leq 1$ and $\sum_j w(i,j) = 1$. In this paper the authors measured the similarity of a square neighborhood $\mathcal{N}_i$ of a fixed size at each pixel i as a decreasing function of the weighted Euclidean distance, $||u(\mathcal{N}_i) - u(\mathcal{N}_j)||^2_{2,a}$, where $j \in I$ and $a > 0$ is the standard deviation of the Gaussian kernel. The weight function is defined as

$$w(i,j) = \frac{1}{C(i)} e^{-\frac{||u(\mathcal{N}_i)-u(\mathcal{N}_j)||^2_{2,a}}{h^2}}, \tag{12}$$

where $C(i)$ is the normalizing constant given by

$$C(i) = \sum_j e^{-\frac{||u(\mathcal{N}_i)-u(\mathcal{N}_j)||^2_{2,a}}{h^2}}. \tag{13}$$

The similar neighborhoods have a very small Euclidean distance which in turn results in larger weight. Since the nonlocal means uses the similarity neighborhood for denoising, it works really well for periodic or patterned textured cases. But though the method is very good in removing noise, it is very slow and hence quite impractical.

2.3 Nonlocal TV Model

In 2008, Gilboa and Osher extended some known PDE and variational techniques in image processing to the nonlocal framework [3]. The motivation behind this was to make any point interact with any other point in the image. Since the classical derivatives were local operators, they had to redefine the required operators in [3] following the ideas of Zhou and Scholkopf [10, 11]. Gilboa and Osher proposed the nonlocal TV (NLTV) model based on the NLTV operator ($\int_\Omega |\nabla_{\mathrm{NL}} u|$), as follows:

$$\begin{aligned} \frac{\partial u}{\partial t} &= \kappa_{\mathrm{NL}}(u) - \lambda(u - f) \\ &= \int_\Omega w(x,y)(u(y) - u(x))(|\nabla_{\mathrm{NL}} u|^{-1}(x) + |\nabla_{\mathrm{NL}} u|^{-1}(y))dy - \lambda(u - f), \end{aligned} \tag{14}$$

where $\nabla_{\mathrm{NL}}(\cdot) =$ nonlocal gradient, $div_{\mathrm{NL}}(\cdot) =$ nonlocal divergence and $\kappa_{\mathrm{NL}}(\cdot) = div_{\mathrm{NL}} \frac{\nabla_{\mathrm{NL}}}{|\nabla_{\mathrm{NL}}|}$ denotes the nonlocal curvature. Since the steepest descent scheme (14) is very slow, a faster numerical scheme was introduced by Bresson in [1]. The scheme is based on the Split Bregman method introduced in [4]. It was proved to be very fast for regular TV.

3 Nonlocal TV Based Speckle Denoising Model

We now present our new model for speckle denoising. We extend the idea presented in Sect. 2.3 and develop the following nonlocal TV based speckle denoising model:

$$\min_u F(u), \quad F(u) = \int_\Omega \left[|\nabla_{\mathrm{NL}} u| + \lambda \left(\frac{f-u}{\sqrt{u}} \right)^2 \right] d\mathbf{x} \tag{15}$$

where ∇_{NL} is a nonlocal gradient defined in [3]. For faster computation, we adopt Split Bregman scheme [4] for finding a solution to our new model (15). We first develop in Sect. 3.1 the Split Bregman scheme for Krissian et al. [6].

3.1 Split Bregman Scheme for Krissian et al. Model

As discussed in Sect. 2.1, the minimization functional was given by (8). For Split Bregman scheme, we introduce $d = \nabla u$ and construct the unconstrained functional as

$$\min_u F(u), \quad F(u) = \int_\Omega |d| + \lambda \frac{(f-u)^2}{u} d\mathbf{x} + \frac{\beta}{2} \|d - \nabla u - b\|_2^2. \tag{16}$$

Here β is a penalty parameter. We can now split this in two subproblems of u and d given as:

$$\min_u \int_\Omega \lambda \frac{(f-u)^2}{u} d\mathbf{x} + \frac{\beta}{2} \|d - \nabla u - b\|_2^2, \tag{17}$$

$$\min_d \int_\Omega |d| d\mathbf{x} + \frac{\beta}{2} \|d - \nabla u - b\|_2^2. \tag{18}$$

Now,

$$\begin{aligned}\frac{\partial}{\partial u}\left(\lambda\frac{(f-u)^2}{u}\right) &= \lambda\left[\frac{-2u(f-u)-(f-u)^2}{u^2}\right]\\ &= \lambda\left[\frac{(u-f)(2u-u+f)}{u^2}\right]\\ &= \frac{\lambda}{u^2}(u-f)(u+f).\end{aligned} \tag{19}$$

Therefore the optimality condition for u in (17) gives us

$$\begin{aligned}&\frac{\lambda(u+f)}{u^2}(u-f)+\beta\div(d-\nabla u-b)=0\\ \Rightarrow &\left(\lambda-\beta\frac{u^2}{u+f}\Delta\right)u=\lambda f-\beta\frac{u^2}{u+f}\div(d-b).\end{aligned} \tag{20}$$

The optimality condition can be discretized using the definition of discrete gradient, divergence and Laplacian as discussed in [3]. This gives

$$\begin{aligned}\left(\lambda+4\beta\frac{u_{ij}^2}{u_{ij}+f_{ij}}\right)u_{ij} = \lambda f_{ij}+\beta\frac{u_{ij}^2}{u_{ij}+f_{ij}}\Big(&u_{i+1,j}+u_{i-1,j}+u_{i,j+1}+u_{i,j-1}\\ &+d_{x,i-1,j}-d_{x,ij}+d_{y,i,j-1}-d_{y,ij}\\ &-b_{x,i-1,j}+b_{x,ij}-b_{y,i,j-1}+b_{y,ij}\Big).\end{aligned} \tag{21}$$

Thus, we get

$$\begin{aligned}u_{ij}^{k+1} = \frac{1}{\left(\lambda+4\beta\frac{u_{ij}^{k\,2}}{u_{ij}^k+f_{ij}}\right)}&\left[\lambda f_{ij}+\beta\frac{u_{ij}^{k\,2}}{u_{ij}^k+f_{ij}}\Big(u_{i+1,j}^k+u_{i-1,j}^k+u_{i,j+1}^k+u_{i,j-1}^k\right.\\ &\left.+d_{x,i-1,j}^k-d_{x,ij}^k+d_{y,i,j-1}^k-d_{y,ij}^k-b_{x,i-1,j}^k+b_{x,ij}^k-b_{y,i,j-1}^k+b_{y,ij}^k\Big)\right].\end{aligned} \tag{22}$$

Since there is no coupling between elements of d, the optimal value of d is computed using the shrinkage operator described in [7, 9]:

$$d_{ij}^{k+1} = \frac{\nabla u_{ij}^{k+1}+b_{ij}^k}{|\nabla u_{ij}^{k+1}+b_{ij}^k|}\max\{|\nabla u_{ij}^{k+1}+b_{ij}^k|-1/\beta,0\}. \tag{23}$$

The variable b is initialized to zero and is updated after each Bregman iteration as:

$$b_{ij}^{k+1} = b_{ij}^{k} + \nabla u_{ij}^{k+1} - d_{ij}^{k+1}. \tag{24}$$

The scheme (22)–(24) provides a much faster solution for the TV based speckle denoising model introduced by Krissian et al. [6].

3.2 Split Bregman scheme for nonlocal TV speckle model

For our new model (15) the Split Bregman functional will be of the form

$$\min_u F(u), \quad F(u) = \int_\Omega |d| + \lambda \frac{(f-u)^2}{u} + \frac{\beta}{2}\|d - \nabla_{\mathrm{NL}} u - b\|_2^2. \tag{25}$$

Then the optimality condition for u gives us the equation

$$\left(\lambda - \beta \frac{u^2}{u+f} \Delta_{\mathrm{NL}}\right) u = \lambda f - \beta \frac{u^2}{u+f} div_{\mathrm{NL}}(d-b). \tag{26}$$

Denoting the discretized points $x, y \in \Omega \times \Omega$ by i and j, and using the discrete definition for div_{NL} and $\Delta_{badhbox}$ [1, 3], the discrete minimization scheme is given as:

$$\begin{aligned} &\lambda u_i - \beta \frac{u_i^2}{u_i + f_i}\left(\sum_j w_{ij}(u_j - u_i)\right) \\ &= \lambda f_i - \beta \frac{u_i^2}{u_i + f_i}\left(\sum_j \sqrt{w_{ij}}(d_{ij} - d_{ji} - b_{ij} + b_{ji})\right). \end{aligned} \tag{27}$$

Hence, the iteration steps are:

$$\begin{aligned} u_i^{k+1} = &\frac{1}{\lambda + \beta \frac{u_i^2}{u_i+f_i}\sum_j w_{ij}} \left[\beta \frac{u_i^2}{u_i+f_i} \sum_j w_{ij} u_j^k + \lambda f_i \right. \\ &\left. - \beta \frac{u_i^2}{u_i+f_i}\left(\sum_j \sqrt{w_{ij}}(d_{ij} - d_{ji} - b_{ij} + b_{ji})\right)\right], \end{aligned} \tag{28}$$

Table 1 Summary of numerical results

	Krissian et al.		Nonlocal means		NLTV	
Images	Time (s)	PSNR	Time (s)	PSNR	Time (s)	PSNR
Lenna (*PSNR* = 24.51)	0.72	26.39	11.46	28.21	4.67	29.10
Gallstones	0.14	–	11.92	–	4.62	–

$$d_{ij}^{k+1} = \frac{\sqrt{w_{ij}}(u_j^{k+1} - u_i^{k+1}) + b_{ij}^k}{\sqrt{\sum_j w_{ij}(u_j^{k+1} - u_i^{k+1})^2 + b_{ij}^{k\,2}}} \max\left(\sqrt{\sum_j w_{ij}(u_j^{k+1} - u_i^{k+1})^2 + b_{ij}^{k\,2}} - \frac{1}{\beta}, 0\right), \tag{29}$$

$$b_{ij}^{k+1} = b_{ij}^k + \sqrt{w_{ij}}(u_j^{k+1} - u_i^{k+1}) - d_{ij}^{k+1}. \tag{30}$$

(28–30) provides a fast and accurate nonlocal scheme for TV based speckle denoising model.

4 Numerical Results

The numerical results are displayed in the next pages. We have considered the images of size 128 $\times$ 128 and the summary of results are in Table 1. Other than the visual results, we have used peak-signal-to-noise ratio (PSNR) for synthetic images, which was mentioned in [5], to measure the efficiency of the models.A cleaner image provides a higher PSNR value. We have provided the absolute residual ($|f - u|$) and the speckle noise residual $\left(n = \frac{f-u}{\sqrt{u}}\right)$ for all images. It is evident that the new nonlocal speckle model works best among the models compared here. For this model only 1–2 iterations were enough for all images. The Krissian et al. is not denoising very well compared to the new one. The residual shows presence of edges and fine textures due to the blurring effect of TV. But the Krissian et al. is certainly the fastest one. The nonlocal means is not working well for fine textures and slower than the new model. It is also the slowest of all. Nonlocal speckle model maintains texture and removes noise very well. The noise residuals show that nonlocal speckle model has picked up more noise compared to the other ones.

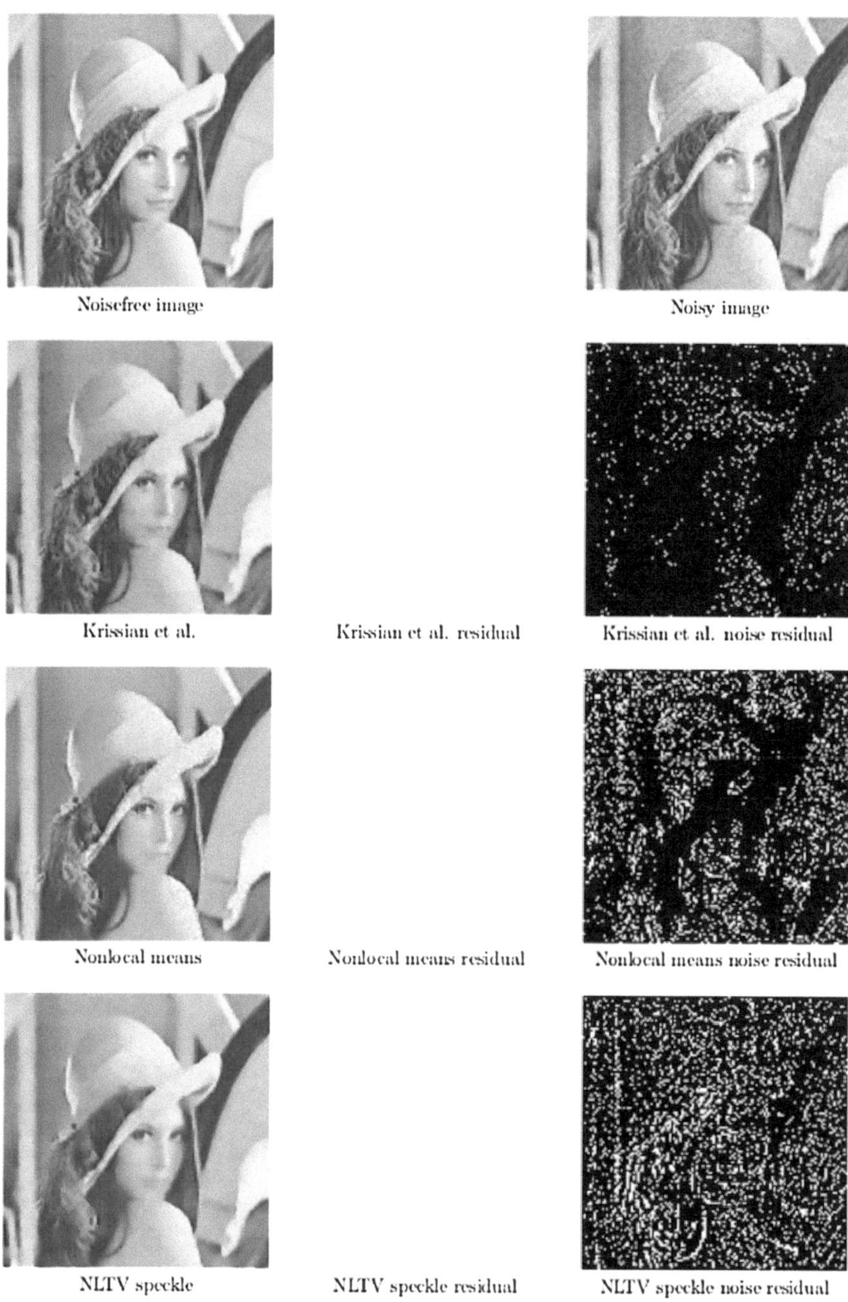

Synthetic image result (Lenna)

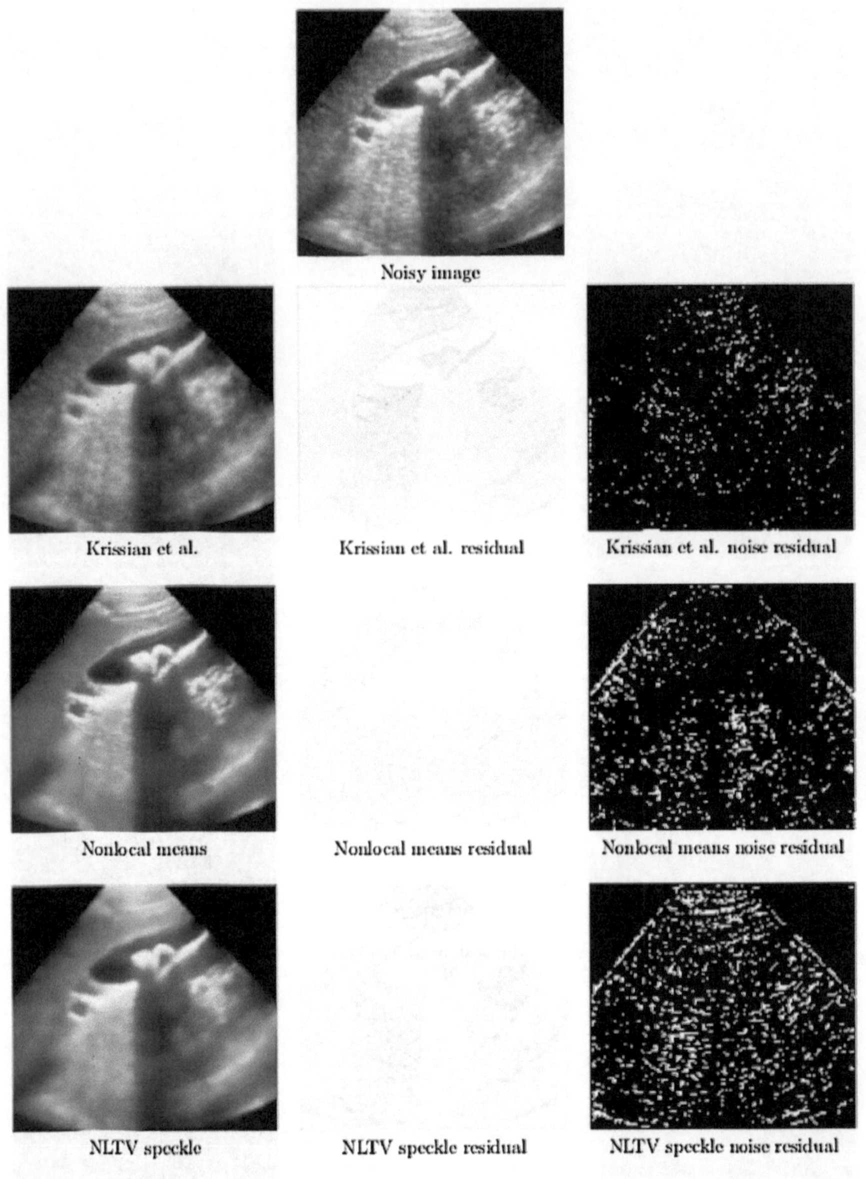

Ultrasound image result (Gallstones)

5 Conclusion

In this paper we have incorporated the positive aspects of both PDE and filtering models. The PDE based TV model is very fast but has a blurring effect. Nonlocal means has very effective denoising properties but it is very slow. Therefore, have developed an accurate model using both of them. We have extended the existing Krissian et al. speckle denoising model [6] to a nonlocal framework, and provided an efficient nonlocal TV based speckle denoising model. We have also developed a numerical scheme for solving this new model. The numerical scheme is based on the Split Bregman method introduced by Gilboa and Osher. The results show the new model is more accurate than both the Krissian et al and nonlocal means model. It is also much faster than nonlocal means.

References

1. Bresson X (2009) A short note for nonlocal tv minimization. Technical report
2. Buades A, Coll B, Morel JM (2005) A non-local algorithm for image denoising. In: IEEE computer society conference on computer vision and pattern recognition, CVPR 2005, vol 2, pp 60–65. doi:10.1109/CVPR.2005.38
3. Gilboa G, Osher S (2008) Nonlocal operators with applications to image processing. Multiscale Model Simul 7(3):1005–1028. http://dblp.uni-trier.de/db/journals/mmas/mmas7.html#GilboaO08
4. Goldstein T, Osher S (2009) The split bregman method for l1-regularized problems. SIAM J Imaging Sci 2:323–343. doi:10.1137/080725891, http://portal.acm.org/citation.cfm?id=1658384.1658386
5. Kim S, Lim H (2007) A non-convex diffusion model for simultaneous image denoising and edge enhancement. Electron J Differ Equ Conference 15:175–192
6. Krissian K, Kikinis R, Westin CF, Vosburgh K (2005) Speckle-constrained filtering of ultrasound images. In: CVPR '05: Proceedings of the 2005 IEEE computer society conference on computer vision and pattern recognition (CVPR'05), vol 2. IEEE Computer Society, Washington, DC, pp 547–552. doi:10.1109/CVPR.2005.331
7. Li C (2003) An efficient algorithm for total variation regularization with applications to the single pixel camera and compressive sensing. Master's thesis, Rice University, Houston, Texas
8. Rudin LI, Osher S, Fatemi E (1992) Nonlinear total variation based noise removal algorithms. Physica D: Nonlinear Phenom 60(1–4):259–268. doi:10.1016/0167-2789(92)90242-F
9. Wang Y, Yin W, Zhang Y (2007) A fast algorithm for image deblurring with total variation regularization. Technical report, Department of Computational and Applied Mathematics, Rice University, Houston, Texas
10. Zhou D, Schlkopf B (2004) A regularization framework for learning from graph data. In: ICML workshop on statistical relational learning, pp 132–137
11. Zhou D, Schlkopf B (2005) Regularization on discrete spaces. In: Pattern recognition. Springer, Heidelberg, pp 361–368

Foot Motion Recognition for Human–Computer Interaction

Hyun-Min Choi, Yeong-Seob Jeong, Seung-Jun Son and Seung Eun Lee

Abstract In this paper, we introduce a foot motion recognition system for the human–computer interaction. Our system consists of five pressure sensors in a shoe, 3-axis acceleration sensor in the front of the shoe, and signal processing unit on FPGA. Bluetooth establishes the wireless communication channel between our system and a host PC. Experimental results demonstrate the feasibility of our proposal for human–computer interaction, by providing full control of the host PC though the foot motion recognition.

Keywords Human–computer interaction · User experience

1 Introduction

Over the last decade, smart interface becomes a major issue that is how to support more natural and immersive user interface. The user interface has evolved from simple devices such as a keyboard and a mouse, to smart equipments employing gesture recognition, motion tracking, and speech recognition. Among the devices, a mouse is the most popular one. However, many people suffer pain in their hands to operate computer or machine by using a mouse. These people also suffer pain in their arm. It finally causes a new syndrome called "the mouse elbow" that people who play game suddenly suffer pain in the fingers and then massage their hand unwittingly. In this case, it can be considered as Repetitive strain injury (RSI). RSI

H.-M. Choi · Y.-S. Jeong · S.-J. Son · S. E. Lee (✉)
Department of Electronic Engineering, Seoul National University of Science and Technology, Seoul, South Korea
e-mail: seung.lee@seoultech.ac.kr

Mohan S. and S. S. Kumar (eds.), *Proceedings of the Fourth International Conference on Signal and Image Processing 2012 (ICSIP 2012)*, Lecture Notes in Electrical Engineering 221, DOI: 10.1007/978-81-322-0997-3_47,

is an injury of the musculoskeletal and nervous systems that may be caused by repetitive tasks, forceful exertions, vibrations, mechanical compression pressing against hard surfaces, or sustained. The types of RSIs that affect computer users may include non-specific arm pain or work related upper limb disorder. So there is a growing need to introduce alternative mouse in preparation for people who have difficult to use their hands to operate the devices. Recently, wearable alternative devices were introduced [1, 2]. In addition, systems to capture the human movements adopting a shoe were proposed [3–5].

In this paper, we propose a foot motion recognition system for the human–computer interaction. Bluetooth establishes the communication channel between the proposed sensing equipment and a host PC. Experimental results demonstrate the feasibility of our proposal for a alternative human–computer interaction, by providing full control of the host PC through the foot motion recognition. Our system is designed to facilitate the computer especially for people who have difficulty in using their hands to operate devices. We expect that portable, wearable, and compact features of our system could help the people who have a problem in their hands and make efforts to promote their handicapped welfare.

The reset of paper is organized as following: Sect. 2 describes our system overview. Section 3 shows the experimental results, verifying the functionality of our system. We conclude in Sect. 4 by outlining the direction for the future work on this topic.

2 Overview

The proposed system for the foot motion recognition utilizes five pressure sensors, one acceleration sensor, signal processing unit on FPGA, Bluetooth communication module, and a host PC (See Fig. 1). First, the five inputs coming from the FSR pressure sensor that are attached to the soles of shoes enter the amplifier. The amplified five analog signals and three axis acceleration sensor x, y, z inputs go into the external AD converter. In other words, a total of eight analog signals enter the external A/D converter and they are converted to 8bit digital signals. After the 8 bit signals go into the FPGA, they are forwarded to a RS232 serial port at a speed of 9,600bps. The data values are sent with Bluetooth communication and monitored on the Host PC. A Labview program in the Host PC provides the way for user to control the mouse using the foot.

3 Experimental Results

In order to verify the functionality of our system, we established the experimental environment which connects to a Host PC. We extracted the received data and verified that the mouse cursor is rightly moved in accordance with the date value.

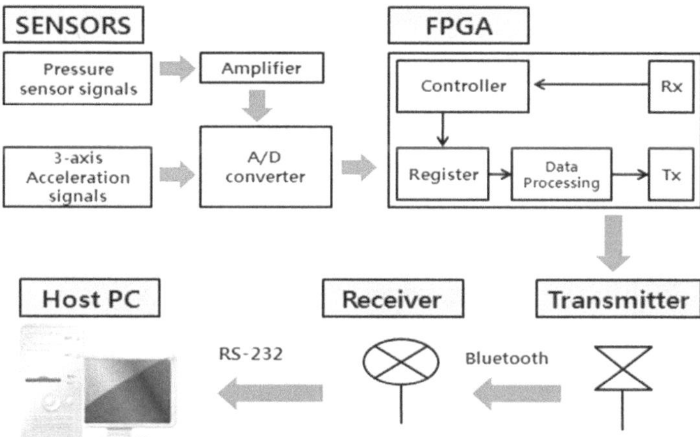

Fig. 1 System flow of foot motion recognition system

3.1 Cursor Movement Recognition

Positions of the five force sensor embedded in a shoe were selected by experiment using molding technique. The force sensors are installed at the toe, left side of the sole, right side of the sole and the heel. When a user presses each of the force sensor located in sole, IPFM system enables the motion of the mouse-cursor and force intensity is shown in the real-time grape on the screen. Figure 2 displays a sample output waveform of the each force data when the user uses the IPFM system on computer. Each part of performance property in the IPFM is as in the following. If the part in the front of shoe is pressed, the IPFM makes the mouse-cursor go up. If the part on the left side of the shoe is pressed, the mouse-cursor will be moved to the left. The other performance property of going right and down are based on the same principle as above. Our system has absolutely no noises at all because our system does not need to drag the shoes.

3.2 Click Recognition

Accelerometer at the top of the shoe is adapted for a single click. The accelerometer sensor can output x-axis acceleration, y-axis acceleration and z-axis acceleration. Among these outputs, we choose the z-axis acceleration for click function, because it reacted sensitively among the outputs of acceleration sensor when we shook the foot in the air and it was easy to control. Figure 3 displays a sample output wave form of z-axis acceleration when the shoe shook up and down. If the acceleration is bigger than the threshold value (200), the IPFM system has the click function of the left button.

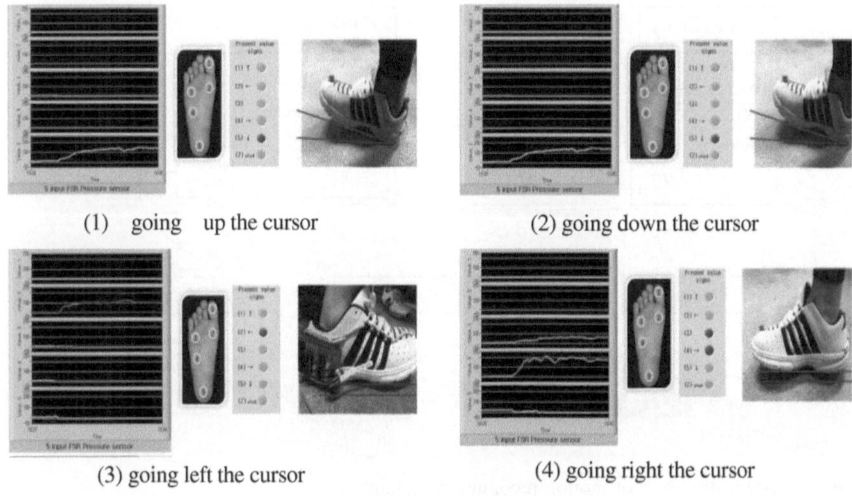

Fig. 2 The acquired signals from sensors according to the motion

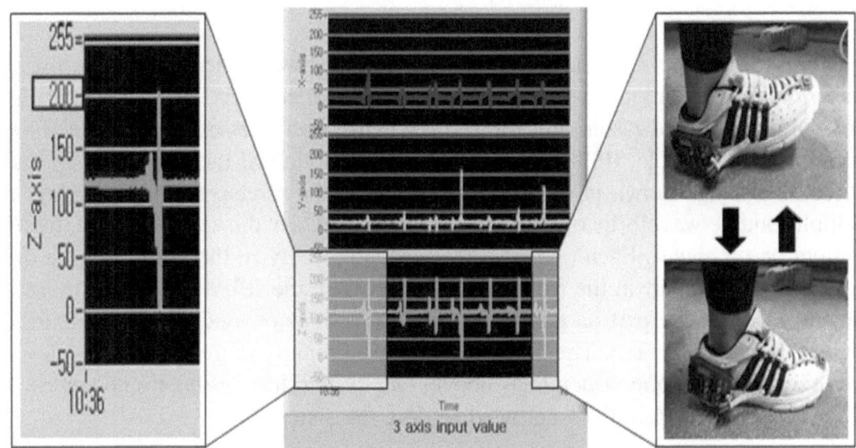

Fig. 3 The acquired signals from sensors when user shook the shoe

4 Conclusions

In this paper, we presented a wearable sensor shoes for a mouse that is operated without hands. The data from the shoe is transmitted to Host PC and make mouse move on the screen, establishing control channel between shoe and the computer.

Our system has advantage that does not make any noise even when the function of click takes effects. Also it is portable, light and easy to use due to its simple usage. Thus, it enables the system to be used regardless location and age. Also this system is economical and suitable for everyday life because it embeds pressure sensors in a pair of shoes which allows users to wear them without any interference.

Acknowledgments This study was financially supported by the Seoul National University of Science and Technology, Seoul, Korea.

References

1. Kim BS, Park M, Anumas S, Yoo J (2010) Head mouse system based on gyro and opto sensors. In: Proceedings of the 3rd international conference on biomedical engineering and informatics, Oct 16–18
2. Savioz G, Markovic M, Perriard Y (2011) Towards multi-finger haptic devices: a computer keyboard with adjustable force feedback. In: Proceedings of international conference of electrical machines and systems, pp 1–6
3. Vel'azquez R, Baz'an O (2009) Marco magana: a shoe-integrated tactile display for directional navigation. Intell Robots Syst 1235–1240
4. Fujimoto M, Fujita N, Takegawa Y, Terada T, Tsukamoto M (2009) A motion recognition method for a wearable dancing musical instrument. International symposium on wearable computers, pp 11–18
5. Rasool B, Masoomeh A (2009) A power generating system for mobile electronic devices using human walking motion. International conference on computer and electrical engineering, pp 385–388

Content Based Image Retrieval Using Self Organizing Map

Purohit Shrinivasacharya and M. V. Sudhamani

Abstract From the recent literature it is observed that information storage and retrieval through the Internet has made impressive progress. Practical searching for information still confronts us with retrieval systems that are present. A Content Based Image Retrieval (CBIR) system provides an efficient way of retrieving related images from image collections. In this paper we present a new feature extraction techniques and clustering of the features to achieve better performance in image retrieval system. The proposed method uses an approach which combines edge information and median filtering technique to extract the features from the image. Self Organizing Map (SOM) technique is used for clustering the extracted image features. The median filtering technique is applied to the original image to get a smooth image. The edge information can be extracted from the image using Bi-directional Empirical Mode Decomposition (BEMD) technique. Then replace only the values of edge position of smooth image with the detected edge image values by BEMD and extracted only 64 bins gray features. These extracted features are supplied as input to the SOM neural network for clustering where features are clustered into nine different groups. Finally query image features are feed to the neural network to identify the cluster to which the query image belongs. The surrounded clustered features are compared with the query image features and display the similar resultant images. The experiment is carried out on a ground truth database which has 1000 images of different categories. The experimental results have been compared with the conventional Median filter histogram technique. Here performance of the retrieval system is good because of combination of

P. Shrinivasacharya (✉)
Department of ISE, Siddaganga Institute of Technology, 03 Tumkur, India
e-mail: purohitsn@gmail.com

M. V. Sudhamani
Department of ISE, RNS Institute of Technology, 61 Bengaluru, Karnataka
e-mail: mvsudha_raj@hotmail.com

Mohan S. and S. S. Kumar (eds.), *Proceedings of the Fourth International Conference on Signal and Image Processing 2012 (ICSIP 2012)*, Lecture Notes in Electrical Engineering 221, DOI: 10.1007/978-81-322-0997-3_48, © Springer India 2013

median, edge and SOM techniques. It gives an average precision 2.37 % and recall 2.82 % improvement compared with an existing system.

Keywords CBIR · BEMD · Indexing · Image database · Histogram · SOM

1 Introduction

In the past few years, there has been enormous amount of images which are being added every minute to the World Wide Web (WWW). As it is the need for effective and efficient image retrieval system. Retrieving an image having some characteristics in a big database is a crucial task. Searching for an image among a collection of images can be done by different approaches. Text based search uses the surrounding information of the images like image name, the web page surrounding text etc. and it requires humans to personally describe every image information in the database. The limitations of the text based search are moving the trend towards the CBIR technique. The CBIR uses some features of the image content instead of surrounding text for searching the image. The main intention of CBIR technique is to automatically retrieve using some image features from the image database. This is very impractical for large amount of image databases. It is possible to miss images which use different synonyms in their descriptions.

A CBIR is an alternative or complementary method for the textual indexing search. CBIR is an important part of multimedia information retrieval, while image feature extraction and expression is the basement of CBIR. The process of retrieving images from a database on the basis of features that are extracted from the images themselves is called CBIR. This paper presents a CBIR system which accepts a query image as input and relevant images are retrieved based on the similarity of the features of the query image and features of the individual images stored in database. The proposed method uses the edge using BEMD [1, 2], Median filtering [3] histogram [4] and SOM [5, 6] clustering techniques to build new system. The proposed system uses the different category images. Ten categories have been chosen for experiment to show that the proposed method performance is good and the categories are as shown in the following Fig. 1.

The detailed description of the work will be organized as follows: Sect. 2 describes the general clustering method. The proposed CBIR architecture and different phase of the architecture are presented in Sect. 3. Section 4 gives the details of feature extraction and clustering using edge detection using BEMD and training the network using batch algorithm for SOM. Experimental results and graphs are presented in Sect. 5. Finally, Summary and conclusions of this study and future work are mentioned in Sect. 6.

Fig. 1 Sample database of CBIR system

2 Clustering

A clustering 'P' is a partitioning of a data set into a set of clusters P_i, $i = 1, 2, \ldots, m$ in such a way that each data sample exactly belongs to one cluster. The clustering can be carried out using two-level approach, where the data set clustered using the SOM. The important benefit of SOM is that computational load decreases considerably, building it is possible to cluster large amount data sets and to consider several type of preprocessing strategies in a limited time.

Self organizing maps were developed by Kohonen [5] in 1980. It is a type of artificial neural network that is trained with unsupervised learning to produce a low dimensional discretized demonstration of the input space of training samples, called a map. The SOM consists of small working components called neurons. Associated with each neuron is a weight vector of the same dimension as the input information vectors and a position in the map space. The weights of the neurons are initialized either to random values or sampled lightly from the subspace spanned by the two largest principal component of eigenvectors. Internally SOM uses the Euclidean distance metric to make the cluster. The main features of the SOM are

- Clustering of high dimensional data
- Resulting clusters are arranged on a grid.

3 Proposed Content Based Image Retrieval (CBIR) System

CBIR system mainly consists of following two phases.

Offline Phase: In this phase each image feature vectors are extracted from the collection of images, store features and their index in the database. These selected features are used to create cluster using the SOM neural network. The clustering process will be repeated, if any new images added to the system. This phase is the top portion in Fig. 2.

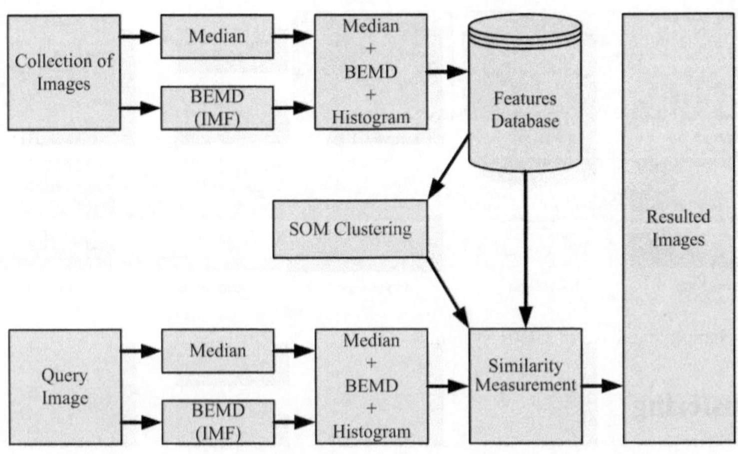

Fig. 2 Proposed CBIR system architecture

Online Phase: In this phase user will give query image for extracting the similar images from the database. Query image feature vector are extract in the same manner and is given to the SOM neural network to identify the cluster number to which it belongs. Identified cluster and its surrounding clusters feature vectors are compared to identify the similar images. The selected similar images are displayed on the Graphical User Interface (GUI). This phase is the bottom part of the Fig. 2.

4 Feature Extraction

4.1 Edge Detection Using BEMD

In this paper, empirical mode decomposition algorithm is used to detect the edge of the image. When the original image is decomposed using BEMD, first Intrinsic Mode Frequency (IMF) shows fine edge characterization. The clear edge image is obtained from the first IMF by applying suitable threshold. Extracting the IMF's from the image is called process of shifting is as follows.

Assume that $Y(t)$ is the original signal and let $S(t) = Y(t), k = 0$ and $i = 0$

1. Find the local minima and maxima of $S(t)$
2. Find the lower envelop $LE(t)$ by interpolating between minima and find the upper envelop $UE(t)$ with maxima.
3. Compute the mean envelop as an approximation to the local average

$$M(t)_1 = (UE(t) + LE(t))/2 \quad (1)$$

4. Let $i = i + 1$ and define the intermediate function as

$$IM_i(t) = S(t) - M(t) \tag{2}$$

5. Repeat steps 1 to 4 on $IM_i(t)$ until it is an IMF, then record the IMF

$$C_1 = IM_i(t) \tag{3}$$

6. Let $S(t) = S(t) - C_k(t)$, if stopping criteria is reached then stop the shifting process otherwise $k = k + 1, i = 0$, and goto step 1.

After completing the shifting process the original signal $Y(t)$ can be represented using extracted IMF's as follows

$$Y(t) = \sum_{j=i}^{n} (C_n(t) + S_n) \tag{4}$$

where $C_n(t)$ is the nth IMF and $S(t)$ is the residue.

The above process is used for signal it is like single dimension, but image is two dimensions the following standard division criteria for stopping the sifting process [2] is used

$$SD_k = \sum_{m=0}^{N} \frac{|IM_{i-1}(m) - IM_i(m)|^2}{(IM_{i-1}^2(m))} \tag{5}$$

4.2 Feature Extraction Using Edge and Median Filtering

The following steps are carried out for generating feature vector for the image.

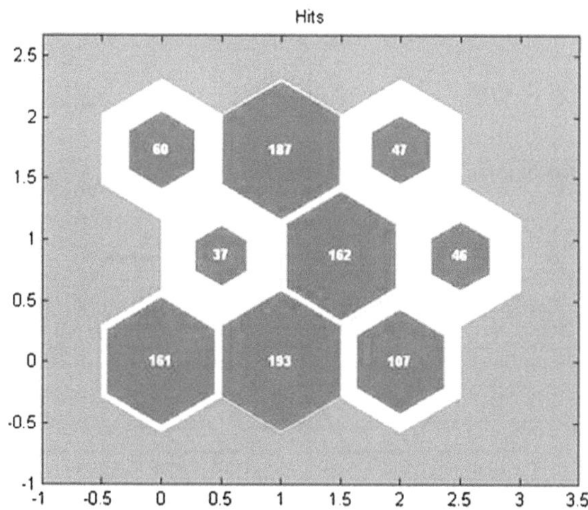

Fig. 3 Feature points associated with neuron

1. The image is converted to gray scale image.
2. Histogram Equalization is applied for gray scale image.
3. Extract edge using empirical mode decomposition.
4. Median filtering is applied to the histogram equalized gray scale image block of 3×3.
5. Replace the values of edge position of median filtering image with detected edge values by BEMD.
6. Extract 64 bins vector and is stored in the database.

4.3 Training with the Batch Algorithm

In this approach SOM neural network will be created and trained using batch training algorithm. The neural network consists of 3-by-3, two-dimensional map of 9 neurons. During training, 200 iterations of the batch algorithm will be run for making the cluster, the SOM neural network will be distributed to all the image features space into 9 neurons.

SOM Algorithm
1. Initialization: Chose any random values for the initial weight vectors W_j.
2. Sampling: Take a sample training input vector X from the input space.
3. Matching: Find the winning neuron $WN(X)$ with weight vector closest to input vector.
4. Updating: Apply the weight update using equation $$\Delta W_{ji} = \eta(t)\ T_{ji}\ X(t)(X_i - W_{ji})$$
5. Continuation: Repeat steps 2–4 until the feature map stops changing.

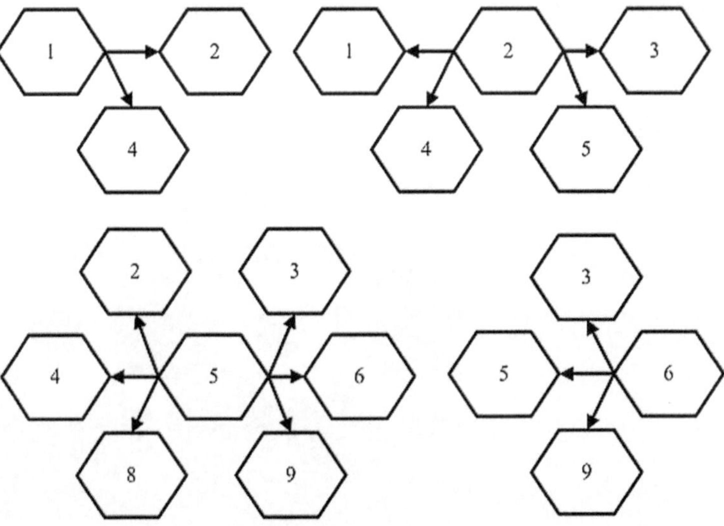

Fig. 4 Neuron and its nearest neurons

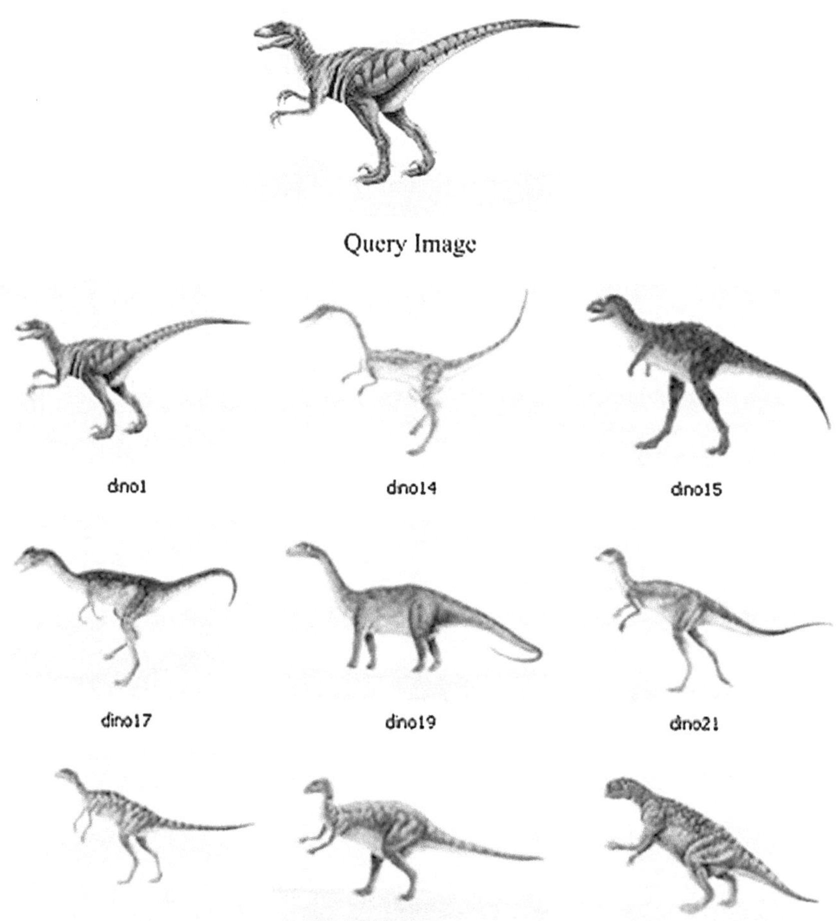

Fig. 5 Query image and its resulted images for Dinosaurs and Buses category

Figure 3 shows total number of image feature points are associated with each neuron.

4.4 SOM Algorithm

The following steps describe the SOM algorithm to make the clusters.

Query Image

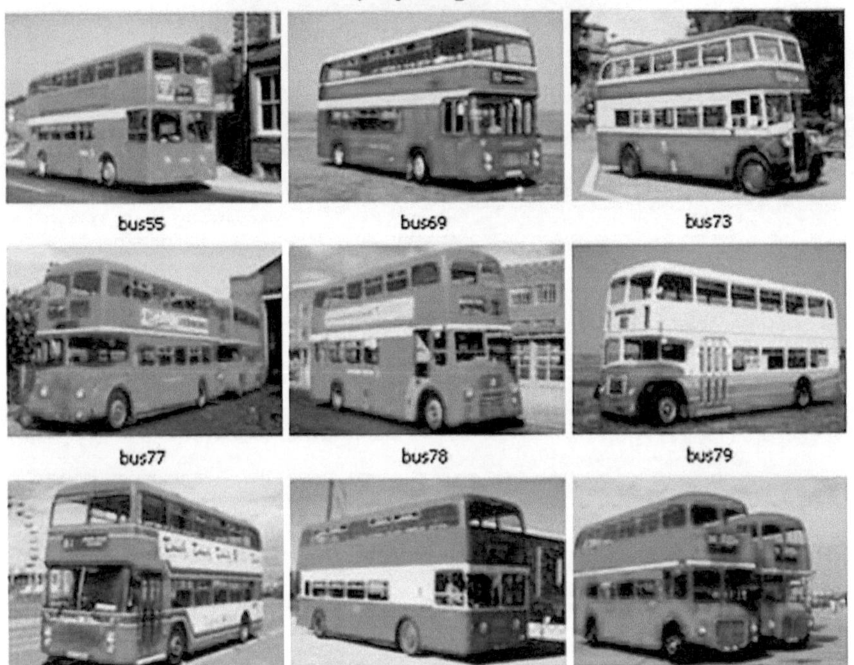

Fig. 5 continued

4.5 Similarity Measure

Figure 4 shows the structure of each neuron and its neighbor neuron. The query features are submitted to SOM neural network to identify the cluster number. Once the cluster number is identified, then the cluster and its surrounding cluster features are extracted from the database and compared with query features by using Euclidean distance technique. The smallest distance will be selected and the corresponding image will be displayed as result.

Let **'Q'** be a query image and **'A'** be an image in database and $Q(n)$ and $A(n)$ be the average value of pixels of each bin, the difference between the value of each

Query Image

Fig. 6 Query image and its resulted images for Buildings and Roses category

bin is calculated as $diff(n) = |A(n) - Q(n)|$, *where* $n = 1, 2, \ldots, 64$. The average value of $diff(n)$ is stored in the array **'SI'**. Finally chosen images difference $diff(n)$ is arranged in the ascending order to display most nearby images on top.

5 Experimental Results

The performance estimation of the CBIR system is done by submitting query image to retrieve similar image from various categories of database images. The experiments are conducted on the ground truth database provided by James S.

Fig. 6 continued

Wang et al. [7, 8]. The ground truth database consists of 1000 images of 10 different categories and each category has 100 related images. The sample query image and its corresponding retrievals are shown in Figs. 5 and 6. Here, only top nine similar images of results are shown.

The Precision and Recall are two generally used metrics for evaluating the accuracy of CBIR system. Precision and recall can be calculated as follows:

$$Precision = \frac{\{relevant\ images\}}{\{retrieved\ images\}}$$

$$Recall = \frac{\{relevant\ images\}}{\{relevant\ images\ in\ the\ DB\}}$$

Table 1 Precision and recall for existing and proposed method

Sl. No.	Category	Existing precision	Method recall	Proposed precision	Method recall
1	Buildings	47.25	20.00	49.25	22.14
2	Africans	73.33	42.00	74.33	44.23
3	Buses	48.33	33.00	55.00	30.10
4	Dinosaurs	99.00	98.00	93.30	90.00
5	Elephants	56.00	35.00	63.70	35.66
6	Food	57.00	33.00	62.00	31.30
7	Flowers	74.82	52.00	75.60	51.00
8	Horses	89.40	40.00	87.10	44.20
9	Mountains	28.00	19.00	38.00	21.00
10	Sunset	48.25	30.00	55.50	28.56
	Average	63.00	37.00	65.37	39.82

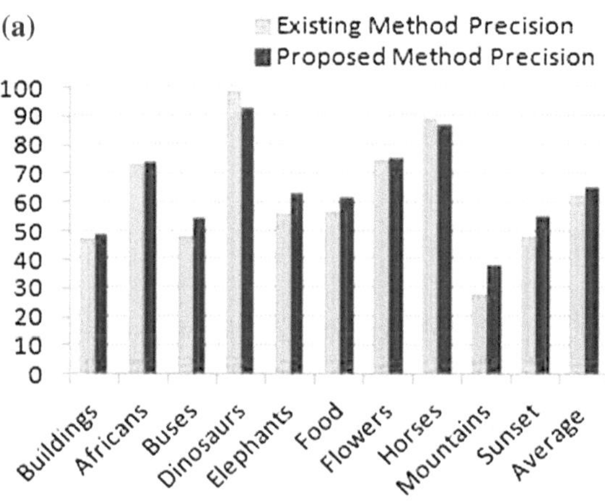

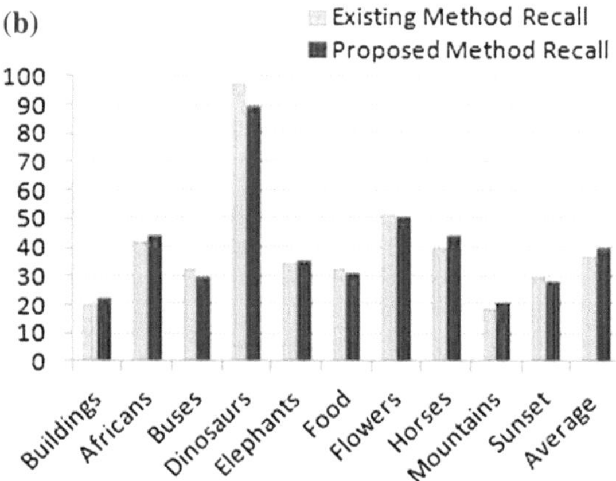

Fig. 7 Each category Precision and recall for existing and proposed method

The precision and recall values of the existing method and the proposed system are shown in the Table 1. and its graph in Fig. 7. From the experimental results it is observed that a substantial improvement in the average value of precision 65.37 % and recall value of 39.82 % compared to existing system [10] values of 63 % and 37 % respectively. The performance is also increased because of less number of compressions of database features.

6 Conclusion

In this paper we have presented a novel approach for image retrieval by combining edge, median histogram with SOM neural network approach for features classification. The techniques were implemented and tested for 500 queries on image database of 1000 images with 10 different categories. The experimental results shows that there is a substantial improvement in the performance of image retrieval system in respect of precision and Recall. The system performance can be enhanced by exploring different techniques which is our current research focus.

References

1. LingFei L, ZiLiang P (2008) An edge detection algorithm of image based on empirical mode decomposition. Second international symposium on intelligent information technology application. In: Proceedings of IEEE, vol 1. pp 128–132
2. Nunes JC (2005) Texture analysis based on the bidimensional empirical mode decomposition. Mach Vis Appl, Guwahati 16(3):177–188
3. Hui Z, Pankoo K, Jongan P (2009) Feature analysis based on edge extraction and median filtering for CBIR. In: 11th International Conference on Computer Modelling and Simulation, vol 48. pp 245–249
4. Sizintsev M, Derpanis KG, Hogue A (2008) Histogram-based search: a comparative study. In: Proceedings of IEEE, CVPR, pp 1–8
5. Kohonen T (1990) The self organizing map. Proc IEEE 78(9):1464–1480
6. Juha V, Esa A (2000) Clustering of the self-organizing map. EEE Trans Neural Netw 11(3):1464–1480
7. Li J, Wang JZ (2003) Automatic linguistic indexing of pictures by a statistical modeling approach. IEEE Trans Pattern Anal Mach Intell 25:1075–1088
8. Wang JZ, Li L, Wiederhold G (2000) SIMPLIcity: semantics-sensitive integrated matching for picture libraries. In: Advances in visual information systems: 4th international conference VISUAL. Loyn, France
9. Ait Aoudia S, Mahiou R, Benzaid B (2010) YACBIR-Yet another content based image retrieval system. In: 14th international conference information visualisation, pp 570–575
10. Shrinivasacharya P, Kavitha H, Sudhamani MV (2011) Content based image retrieval by combining median filtering and BEMD technique. Int Conf Data Eng Commun Syst (ICDECS) 1(2):231–236

Moving Object Tracking in Occluded and Cluttered Backgrounds Using Adaptive Kalman Filtering

Mohammed Ahmed, Youngshin Ahn and Jaeho Choi

Abstract This paper considers the problem of object tracking when a moving object undergoes partial or complete occlusion by the cluttered and noisy background. The presented algorithm is based on the Kalman filter and background checking combined with the mean shift algorithm. First, a rectangular region is defined surrounding the object of interest and the region is searched for a similar histogram distribution of that of the object of interest. Then, the model of the Kalman filter is constructed. Using the mean shift algorithm, the centroid of the object is predicted. The predicted values are fed into the Kalman filter. Interactively, the resulting parameter estimates of Kalman filtering are fed back to the mean shifting processor. The verification on the performance of the proposed method shows us that the proposed method can successfully track a moving object under complete or partial occlusion, even when the object has a similar color and texture with the background.

Keywords Moving object tracking · Mean shift · Adaptive Kalman filtering · Cluttering and occlusion

M. Ahmed (✉) · Y. Ahn · J. Choi
Department of Electronic Engineering, CAIIT Chonbuk National University Chonju, Chonju, Chonbuk, Republic of Korea
e-mail: gettoahmed@gmail.com

Y. Ahn
e-mail: math1532@nate.com

J. Choi
e-mail: wave@jbnu.ac.kr

Mohan S. and S. S. Kumar (eds.), *Proceedings of the Fourth International Conference on Signal and Image Processing 2012 (ICSIP 2012)*, Lecture Notes in Electrical Engineering 221, DOI: 10.1007/978-81-322-0997-3_49,

1 Introduction

Tracking of moving objects in video sequences is a very active research area and there are many approaches which attempt to develop robust techniques for varying video conditions. Among numerous object tracking algorithms, mean shift (MS) object tracking has recently received growing interest since it was introduced by Comaniciu, Ramesh, and Meer [1]. In this approach, objects of interest are characterized by the probability density functions of their color or texture features. By masking the distribution with a monotonically decreasing kernel, spatially-smooth similarity functions is defined and mean shift iteration can use the gradient of this similarity function as an indicator of the direction of target's movement. The similarity is expressed in terms of Bhattacharyya coefficient. Though the MS object tracking algorithm performs well on sequences with relatively small object displacement, its performance is not guaranteed when the objects move fast, undergo partial or full occlusion or have similar color histogram with the background or surrounding objects. To overcome this disadvantage of the MS tracking method, an improved MS object tracking algorithm was proposed.

In this work a novel approach using Kalman filter and segmentation with a mean shift is proposed. First, a rectangular region (RR) is defined surrounding the object of interest (OI) and we search in RR for object with similar color histogram. Then the model of Kalman filter is constructed. The estimate parameters of KF are adjusted by Bhattacharyya coefficients and background similarity information.

The paper is organized as follows. The related works are summarized in Sect. 2, and the review on the mean shift algorithm is given in Sect. 3. In Sect. 4, the Kalman filtering algorithm is given, and in Sect. 5, the proposed object tracking algorithm is presented. The experimental results are given in Sect. 6, and the paper concludes in Sect. 7.

2 Related Works

The mean shift algorithm and Kalman filtering for object tracking is a very active research area and there are many approaches that attempt to develop robust techniques for varying video conditions such as partial or complete occlusions, clutter, and noise. Comaniciu and Ramesh [2] initiated the fusion of the mean shift technique with Kalman filtering. Also, Lee [3] proposed a method in which the exact target center is obtained by combining the two estimated target centers obtained by KF and MS algorithms together. Both of those methods fail to adapt to different environment. Since the prediction and measurement errors of KF are set constant, these algorithms are not robust enough. Later, another object tracking scheme was proposed by Babu [4] that combines the sum-of-squared-differences method and the MS object tracking method in the KF framework. In this method, to handle a partial occlusion, the whole object is represented by a number of elementary MS modules embedded within the object, rather than using a single global MS tracker. Therefore, this scheme is time consuming. On the other hand, Li [5] proposed a method that calibrates the Kalman filter by setting the measurement noise and the system noise and both of the noises depend on Bhattacharyya coefficients. This method however missed the generality of the Kalman filtering algorithm by losing the independency between measurement and state noises. It fails to capture the moving object when the target has a similar color texture with the background. Note that none of the above mentioned methods handle the moving object tracking problem in the cluttered environment, i.e., similar background color texture; and none of them are adaptable to varying amount of occlusion between the background and the moving object in the same Kalman filtering framework.

3 Mean Shift Tracking

3.1 *Target Model*

The target color probability density function (pdf) is approximated by a histogram of m bins $\hat{q} = \{\hat{q}_u\}_{u=1\ldots m}, \sum_{u=1}^{m} \hat{q}_u = 1$ with $\hat{q}_u$ being the u-th bin. To form the histogram, only the pixels inside a rectangular region surrounding the object are taken into account. The objects' density estimates (i.e. histograms) were weighted by a monotonically decreasing the Epanechnikov kernel. The kernel is defined as follows:

$$K_E = {}^1\!/_2 c_d^{-1}(1 - d)\left(1 - \|x^2\|\right) \tag{1}$$

where $c_d$is the volume of the d-dimensional sphere and x is the normalized pixel coordinates within the target, relative to the center; and $\|x^2\|$ is the squared Euclidean distance of each pixel from the center of the target. Since we were dealing with a two-dimensional image space, the kernel function can be expressed as follows:

$$K_E = {}^2\!/_\pi\left(1 - \|x^2\|\right) \tag{2}$$

The rationale for using a kernel is to assign the smaller weights to the pixels farther from the center. It assumes that those pixels are the least reliable, since they are the ones more affected by occlusion or interference from the background, i.e., color-bleeding due to a low resolution, transmission errors, and noise. A kernel with the Epanechnikov profile is essential to obtain a smooth similarity function between the distributions. Since its derivative is constant, the kernel masking leads to a function suitable for a gradient optimization; and the gradient provides us the direction of the target's movement. The search for the matching target candidate in that case is restricted to a smaller area, and therefore, the searching becomes faster than the exhaustive search.

3.2 Distance Minimization

Based on the fact that the probability of a classification error is directly related to the similarity of the two distributions, the choice of the similarity measure in [1] was such that it was supposed to maximize the Bayes error arising from the comparison between the target and candidate pdf's. Being a closely related entity to the Bayes error, the Bhattacharyya coefficients were calculated and its maximum was searched for estimating the target localization. The set of Bhattacharyya coefficients of the two statistical distributions is defined as follows:

$$\rho(y) = \rho[p(y), q] = \sum_{u}^{m} [p_u(y) q_u]^{1/2} \tag{3}$$

where $p_u$and q_u represent the histograms of the target and the candidate, respectively.

3.3 Mean Shift

In order to find the best match of the moving target in the sequential frames, one needs to maximize the Bhattacharyya coefficient given by Eq. 3. In other words, as explained in [1], it needs to maximize the term defined as follows:

$$\sum_{i=1}^{n} \omega_i k(\|(y - x_i)/h^2\|) \tag{4}$$

where h is the kernel's smoothing parameter, or bandwidth, and w_i is as follows:

$$\omega_i = \sum_{u=1}^{m} \sqrt{\frac{q_u}{p_u(y_0)}} \delta[b(X_i) - u] \tag{5}$$

and δ is the Kronecker delta function, and it is 1 at the particular bin u. The terms $q_u$and p_u are the values of the target and candidate histograms corresponding to pixel x_i of the candidate object. Note that the sum in Eq. 4 is actually a density estimate, i.e. a histogram, of the object centered at y in the current frame, computed with a kernel profile k(x) and weighted by ω_i. The maximum of this density in the local neighborhood, starting from the last known position of the target, gives us the most probable target position in the current frame; and it can be obtained d by using the mean shift procedure. During this procedure, the center of the target candidate is successively shifted as follows:

$$Y = \frac{\sum_{i=1}^{n} X_i \omega_i \, g\left(\left\|\frac{Y_0 - X_i}{h}\right\|^2\right)}{\sum_{i=1}^{n} \omega_i \, g\left(\left\|\frac{Y_0 - X_i}{h}\right\|^2\right)} \tag{6}$$

where y_0 is the current location of the candidate center and g(x) is the derivative function. Since the derivative of the Epanechnikov kernel profile is constant, the above expression reduces to a weighted distance average and it can be rewritten as follows:

$$Y = \frac{\sum_{i=1}^{n} X_i \omega_i}{\sum_{i=1}^{n} \omega_i} \tag{7}$$

4 Kalman Filter

4.1 Kalman Filter Model

In general, it is assumed that there is a linear process governed by an unknown inner state producing a set of measurements. More specifically, there is a discrete time system and its state at the time k−1 is given by the vector x_{k-1}. The state in the next time step k is as follows:

$$x_k = Ax_{k-1} + w_{k-1} \tag{8}$$

where *A* is the transition matrix from state $x_{k-1}$to x_k and w_{k-1} is the additive white Gaussian noise (AWGN) with zero mean and the covariance matrix S_w^{k-1}. The measurement vector y_k is as follows:

$$y_k = Cx_k + \boldsymbol{\mu}_k \tag{9}$$

where *C* is the measurement matrix and μ_k is the AWGN with zero mean and the covariance matrix S_z^{k-1}. In Eq. 9, the measurement y_k depends only on the current state x_k and the noise vector μ_k is independent of the noise w_{k-1}.

The Kalman filter computes the minimum mean-squared error estimate of the state x_k given the measurement vectors y_k. The solution is obtained by a recursive procedure described in the sub-section below.

4.2 Kalman Filtering Algorithm

The Kalman filtering is basically consisted of three steps as follows:

1. Initialize

$$x_0^s = E(x_0)$$

$$P_o = E(x_0 - E(x_0))(x_0 - E(x_0))^{\mathbf{t}}$$

2. Prediction

a. Innovation vector:

$$Innov = y - x_k^s$$

b. Covariance of innovation

$$s = CPC' + S_z^{k-1}$$

$$P = APA' + S_w^{k-1}$$

c. Kalman gain

$$K = APC's^{-1}$$

d. New state estimate

$$x_k^s = x_k^s + K\, Innov$$

3. Estimation

a. Covariance of the estimation error

$$P = APA' - APC's^{-1}CPA' + S_w^{k-1}$$

b. Go back to the prediction steps for the next prediction.

5 The Proposed Method

5.1 The Proposed Kalman Filter Model

The section considers a dynamic scheme for the Kalman filter as the elements of its state matrix are updated on-line depending on a measure evaluating the quality of the observation.

By these means the tracking procedure may be significantly accelerated. In the proposed Kalman filtering process, the state transition matrix A is defined as follows:

$$A = \begin{bmatrix} 1 & \Delta t & 0.5\Delta t^2 \\ 0 & 1 & \Delta t \\ 0 & 0 & 1 \end{bmatrix} \tag{10}$$

where Δt is the frame rate. This enables to capture the acceleration and position of the moving object for various frame rates. The initial state estimate for the recursive Kalman filtering is defined as follows:

$$x = \begin{bmatrix} x \\ v \\ a \end{bmatrix} = \begin{bmatrix} \mu_{mlp}^{N-1} + \frac{1}{N}\left(x_N - \mu_{mlp}^{N-1}\right) \\ \mu_{mlv}^{N-1} + \frac{1}{N}\left(v_N - \mu_{mlv}^{N-1}\right) \\ \mu_{mla}^{N-1} + \frac{1}{N}\left(a_N - \mu_{mla}^{N-1}\right) \end{bmatrix} \tag{11}$$

where N represents the number of recursive calls to KF. $\mu_{\mathrm{mlp}}^{\mathrm{N-1}}$, $\mu_{\mathrm{mlv}}^{\mathrm{N-1}}$, and $\mu_{\mathrm{mla}}^{\mathrm{N-1}}$ are the mean values of predictions obtained by the Kalman filter for the position, velocity and acceleration, respectively. $\mathrm{x_N}, \mathrm{v_N},$and $\mathrm{a_N}$ are the current measurements of position, velocity, and acceleration, respectively see Fig. 1; and they are defined as follows:

$$x_n = MS_n$$

$$\mathrm{v_N} = \frac{\mathrm{MS_n}}{\Delta \mathrm{t}}$$

$$a_N = \frac{v_N - v_{N-1}}{\Delta t} \tag{12}$$

The rationale for using this initial state estimate is that on each recursive call to the Kalman filtering procedure, the system model will tuned by estimating the system model parameters, sequentially and adaptively.

On the other hand, the object motion is described by its centroid, hence, the measurement matrix is defined as follows:

$$C = [100] \tag{13}$$

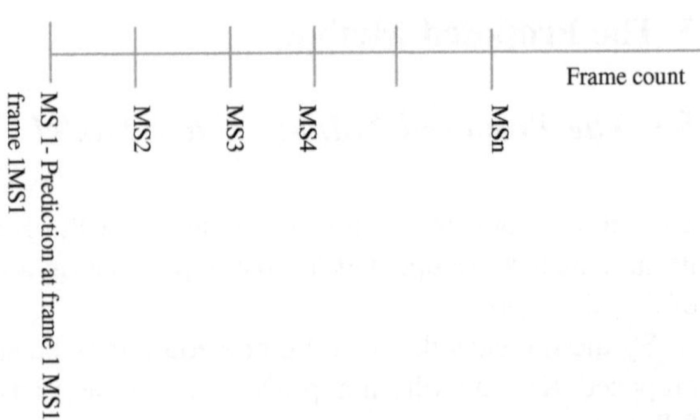

Fig. 1 Initial state estimation for KF

From the above discussion we can see that the Kalman filter is applied for the vertical and horizontal position estimates, separately. The process and measurement noises are assumed to be AWGN and they are defined as follows:

$$S_w = \begin{bmatrix} Pnoise + \frac{1}{Factor} & 0 & 0 \\ 0 & (0.01) * \left| \left(\frac{v_N - \mu_{mlv}}{|v_N| + |\mu_{mlv}|} \right) \right| & 0 \\ 0 & 0 & (0.1) \left| \left(\frac{a_N - \mu_{mla}}{|a_N| + |\mu_{mla}|} \right) \right| \end{bmatrix} \tag{14}$$

where

$$\mu_z = (1 - Bhattacharyya\ coefficient) * Factor$$

Using a series of experiments, various expressions for the term 'Pnoise' have been examined. Also, an experimental assumption is made that the difference between x_{N-1} and x_N cannot be more than four times the mean position estimate. Hence, we can consider that if the prediction is larger than 4 μ_{mlp}^{N-1} the measurement is erroneous, and we only trust the predicted value. Finally, we detect a partial or complete occlusion based on the Bhattacharyya coefficient. If Bhattacharyya coefficient falls below some threshhold value, one can guess that a partial or complete occlusion has occurred. By controlling a decaying 'Factor' term in Eq. 14, one can control the duration of occlusion, and furthermore, the same setting can be applied to various durations of occlusion problems.

5.2 Summary of Moving Object Tracking in Occlusion

One of the problematic elements in moving object is the cluttered background. This problem is a challenging task since it is always desirable to keep the computation complexity in the mean shift procedures. Here, a novel method is proposed to overcome this problem by using the same configuration of the proposed Kalman filtering algorithm discussed above without any modification. In order to effectively reduce the computation, a rectangular region is used around the object of interest.

The size of the RR depends on the average prediction of the mean shift (i.e. the speed of the target). Then, the existence of a similar target distribution is searched within the RR after observing the prediction from the mean shift procedure using the Bhattacharya coefficient. This can be done by applying a mask to the target to avoid rediscovering the target see Fig. 3d. The above outlined procedure can be summarized as follows:

1. Read input image
   a. Apply mask (i.e. to decrease the number of iterations in the mean shift)
   b. Extract the target
2. Read new image
   a. Extract the candidate
   b. Apply mask
3. Apply the mean shift algorithm as described in Sect. 3.3
4. Apply the Kalman filtering algorithm as described in Sect. 4.2
5. Check the background similarity
   a. If true go to step 4
   b. Else go to step 2.

6 Experimental Results

In order to evaluate the proposed algorithm, we have performed comparisons with the standard mean shift algorithm [1]. Various test sequences were employed in the evaluation. These sequences consist of complete occlusion, partial occlusion, and cluttered background scenarios. First, the target model is constructed 16-quantization levels, i.e., the histogram size of $16 \times 16 \times 16$, as shown in Fig. 2. The Epanechnikov kernel profile is used for the derivation of the smooth similarity function between the distributions. Since its derivative is constant; thus, the kernel masking leads to a function suitable for gradient optimization, which provides us the direction of the target's movement. The search for the matching target candidate in that case has been restricted to a much smaller area, and therefore, the computational complexity can be reduced in comparison to the exhaustive search. The Epanechnikov kernel profile is shown in Fig. 2. On the other hand, the target was initialized by detecting the object entering the camera range. The representative frames are shown in Fig. 3. The proposed method was tested on a 1.6 GHz machine with 1 GB of memory: The test video contains 185 video frames and is implemented within a MATLAB environment. The result of our experiment and comparison with other implementation [5] and [1] is shown in Fig. 4.

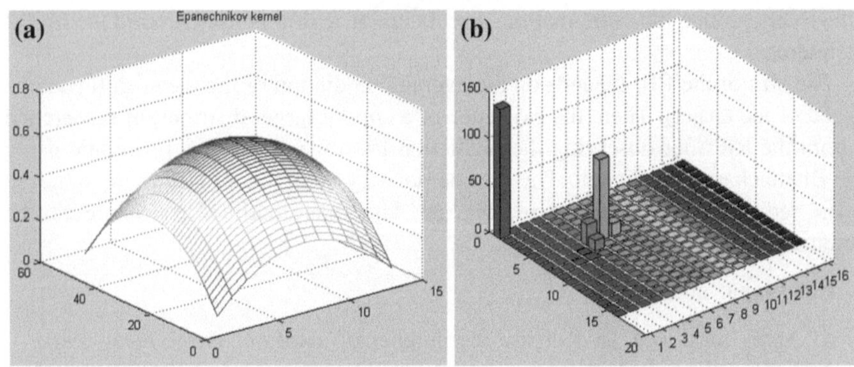

Fig. 2 **a** Epanechnikov kernel. **b** Target histogram

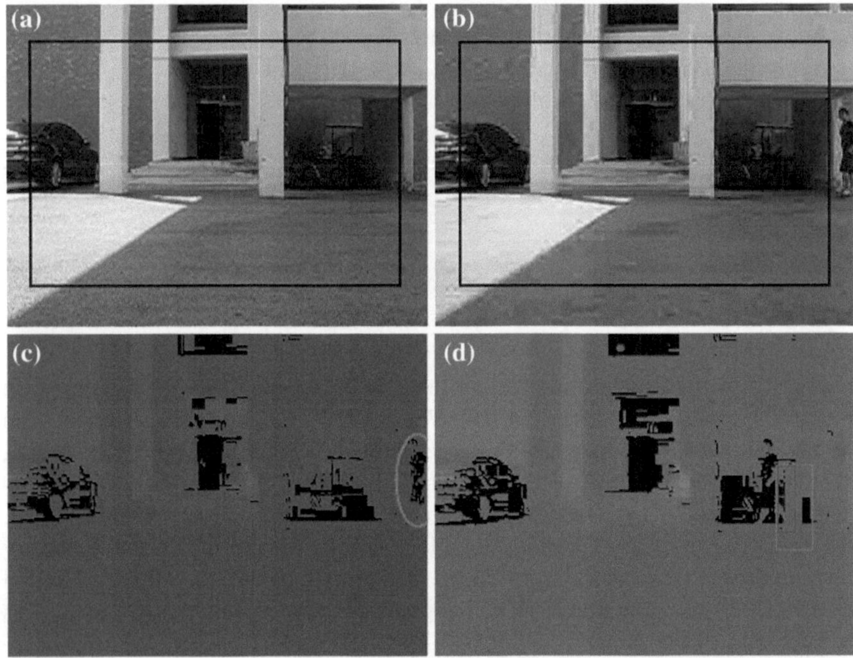

Fig. 3 **a** It shows the detection area boxed with *black*. **b** The target begins to come into the object detection area. **c** The normalized RGB of the image after applying the mask and **d** The target centroid in *blue* and the RR in *red* around the target in moving object tracking

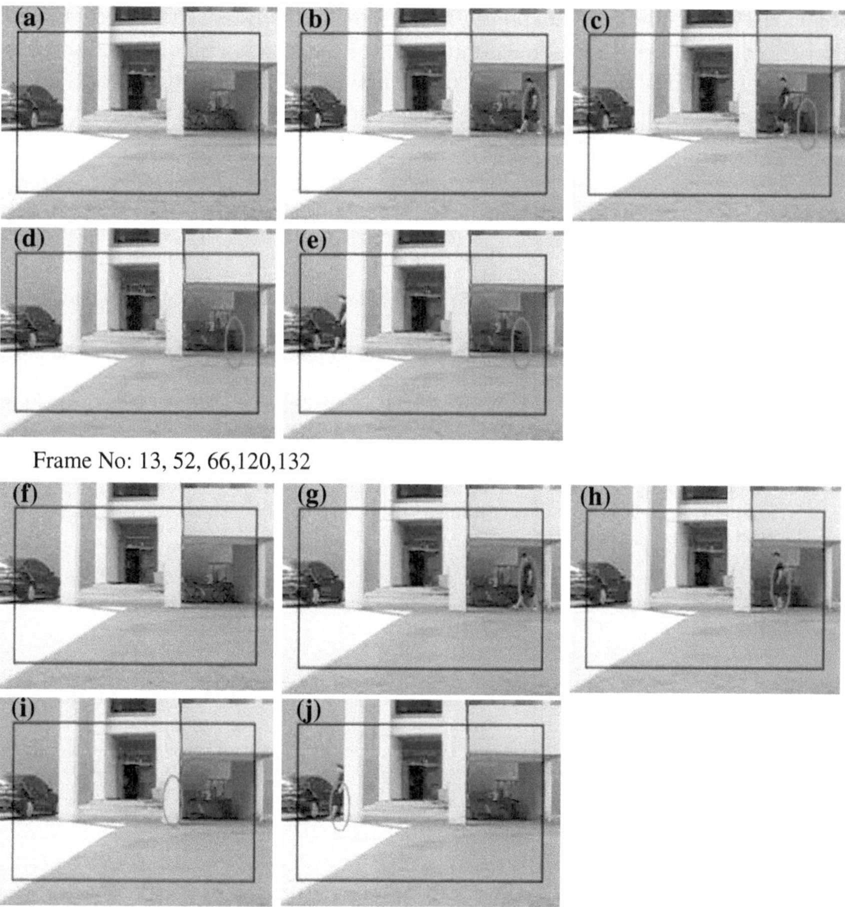

Frame No: 13, 52, 66,120,132

Fig. 4 The algorithm in the literature [5] is implemented and the results are shown in the representative image sequences (a, b, c, d, e); while the results for the proposed method is shown in (f, g, h, i, j, k). As shown, the proposed method is successful but [5] fails due to inability to adapt to cluttered background

7 Conclusion

In this paper we have presented resolvable moving object detection and tracking algorithm and solved the problem of cluttered background and occlusion problem in the same Kalman filter framework. The MS object tracking method is integrated with the KF framework; and an adaptive KF algorithm has been discussed. The MS is initialized by the predicted value of KF and it is used to track the target position. Then the tracking result of the MS is fed back as to the KF as the measurement, and the estimate parameters of the KF are adjusted by the

Bhattacharyya coefficients, adaptively. According to the Bhattacharyya coefficient, the KF can be adjusted to estimate the center of the moving object. The proposed method successfully detects and tracks the moving object that undergoes a complete or partial occlusion in a clutter background.

References

1. Comaniciu D, Ramesh V, Meer P (2003) Kernel-based object tracking. IEEE Trans Pattern Anal Mach Intell 25:564–577
2. Comaniciu D, Ramesh V (2000) Mean shift and optimal prediction for efficient object tracking. In: Proceedings IEEE, international conference image processing, Canada
3. Lee W, Chun J, Choi B, Yang Y, Kim S (2009) Hybrid real-time tracking of non- rigid objects under occlusions. In: Proceedings SPIE 7252
4. Babu R, Perze P, Bouthemy P (2007) Robust tracking with motion estimation and local kernel-based color modeling. Image Vis Comp 25:1205–1216
5. Li X, Zhang T, Shen X (2010) Object tracking using an adaptive Kalman filter combined with mean shift. In: Proceedings SPIE 7252

Adaptive Enhancement of Underwater Images

Jharna Majumdar, Aparna Manikonda and G. M. Venkatesh

Abstract Underwater images usually suffer from low contrast and non-uniform lightning. To overcome this histogram equalization is the basic technique for enhancement due to its simple function and effectiveness. This method tends to change the brightness of an image and therefore, not suitable for consumer electronic products, where preserving the original brightness is essential to avoid annoying artifacts. A number of techniques have been developed over a period of time to overcome these undesirable effects. But none of the technique is found suitable for enhancement of image under poor illumination conditions, which preserve the brightness of the original image. In this article we present a survey of different techniques that are based on histogram equalization, and also applied those techniques on underwater images. We also made a comparison of the processed images by a set of ten quality metrics.

Keywords Adaptive enhancement · Histogram equalization · Quality metrics

1 Introduction

Under water imaging requires that vehicles carry their own light sources with them as ambient light is nonexistent. The images are affected with the transmission of

J. Majumdar (✉) · A. Manikonda · G. M. Venkatesh
Nitte Meenakshi Institute of Technology, Bangalore 560064, India
e-mail: jharna.majumdar@gmail.com

A. Manikonda
e-mail: aparna.subhadra@gmail.com

G. M. Venkatesh
e-mail: venkatesh.gm89@gmail.com

Mohan S. and S. S. Kumar (eds.), *Proceedings of the Fourth International Conference on Signal and Image Processing 2012 (ICSIP 2012)*, Lecture Notes in Electrical Engineering 221, DOI: 10.1007/978-81-322-0997-3_50,

limited range of light, disturbance of lightening, low contrast and blurring. Research work is in progress to compensate these artifacts and improve the visibility of underwater images [1–4]. In this paper we have proposed few image enhancement techniques based on histogram equalization. The effectiveness of various algorithms is accessed using set of quality metric parameters.

In conventional methods of image enhancement, human viewers choose the method of enhancement that is most appropriate for a given input image. That is to say, the method is chosen in an ad-hoc manner. The assessment of the algorithm is done in a subjective manner by human beings. While a human image processing expert may select the best method on a case-to-case basis based on visual inspection, such a human intervention may not be feasible in practice. For applications in which images are ultimately to be viewed by human beings, the only "correct" method of visually quantifying the image is through subjective evaluation. In practice however, subjective evaluation is usually too inconvenient, time-consuming unreliable and expensive. It is therefore necessary to devise methods to automatically select the suitable enhancement routine for any given input image. The adaptive methods of enhancement are modification of classical methods. Depending on the characteristics of the input image, adaptive method decides whether to increase the dynamic range of the image or to enhance the details of the dark regions of the image without affecting mid and bright pixels.

Image enhancement techniques have been used in various applications where the subjective quality of images is very important. Contrast is an important factor in image quality estimation. The term refers to the amount of gray scale differentiation that exists between various image features while working on gray scale images. Images having higher contrast level display a larger grayscale difference than those of lower contrast. The contrast variations affect the ultimate form of the image. There are many approaches for enhancing the contrast of images. Techniques using histograms are most common in the Contrast problem. Among these, Histogram Equalization (HE) is the classical method due to its simplicity and effectiveness. HE uses histogram information of the image and turns them into images with uniform histogram distributions. However, this technique is less effective when the contrast characteristics vary drastically across the image as in backlight conditions. That is the bright area becomes saturated in the resultant image due to the compensation taken place in the dark area. Moreover, the resultant image may have regions of decreased local contrast. This is because HE only uses global information (from the whole image) and does not consider local information of luminance variation within neighborhoods of each pixel. To overcome this many researchers have proposed techniques [5–15] for histogram equalization using local information.

In this article we compare each of these different techniques of histogram equalization and their effectiveness while processing the underwater images. We have used some of the quality parameters to analyze the results while finding the suitability of the algorithm.

2 Techniques Based on Histogram Equalization

In general, a histogram is the estimation of the probability distribution of a particular type of data. An image histogram represents a graphical tonal distribution of the gray values in a digital image. By viewing the image's histogram, we can analyze the frequency of appearance of the different gray levels contained in the image. A good histogram is that which covers all the possible values in the gray scale used. This type of histogram suggests that the image has good contrast and that details in the image may be observed more easily.

2.1 Histogram Equalization

Histogram equalization is a straightforward enhancement technique to achieve better quality images in grey scale. The histogram equalization redistributes intensity values along the total range of values in order to achieve higher contrast. This method is especially useful when an image is represented by close contrast values, such as images in which both the background and foreground are bright at the same time, or else both are dark at the same time [5].

2.2 Brightness Preserving Bi-Histogram Equalization

Brightness Preserving Bi-Histogram Equalization (BBHE) has been proposed by Kim [6]. The BBHE firstly decomposes an input image into two sub-images based on the mean of the input image. One of the sub-images is the set of samples less than or equal to the mean whereas the other one is the set of samples greater than the mean. Then the BBHE equalizes the sub-images independently based on their respective histograms with the constraint that the samples in the formal set are mapped into the range from the minimum gray level to the input mean. The samples in the latter set are mapped into the range from the mean to the maximum gray level.

2.3 Recursive Mean-Separate Histogram Equalization

Recursive Mean-Separate Histogram Equalization (RMSHE) has been proposed by Chen and Ramli [7]. This method is a generalization of BBHE referred to as to provide not only better but also scalable brightness preservation. BBHE separates the input image's histogram into two based on its mean before equalizing them independently. While the separation is done only once in BBHE, this method proposes to perform the separation recursively; separate each new histogram further based on their respective mean. Besides, the recursive nature of RMSHE also allows scalable brightness preservation, which is very useful in consumer electronics.

2.4 Bilinear Interpolation Dynamic Histogram Equalization

Bilinear Interpolation Dynamic Histogram Equalization (BIDHE) has been proposed by Xu and Liu [8]. First, the original image is divided into some same size sub-images. Second, the histogram of each sub-image is partitioned into sub-histograms without domination. Third, the new dynamic ranges are allocated for sub-histograms. Finally, HE and Bilinear Interpolation are respectively implemented to the image.

2.5 Brightness Preserving Dynamic Histogram Equalization

Brightness preserving dynamic histogram equalization (BPDHE) has been proposed by Ibrahim and Kong [9], which is an extension to HE that can produce the output image with the mean intensity almost equal to the mean intensity of the input, thus fulfill the requirement of maintaining the mean brightness of the image. First, the method smoothes the input histogram with one-dimensional Gaussian filters, and then partitions the smoothed histogram based on its local maximums. Next, each partition will be assigned to a new dynamic range. After that, the histogram equalization process is applied independently to these partitions, based on this new dynamic range. For sure, the changes in dynamic range, and also histogram equalization process will alter the mean brightness of the image. Therefore, the last step in this method is to normalize the output image to the input mean brightness.

2.6 Multipeak Histogram Equalization with Brightness Preserving

The Multipeak histogram equalization (MHE) has been proposed by Wongsritong et.al. [10]. Here, each detected peak of histogram is independently equalized. The effect of brightness saturation can be defeated, as the perceptibility can be improved. By using histogram equalization method, the output image can be preserved the mean value brightness of the input image.

2.7 Minimum Mean Brightness Error Bi-Histogram Equalization

Minimum Mean Brightness Error Bi-Histogram Equalization (MMBEBHE) has been proposed by Chen and Ramli [12] to provide maximum brightness preservation. BBHE separates the input image's histogram into two based on input mean before equalizing them independently. This method proposes to perform the separation based on the threshold level, which would yield minimum Absolute Mean Brightness Error (AMBE—the absolute difference between input and output mean). An efficient recursive integer-based computation for AMBE has been formulated to facilitate real time implementation.

2.8 Dualistic Sub-Image Histogram Equalization

Dualistic Sub-Image Histogram Equalization (DSIHE) has been proposed by Yu Wan et al. [11]. Here, first, the image is decomposed into two equal area sub-images based on its original probability density function. Then the two sub-images are equalized respectively. At last, we get the result after the processed sub-images are composed into one image.

2.9 Adaptively Modified Histogram Equalization

Adaptively Modified Histogram Equalization (AMHE) is proposed by Hyoung-Joon Kim et al. [13] which is an extension of typical histogram equalization. To prevent any significant change of gray levels between the original image and the histogram equalized image, the AMHE scales the magnitudes of the probability density function of the original image before equalization. The scale factor is determined adaptively based on the mean brightness of the original image.

2.10 Weighted and Threshold Histogram Equalization

Weighted and threshold histogram equalization (WTHE) is proposed by Wang and Ward [14]. In this method, the Probability distribution function of an image is modified by weighting and thresholding before the histogram equalization (HE) is performed. This method provides a convenient and effective mechanism to control the enhancement process while being adaptive to various types of images.

2.11 Average Luminance with Weighted Histogram Equalization

Average Luminance with Weighted Histogram Equalization (ALWHE) has been proposed by Tai et al. [15] for Low Dynamic Range Images (LDRI). Automation of parameter selection can easily be done by iteratively using the proposed method. The proposed method overcomes the disadvantage of HE like luminance shifting and washed-out looking.

3 Quality Metrics

Image quality measurement is crucial for most image processing applications. The best way to assess the quality of an image is perhaps visual observation called subjective quality measurement, since human eye is the ultimate receivers in most image processing environment. The subjective quality metric, Mean Opinion Score (MOS), although used for many years, found to be too inconvenient, slow and expensive for practical usage. The objective image quality metrics can predict perceived image quality automatically using a set of Quality Metric Parameters.

In this paper, we have proposed a number of quality metric parameters and used them to study the performance of the proposed enhancement algorithms for a set if input images.

The quality parameters are: Entropy, Entropy Error Rate (EER), Grey level Energy (GLE), Spatial Frequency (SF), Global Contrast (GC), Quality Index (QI), Absolute Mean Brightness Error (AMBE), Relative Entropy (RE), Structural Content (SC) and Maximum Difference (MD).

A brief description of each of these metrics is given below:

3.1 Entropy

The entropy [16] is the measure of information content in an image and is given by

$$\text{Entropy} = -\sum_{k=0}^{255} p(k)\log_2(p(k)) \tag{1}$$

If the enhanced image is higher value of entropy than the input image, then we can say that the image is truly enhanced. Hence, the entropy ratio for image enhancement is given by

$$\text{ER} = \frac{\text{Entropy(OutputImage)}}{\text{Entropy(InputImage)}} \tag{2}$$

3.2 Entropy Error Rate

The entropy error rate (EER) [17] is a measure of distribution of image information. EER of an image is given as

$$\mathrm{EER} = \frac{\overline{\mathrm{HD}} - \overline{\mathrm{HB}}}{\mathrm{S}} \tag{3}$$

where $\bar{\mathrm{H}}_{\mathrm{D}}$ and $\bar{\mathrm{H}}_{\mathrm{B}}$ are the average entropy for the darker pixels and brighter pixels, and S is a statistic that estimates the relative position of the mean within the intensity histogram. If EER of an image has a relatively large positive value, then the image is enhanced.

3.3 Gray Level Energy

The grey level energy (GLE) [16] indicates how the grey levels are distributed. If an image has a GLE value approaching to 1, then the image is said to be enhanced. The value of GLE is given by the following expression:

$$\mathrm{E(x)} = \sum_{0}^{255} \mathrm{p(i)}^2 \tag{4}$$

where E(x) refers to the gray level energy with 256 bins and P(i) refers to the probability distribution function of the histogram.

3.4 Spatial Frequency

The spatial frequency (SF) measure indicates the overall activity level in an image [18]. SF is defined as follows:

$$\mathrm{SF} = \sqrt{R^2 + C^2} \tag{5}$$

where

R is column frequency and is given by $\mathrm{R} = \sqrt{\frac{1}{MN} * \sum_{j=1}^{M} \sum_{k=2}^{N} (X_{j,k} - X_{j,k-1})^2}$

C is column frequency and is given by $\mathrm{C} = \sqrt{\frac{1}{MN} * \sum_{k=1}^{N} \sum_{j=2}^{M} (X_{j,k} - X_{j-1,k})^2}$

Xj, k denotes the pixel intensity values of image, M and N are number of horizontal and vertical directions. The value of spatial frequency is high for a good enhanced image.

3.5 Global Contrast

The global contrast (GC) [19] value of an image is defined as the second central moment of its histogram divided by the total number of pixels in the image and is given by

$$\text{Global Contrast} = \frac{\sum_{i=0}^{L} (\mathrm{i} - \mu)2 * Hist(i)}{\mathrm{N}} \tag{6}$$

where μ is the average intensity of the image, Hist (i) is the number of pixels in the image with intensity value i, L is the highest intensity value.

3.6 Quality Index

The Image quality index (QI) is suggested according to the statistical features gray level histogram of an image and is given as

$$\mathrm{Q} = \frac{1}{\Delta f}\sqrt{\frac{\sum_{i=0}^{L} (\mathrm{p_j} - \bar{\mathrm{p}})^2}{\mathrm{L}}} \tag{7}$$

where Q is image quality, Δf is the dynamic range, $\mathrm{p_j}$ is the histogram, $\bar{\mathrm{p}}$ is the average of the histogram and L is the number of gray levels.

3.7 Absolute Mean Brightness Error

Absolute Mean Brightness Error (AMBE) [20] measures the deviation of the processed image mean from the input image mean. The AMBE value provides a sense of how the global appearance of the image has changed with respect to lower values. The expression for AMBE is given as:

$$\text{AMBE} = |\mu\mathrm{p} - \mu\mathrm{i}| \tag{8}$$

where μ_p is the processed image and $\mu_i$is the input image.

3.8 Relative Entropy

The relative entropy (RE) of probability distribution of the input image with respect to the probability distribution of the output image is defined as the summation of all the possible states of the system. The lower value of RE indicates good enhanced image. The expression for RE is given as:

$$\mathrm{RE} = -\sum_{\mathrm{k}=0}^{255} \mathrm{p(k)log_2}\left(\frac{\mathrm{p(k)}}{q(k)}\right) \tag{9}$$

3.9 Structural Content

Structural content (SC) of an image is defined as the ratio between the sum of the pixels in the original image at location (m, n) to the sum of the pixels in the processed image at the same location (m, n) and is given as:

$$\mathrm{SC} = \frac{\sum_{m=1}^{M}\sum_{n=1}^{N} X(m,n)^2}{\sum_{m=1}^{M}\sum_{n=1}^{N} Y(m,n)^2} \tag{10}$$

where M is the number of columns, N is the number of rows, x(m, n) denote the pixel values of the original image at location (m, n), y(m, n) denote the pixel values of the processed image at location (m, n).The value of structural content will be high for a good enhanced image.

3.10 Maximum Difference

Maximum difference (MD) of an image is defined as the absolute difference between the original image at location (m, n) and the processed image at the same location (m, n). The value of MD will be high for a good enhanced image. The expression for MD is given as:

$$\mathrm{MD} = \mathrm{Max}\ (|X(m,n) - Y(m,n)|) \tag{11}$$

where X(m, n) denote the pixel values of the original image at location (m, n), Y(m, n) denote the pixel values of the processed image at the same location (m, n).

4 Discussion

HE [5] is a technique commonly used for image contrast enhancement, since HE is computationally fast and simple to implement. HE performs its operation by remapping the gray levels of the image based on the probability distribution of the input gray levels. Brightness preserving Bi-Histogram Equalization (BBHE) [6], Recursive Mean Separate HE (RMSHE) [7], Brightness preserving Dynamic Histogram Equalization (BPDHE) [9] and AMHE [13] are the variants of HE based contrast enhancement techniques. BBHE divides the input image histogram into two parts based on the mean of the input image and then each part is equalized independently. This method tries to overcome the problem of brightness preservation. RMSHE [7] is an improved version of BBHE. However, it is also not free from side effects. For this purpose, Ibrahim and Kong proposed Brightness Preserving Dynamic Histogram Equalization (BPDHE) [9], this method partitions the image histogram based on the local maxima of the smoothed histogram. It then assigns a new dynamic range to each partition. Finally the output intensity is normalized to make the mean intensity of the resulting image equal to the input one. Adaptively Modified Histogram Equalization [13] scales the magnitudes of the probability density function of the original image before equalization. To resolve the fog-degraded image problem, fog removal method based on bilinear interpolation dynamic histogram equalization (BIDHE) [8] was proposed. The original image was split into some sub-images with same size, and the histogram of each sub-image was partitioned by the local minima. Minimum Mean Brightness Error Bi-Histogram Equalization (MMBEBHE) [11] is a variation of BBHE which perform the separation based on the threshold level, which would yield minimum Absolute Mean Brightness Error. The effect of brightness saturation can be defeated using Multipeak Histogram Equalization [10] which is based on global histogram equalization technique. However, luminance shifting and washed out looking can be occurred on the images after processing by this histogram based methods. To overcome this Average Luminance with Weighted Histogram Equalization (ALWHE) has been proposed by [15] where the automation of parameter selection can easily be done by iteratively using the proposed method.

5 Results and Analysis

A set of three underwater images were taken for the above study. The enhancement algorithms proposed in the Sect. 2 of the paper is applied to all the images and the results were analyzed using the set of quality parameters described in Sect. 3. The detail results are shown in Appendix-1. Figures 1, 2 and 3 shows the results of enhancement using each of these algorithms. Tables 1, 2 and 3 shows the analysis using quality parameters. The quality parameters across each algorithm

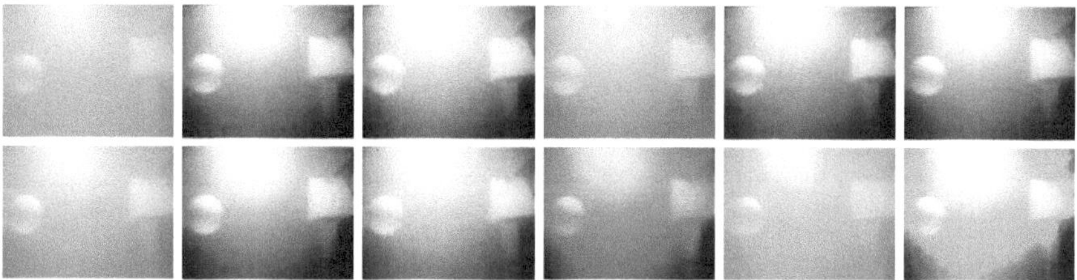

Fig. 1 Image courtesy of Bazeille et al. [12]. From *left* to *right* input Image, HE, BBHE, RMSHE, MMBEBHE, DSIHE, AMHE, WTHE, ALWHE, BIDHE, BPDHE, MHE

Fig. 2 Image courtesy of Bazeille et al. [12]. From *left* to *right* input Image, HE, BBHE, RMSHE, MMBEBHE, DSIHE, AMHE, WTHE, ALWHE, BIDHE, BPDHE, MHE

Fig. 3 From *left* to *right* input Image, HE, BBHE, RMSHE, MMBEBHE, DSIHE, AMHE, WTHE, ALWHE, BIDHE, BPDHE, MHE

for a given input image are studied. Since the parameters GLE and QI do not vary much across different algorithms, we have discarded them for further analysis.

A study of the above three sets of results show that the few algorithms of adaptive enhancement gives the desired value of quality parameters. Analysis show that HE, BBHE, DSIHE and BIDHE gives comparatively higher value of Entropy, Entropy Error Rate, Spatial Frequency, Structural Content, Maximum Difference and lower value of Relative Entropy.

Table 1 Results for image 1

Quality metric	Original	HE	BBHE	RMSHE	MMBEBHE	DSIHE	AMHE	WTHE	ALWHE	BIDHE	BPDHE	MHE
Entropy	6.6354	6.5244	6.5036	6.4579	6.5348	6.5349	6.6191	6.5055	6.6324	7.3265	6.4017	6.4933
EER	0.0059	−0.0020	−0.0110	0.0049	−0.0022	−0.0028	0.0017	−0.0025	−0.0145	0.0081	−0.0118	−0.0191
GC	770.41	5,400.08	5,543.14	1,680.43	5,354.27	5511.38	1,400.16	5,285.62	2,598.29	2,164.19	927.78	3,348.03
SF	13.80	25.07	24.41	16.57	24.95	25.15	16.16	24.49	20.43	27.29	13.66	21.29
GLE	0.0118	0.0120	0.0121	0.0124	0.0120	0.0120	0.0118	0.0121	0.0118	0.0082	0.0139	0.0126
QI	1.5989	1.115	1.110	1.6666	1.1171	1.1118	1.4456	1.1532	1.0867	0.8891	1.4091	1.1314
RE	–	0.1530	0.1610	0.1925	0.1600	0.1476	0.1921	0.1613	0.1342	-1.6288	0.1846	0.1840
AMBE	–	36.97	22.39	8.0954	35.98	34.94	6.90	39.36	16.43	59.59	0.4552	3.0195
SC	–	1.2848	1.0821	1.0644	1.2726	1.2484	1.0605	1.3285	1.1346	2.1046	0.9998	0.9466
MD	–	116	113	36	114	115	26	117	100	255	52	105

Table 2 Results for image 2

Quality metric	Original	HE	BBHE	RMSHE	MMBEBHE	DSIHE	AMHE	WTHE	ALWHE	BIDHE	BPDHE	MHE
Entropy	6.9516	6.9227	6.9211	6.7692	6.9203	6.9286	6.9456	6.9084	6.9503	7.2217	6.9076	6.8041
EER	0.0005	−0.0019	0.0084	−0.0146	−0.0028	−0.0025	0.0003	−0.0027	0.0085	0.0341	−0.0028	0.0062
GC	1,299.08	5,400.88	5,417.80	1,499.89	4,166.45	5,396.63	2,330.61	5,213.55	4,648.34	2,094.04	4,052.509	2,555.74
SF	4.5849	10.48	10.08	5.95	9.1946	10.48	6.27	10.38	8.69	13.91	9.0907	11.8101
GLE	0.0088	0.0088	0.0088	0.0099	0.0088	0.0088	0.0088	0.0089	0.0088	0.0082	0.0089	0.0096
QI	0.8852	0.5212	0.5200	0.7930	0.5928	0.5203	0.7097	0.5359	0.5168	0.5364	0.6062	0.5623
RE	–	0.0124	−0.1047	0.1352	−0.0475	0.0308	0.0647	0.0033	−0.0854	−1.3077	−0.0364	0.1545
AMBE	–	15.71	3.61	0.7493	0.5430	16.19	9.1211	13.29	6.4128	36.31	0.5183	3.0615
SC	–	0.6388	0.7382	0.9742	0.8234	0.6354	0.8149	0.6632	0.7426	1.7674	0.8409	0.9602
MD	–	75	75	43	50	75	30	70	73	184	50	73

Table 3 Results for image 3

Quality metric	Original	HE	BBHE	RMSHE	MMBEBHE	DSIHE	AMHE	WTHE	ALWHE	BIDHE	BPDHE	MHE
Entropy	7.0438	6.9844	6.9883	6.8740	6.9899	6.9673	7.0355	6.9798	7.0426	7.3817	6.9115	6.9300
EER	−0.0090	0.0016	−0.0069	−0.0065	0.0016	0.0097	−0.0041	0.0019	−0.0065	0.0238	0.0025	−0.0017
GC	1,278.54	5,448.03	5,382.94	1,787.48	5,407.15	3,984.35	2,290.77	5,323.44	3,381.54	2,692.08	5,558.22	3,163.95
SF	4.5793	8.7912	8.6880	6.6375	8.7712	7.5359	5.9147	8.6834	7.2813	17.9848	29.75	8.5484
GLE	0.0084	0.0085	0.0085	0.0092	0.0085	0.0086	0.0084	0.0085	0.0084	0.0073	0.0088	0.0089
QI	1.2098	0.8521	0.8365	0.9052	0.8494	0.8781	1.0389	0.8738	0.8324	0.8595	0.8825	0.8660
RE	–	0.1474	0.0835	0.1518	0.1417	0.2229	0.1607	0.1615	0.0913	−1.8994	0.0641	0.1274
AMBE	–	4.8258	6.2111	0.6747	3.8008	22.92	3.3455	7.9872	2.4734	49.8495	6.4832	9.4111
SC	–	0.8668	0.7657	0.9649	0.8580	1.1779	0.9930	0.9051	0.8725	1.9753	0.8793	0.8088
MD	–	62	63	47	62	57	23	57	44	199	189	61

6 Conclusion

The paper discusses a number of techniques for enhancement of underwater images, all of which, in principle, are based on classical histogram equalization techniques. Based on results obtained, the methods have proven to improve several distinctive obstructions in the underwater images. The analysis of the results concludes that HE, BBHE, DSIHE and BIDHE can preserve the mean brightness of underwater images better in comparison to the other methods. The study also shows that only some of the quality parameters can be chosen for the purpose of comparison of different algorithms.

Besides the good features of the algorithms, there still exist some issues and areas of improvement for the proposed techniques. First, the good performance of the algorithm depends on the wise selection of few input parameters. It might be possible to find a set of parameters that optimize the performance however, it requires extensive tests and experiments with different inputs. At the moment, the method for removing outliers is not stable because sometimes it removes too much information which is not outliers and sometimes it misses some artifacts. In terms of algorithms, we can always update with new techniques of image enhancement. There are more and more interest in underwater imaging and many good techniques have been proposed in recent years.

References

1. Kolar R, Odstrcilik J, Jan J, Harabis V (2011) Illumination correction and contrast equalization in colour fundus images. In: 19th European signal processing conference (EUSIPCO 2011) Barcelona, Spain, August 29–September 2 2011
2. Bazeille S, Quidu I, Jaulin L, Malkasse J (2006) Automatic underwater image pre-processing CMM'06—caracterisation du milieu marin, 16 –19 Octobre 2006. Empirical mode decomposition based visual enhancement of underwater images, Aysun Taşyapı Çelebi and Sarp Ertürk
3. Schettini R, Corchs S (2010) Underwater image processing: state of the art of restoration and image enhancement methods. EURASIP J Adv Sig Process 2010, Article id 746052 (Hindawi Publishing Corporation)
4. Iqbal K, Abdul Salam R, Osman A, Zawawi Talib A (2007) Underwater Image Enhancement Using an Integrated Colour Model. IAENG Int J Comput Sci 34:2 IJCS_34_2_12
5. Gonzalez RC, Woods RE (2002) Digital image processing, 2nd edn. Prentice-Hall, New Jersey
6. Kim YT (1997) Contrast enhancement using brightness preserving bi-histogram equalization. IEEE Trans Consum Electron 43(1):1–8
7. Chen S, Ramli AR (2003) Contrast enhancement using recursive mean-separate histogram equalization for scalable brightness preservation. IEEE Trans Consum Electron 49(4):1301–1309
8. Xu Z, Liu X (2010) Bilinear interpolation dynamic histogram equalization for fog-degraded image enhancement. J Inf Comput Sci 7(8):1727–1732
9. Ibrahim H, Kong NSP (2007) Brightness preserving dynamic histogram equalization for image contrast enhancement. IEEE Trans Consum Electron 53(4):1752–1758

10. Wongsritong K, Kittayaruasiriwat K, Cheevasuvit F, Dejhan K, Somboonkaew A (1998) Contrast enhancement using multipeak histogram equalization with brightness preserving. IEEE Asia Pac Conf circuits syst 452–458
11. Wang Y, Chen Q, Zhang B (1999) Image enhancement based on equal area dualistic sub-image histogram equalization method. IEEE Trans Consum Electron 45(1):68–75
12. Chen S, Ramli AR (2003) Minimum mean brightness error bi-histogram equalization in contrast enhancement. IEEE Trans Consum Electron 49(4):1310–1319
13. Chang LW, Lie W-N, Chiang R (eds) (2006) PSIVT 2006, LNCS 4319. © Springer, Berlin Heidelberg, pp 1150–1158
14. Wang Q, Ward R (2007) Fast Image/Video Contrast Enhancement Based on WTHE. In: Proceedings of IEEE on MSP 2007. pp 339–342
15. Tai SC, Chang YY, Li KM, Tsai TC (2009) Contrast enhancement method based on average luminance with weighted histogram equalization. In: Proceedings of IEEE on IAS 2009. pp 555–558
16. Ye Z (2009) Objective assessment of nonlinear segmentation approaches to gray level underwater images. ICGST-GVIP J 9(2) ISSN 1687-398X
17. Lizuo J, Shin'ichi S, Masao S (2004) A novel adaptive image enhancement algorithm for face detection. In: Proceedings of the 17th international conference on pattern recognition (ICPR'04)
18. Grgi S, Grgic M, Mrak M (2004) Reliability of objective picture quality measures. J Electr Eng 55(1, 2):3–10
19. Leu JG (1992) Image contrast enhancement based on the intensities of edge pixels. CVGIP: Graph Models Image Process 54(6):497–506
20. Jafar I, Ying H (2007) A new method for image contrast enhancement based on automatic specification of local histograms. IJCSNS Int J Comput Sci Netw Secur 7(7):1–10

Real-Time Automatic Detection and Recognition of Hamming Code Based Fiducial Marker

Sundar Raj Deepika and N. Avinash

Abstract A fiducial marker is an object used in the field of view of imaging system which appears in the image produced, for use as a point of reference or measure. This paper explains a novel fiducial marker generation and also technique to automatically detect and recognize them in different environments. Hamming code technique is used to encode and decode a fiducial marker, and for the detection process we make use of grey scale segmentation algorithm and the methods have been validated. In the experiment result we compare the performance of the algorithm under different lighting condition and we also calculate its efficiency.

Keywords Fiducial marker system · Augmented reality · Robot navigation

1 Introduction

A fiducial marker is an object used in the field of view of an imaging system. This appears in image produced, for use as a point of reference or a measure, which may also be placed into the imaging subject. In view of its planar nature it can be placed on floor, ceiling, wall and so on, which will be detected by the camera without any recalibration. By viewing the basic shape of fiducial, machine vision technique make use of perspective and affine distortion to maintain recognition.

S. R. Deepika (✉) · N. Avinash
PES Institution of Technology, Bangalore, Karnataka, India
e-mail: deepisr27387@gmail.com

N. Avinash
e-mail: avinashn@pes.edu

Mohan S. and S. S. Kumar (eds.), *Proceedings of the Fourth International Conference on Signal and Image Processing 2012 (ICSIP 2012)*, Lecture Notes in Electrical Engineering 221, DOI: 10.1007/978-81-322-0997-3_51,

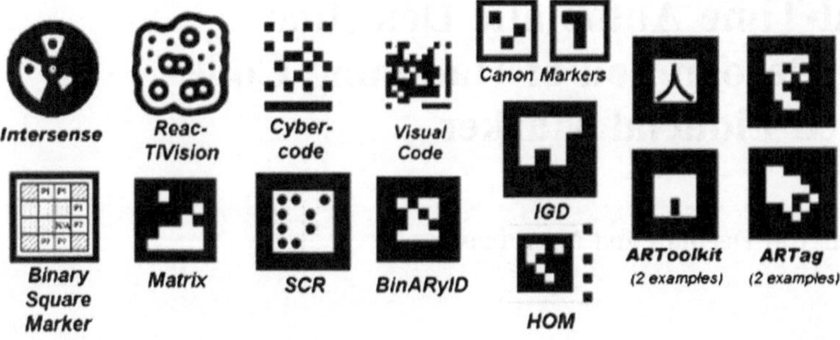

Fig. 1 Several fiducial marker systems

These markers are useful for Augmented Reality, robot navigation, photogrammetry and in application where relative pose between a camera and object is required.

Fiducial marker systems are systems consisting of unique pattern, which are automatically detected by digital camera to locate their projection in image using necessary algorithms. The fiducial marker patterns should be distinct enough to avoid confusion with respect to the environment. Ideally there should be a library of unique markers to distinguish from one another and robust enough to find them in uncontrolled lighting, image noise and blurring, unknown scale or partial occlusion.

The Fig. 1 below shows some of the fiducial marker system such as HOM, SCR, IGD, ARTag, ARToolkit etc. In HOM [1], the original version of marker system are used for processing static image and is compiled into a software library by Siemens Corporate Research(SCR) for processing real-time video sequences. In case of IGD marker system [1] feature extraction and processing speed is considered to be accurate and reasonable for real-time video sequence. ARToolKit [2], ARTag [3–5], ARToolKitPlus [6] systems are mainly used for augmented reality, robot navigation, photogrammetry etc. In ARToolKit [2] system the detection of marker is threshold based and the verification and identification mechanism is done using correlation and the processing time also depends on the library size. The ARTag [3–5] make use of edge based approach for detection and Forward Error Correction (FEC) and robust digital techniques of checksums are used during verification and identification. ARToolkitPlus [6] uses the same unique feature detection step, and only differs from ARToolkit in the verification/identification step.

The above marker systems are commercially available in market. Since these markers are expensive, it is not appropriate to use them for simple application such as calculating the angle of an object, position of an object etc. To handle these kind of issue we aim to develop fiducial marker system using simple technique, which can be applied to develop these kind of application in cost efficient way.

This paper focuses on the algorithms used for generation of fiducial marker, segmentation and recognition of fiducial marker and their performance under different constraints.

2 Methodology

Two different systems developed are explained in this section namely

- Generation of Fiducial Marker
- Segmentation and Recognition of Fiducial Marker

In the generation of fiducial marker system we design the fiducial marker using Hamming code technique based on the decimal value provided as input. Where as in the segmentation and recognition stage we first perform segmentation operation on the input image to separate the fiducial marker from the input image, and then make use of Hamming code technique to decode the marker. Both the stages are explained in detail in the Sects. 2.1 and 2.2

2.1 Generation of Fiducial Marker Algorithm

The steps followed for generation of fiducial marker is as follows

Algorithm 1 *Algorithm explaining the generation of fiducial Marker*

Step 1. Input a decimal value.
Step 2. Obtain a 10 bit binary number for the given input value.
Step 3. For every consecutive 2 bit of binary number, create corresponding Hamming Code and store the value as a 2 dimensional matrix.
Step 4. Create a binary black image.
Step 5. Markers are generated based on the value stored in the matrix.

In step 1, to generate a fiducial marker first select a decimal value between 0 and 1,023. This range is selected because a 5×5 matrix is used to generate fiducial marker, where each row of the matrix represents a hamming code of 2-bits binary number. Since a 5×5 matrix can hold a maximum of 10-bit binary number, i.e. a maximum number of 1,024 different numbers. In step 2, the decimal input value is converted to a 10-bit binary number. To this binary number add 2 zeros at both end which results in a 14-bit binary number. In step 3, for each of the 2-bits of the 14-bit binary number a hamming code value is calculated, which is in-turn stored in a 7×7 matrix from 2nd to 6th position and the remaining position are filled with zeros as shown in Fig. 2a below. In step 4 a black binary image is created, which is used along with hamming code matrix to generate fiducial marker. In step 5, the binary image is first divided into 7 equal parts both horizontally and vertically. Based on the value present in hamming code matrix corresponding sub region within the binary image is filled with either black or white colour (0 for black and 1 for white). After the execution of the above steps the output obtained results in fiducial marker. The design of fiducial marker depend on the input value selected

2.2 Segmentation Algorithm

The step followed in segmentation algorithm is as follows

Algorithm 2 *Algorithm explaining the segmentation of image*

Step 1. Input to the system is an image of a scene in which a Fiducial Marker is present
Step 2. Apply contrast stretching on the Grey scale image
Step 3. Apply Bimodal Thresholding technique to obtain the threshold value of the image
Step 4. Apply Contour Detection to obtain the possible ROI (Region of Interest) location of marker within the image
Step 5. Reduce the number of false ROI s detected based on heuristic rules

In step1, the Images with one or more fiducial markers, streamed from a camera (e.g. built-in webcam of Laptop) are used as input. This input image is converted to gray scale (shown in Fig. 5a). In step 2, Contrast-stretching [7] is performed on the gray image. The Fig. 3 shows the transformation function where 'r' represents the input gray level and 's' represents the output gray level, where the r value is considered to be L/2 where L value is considered as 256. To determine the minimum and maximum contrast stretching required, initially the 'r' value (both r1, r2) is saturated to 127 and the s1 value ranges between 0 and 127 and s2 value ranges between 127 and 255. An experiment is performed to determine at what values of s1 and s2, the marker detection rate is high (shown in Fig. 4 where the x coordinates represents the output grey level s and the y coordinates the percentage of marker detected). From the experiment we come to know that at a value of 50 in case of s1 and 200 in case of s2 the marker detection rate is high. Therefore we set (r1,s1) and (r2,s2) value to (127,50) and (127,200), in order to make the bright region of the input image more brighter and the dark region more darker, to increase the contrast of the image, which in turn aid in obtaining a high quality.

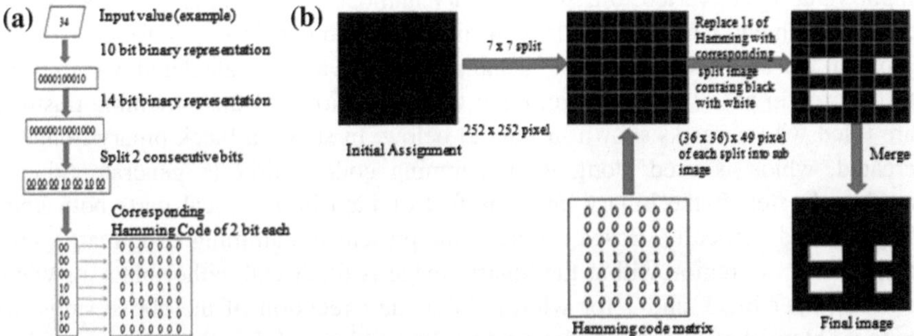

Fig. 2 **a** Hamming code matrix generation. **b** Conversion of hamming code matrix into fiducial marker

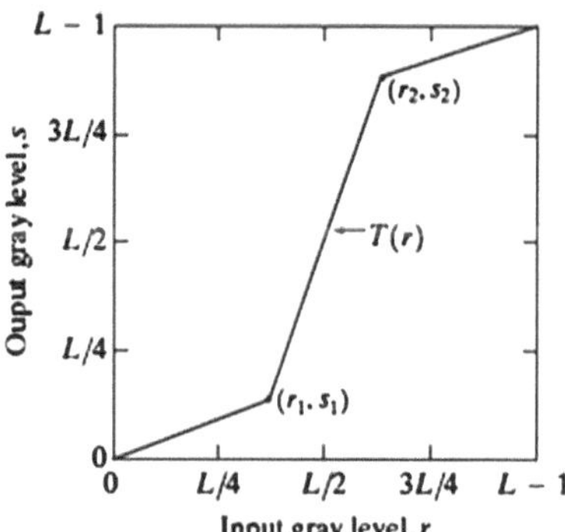

Fig. 3 Form of transformation function

During Bimodal thresholding [8], the local minimum value obtained, will be considered as the threshold value, which is applied on the input image (contrast stretched image) in order to obtain the binarized image as shown in Fig. 5c.

In step 4, contour detection is applied on the binarized image obtained in previous step. In the contour detection process all the contours [9] within the image are detected and highlighted to view the location of contour within image (as shown in Fig. 5d). In step 5, the contour detected in previous step represents the region of the image in which the fiducial marker may or may not be present. So we make use of certain heuristic rules such as area of fiducial marker, aspect ratio of the marker, to select the region of image in which the marker is present, which in turn helps in identifying the location of fiducial marker in image (as shown in Fig. 5e) In step 6, after obtaining the location of fiducial marker, plane to plane homography is used to perspectively undistort the regions in the image [10] This is solved as explained in equation below.s

$$A = T * B \tag{1}$$

where A is given as

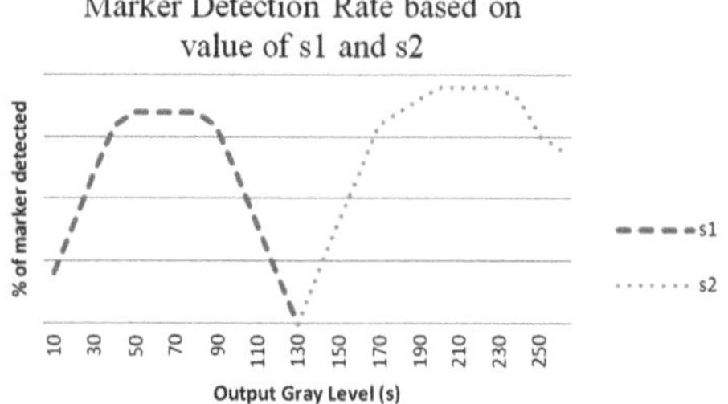

Fig. 4 Marker detection rate based on s1 and s2

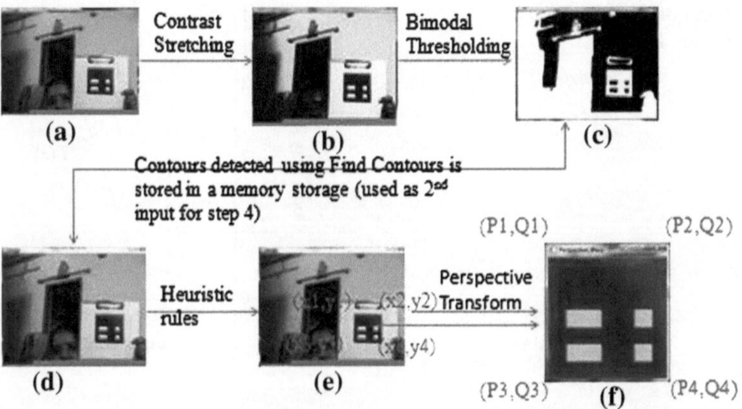

Fig. 5 Segmentation algorithm

$$
\begin{bmatrix}
-x1 & -y1 & -1 & 0 & 0 & 0 & P1x1 & P1y1 \\
0 & 0 & 0 & -x1 & -y1 & -1 & Q1x1 & Q1y1 \\
-x2 & -y2 & -1 & 0 & 0 & 0 & P2x2 & P2y2 \\
0 & 0 & 0 & -x2 & -y2 & -1 & Q2x2 & Q2y2 \\
-x3 & -y3 & -1 & 0 & 0 & 0 & P3x3 & P3y3 \\
0 & 0 & 0 & -x3 & -y3 & -1 & Q3x3 & Q3y3 \\
-x4 & -y4 & -1 & 0 & 0 & 0 & P4x4 & P4y4 \\
0 & 0 & 0 & -x4 & -y4 & -1 & Q4x4 & Q4y4
\end{bmatrix}
$$

and the vectors

$$T = (a_{11}, a_{12}, a_{13}, a_{21}, a_{22}, a_{23}, a_{31}, a_{32})^{T} \tag{2}$$

$$B = (P_1, Q_1, P_2, Q_2, P_3, Q_3, P_4, Q_4)^{T} \tag{3}$$

in which (x1,y1), (x2,y2),(x3,y3) and (x4,y4) represents the location of the marker in the image and (P1,Q1),(P2,Q2),(P3,Q3), and (P4,Q4) represents the corners of a quadrilateral. The detail description of the process is shown in Fig. 5f.

2.3 Algorithm for Recognition of Fiducial Marker

The steps followed for decoding of fiduicial marker is as follows

Algorithm 3 *Algorithm explaining the recognition of fiducial marker*

Step 1. ROI of Fiducial Marker is divided into 7 × 7 sub image
Step 2. All 7 × 7 sub image are binarized

Step 3. In each sub image the maximum contribution of white/black region are considered and a 7 × 7 matrix is formed where each row and coloum depicts the maximum value of either white or black correspondingly. This is the Hamming Code Matrix

Step 4. From the Hamming code matrix corresponding binary value are formed for each row of Hamming code matrix

Step 5. The binary value obtained are arranged sequentially

Step 6. Obtain a 10 bit binary value from the 14 bit binary value

Step 7. Corresponding decimal value of 10 bit binary number is obtained

In step1, the output image (ROI of fiducial marker) obtained after performing segmentation process is used as input, which is divided into 7 equal parts both horizontally and vertically as shown in Fig. 6a. In step 2, all the 7 × 7 sub image are binarized i.e. the black color in the image is replaced by 0 and the white color is replaced by 1. In step 3, in each of the sub image, maximum contribution of

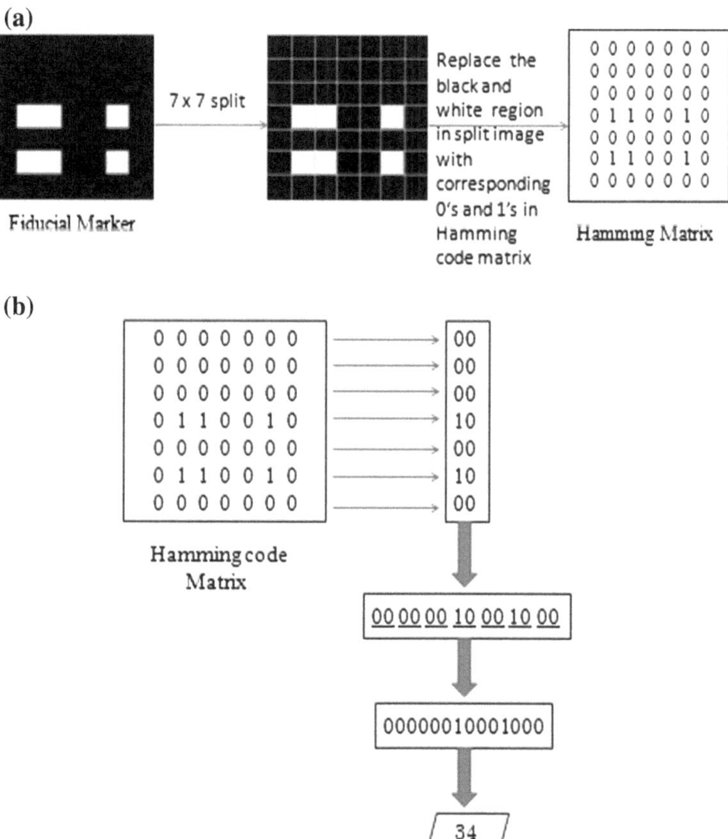

Fig. 6 **a**, **b** Explains stage wise recognition of fiducial marker

white/black region are considered and a 7 × 7 matrix is formed where each row and column depicts the maximum value of either white or black correspondingly (where 0 in the matrix represents black and 1 represents white)

This is how the Hamming Code Matrix is generated (as shown in Fig. 6a). In step 4, in the Hamming code matrix the element from 2nd to 6th position in each row represents a hamming code value for a 2 bit binary number, which results in 7 pair of 2-bits binary number as show in Fig. 6b. In step 5, the 7 pairs of 2-bit binary number are sequenced to obtain a 14-bit binary number, where the 2 zeros at both the binary number, the 2 zeros at both the ends are eliminated to obtained a 10-bit binary ends represent the border of the marker and the remaining 10-bits represents the actual information as shown (Fig. 6b). In step 6 among the 14-bits in the 14-bit number. In step 7, the 10-bits binary number is in-turn converted to its equivalent decimal value, which is the output of the decoding process.

3 Measuring Current Fiducial Marker

A fiducial marker detection system must be reliable and any failure includes poor tolerance/lighting, false reporting objects of similar in appearance is encountered. A non robust marker will provide false measurement. The robustness of fiducial marker system depends on qualitative observation of false positive rate, false negative probability, minimal marker size, library size, immunity to lighting, and the perspective support with low cost computing power.

False positive rate reports falsely the presence of a marker. The intermarker confusion will be due to wrong id. The false negative probability when the fiducial marker is present in an image but not reported. The minimal marker size in pixel affects the reliable detection. The library size should be supported by the system. Immunity to lighting is important to recognize the fiducial marker pattern. Perspective support is necessary with at least 4 salient points.

In this current work we consider the proof or concept as software system to test the performance on light condition, and perspective support as the very important factors for software system.

4 Experimental Results

In experiment 1, the fiducial marker system is tested under different light condition (i.e. from normal to very bright progressively), for which we consider a video of 7,000,000 frames as input. For statistical analysis purpose we have considered only three conditions namely normal, bright, very bright, which are in turn subjected to different test case. In each of the test case, the 2nd peak in bimodal thresholding [8] is calculated by considering the highest peak in the histogram that appear between a particular bin value and 255, where the bin value ranges from 0 to 255,

Table 1 Marker detection rate under different light conditions

Light condition	Test 1 (%)	Test 2 (%)	Test 3 (%)
Normal condition	49	95	92
Bright condition	82	80	78
Very-bright condition	52	52	55

Table 2 Average of 50 experiments of fiducial marker detected rate at different pan angle

Pan angle	−50	−40	−30	−20	−10	0	10	20	30	40	50
Fiducial marker detection rate (%)	51.6	66	78.7	82.9	89.2	95	89.2	82.9	78.7	66	51.6

where as the 1st peak remains the same as explained in the algorithm. For statistical purpose we have considered 3 test cases test1, test2, test3 where the 2nd peak of bimodal thresholding [8] is considered between the region 0–255, 35–255, 50–255 of the histogram The Table 1 and Fig. 7 below show the percentage of marker detected in different light condition for each test case, where the x coordinates in Fig. 7 represents the bin value and the y coordinates represents the percentage of marker detected.

From the Table 1 and Fig. 7, we can infer that at the bin value of 35, the marker detection rate is 95 and 80 % in normal and bright condition, and acceptable in very bright condition when compared with other bin value, which is used in further experiments. In experiment 2, we calculate the pan angle of fiducial marker under different light condition where the fiducial marker detection rate is 95 %, for

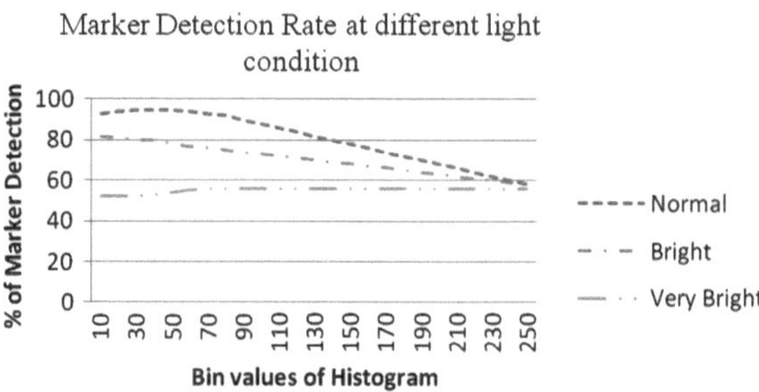

Fig. 7 Marker detection in different light condition

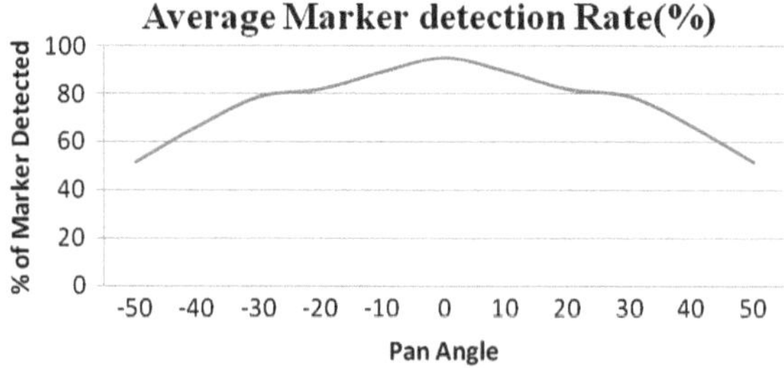

Fig. 8 Marker detection with respect to pan angle

which we consider different videos of 3,000,000 frames as input. These videos are captured by placing the markers at different angle(ranging from −50 to 50) on the calibrated table, whose results are shown in the Table 2 and Fig. 8 below, where the x coordinates of the Fig. 8 represents pan angle and y coordinates represents the percentage of fiducial marker detected.

From Table 2 and Fig. 8 we can infer that at a pan angle of −30 to +30° the marker detection is above 80 %.

5 Conclusion

In the paper we present two algorithms namely generation of fiducial marker and the 2nd is automatic detection and recognition of fiducial marker. In the generation of fiducial marker, the fiducial marker is generated based on the input decimal number. In detection process a natural image is captured which contains a fiducial marker in it. From the given input image the fiducial marker is segmented and recognized for its decimal (marker id). Along with this we perform experiment to determine the marker detection rate at different light conditions and also the maximum permissible pan angle of the fiducial marker.

From the above two experiments we come to a conclusion that the fiducial marker can be rotated up to a pan angle of 30° (both clockwise and anticlockwise) under normal condition to obtain detection rate to be greater than 70 %.

References

1. Zhang X, Fronz S, Navab N (2002) Visual marker detection and decoding in AR systems: a comparative study. In: International symposium on mixed and augmented reality ISMAR 2002
2. Poupyrev I, Kato H, Billinghurst M (2000) ARToolkit user manual, version 2.33. Human Interface Technology Lab, University of Washington, Washington
3. Fiala M (2005) ARTag, a fiducial marker system using digital techniques. IEEE Comput Soc Conf Comput Vis Pattern Recogn 2:590–596
4. Fiala M (2010) Designing highly reliable fiducial marker. IEEE Trans Pattern Anal Mach Intell 32:1317–1324
5. Fiala M (2004) Artag revision 1, a fiducial marker system using digital techniques. In National Research Council Publication, Canada
6. Wagner D, Schmalstieg D (2007) ARToolKitPlus for pose tracking on mobile devices. In: Proceedings of 12th computer vision winter workshop 2007
7. Gonzalez RC, Woods RE (2002) Digital image processing, 2nd edn. Prentice Hall, Upper Saddle River
8. Shen H, Johnson CR (1994) Semiautomatic image segmentation: a bimodal thresholding approach. University of Utah, Salt Lake
9. Bradski G, Kaehler A Learning open CV
10. Wang X, Klette R, Rosenhahn B (2005) Geometric and photometric correction of projected rectangular pictures. The University of Auckland, New Zealand

Automated Video Surveillance for Retail Store Statistics Generation

N. Avinash, M. S. Shashi Kumar and S. M. Sagar

Abstract Video Surveillance Systems have gained immense popularity in the recent past because of the fact that it can be used in numerous real-world scenario applications. Monitoring the people flow pattern as well as counting them serves as valuable information in many surveillance related applications. In this paper we propose a system that is used for counting the number of people passing through the camera field of view. A single overhead camera is used to get a clear top-view which avoids occlusions. For background subtraction, running Gaussian approach has been used as a preprocessing step, to facilitate the further segmentation and tracking procedures. Connected component analysis is used to group the similar blobs together followed by intensity based correlation for blob matching followed by Kalman tracking. The percentage of blobs that crosses a reference line is recorded. Two counters are incremented depending on the direction of movement of the blobs and the algorithm is able to count the number of people moving up/down the scene.

Keywords Running Gaussian approach · Kalman tracker · Connected component analysis · Intensity based correlation

N. Avinash (✉) · M. S. Shashi Kumar
Wittybot Technologies, Bangalore, India
e-mail: avinash@wittybot.com

M. S. Shashi Kumar
e-mail: shashikumar@wittybot.com

S. M. Sagar
Manipal Centre for Information Science (MCIS), Manipal, India
e-mail: sagggy.s@gmail.com

Mohan S. and S. S. Kumar (eds.), *Proceedings of the Fourth International Conference on Signal and Image Processing 2012 (ICSIP 2012)*, Lecture Notes in Electrical Engineering 221, DOI: 10.1007/978-81-322-0997-3_52, © Springer India 2013

1 Introduction

Automatic People Counting systems are stand-alone devices which are mounted to the ceiling that counts the number of people passing through its field of view. These devices can be used in many public places (retail shop, stadium, airport, bus, train) for monitoring and controlling human traffic. Having an accurate count of the number of people can prove to be very useful in developing a suitable marketing strategy, crowd management, safety and security purposes etc.

Counting people manually has proved to be monotonous, erroneous, and ineffective. Mechanical counters such as turn-stiles are more accurate, but it counts only one person at a time, causing congestion in crowded places. Beam sensors can also be used to determine the number of people, but it doesn't deal with the occlusion problem. Hence, many image-processing based approaches [1–7] have been put forward to address the above problems. It is a challenging task, because of the variation in the size and shapes of humans and also the color, which depends on their clothing and appearance.

There are two kinds of approaches existing in literature, one is, the count happens on a frame by frame basis [8, 9] where features are extracted from every single frame and doesn't consider the correspondence of the same person in multiple frames, because of which the count obtained may be inaccurate, when taken over a period of time. The second is based on detection and tracking [1, 5], which happens in every frame, before a particular person is counted. Here the count is not dependent on a single frame but on a sequence of frames.

Among many techniques used to separate the moving parts from the background, frame differencing [1, 4, 10] and background subtraction [1, 5] were most common ones found in the literature. There are different methods followed by the authors for object detection, like, Georgino et al. [3] used a self-organizing map for the detection of head and non-head pixels. Hsieh et al. [11] described an intelligent head-shoulder region detector for extracting all possible pedestrian candidates. Zhao et al. [12] modeled different parts of a human body for a given pose, where different poses were stored in the database and when a blob was detected it was compared with the existing poses in the database to find a match. This method is only confined to the information available in the database. Chen [4] proposed a hue saturation color value model for detection of foreground regions.

The tracking method employed depends on the kind of features extracted. Over various methods found in literature, Kalman filter [6] and mean-shift [7] based tracking are used to track the trajectories of the detected objects. Aik et al. [2] talks about a curve-analysis method, where a curve is generated when an object is found to be in the counting area. Virtual lines [4] drawn at the centre of the images are used for the counting purpose.

In this paper, we propose a bi-directional people counting algorithm, where background subtraction method is implemented using a running Gaussian average [13], to extract the foreground regions which is followed by noise removal using morphology filtering [10]. These foreground objects are detected using blob analysis [3, 6, 10] followed by shadow correction [6, 14]. Features extracted from these objects of interest were used to filter the blobs. These blobs are tracked using Kalman filter [6, 7] and matched using a constructed match matrix. A counter is incremented whenever a blob crosses over the reference line at the center of the image.

The next section gives an introduction about the system overview. Section three explains about the methodology part consisting of background subtraction, then the segmentation module and finally the tracking and matching of the segmented blocks. Section four describes the counting technique, which is followed by results in section five and conclusion in section six.

2 System Overview

This system is divided into three main blocks which constitutes the people counter engine as shown in Fig. 1. The blocks are as follows, Background subtraction, Segmentation and the Tracking and Matching block. The images are captured from an over-head fixed camera. The frame rate of the camera is 25 fps with a resolution of 320 × 240 pixels. The captured frames are processed in the background subtraction module, where the foreground objects are extracted using a dynamic background model. This background model is created using Running Gaussian average [13]. Morphology filtering [10] is used for noise removal. The filtering process is followed by the segmentation stage, where blob detection is used which works on connected component analysis. Once the blobs are detected, they are filtered based on their attributes. After removing the blobs which did not fit into the criteria, the remaining blobs are tracked using Kalman filter. To find the correspondence between the blobs a match matrix is formulated using Euclidian distance and intensity correlation. A reference line at the center of the image is used to find out how much percentage of a certain blob crosses it, counting is based on this information. Two different counters are incremented depending on the direction of movement of the blobs. The camera set up is as shown in the Fig. 2.

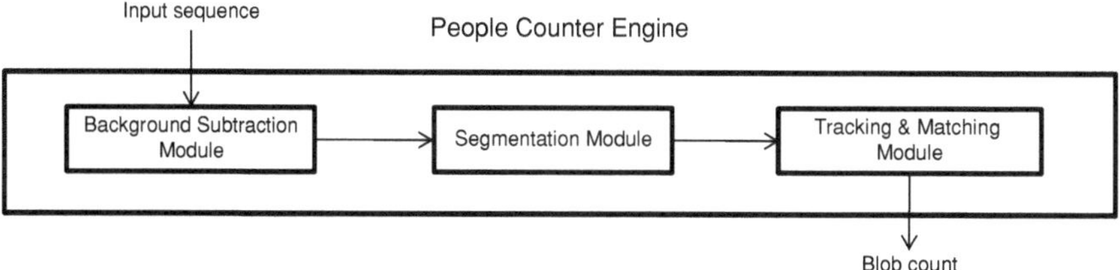

Fig. 1 Camera setup

3 Methodology

A dynamic adaptive background model is formed using Gaussian averaging method for background subtraction, which gives us a set of frames with only the foreground objects in it. The subtraction technique has been explained in detail below.

3.1 Background Subtraction

In most of the surveillance applications, segregating the foreground objects from the static background is the first and foremost step. In our approach Gaussian averaging [13] is used to create an adaptive background model which separates the foreground pixels form the background.

An adaptive background model is created where every pixel in the image is modeled based on the Gaussian probability density function. Here each pixel is being modeled independently based on the previous n frames. Initially the camera is run in the warm-up mode for a few seconds (3–5 s) where the mean and variance are calculated for each frame at pixel level. When it is done with warming up (learning), classification of each pixel into foreground or background begins.

A certain pixel is termed as foreground/background based on the inequality equation mentioned below, and if a pixel is classified as foreground it will be removed from the background model. This method is unaffected by the slight variations in the illumination, which happens to be a common problem out-door surveillance applications.

$$\mu_{t+1} = \alpha I_t + (1 - \alpha)\mu_t \tag{1}$$

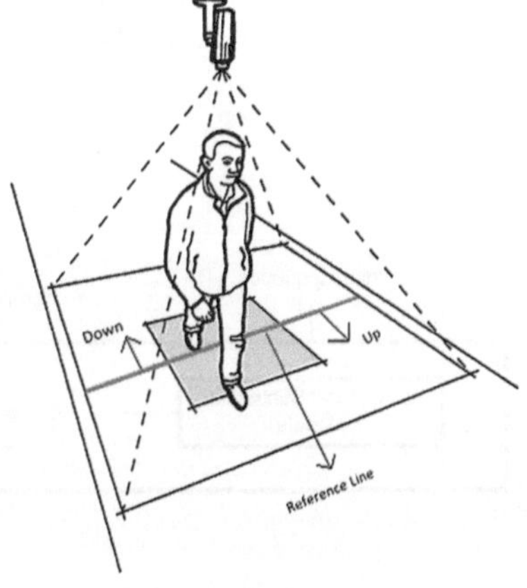

Fig. 2 People counter engine

$$\sigma_{t+1}^2 = \alpha(I_t - \mu_t)^2 + (1 - \alpha)\sigma_t^2 \tag{2}$$

$$|I_t - \mu_t| > Th \tag{3}$$

where μ is the mean, α is the learning factor which is chosen accordingly, I_t is the current pixel value and σ^2 is the variance. The system's learning ability (fast/slow) depends on α.

Once we have obtained the subtracted images, we use a filtering process to remove the tiny bright foreground pixels (noise) using morphology. The binary frames obtained from the classification stage are subjected to open based operation of the morphology [10], where the images are eroded and the resulting images are then dilated. This effect helps in the removal of noise caused by the bright foreground pixels.

Opening of an image 'I' by a structuring element 'S' is given by,

$$I \bigcirc S = (I \ominus S) \oplus S \tag{4}$$

Where $\bigcirc$ represents open operation, $\ominus$ represents erosion and $\oplus$ represents dilation.

A 3 × 3 mask as shown in Fig. 3 was used as a Structuring Element for performing erosion and dilation, which helped in filtering out the objects (pixels) which are smaller in area compared to the SE.

The Fig. 4 shows the output of background subtraction and morphology filtering.

The Fig. 5 detailed block diagram showing the first two blocks, which correlate the Fig. 2.

Fig. 3 Structuring element

1	1	1
1	1	1
1	1	1

Fig. 4 Original frame and subtracted frame

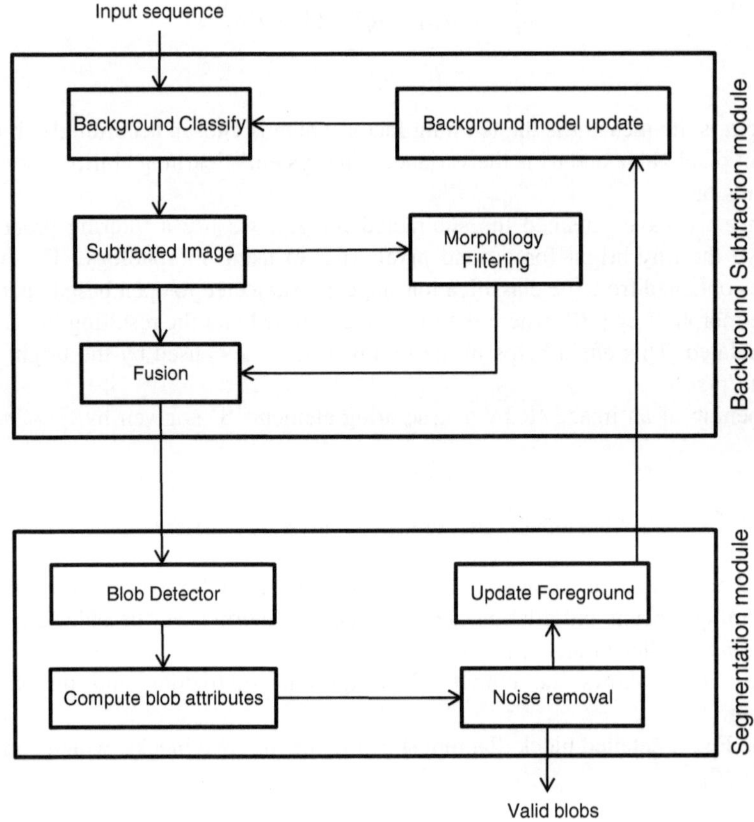

Fig. 5 First two blocks of Fig 2, in detail

3.2 Blob Detection

Morphological filtering is followed by blob detection. The actual technique involved in the blob detection process is the Connected Component Analysis [4], which connects the pixels with similar attributes together as one component based on their connectivity. A connected component can be defined as a cluster of pixels connected to each other. The images are scanned from left to right and from top to bottom. A four-connectivity method is used where the current pixel is compared with its neighbors before it is assigned a label. Thus, several groups of connected pixels will be labeled accordingly giving rise to many individual blobs.

For a pixel 'm' with co-ordinates (x, y), adjacency criterion between pixels in a four-connected method,

$$N_4(m) = \{(x+1, y)(x-1, y)(x, y+1)(x, y-1)\} \tag{5}$$

After the detection of the blobs, the images are sub-sampled to enhance connectivity between blobs. Shadows may be introduced in some parts of the scene when an object comes in the way of light. Presence of shadows in the segmented frames causes problems, like shadows of particular objects may fall on others leading to wrong object detection (shadow detected as object), merging of objects or even object losses. To overcome this problem, a shadow correction technique [6, 14] is applied to remove the shadow inside the blob. Once the shadow within the blob was removed, it was enclosed with a bounding box and blob statistics were calculated. Figure 6 shows a bounding box indicating a detected blob.

The enclosure of the bounding box was followed by the calculation of their attributes like, its width, height, area, and centroid. All the detected blobs are filtered based on their features (area, compactness), this forms the second phase of noise removal after morphology filtering. Filtering will remove the smaller blobs and retain a set of blobs which are termed as valid; these blobs will be assigned an id. For every few frames after the blob detection and filtering stages the background model gets updated. Tracking of the valid blobs are explained in the following module.

3.3 Tracking and Matching

The speed at which a particular blob is moving is calculated, which is used to estimate its location in the following frames. In this tracking process, estimation and prediction happens simultaneously. Prediction begins after a few frames, once a particular threshold is crossed. Meta data block is also updated every time with the estimated and predicted values. These values are sent to the object matcher. The condition box will check if the object is broken down into many objects (split) or has it merged as one or if it is a matched object, based on which a corresponding flag will be set (Fig. 7).

In case of merging, two different blobs having unique trials may merge in the counting area and be counted as one, In such cases the matcher checks for the

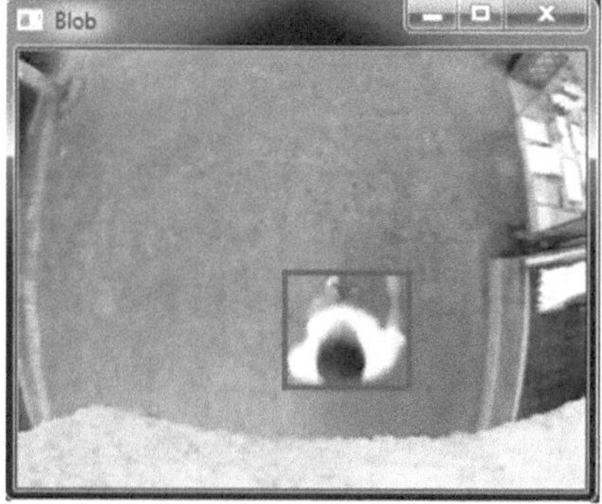

Fig. 6 Detected blob

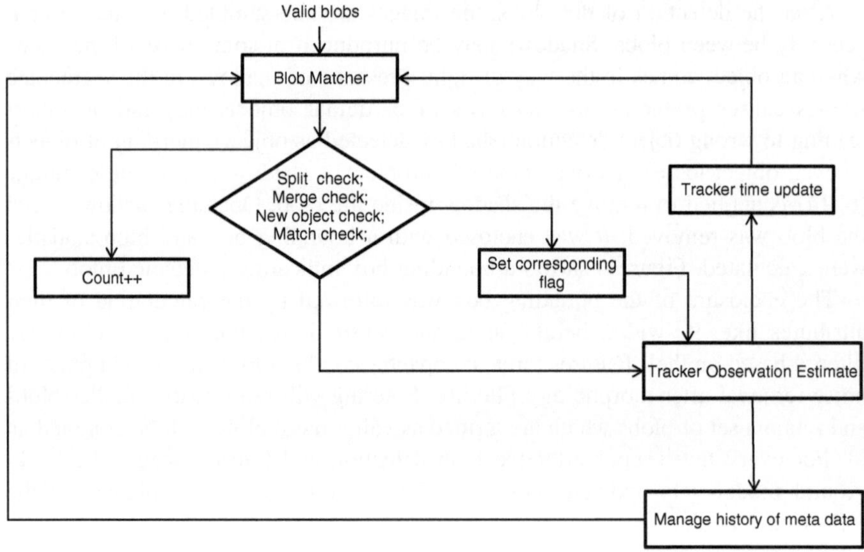

Fig. 7 Tracking module

history of those two blobs stored in the meta data block and then corrects the count. Likewise split and new object detection is also addressed depending on the flag set.

Prediction normally takes over the actual estimation when the object abruptly disappears from a scene. Major part of the predicted value and some part of the estimated value is considered in this case.

Kalman filter [6, 7] is an optimal recursive estimator. They work by predicting and updating the state values, and comparing them with the actual measurements and then modifying its estimated value based on the comparison.

Using the set of equations mentioned below the filter predicts the state of a vector at a future time k + 1, and then corrects it based on all the measurements it has.

Prediction:

$$\overline{X_{k+1}} = A_k X_k \tag{6}$$

$$\overline{P_{k+1}} = A_k P_k A_k^T + Q_k \tag{7}$$

Correction:

$$K_k = \overline{P_k} H_k^T \left(H_k \overline{P_k} H_k^T + R_k \right)^{-1} \tag{8}$$

$$X_k = \overline{X_k} + K_k (z_k - H_k \overline{X_k} \tag{9}$$

$$P_k = (1 - K_k H_k) R_k \tag{10}$$

Where, A and H are the transition and measurement matrix respectively. X_k represents the state at time instant k, z is the measurement and K is the Kalman gain. P is the error covariance, Q and R are the noise co variances. The values with a bar on top are predicted values.

Blob matching is a process which tries to find an exact match between every corresponding blob. A match matrix is formed to find the correspondence between blobs in the following frames. The entry in the match matrix consists of the Euclidean distance and intensity correlation of every tracked blob which is used as a measure to find the matching blob in the next frame. Match matrix is as shown in Table 1, it is a data structure in the form of a list containing the history of the frames and versus number of blobs in it. Euclidean distance is calculated between blobs of the current frame and previous frames which is used to identify that the blob belongs to the same person. Histogram of every valid blob was found, which was used in the intensity correlation. The values obtained from the Euclidean distance and intensity correlation were averaged out to find the exact correspondence between the blobs in the current frame and the tracks from the previous frames.

3.4 Counting

In the last stage, the number of valid blobs which crossed a reference base line is counted. This line was drawn at the center, and if more than three-fourths of a blob crossed this line, a counter is incremented. Tracking provides cues of the trial history which marked by the blob which is crossing over and the counter is incremented based on number of trials that blob is associated with. This provides the number of people passing through the camera's field of view.

4 Experiments and Results

This approach is tested on six videos of ten minute duration captured form an overhead mounted camera. All the videos had a frame rate of 25 fps and a resolution of 320 × 240. The videos were taken at uniform lighting conditions. However, shadows were encountered in a few videos, which were removed using shadow correction. The actual number of people passing through in video 1 is,

Table 1 Match matrix; the numbers in block showing the closest match

	Blob 1	Blob 2	Blob 3
Predicted 1	**0.9986**	0.0001	0.0012
Predicted 2	−1.0345	0.0021	**0.8345**
Predicted 3	−0.3455	**0.9124**	−0.8873

Table 2 Experimental analysis

	Accuracy (%)	Actual count	Algo count
Video 1	90	222	200
Video 2	91	137	125
Video 3	88	156	138
Video 4	89	129	115
Video 5	88	182	157
Video 6	94	163	154

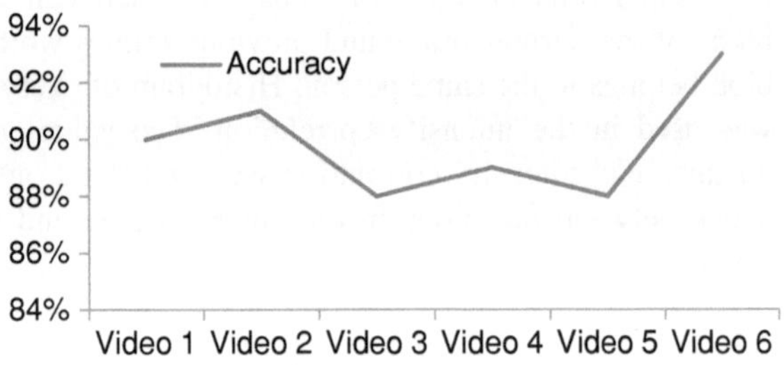

Fig. 8 Graph showing videos versus accuracy

up = 133 and down = 89, the algorithm count is given as up = 121 and down = 79. Accuracy is around 90 % on an average as seen from all the six videos we have tested (Table 2, Fig. 8).

From the experimental analysis, it is found that, if two objects appeared very close to each other and moved together in the same way until they crossed the reference line, it was counted as one and the algorithm had problems in segmenting persons who wore dark clothes similar to the background, and due to this further detection was also limited.

But the algorithm performed correctly if the objects had different trials initially, and if they appeared to be merged as one in the counting zone (Fig. 9a–h).

5 Future Work

The problem we faced, where two persons entered the scene appearing very close to each other and continued in the same way until they left the field of view can be addressed by performing a shape based or color based retrieval for differentiating between them and getting an exact count.

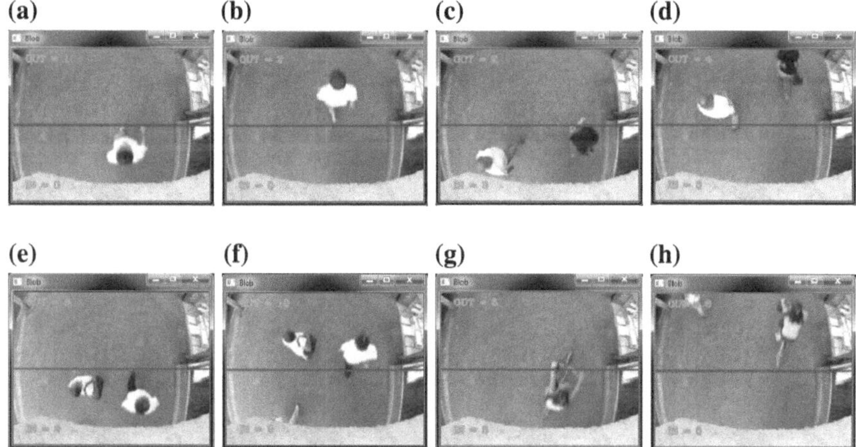

Fig. 9 **a**, **b** A person crossing the reference line, counter incremented accordingly. **c**, **d**, **e**, **f** People passing through the reference line and the count incremented as shown. **g**, **h** A person on a bicycle crossing the line and the counter incremented

6 Conclusion

In this paper we have proposed an efficient people counting algorithm, which counts people moving in both directions. The people counter engine is divided into three blocks, namely, background subtraction, segmentation and tracking and matching. The initial step involves extraction of foreground regions using a running Gaussian average which works faster compared to other background subtraction techniques as it consumes lesser memory. These foreground regions are detected and their trajectories are tracked. A virtual line is used as a reference to count the number of people passing in the scene. We have tested the algorithm on a set of videos of people passing through a street.

References

1. Hayman JA (2003) Computer vision based people tracking for motivating behavior in public spaces. Massachusetts Institute of Technology, Department of Electrical Engineering and Computer Science, pp 121–125
2. Aik LE, Zianuddin Z (2009) Real-time people counting system using curve analysis method. Int J Comput Electr Eng 1(1):1793–8198
3. Adriano GP, Mendoza SIV, Montinola FNJ, Naval PC (2005) APeC: automated people counting form video. In: Computer vision and machine intelligence group, Department of Computer Science, College of Engineering, University of the Philippines-Diliman
4. Chen C-H, Chen T-Y, Wang D-J, Chen T-J (2012) A Cost-effective people counter for a crowd of moving people based on two-stage, segmentation. J Inform Hiding Multimedia Signal Process 3(1):2073–4212

5. de Almeida SS, de Melo VHC, Menotti D An Evaluation of two people counting system using zenithal camera
6. Berg R-E (2007) Real-time people counting system using video camera
7. Lu H, Zhang R, Chen Y-W (2008) Head detection and tracking by mean-shift and Kalman filter. Department of Electronics, Dalian University, Dalian
8. Lempitsky V, Zisserman A (2010) Learning to count objects in images. In: Proceedings of NIPS, pp 1324–1332
9. Conte D, Foggia P, Percannella G, Tufano F, Vento M (2010) Counting moving people in videos by salient points detection. In: Proceedings of ICPR, 2010
10. Lefloch D (2007) Real-time people counting system using video camera. Department of Computer Science and Media Technology, Gjøvik University College, Norway
11. Hsieh J-W, Fang F-J, Lin G-J, Wang Y-S (2012) Template matching and Monte Carlo Markova chain for people counting under occlusions. In: Proceedings of 18th international conference on advances in multimedia modeling, pp 761–771
12. Zhao T, Nevatia R (2003) Bayesian human segmentation in crowded situations. In: Computer vision and pattern recognition, IEEE, pp 459–466
13. Piccardi M (2004) Background subtraction techniques: a review. In: IEEE international conference on systems, man and cybernetics, 2004
14. Ibrahim MM, Anupama R (2005) Scene adaptive shadow detection algorithm. In: Proceedings of world academy of science, engineering and technology, 2005

Cataloging of Happy Facial Affect Using a Radial Basis Function Neural Network

M. Nachamai and Pranti Dutta

Abstract The paper entitled "Cataloging of Happy facial Affect using a Radial Basis Function Neural Network" has developed an affect recognition system for identifying happy affect from faces using a RBF neural network. The methodology adapted by this research is a four step process: image preprocessing, marking of region of interest, feature extraction and a classification network. The emotion recognition system has been a momentous field in human–computer interaction. Though it is considerably a challenging field to make a system intelligent that is able to identify and understand human emotions for various vital purposes, e.g. security, society, entertainment but many research work has been done and going on, in order to produce an accurate and effective emotion recognition system. Emotion recognition system can be classified into facial emotion recognition and speech emotion recognition. This work is on facial emotion recognition that identifies one of the seven basic emotions i.e. happy affect. This is carried out by extracting unique facial expression feature; calculating euclidean distance, and building the feature vector. For classification radial basis function neural network is used. The deployment was done in Matlab. The happy affect recognition system gave satisfactory results.

Keywords Image processing · Radial basis function network · Emotion recognition

M. Nachamai (✉) · P. Dutta
Department of Computer Science, Christ University, Bangalore, INDIA
e-mail: nachamai.m@christuniversity.in

P. Dutta
e-mail: prantid@gmail.com

Mohan S. and S. S. Kumar (eds.), *Proceedings of the Fourth International Conference on Signal and Image Processing 2012 (ICSIP 2012)*, Lecture Notes in Electrical Engineering 221, DOI: 10.1007/978-81-322-0997-3_53,

1 Introduction

Human computer interaction (HCI) is the way by which human can communicate with the computers and thus paves way for a computer to successfully interact with human beings. According to the association for computing machinery (ACM) special interest group on computer human interaction (SIGCHI) the definition of HCI is "HCI is a discipline concerned with the design, evaluation and implementation of interactive computing systems for human use and with the study of major phenomena surrounding them". But it is not amicable to work with the communication between human and computers as a machine is involved. Though this interaction is already possible in and around the peripheral devices, still there exists lot of scope for further improvement. From every aspect, new techniques and interactive devices are being explored; among them one of the most popular areas is visualization that comes under the field of image processing. Image processing is a processing technique that converts an input image to an output image of better quality or as per the requirement i.e. obtains an image in a readable format. The importance of image processing is to humanize the visual appearance of an image to a human viewer along with various applications across various fields such as computer vision, face detection, medical imaging, microscope image processing, raster operation, etc. Communication between human and computer through faces is one of the major cruxes of face recognition. In image processing, facial images can be conceived as vectors. Facial expression plays an important role in daily life which led to concentrating on the need of studying the emotions in the realm of interactive design. This is the most challenging task to detect a face automatically. There are many factors affecting the images like image background, image quality, noise, lighting effect, person's pose and facial expressions. Digital image processing is extensively applied to improve quality and remove noise from images. The word emotion itself is a very complex psychology in human which involves different states of unstable mind which includes observable behaviors, expressed feelings, sensibility and changes in the body language. Emotions are classified as basic and complex. Happy, interest, anger, sadness, fear, surprise, disgust and distress, are all categorized as "basic emotions". Embarrassment, pride, guilt and shame are categorized as "complex emotions". This work is focused on one of the emotions 'happy affect'. It is a very challenging task to recognize and speculate a face with emotions via computer systems from the user image. Facial expressions are generated by contractions of facial muscles, which result in temporally malformed facial features such as eye lids, eye brows, nose, lips and skin textures, often revealed by wrinkles and bulges. Therefore for face recognition or facial expression recognition one of the most important initial steps is feature extraction. There are many approaches for extracting facial points from the images. Some of them are Geometry-based, Template-based, Color segmentation techniques and Appearance-based approaches. Methods such as principal component analysis (PCA), independent component analysis (ICA), Gabor wavelets and a statistical model based approach named active appearance model (AAM) etc. are used to extract the feature vector. It is important in order to recognize a special affect on face, firstly a particular region from a face should be located and then only extraction of facial features can be done. It is a very different mechanism to recognize a particular affect in a face. After identifying feature points from a particular region, it is important to determine the distances between the points. Metric is a measure of the similarity or dissimilarity between two data objects and it is used to form data points into clusters based on two classes of distance (correlation coefficients and distances metric). Euclidean distance is the most commonly used distance procedure in comparison to others like squared Euclidean distance, correlation-based distance, statistical distance and Manhattan distance (also known as city-block). Adopting Euclidean distance for the proposed system is due to its accuracy i.e. Euclidean distance measures an actual distance between the two points of interest; it corresponds to the geometric distance into the multidimensional space. Once the features are extracted and selected, the next step is to classify the image. Artificial neural networks (ANNs) are a popular tool in pattern recognition. Multilayer perceptron (MLP) is used to depict any general feed forward (no recurrent connections) network. In the field of neural network, there are many methods for facial expression recognition system based on learning. Learning is a process to optimize a performance criterion using example data or past experience. There are two types of learning process namely supervised and unsupervised learning. Based on learning methods, there are lots of neural network models like back propagation neural network (BPNN), self-organizing map (SOM), adaptive resonance theory (ART), probabilistic neural networks (PNN), general regression neural networks (GRNN) and radial basis function neural network (RBFNN). In the proposed system RBFNN is used for classification due to excellent approximation capabilities, nonlinear approximation, faster as it is capable of modeling complex mappings.

2 Literature Review

Over the last few decades, the ways of interaction between a human and a computer, popularly known as HCI has seen remarkable advancements. Research on HCI took center stage in between 1980 and 1990s and was most prominently presented by Card, Moran and Newell in their book, The Psychology of Human–Computer Interaction. In technical terms, human is seen as an information processor with inputs, mental processing and outputs which in turn 'input' information or data into the computer [1]. Human eye can process huge amount of visual information very quickly differentiating colors, shapes, size and different faces. Image processing is a technique which involves changing the nature of an image either to improve its pictorial information for human interpretation as well as render it more suitable for autonomous machine perception. A digital image differs from a normal image by some arrays of discrete values and involves a computer to change the nature of the digital image [2]. Face recognition, expression recognition as well as object recognition are the core research areas in HCI. A field that has further emerged within face recognition is the expression recognition. Emotional expressions add further meaning to faces as they reveal the state a person is in at that moment. To automate recognition of emotional state, machines must be taught to understand facial gestures [3]. For recognizing a particular expression within a human face it is necessary to extract facial features from that particular face. Human facial features play a significant role for face recognition and the eyes, mouth, and nose are amongst the most important features for recognition [4, 5]. Various approaches have been proposed to extract these facial points from images or video sequences of faces [6]. The general approach to automatic facial expression recognition consists of three steps—face acquisition, facial feature extraction and representation and facial expression recognition [7]. To speed up processing procedure and to get an accurate output from an image, normalization factors like scaling, cropping and rotating are used [8]. Methods such as Gabor wavelets, ICA, AAM and PCA etc., are used to extract the feature vector [9–11]. After extracting and selecting the features, the next step is to classify and train the image for recognition. A successful face recognition methodology depends heavily on the particular choice of the features used by the pattern classifier. ANN is most widely used mathematical model or computational model that tries to simulate the structure and/or functional parameters of biological neural networks. In training, changes based on the weight of the network parameters and the network is modified according to the size of these parameters. There are lots of classifier models based on supervised and unsupervised learning such as BPNN, SOM, ART, PNN, GRNN and RBFNN. Among all many researcher use radial basis function (RBF) as a neural network method to develop facial expression recognition because of its advantages. The RBF calculations are simple within and are overall very resistant. The performance of RBF network does not decrease considerably despite alterations in orientation, scale, pose and lighting. This helps it to detect faces in real time. RBFNN comprises of three layers; input layer, hidden and output layer. The hidden layer may consist of many intermediate layers. The two common RBF designs are the forward selection and Gaussian mixture model [12, 13]. Designing and simulation of neural networks can be done in MATLAB. MATLAB stands for MATrix LABoratory, is a software package developed by MathWorks, Inc. The concept of neural networks has evolved from the biological nervous system and the connections between the elements determine the network function [14].

3 Methods and Materials

3.1 Database and Tool Used

The JAFFE data set is used for the experiments. The gallery consists of 215 images of seven facial expressions which includes six basic facial expressions and one neutral posed by ten Japanese female models. Each image has been rated on six emotion adjectives by sixty Japanese subjects. All images in the data set are grayscale images, with .tiff extension. It is important to use an effective and applicable programming language that has to be chosen carefully. The selected programming language for coding the essential practicality of the approaches chosen is Matlab 7.10.0 (R2010a). The whole recognition system is carried out in four steps; preprocessing, marking region of interest, features extraction and RBFNN for classifier.

3.2 Preprocessing

Preprocessing stage consists of two steps, first, identifying the facial region of the image; second, improving the quality of the image and removing the noise. This stage would pave way for better results in the emotion recognition system but it is very important to empower the system to make it apposite to apply any data set of facial images. Original JAFFE images were first spatially (geometrically) transformed to get the fixed points from the mouth region. The accurate region of interest was calculated using cropping function and all images were then cropped to 224 × 128 pixels (using coordinate values) and a rectangular mask used to further eliminate the background. Finally, image pixel values were histogram equalized to the range of values from 0 to 255 and then adjusted the points using the function imadjust. These preprocessing steps were carried out on all images prior to performing the experiments. This is shown in Figs. 1 and 2 with sample images.

The mouth portion (region of interest) was acquired using crop function by defining the length of x-coordinate and y-coordinate. This is shown in Fig. 3.

3.3 Proposed RBF Network

The proposed network consists of nine nodes. The input layer, hidden layer and output layer includes four nodes, three nodes and two nodes respectively. In the hidden layer, the three nodes are not in the same level. X_i, Y_i, X_j and Y_j are inputs given to the network taken as four coordinates points. Euclidean distances (ED) is calculated from the coordinate's points of the input layer and compared with

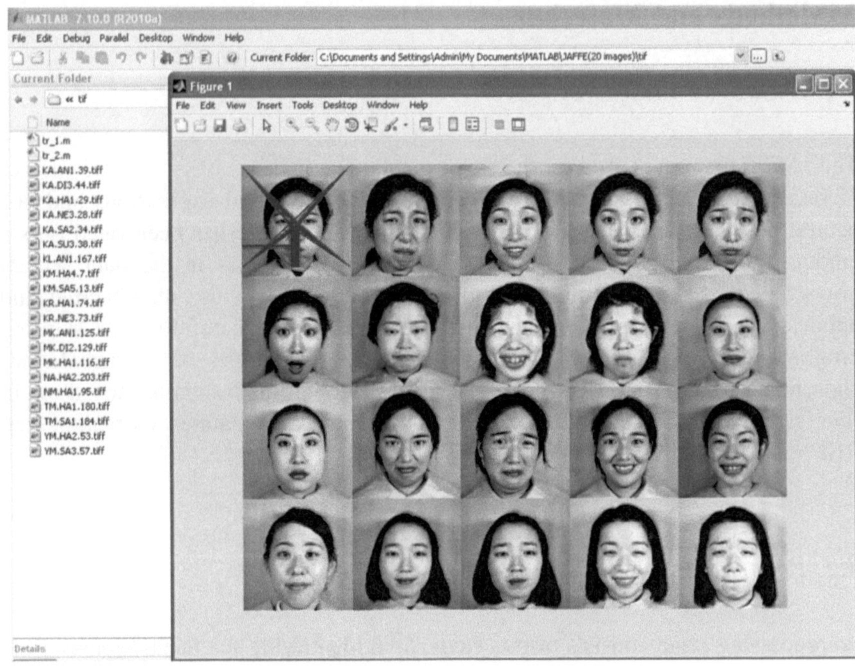

Fig. 1 Sample images before preprocessing

biased threshold (BT), which in turn gives the necessary output. In the proposed system, the mean value of all ED from the neutral faces of the JAFFE database is taken as BT. The architecture of the proposed RBF network is shown in Fig. 4. A portion of images are taken from the database for training and testing by the network.

3.4 Algorithm for the Proposed System

Step1: An image is read from the database.

Step2: The image is preprocessed. Preprocessing involves image equalization, image adjusts, histogram check and sobel edge detection is done to convert the image into black and white; with white discontinuities as edges of images.

Step3: Calculating the centroid of the image with cross hair technique.

Step4: Cropping the region of interest, the mouth portion (e.g. imcrop (image,[x y 3 4]) or imcrop (image, [1 2 3 4])), where x/1 is first point on x-axis, y/2 is first point on y-axis and 3 is extension from x i.e. $x + 3$ towards right, if -3 then $x - 3$ towards left. Similar +3 is downwards (only in images), -3 is upwards.

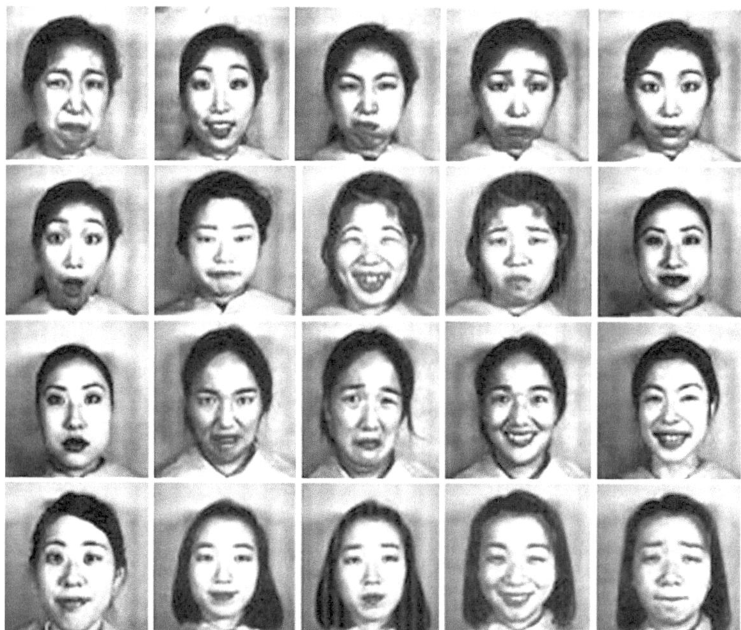

Fig. 2 Sample images after preprocessing

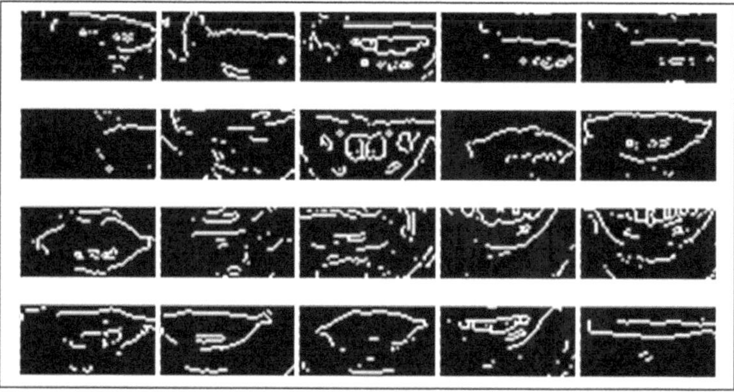

Fig. 3 Sample images after cropping and marking the ROI

Step5: Extraction of facial feature points in the form of coordinates values as $x1$, $x2$, $y1$ and $y2$ which is made as inputs to the RBF network.

Step6: Train the image after training, the RBF output will be stored in the external file, which will give the final weight and bias.

Step7: With the feature vector i.e. coordinate points the euclidean distance is calculated as (d = pdist(z,'euclidean')).

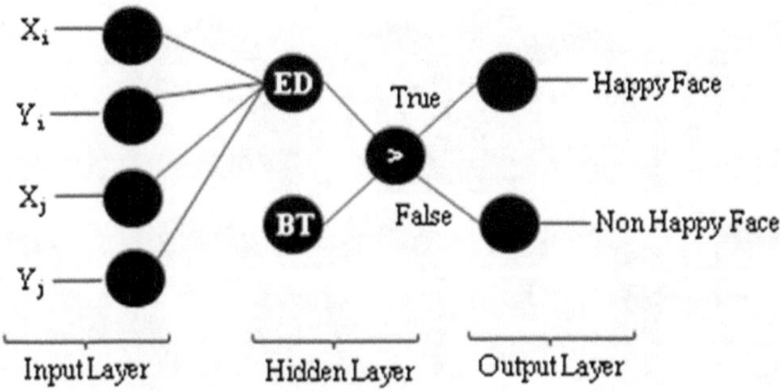

Fig. 4 Architecture of the proposed RBF network

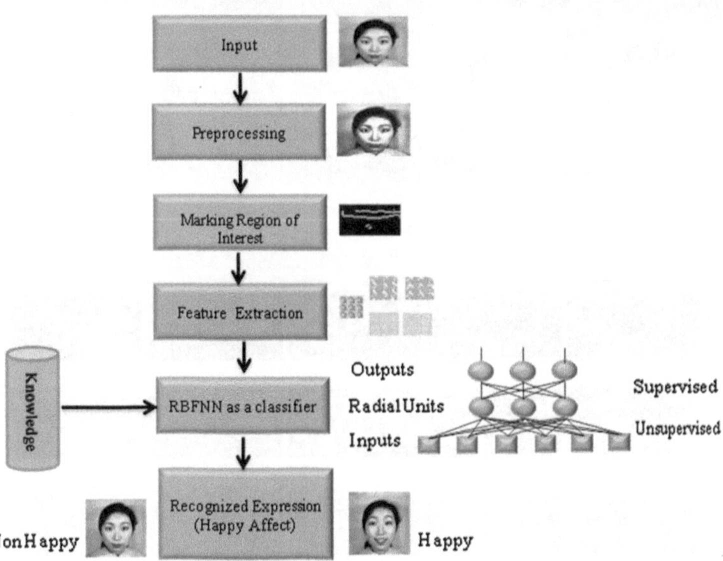

Fig. 5 Flow of the system

Step8: A bias threshold value (calculated on basis of all neutral faces) is compared with the euclidean distance of the loaded/test image. The RBF classifier will identify it as a happy or non-happy face.

On the basis of the proposed algorithm, the flow of the system is provided in Fig. 5.

4 Experimental Results and Discussion

For cross validation, the system has been tested with different data groupings: testing dataset-1 and testing dataset-2. In the dataset-1, the number of training images was 173 and a set of 42 images has been used for testing purposes. In the testing stage, number of happy images was 6 and number of non-happy images was 36. Confusion matrix of the classification is given in Table 1.

In the dataset-2, the number of training images was 134 and the number of testing images was 81. In the testing stage, number of happy images was 30 and number of non-happy images was 51. Confusion matrix of the classification is given in Table 2.

On consolidation of accuracy percentage from both the data groups of images (dataset-1 and dataset-2), the results were evaluated with different network spread values. As tabulated in Table 3, the accuracy percentage shows variations with different spread values. With increase in spread value there is a change in the performance noted for datagroups with more true positive samples. For a less true positive sample size the spread size does not have a change in accuracy.

A comparison of classification rate using various feature extraction methods and classifiers based on JAFFE database is given in Table 4.

Table 1 Confusion matrix (dataset-1)

		Predicted class	
		Happy	Non-happy
Actual class	Happy (6)	6	0
	Non-happy (36)	20	16

Table 2 Confusion matrix (dataset-2)

		Predicted class	
		Happy	Non-happy
Actual class	Happy (19)	16	3
	Non-happy (62)	20	42

Table 3 Spread value comparison

Dataset	RBF spread value	No. of happy faces recognized	Accuracy in %
Dataset-1 (happy faces-6)	2.1	5	83.3333
	3.1	5	83.3333
	4.1	5	83.3333
Dataset-2 (happy faces-19)	2.1	14	73.6842
	3.1	15	78.9473
	4.1	15	78.9473

Table 4 Comparison of classification rate (%)

Methods Used	Accuracy rate (%)	Size of the dataset	Reference (paper name)	Year
PCA reconstruction	78.95	10	Facial expression classification using PCA and hierarchical RBFN	2006
RBFN with PCA	68.42	13	Facial expression classification using PCA and hierarchical RBFN	2006
Artificial neural network	73.3	30	Rapid facial expression classification using artificial neural network	2007
Multi Artificial Neural Network	83	213	Facial expression classification using multi artificial neural network	2010
Canny_PCA_ANN	85.7	213	A facial expression system classification integrating Canny, PCA and ANN	2011
Gabor filters and naïve Bayesian classifier	98.5	60	Feature extraction for facial expression recognition based on hybrid face regions	2009
Appearance-based local approach, radial symmetry transform and edge projection analysis	83	150	Facial expression recognition	2010
Gabor filters, Haar classifier	80	213	Facial expression recognition system using weight-based approach	2010
Gabor wavelet-based method, cosine similarity measure, maximum correlation classifier	83.50 (CSM-LFR), 80 (CSM-HFR), 83.5 (MCC-LFR), 80.80 (MCC-HFR)	213	Recognition of facial expressions in presence of partial occlusion	2003
Fit forward back propagation ANN	90.2	213	An Automatic Region-based Methodology for Facial Expression Recognition	2008
RBFNN	**83.3333**	**81/42**	**Delineating happy affect face using a Reckoning neural network**	**2012**

5 Conclusion

The experiments examine the efficacy of the emotion recognition system based on the facial expressions. The project is examined on two groups of static images. The two groups of images are categorized from the JAFFE database as dataset-1 (80–20 %) and dataset-2 (60–40 %). The experimental results demonstrate the accuracy of the dataset-1 based on happy images is 100 % that is higher than that of the dataset-2 giving an accuracy of 84.21 %. The accuracy of the dataset-1 for all the images is 52.83 and 71.60 % for dataset-2. The system works well and satisfactorily with the two groups of images. For cross evaluation, both the data groups was evaluated on different RBF spread values. The change in the RBF spread value showcased a consistent accuracy for dataset-1. For the spread values 2.1, 3.1 and 4.1 of the network the accuracy was 83.33 % respectively. A change was observed when the dataset-2 was tested with a spread value 3.1. For 2.1 the accuracy was 73.68 %, a good improvement was seen when spread value was 3.1 and 4.1 respectively. The accuracy came to 78.94 %. The spread value also plays an important role in the performance of the system.

6 Future Work

The proposed emotion recognition system is successful in identifying happy affect in an image. But this paper is limited to 'mouth region' as the region of interest. The extraction of facial feature points within the mouth region only i.e. the identification of happy affect is based on feature point extraction within the mouth region. However to make the system capable to enhance further accuracy, the number of extracted facial features should be increased to cover the full face like eyes region, eye brows and forehead region. Then apply the PCA algorithm for each part individually and learn the system. The use of hybrid classifier system will be more effective for the extracted features. It can also be developed into a more complicated emotional recognition system proficient of distinguishing between more basic emotions. It will lead to a satisfactorily flexible and improved emotion recognition system which can progress the process easily.

Finally, to make it more efficient and capable of distinguishing between more than one emotion to encourage its accuracy based on different databases like different aged images, male databases, real time images, colored images, different poses images, and occluded images.

References

1. Harper R et al (2008) Being human: human-computer interaction in the year 2020. Microsoft Research Ltd.7JJ Thomson Avenue, Cambridge, CB3 0FB, England
2. McAndrew A (2004) An introduction to digital image processing with matlab. Notes for SCM2511 image processing 1 semester 1. Victoria University of Technology, Melbourne
3. Arumugam D (2011) Emotion classification using facial expression. Int J Adv Comput Sci Appl (IJACSA) 2(7):92–98
4. Nagpal A, Garg A (2011) Recognition of expressions on human face using AI techniques. Int J Comput Sci Manage Stud (IJCSMS) 11(2):165–169
5. Sohail ASM, Bhattacharya P (2006) Detection of facial feature points using anthropometric face model. In: Proceedings fo IEEE international conference signal-image technology and internet-based systems, pp. 656–665
6. Bagherian E, Rahmat RW, Udzir NI (2009) Extract of facial feature point. IJCSNS Int J Comput Sci Netw Secur 9(1):49–53
7. Shan C et al (2009) Facial expression recognition based on statistical local features. Universal short title catalogue book, Chapter 4
8. Jansson M, Johansson J (2003) Interactive visualization of statistical data using multidimensional scaling techniques. Thesis, Institute of Technology, Linkopings University, Sweden
9. Praseeda LV, Sasikumar M (2008) A neural network based facial expression analysis using Gabor wavelets. World Acad Sci Eng Technol 42:563–567
10. Neggaz N, Besnassi M, Benyettou A (2010) Application of improved AAM and probabilistic neural network to facial expression recognition. J Appl Sci 10(15):1572–1579
11. Hiremath V, Mayakar A (2009) Face recognition using Eigenface approach. In: IDT workshop on interesting results in computer science and engineering, Sweden
12. Fajaryanti J (2010) Implementation radial basis function neural network for training process in face recognition system. In: Proceedings of faculty of industrial technology. Gunadarma University, Indonesia
13. Bors AG (1996) Introduction of the radial basis function (RBF) networks. In: Online Symposium of electronic engineers
14. Arulmozhi V (2011) Classification task by using matlab neural network tool box—a beginner's view. Int J Wisdom Based Comput 1(2):59–60

Color Invariant Feature Detection and Matching in Underwater Stereo Images

C. J. Prabhakar and P. U. Praveen Kumar

Abstract In this paper, we present an approach to find correspondences in underwater stereo images based on detection and matching of feature points, which are invariant to photometric variations. In underwater environment, the problem of finding correspondences in stereo images is specific step in order to estimate the motion of an underwater vehicle. The current state-of-the-art feature detectors have been proven to be the most robust to geometric variations and avoid dealing with color images due to color constancy problem. The propagation property of light in the underwater causes variations in color information between two underwater images taken under same imaging conditions. To render the color values changed by the various radiometric factors of underwater environment, we use comprehensive color image normalization method to normalize the color image. Our technique uses SIFT to detect interest points from the normalized image. In order to establish correspondences between images, we use window-based correlation measure instead of feature-based correlation techniques. The underwater images are low contrast in nature and lack of image features cause the feature-based techniques matching procedure to fail. Our approach is evaluated extensively to verify its effectiveness with data sets acquired in underwater environment. A new approach based on color invariant feature detection and window-based correlation matching significantly improves the matching reliability.

Keywords Underwater stereo images · Color image normalization · SIFT · Cross-correlation · Similarity measure

C. J. Prabhakar (✉) · P. U. Praveen Kumar
Department of P.G. Studies and Research in Computer Science, Kuvempu University, Shankaraghatta 577451 Karnataka, India
e-mail: psajjan@yahoo.com

P. U. Praveen Kumar
e-mail: praveen577302@gmail.com

Mohan S. and S. S. Kumar (eds.), *Proceedings of the Fourth International Conference on Signal and Image Processing 2012 (ICSIP 2012)*, Lecture Notes in Electrical Engineering 221, DOI: 10.1007/978-81-322-0997-3_54,

1 Introduction

Feature detection and matching of stable and discriminative features are fundamental research problems in many computer vision and machine learning applications. Many feature detection techniques have been proposed during the last few decades for various applications such as, image stitching, image registration, object detection, object localization and object recognition. The significant point of feature point detection is stability and distinctiveness, i.e. the estimated feature point in the successive images should be stable with respect to geometric and photometric variations and it should be distinctive, which means that the extracted features should have minimum information to distinguish between the object which they describe and other objects [1]. Unmanned underwater vehicles (UUVs) are ideal tools to accomplish periodical inspections of jacket structures of offshore platforms [2]. A stereo vision based method is used to navigate the vehicle to locate itself with the aid of environmental reference information. With the aid of the location information, motion of the vehicle can be estimated and based on the motion of the vehicle the pilot can navigate the vehicle in a more efficient and safer way. Implementation details include several steps such as feature extraction and tracking, stereo correspondence and robust motion estimation. In this paper, we focus on performing one important step such as feature detection and matching in underwater stereo images in order to estimate the motion of the vehicle. Suppose the features have been successfully detected in the left image; the next step is to find the corresponding features in the right image.

In an underwater environment, detecting the invariant features from the images is a challenging task due to optical properties of the water, which varies the same feature in sequence of images. When imaging underwater, even an image of the same object can be drastically different due to varying water conditions [3]. The added complexities of impure water introduce issues such as turbulence, air bubbles, particles (such as sediments) and organic matter that absorb and scatter light, which can result in a very blurry and noisy image. Moreover, many parameters can modify the optical properties of the water, and underwater images show large variations in color rather than geometrical variations. As a result, color values of the same point on an object may be completely different between two underwater images taken under same imaging conditions.

Most of the existing approaches use gray geometric-based features. Nearly all geometrical invariant approaches avoid dealing with color images due to color constancy problems. Color is an important component for distinction between objects. If the color information in an object is neglected, a very important source of distinction may be lost, and many objects can be mismatched and misclassiffied [1]. Color invariance features provide high discriminative information cues, as color varies considerably with change in camera viewpoint, object pose and illumination. Underwater images usually suffer from radiometric variations using this raw color information leads in degradation in the performance of feature detection. Color-constancy algorithms often attempt to separate the illumination and the reflectance components on images similar to the human visual system. The most simple and commonly used color-invariant approach for Lambertian surface are the Chromaticity normalization and the gray world assumption. Chromaticity normalization is often used to removal of lighting geometry effects where as gray world assumption is used for removal of illuminant color effects. Neither of the two methods remove the dependency of both lighting geometry and illuminant color simultaneously [4]. Finlayson et al. [5] developed a method called comprehensive color image normalization method, which removes lighting geometry effects and illuminant color effects both iteratively and non-iteratively.

The detection of interest points (keypoints) for a given image is a very challenging task due to variation in scale, orientation and illumination. The interest points are selected based on the distinguishable property of the locations, such as corners, blobs, and T-junctions [6]. The reliability of interest point detectors is evaluated based on repeatability criteria. The repeatability criteria finds the ability of feature detectors to consistently detect the same keypoints in two different images of the same scene under variations such as scale, illumination and view point changes. A wide variety of detectors have been proposed in the literature. The most widely used detectors can be classified into two categories such as Harris-based and Hessian-based detectors [7–11]. The Harris corner detector [7] proposed in 1988 is based on eigen values of the second moment matrix. This method is not scale-invariant due to the fact that the Harris corner detector is very sensitive to changes in image scale, so it does not provide a good basis for matching images of different sizes. To detect the blob-like structures, Lindeberg [12] proposed the concept of automatic scale selection based on Hessian matrix as well as Laplacian. Mikolajczyk and Schmid [10] have further extended this method to create robust and scale-invariant feature detector based on the combination of Harris-Laplacian and Hessian-Laplacian. They used a Harris measure to select the location and Laplacian to select the scale. Lowe [8] proposed to approximate the Laplacian-of-Gaussian (LOG) by a Difference-of-Gaussian (DOG) filter. SIFT proposed by David Lowe [8] is one of the widely used feature detector and descriptor, because of its robustness, distinctiveness and efficient computation time in feature extraction and matching. SIFT uses Harris-based feature detector because it is more stable and repeatable than Hessian-based counterparts which are usually used in SURF [13]. It is also observed that approximations like the DOG used in SIFT can bring speed at a low cost in terms of high accuracy.

In this paper, we introduce an approach to detect color invariant features from underwater left image and to find the corresponding feature points in the right image. We apply comprehensive color image normalization for raw RGB color images in order to render the color values which are changed by various factors such as attenuation and scattering of light in the underwater environment. Color-invariant features are detected from the normalized left image using SIFT feature detection technique i.e. by using Difference-of-Gaussian (DOG) approach. We estimate the candidate matches in the right image for a given interest point in the left image by using window-based correlation measure based on the user defined threshold. For a given, interest point in the left image and the possible candidate matches in the right image, we employ Euclidean similarity measure to find whether there exist true match or false match based on threshold.

The remainder of the paper is organized as follows: the comprehensive color image normalization is discussed in Sect. 2. In Sect. 3, our approach is discussed for finding correspondences. The experimental results on underwater images are presented in Sect. 4. Finally, Sect. 5 draws the conclusion.

2 Comprehensive Color Image Normalization

The raw color recorded by the camera is not a reliable for color invariant feature detection and matching. In order to extract color invariant features, chromaticity normalization is commonly employed to removal of lighting geometry and gray world normalization is employed to remove illuminant color effect. The comprehensive color image normalization combines both chromaticity normalization and gray world assumption to remove both lighting geometry and illuminant color effects [5].

A change in light intensity model can be canceled out by applying chromaticity normalization, which is given by:

$$r_i = \frac{\rho_i R_i}{\rho_i R_i + \rho_i G_i + \rho_i B_i},$$
$$g_i = \frac{\rho_i G_i}{\rho_i R_i + \rho_i G_i + \rho_i B_i}, \quad (1)$$
$$b_i = \frac{\rho_i B_i}{\rho_i R_i + \rho_i G_i + \rho_i B_i},$$

where ρ_i is a scalar value and the new co-ordinates r, $g$and $b$are independent of ρ_i.

To cancel out dependence on illumination color, gray-world normalization is usually employed which is given by:

$$R'_i = \frac{R_i}{\mu(R)}, \quad G'_i = \frac{G_i}{\mu(G)}, \quad B'_i = \frac{B_i}{\mu(B)}, \quad (2)$$

where $\mu(R)$, $\mu(G)$ and $\mu(B)$ are the mean value of red, green and blue channels respectively.

Eq. 1 and 2 remove the effects of lighting geometry and illumination color respectively but neither, by itself, suffices to remove the effect of both factors. Therefore, Finlayson et al. [5] defined a third normalization called comprehensive normalization, which can remove both dependencies. It is defined as:

1. $I_0 = I$ initialization
2. $I_{i+1} = G(C(I_i))$ iteration step
3. $I_{i+1} = I_i$ termination condition

where $G()$ and $C()$ function stands for gray-world and chromaticity normalization. That is, chromaticity and gray-world normalization are applied successively and repeatedly to an image until the resulting image converges to a fixed point.

3 Finding Correspondences

The finding of correspondences begin with rectification of a given pair of normalized images to reduce the searching region along the horizontal axis. The stable and distinctive features of left rectified image are detected using DOG based procedure of SIFT technique. For a given interest point of left image, a set of possible matchings is established in the right image using window-based correlation measure. Finally, the true match is computed for a given interest point in the left image using point–point Euclidean distance measured between given point and its candidate matches based on threshold. The true match is selected as the one with minimum distance which lies within the threshold.

3.1 Uncalibrated Stereo Image Rectification

Given a pair of stereo images, rectification determines a transformation of each image plane such that pairs of conjugate epipolar lines become collinear and parallel to one of the image axes. The important advantage of rectification is that computing stereo correspondences is reduced to a 1-D search problem along the horizontal raster lines of the rectified images. We use Quasi-Euclidean epipolar rectification method for uncalibrated images proposed by Andrea Fusiello et al. [14].

3.2 Detection of Interest Points

Interest points (keypoints) are detected as the steps followed by SIFT feature detection technique proposed by David Lowe [8]. SIFT features are all natural features of images. They are favorably invariant to image translation, scaling, rotation, illumination, viewpoint, noise etc. Good speciality, rich in information, suitable for fast and exact matching in a mass of feature database. Relatively fast speed. The extraction SIFT features from images can be done by applying sequentially steps such as scale-space extrema detection and keypoint localization.

Scale-space extrema detection: Interest or keypoints are detected in this step. This is the first stage of computation searches over all scales and image locations. It is implemented efficiently by using a Difference-of-Gaussian function to identify potential interest points that are invariant to scale and orientation. A Gaussian pyramid is constructed from the input image by repeated smoothing and subsampling, and a Difference-of-Gaussian pyramid is computed from the differences between the adjacent levels in the Gaussian pyramid. Then, interest points are obtained from the points at which the Difference-of-Gaussian values assume extrema with respect to both the spatial coordinates in the image domain and the scale level in the pyramid.

Keypoint Localization: scale-space extrema detection produces too many keypoint candidates, some of which are unstable. A detailed fit to the nearby data for accurate location, scale, and ratio of principal curvatures is done. This information allows points to be rejected that have low contrast (and are therefore sensitive to noise) or are poorly localized along an edge.

3.3 Window-Based Correlation Measure

Correspondence between image pairs is a classical problem in the field of computer vision. Usually correlation measures are used for correspondence matching. We apply window-based correlation measure on the selected interest point in the left image to estimate the candidate matches in right image. Given an interest point p_l in the left image, with the coordinates (x_l, y_l), the correlation window is centered at this point. Then, a search window is defined on the right image, also centered at (x_l, y_l), and the correlation operation is performed. A correlation score is computed in the following way:

$$corr(p_l, p_r) = \frac{\sum_{i=-w}^{w}\sum_{j=-w}^{w} \left(I_L(x_l+i, y_l+j) - \overline{I_l(x_l+y_l)}\right) \cdot \left(I_R(x_r+i, y_r+j) - \overline{I_r(x_r+y_r)}\right)}{w^2\sqrt{\sigma^2(I_L)\cdot\sigma^2(I_R)}}, \tag{3}$$

where $w$is the size of the window considered; $I_L$and I_R are the left and right images, respectively; $\sigma^2(I)$ is the variance of the image computed in the correlation window; $\overline{I(x,y)}$ is the mean of the correlation window in the image.

$$\sigma^2(I) = \frac{\sum_{i=-w}^{w}\sum_{j=-w}^{w} I(x+i, y+j)^2}{w^2} - \overline{I(x,y)}^2 . \tag{4}$$

$$\overline{I(x,y)} = \frac{\sum_{i=-w}^{w}\sum_{j=-w}^{w} I(x+i, y+j)}{w^2} . \tag{5}$$

We define a minimum threshold for estimating possible correspondence matching in the right image for a given interest point in the left image.

3.4 Similarity Measure

We employ Euclidean similarity measure to find whether there exists true match or false match between the pair of feature points based on suitable threshold. Setting the threshold too high or too low results in false matches, i.e. incorrect matches. The mathematical representation is:

$$E(p_l, p_r) = \sqrt{(x_l - x_r)^2 + (y_l - y_r)^2}\,. \tag{6}$$

where $E(p_l, p_r)$ is the Euclidean distance between $p_l$and p_r i.e. left and right image points respectively. (x_l, y_l) and $(x_r, y_r)$are the coordinate values of interest points p_l and p_r respectively.

4 Experimental Results

We have performed several experiments in order to validate our approach for underwater stereo images captured in small water body. The stereo setup consists of two identical digital still cameras, which are Canon D10 water proof. The two cameras are installed on a vertical stereo mount and it is kept in water body to acquire color stereo images. The scene includes several objects at distance [1 m, 2 m], near the corner of the water body. We have captured three sets of stereo images in different water conditions with turbidity levels. Figure 1 shows the captured stereo image datasets. The captured stereo images are suffered from non-uniform lighting, low contrast, blurring and diminished colors. The variations in corresponding color values are very high due to various factors such as, attenuation and scattering of light in underwater environment. The attenuation and scattering of light cause variations in the measured color of each scene point at its projections on the two images. In order to render the color values which are changed by various factors such as attenuation and scattering of light in the underwater environment, we apply comprehensive color image normalization for raw RGB color images.

Uncalibrated stereo rectification technique is applied on the normalized stereo images. Interest points are detected using DOG based feature detection technique to extract stable and distinctive features. The Fig. 2 shows the detected interest points in normalized and rectified left image of dataset1, dataset2 and dataset3 using DOG based feature detection method. The Fig. 3 shows the result of true match by using window-based correlation measure and Euclidean distance employed on dataset1, dataset2 and dataset3 respectively.

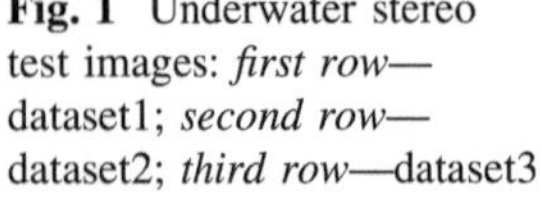

Fig. 1 Underwater stereo test images: *first row*—dataset1; *second row*—dataset2; *third row*—dataset3

Fig. 2 Detected interest points in the left image of dataset1, dataset2 and dataset3 respectively

4.1 Evaluation of Experimental Results

The experimental results are evaluated quantitatively and compared with two standard methods such as SURF [13] and KLT [15, 16] for dataset1 based on the number of feature points detected, total number of matches, repeatability criteria and processing time for interest point detection and matching. In order to evaluate accuracy of the matching procedure, we have used two step approach: first, the interest points have been detected in the left image, and the window-based correlation measure with Euclidean distance has been applied to the right image. Next, a human operator has marked all the visually incorrect matches, in order to know which correspondences have been incorrectly established. The repeatability measurement is computed as a ratio between the number of point-to-point correspondences that can be established for detected points and the mean number of points detected in two images

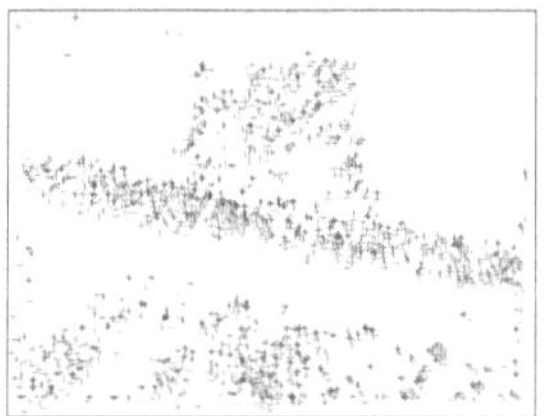

Fig. 3 Correspondence computed for dataset1, dataset2 and dataset3. The *red* colored'+'marks correspond to the interest points in the left image, while *blue* colored '+'marks are their matchings in the right image

Table 1 Comparison of feature detection methods for dataset1

Methods	SIFT	SURF	KLT
Number of feature points detected	840	755	571
Number of matches (human operator)	512	383	267
Repeatability	0.6	0.5	0.44
Processing time (s)	11.85	3.4	15.2

$$R_{lr} = \frac{C(I_l, I_r)}{mean(P_l, P_r)}, \tag{7}$$

where $C(I_l, I_r)$ denotes the number of corresponding matched interest points, P_l and P_r denotes number of interest points found in left and right images respectively. This measurement represents the performance of finding matches.

The processing time and repeatability evaluation is carried out on three feature detection methods, which shows the tendency of the three methods time cost and accuracy of detection and matches. The results are influenced by the factors such as the size and quality of the image types, and the parameters of the algorithm. The results were obtained on an Intel Pentium Core 2 Duo at the speed 1.83 GHz and 2 GB of RAM. The algorithms are implemented in MATLAB environment. The results in Table 1 clearly show that repeatability rate of SIFT is 0.6 where as for SURF and KLT it is 0.5 and 0.44 respectively. SURF and KLT spend approximately 4.5 and 26.62 ms for detection of each feature point where as SIFT spends 14.10 ms. The SIFT feature detection method extract more number of feature points. Even though the processing time of SURF is low compared to SIFT and KLT, the SIFT can efficiently detect more stable, distinctive and repeatable feature points, with increase in the repeatability rate.

5 Conclusion

In this paper, we present a novel approach to detect and match color-invariant features in underwater stereo images. Since there is no benchmark stereo data set is available for underwater images, we conducted experiments using captured stereo images in underwater. In underwater environment, the problem of finding correspondences in stereo images is specific step in order to estimate the motion of an underwater vehicle. The raw RGB images are not suitable for detecting color-invariant feature points, therefore in order to obtain color-invariant features we applied comprehensive color image normalization to normalize the raw RGB values. The interest points are detected in the rectified left image using DOG based feature detection technique of SIFT. The corresponding point in the right image is established for a given interest point in the left image using window-based correlation measure with Euclidean distance. We evaluated our approach for the purpose of suitability for underwater environments and compared with other feature detection methods such as SURF and KLT.

The experimental results show that SIFT based feature detection method is more suitable for real-time application in underwater environment, since it obtains more number of feature points and correspondence matches, its repeatability is also high compared to SURF and KLT based feature detection methods. Even though the processing time of SIFT based feature detection method is high compared to SURF, but the repeatability of SIFT is high compared to SURF. The SIFT method can even efficiently detect more stable and distinctive feature points with illumination and color changes.

Normally, the feature detectors for out-of-water images are compared with other standard feature detectors using benchmark databases to test according to specific criteria. To our knowledge, no such underwater image database exist at present. Therefore, the result of our approach for feature detection and matching is not compared using standard database images. The development of underwater image database could be one of the future research lines from which the underwater community would certainly beneficiate.

Acknowledgments The authors would like to thank the anonymous reviewers for their valuable comments and suggestions. This research was supported by Naval Research Board (Grant No. 158/SC/2008-09), DRDO, New Delhi, India.

References

1. Abdel-Hakim AE, Farag AA (2006) CSIFT: a SIFT descriptor with color invariant characteristics. In: IEEE computer society conference on computer vision and pattern recognition vol 2. pp 1978–1983
2. Garcia R, Batlle J, Cufi X, Amat J (2001) Positioning an underwater vehicle through image mosaicking. In: IEEE international conference on robotics and automation vol 3. pp 2779–2784

3. Mobley C (1994) Light and water: radiative transfer in natural waters. Academic Press, Cleveland
4. Heo YS, Lee KM, Lee SU (2011) Robust stereo matching using adaptive normalized cross-correlation. IEEE Trans Pattern Anal Mach Intell 33(4):807–822
5. Finlayson,GD, Schiele B, Crowley JL (1998) Comprehensive colour image normalization. In: European conference on computer vision pp 475–490
6. Garcia R, Gracias N (2011) Detection of interest points in turbid underwater images. In: IEEE OCEANS pp 1–9
7. Harris C, Stephens M (1988) A combined corner and edge detector. In: proceedings of fourth alvey vision conference pp 147–151
8. Lowe DG (2004) Distinctive image features from scale-invariant keypoints. Int J Comput Vis 2(60):91–110
9. Mikolajczyk K, Schmid C (2001) Indexing based on scale invariant interest points. In: proceedings of eighth international conference on computer vision pp 525–531
10. Mikolajczyk K, Schmid C (2004) Scale and affine invariant interest point detectors. Int J Comput Vis 1(60):63–86
11. Mikolajczyk K, Tuytelaars T, Schmid C, Zisserman A, Matas J, Scha_alitzky F, Kadir T, Gool LV (2005) A comparison of affine region detectors. Int J Comput Vis 65(1–2):43–72
12. Lindeberg T (1994) Scale-space theory: a basic tool for analysing structures at different scales. J Appl Stat 21(2):224–270
13. Bay H, Ess A, Tuytelaars T, Gool LV (2008) SURF: speeded up robust features. Comput Vis Image Underst 110(3):346–359
14. Fusiello A, Irsara L (2010) Quasi-euclidean epipolar rectification of uncalibrated images. Mach Vis Appli 19(4):926–933
15. Shi J, Tomasi C (1994) Good features to track. In: proceedings of the ieee conference on computer vision and pattern recognition pp 593–600
16. Tomasi C, Kanade T (1991) Detection and tracking of point features. Technical Report CMU-CS-91-132, Carnegie Mellon University, PA, Apr

Feature Subset Selection Using PSO-ELM-ANP Wrapper Approach for Keystroke Dynamics

D. Shanmugapriya and G. Padmavathi

Abstract The security of computer access is important today because of huge transactions being carried out every day via the Internet. Username with password is the commonly used authentication mechanism. Most of the text based authentication methods are vulnerable to many attacks as they depend on text and can be strengthened more by combining password with key typing manner of the user. Keystroke Dynamics is one of the famous and inexpensive behavioral biometric technologies, which identifies the authenticity of a user when the user is working via a keyboard. The paper uses a new feature Virtual Key Force along with the commonly extracted timing features. Features are normalized using Z-Score method. For feature subset selection, a wrapper based approach using Particle Swarm Optimization—Extreme Learning Machine combined with Analytic Network Process (PSO-ELM-ANP) is proposed. From the results, it is observed that PSO-ELM-ANP selects less number of features for further processing.

Keywords Keystroke dynamics · Particle swarm optimization · Extreme learning machine · Analytic network process

D. Shanmugapriya (✉)
Department of Information Technology, Avinashilingam Institute for Home Science and Higher Education for Women, Coimbatore, India
e-mail: shanmugapriya.adu@gmail.com

G. Padmavathi
Department of Computer Science, Avinashilingam Institute for Home Science and Higher Education for Women, Coimbatore, India
e-mail: ganapathi.padmavathi@gmail.com

Mohan S. and S. S. Kumar (eds.), *Proceedings of the Fourth International Conference on Signal and Image Processing 2012 (ICSIP 2012)*, Lecture Notes in Electrical Engineering 221, DOI: 10.1007/978-81-322-0997-3_55,

1 Introduction

The confidential information can be secured from unauthorized users by providing authentication. User authentication is defined as the process of verifying the identity claimed by an individual. User Authentication is normally classified into three categories [1] namely, Knowledge based Authentication, Object based Authentication, and Biometric based Authentication. Biometric based authentication provides a very reliable method of authenticating a user and is divided into physiological and behavioral types [2]. Physiological biometrics refer to what the person is or they measure physical parameters of certain parts of the body and Behavioral biometrics shows how the person is using the body for authentication. Keystroke dynamics is a strong behavioral biometric [3] and it is the process of analyzing the way a user types at a terminal by monitoring the keyboard in order to identify the user based on habitual typing rhythm patterns. Unlike other biometric systems, which may be expensive to implement, keystroke dynamics is almost free and it does not require any sophisticated hardware as the only hardware required is the keyboard. Timing features are the one most commonly measured from the keystroke. The huge features often lead to degradation of performance of the system. Hence an attempt is made in this paper to reduce the dimensionality of features using wrapper approaches.

The rest of the paper is organized as follows: the next section discusses the related works done in feature selection, Sect. 3 discusses the methodology and the results are discussed in the Sect. 4. Final section concludes the work with future directions.

2 Related Works

This section discusses the major feature subset selection methods applied in keystroke dynamics. Optimization techniques such as Genetic Algorithm (GA) and Azevedo et al. [4, 5] evaluated the feature subset selection based on Support Vector Machine (SVM) with Genetic Algorithm (GA) and a variation in Particle Swam Optimization (PSO). In this study, wrapper based approach with SVM-PSO variation outperformed SVM-GA with regard to classification error and processing time. Bleha and Obaidat [6] used a reduction technique based on Fisher analysis. However, the technique considered by keeping m − 1 dimension for each vector, with 'm' number of users in the system. Boechat et al. [7] used weighted probability measure to select a subset of 'N' features with the minors of standard deviation. The study eliminates less significant features by keeping 70 % of the features. Experiments are done at Zero False Acceptance Rate. False Rejection Rate reduces when the number of selected features increases. Particle Swarm Optimization (PSO), Genetic Algorithm (GA) and Ant Colony Optimization (ACO) are used by Karnan and Akila [8] for feature subset selection.

Back propagation Neural Network (BPN) has been used for classification. ACO gives better performance than PSO and GA with regard to feature reduction rate and classification accuracy. Sung and Cho [9] adds a uniqueness term in the fitness function of genetic algorithm and used SVM as classifier for classification and Genetic Algorithm (GA) to select the data. The experiments performed better than two phase ensemble selection approach and prediction based diversity term approach. Enzhe Yu and Sungzoon Cho [10] used wrapper based Support Vector Machines (SVM) and Genetic Algorithms (GA) to reduce the feature size. SVM is applied for training and GA is used for searching. GA-SVM wrapper approach gives good accuracy. Feature Selection (FS)—Ensemble is proposed to deal with the over fitting problems.

3 Methodology

The entire methodology is divided into three phases namely, feature extraction phase, normalization phase and feature subset selection phase after obtaining the raw keystroke data (press time and release time). The key function of feature extraction in keystroke dynamics is to extract the fundamental features from the timestamp collected from raw keystroke data for creation of template. The password '.tieRo-an1' has been collected from 100 users each typed the password 10 times. The features Duration or Dwell time, Flight time or Latencies, Digraph, Trigraph and Virtual Key Force [11] are measured. After feature extraction, data normalization using Z-Score [12] is carried out to bring the data in a particular range (0–1) as the time and speed of typing varies for same user even for the same text.

Feature subset selection identifies the most discriminating features. It also reduces the dimensionality of features which improves the assumption accuracy and decreases the computation time. Feature subset selection is applied to high dimensional data before preceding the classification since the increased dimensionality of features makes testing and training of classification method difficult. Feature subset selection is divided into Filter and Wrapper approach [13]. In Filter approach, the feature subset selection is done independently of the learning algorithm. The wrapper approach uses the method of classification to measure the importance of features set.

3.1 Wrapper Based Feature Subset Selection using PSO-ELM-ANP

Particle Swarm Optimization (PSO) is wrapped with Extreme Learning Machine (ELM)—Analytic Network Process to select the subset of features from the extracted features. These proposed wrapper based approaches select an appropriate

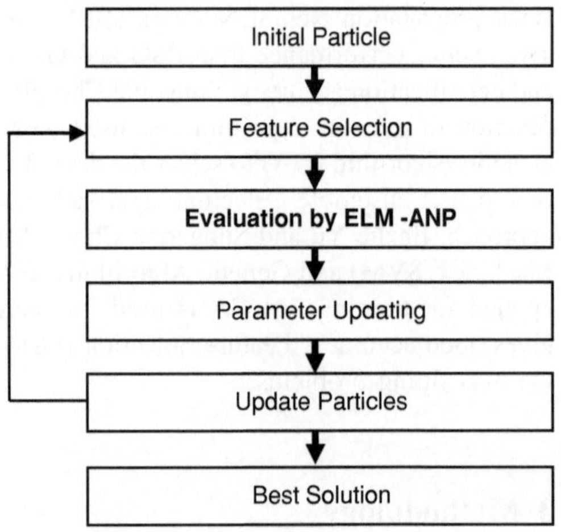

Fig. 1 Flowchart of proposed PSO-ELM-ANP wrapper feature subset selection

subset of features and the rest will not be considered, thus resulting in a more comprehensive model. Flow chart for PSO-ELM-ANP wrapper approach for feature subset selection is given in Fig. 1.

3.1.1 Particle Swarm Optimization

Particle Swarm Optimization (PSO) [14] is a stochastic search technique that aims to optimize an objective function, motivated by social activities of birds gathering or fish schooling. PSO is a population dependent search algorithm in which each individual is indicated as a particle and represents a candidate solution. All the particles in PSO moves through the search space with an adjustable velocity that is dynamically transformed based on its individual moving experience and also the moving experience of the other particles. In PSO each particles attempt to enhance themselves by imitating traits from their successful peers. The position in proportion to the best fitness is known as pbest (Particle best) and the overall best out of all the particles in the population is called gbest (Global best) [14]. The velocity $v_i(t)$and the position $x_i(t)$ of the particles are updated with the following equations.

$$v_i(t+1) = w * v_i(t) + c_1 r_1 (y_i(t) - x_i(t)) + c_2 r_2 (\hat{y}(t) - x_i(t)) \tag{1}$$

$$x_i(t+1) = x_i(t) + v_i(t+1) \tag{2}$$

where i = 1, 2, …, n

$v_i(t)$ Velocity of agent i at iteration t and must lie in the range

$$V_{min} \le V_i(t) \le V_{max} \tag{3}$$

W: Inertia weight factor
c_1, c_2 : Cognitive and social acceleration factors respectively,
r_1, r_2 : Uniformly distributed random number between 0 and 1,
$x_i(t)$: Current position of agent i at iteration t,
$y_i(t)$: Personal best (pbest) and is calculated as

$$y_i(t+1) = \begin{cases} y_i(t), f(x_i(t+1)) \geq f(y_i(t)) \\ x_i(t+1), f(x_i(t+1)) < f(y_i(t)) \end{cases} \tag{4}$$

$\hat{y}(t)$ Global best (gbest) and is calculated using the following equation:

$$\hat{y}_i(t+1) = \operatorname{argmin}\{f(y_i(t+1))\}, i \leq i \leq s \tag{5}$$

where s = number of particles in the swarm.

3.1.2 Extreme Learning Machine

Extreme learning machine (ELM), an emerging technology proposed by [15] overcomes the challenges such as slow learning speed, trivial human intervene and poor computational scalability faced by the computational intelligent techniques such as Back Propagation Neural Network (BPN) and Support Vector Machines (SVMs). ELM works for single-hidden layer feed forward networks (SLFNs). The hidden layer of SLFNs need not be tuned which is an advantage of Extreme Learning Machine. The learning speed of ELM is faster and with least human intervene which gives better performance compared with other computational intelligent techniques. The Algorithm for ELM [15] is narrated below:

Given a training set $\aleph = (x_i, y_i)$, $x_i \in R^{d_1}$, $y_i \in R^{d_2}$ an activation function $f: R \mapsto R$ and the number of hidden nodes H.

- Randomly assign input weights w_i and biases $b_i, i \in [1, H]$..
- Calculate the hidden layer output matrix H.
- Calculate output weight matrix$\beta = H^{\dagger}Y$.

3.1.3 Analytic Network Process

The input weights and the hidden bias are randomly chosen in Extreme Learning machine. The output weights are analytically determined by Moore Penrose inverse. ELM overcomes the problems such as slow learning speed, number of epochs etc., which are being faced by the computational intelligence techniques such as BPN and SVMs and has faster learning rate. However, ELM requires more number of hidden neurons than that of human intervene based algorithms that are mentioned above. The input weights are generated using Analytic Network Process (ANP).

ANP proposed by [16] facilitates for more complex interrelationships among the decision levels and attributes and does not require any strict hierarchical structure. The ANP approach handles interdependent relationships among the elements by obtaining the composite weights developed by the supermatrix. The supermatrix concept contains parallels to the Markov chain process where relative importance weights are adjusted by forming a supermatrix from the eigenvectors of these relative importance weights. The weights are adjusted by determining products of the supermatrix. In this work, the input weights of ELM are determined using ANP technique.

3.1.4 Proposed PSO: ELM: ANP Wrapper Approach Algorithm

In the proposed PSO-ELM-ANP feature subset selection wrapper approach PSO algorithm is wrapped with ELM-ANP approach and used for feature subset selection. Duration, Latency, Trigraph, Digraph and the proposed Virtual Key Force are given as input features. Feature subset selection is done and the selected features are evaluated by Extreme Learning Machine- Analytic Network Process. The process is repeated until the best solution is obtained. The training algorithm used for ELM is modified Levenberg–Marquardt Algorithm [17]. In standard Levenberg–Marquardt (LM) algorithm, the learning parameter μ, is a constant number where modified LM algorithm modifies the learning parameter as

$$\mu = 0.01e^T e \tag{6}$$

where **e** is a k × 1 matrix.

The algorithm of PSO-ELM-ANP Wrapper approach is given as follows:

Step 1: Initialize the number of Iterations, Number of particles, Weight, c_1, c_2, r_1, r_2, $v_i(t)$.

Step 2: Compute the feature values of $x_i(t)$ (Duration, Latency, Digraph, Trigraph, and Virtual Key Force).

Step 3: Evaluate fitness for each feature value using **ELM-ANP**.

Step 4: The following is repeated for number of iterations:

1. Check if p >=pbest then, pbest = p else pbest = pbest.
2. If pbest >=gbest then, gbest = pbest else gbest = gbest.
3. Update velocity by (1) and position is updated by (2), pbest and gbest position are updated using (4) and (5).

Step 4 is repeated until gbest is optimum value.

Table 1 Features selected

Induction Algorithms	PSO			
	W.O. VKF		W. VKF	
	F	FS	F	FS
ELM	60	50	71	53
ELM—ANP	60	46	71	51

Table 2 Features selection percentage

Induction Algorithms	PSO	Feature Selection (%)
ELM	53	74.7
ELM—ANP	51	71.8

4 Experimental Results

In the proposed method, the features are selected by PSO and are given as input vectors to the input layers. The input weights and hidden bias are generated by Analytic Network Process. Extreme Learning Machine analytically determines the output weights and the weights and bias are updated using Modified Levenberg–Marquardt (LM) algorithm which is used to train the network.

The proposed methodology is tested with 100 users typing the password '.tieRoan1' 10 times. The key press and release time of the characters of the passwords are extracted. Timing features such as Duration, Latency and Virtual Key Force are measured. The password contains 11 characters including shift key to press R. Therefore, total of 71 features are measured from the password with VKF and 60 features are measured without VKF. The feature subset selection methods, PSO-ELM and PSO-ELM-ANP approaches with and without the feature Virtual Key Force are experimented. The following Table 1 shows the features selected

F-Features measured, FS-Features Selected, W.O. VKF- Without VKF, W.WKF—With VKF

From Table 1, it is observed that PSO—ELM—ANP wrapper has reduced the existing 71 features to 51 features. The proposed PSO—ELM—ANP wrapper based approach has selected 49.3 % of total 71 features and has reduced 28.2 % (100–71.8 %) of the total features which is shown in the Table 2.

5 Conclusion and Future Directions

The Extreme Learning machine integrated with Analytic Network Process is wrapped with Particle Swarm Optimization to reduce the number of features. The reduced features can be used for further classification. In future more optimization techniques such as Ant Colony Optimization, Genetic Algorithm etc., can be explored.

References

1. O'Gorman L (2003) Comparing passwords. Tokens Biometrics User Authentication Proc IEEE 91(12):2019–2040
2. Francesco B, Gunetti D, Picardi C (2002) User authentication through keystroke dynamics. ACM Trans Inf Syst Secur 5(4):367–397
3. Ahmed AAE, Traore I (2005) Anomaly Intrusion detection based on biometrics. In: Proceedings of 6th IEEE information assurance, workshop, pp. 452–453
4. Azevedo GLF, Cavalcanti GDC, Carvalho Filho ECB (2007) Hybrid solution for the feature selection in personal identification problems through keystroke dynamics. In: International joint conference on neural networks, pp 1947–1952
5. Azevedo GLF, Cavalcanti GDC, Filho ECB (2007) An approach to feature selection for keystroke dynamics systems based on PSO and feature weighting. In: IEEE Congress on, Evolutionary Computation, pp 3577–3584
6. Bleha S, Obaidat M (1991) Dimensionality reduction and feature extraction applications in identifying computer users. IEEE Trans Syst Man Cybern 21(2):452–456
7. Boechat G, Ferreira J, Carvalho E (2006) Using the keystrokes dynamic for systems of personal security. Proc World Acad Sci Eng Technol 18:200–205
8. Karnan M, Akila M (2010) Personal authentication based on keystroke dynamics using soft computing techniques. In: Second international conference on communication software and networks. pp 334–338
9. Sung KS, Cho S (2006) GA SVM wrapper ensemble for keystroke dynamics authentication. In: Proceedings of the international conference on biometrics, China, vol 3832, pp 654–660
10. Yu E, Sungzoon Cho (2003) GA-SVM wrapper approach for feature subset selection in keystroke dynamics identity verification. In: Proceedings of the international joint conference on neural networks, vol 3, pp 2253–2257
11. Shanmugapriya D, Padmavathi G (2011) Virtual key force—a new feature for keystroke. Int J Eng Sci Technol 3(10):7738–7743
12. Jain A, Nandakumar K, Ross A (2005) Score normalization in multimodal biometric systems. Pattern Recogn 38:2270–2285
13. Yu E, Cho S (2004) Keystroke dynamics identity verification: its problems and practical solutions. Comput Secur 23:428–440
14. Panda S, Padhy NP (2007) Comparison of particle swarm optimization and genetic algorithm for TCSC-based controller design. World Acad Sci Eng Technol 27:497–505
15. Huang GB, Zhu QY, Siew CK (2006) Extreme learning machine: a new learning scheme of feedforward neural networks. In: Proceedings of 2004 international joint conference on neural networks
16. Saaty TL (1996) Decision making with dependence and feedback, the analytic network process. RWS Publications, Pittsburgh
17. Suratgar AA, Tavakoli MB, Hoseinabadi A (2005) Modified Levenberg-Marquardt method for neural networks training. World Acad Sci Eng Technol 6:46–48

Age Estimation Using Specific Gabor Filter on Upper Facial Area

Hadi Affendy Dahlan, Syamsiah Mashohor, Sharifah Mumtazah Syed Ahmad Abdul Rahman and Wan Azizun Wan Adnan

Abstract There is a growing research interest in facial aging estimation where the whole facial area was used for age estimation. However, in this paper, we propose a different approach for age estimation whereby we perform feature extraction at the specific part which is called the Region of Interest (ROI) on the upper facial area by employing a specific orientation and scales of Gabor filter. The proposed multi-Support Vector Machine (SVM) was used as the classification, and tested on two databases which are the captured face images of Malaysian citizen, and the FG-NET database. For the scheme, the Leave One Picture Out (LOPO) is used for training and testing according to age groups. The overall results of the proposed method show that the upper ROI performances are better for both Malaysian citizen and FG-NET database than the full facial ROI. Moreover, the Mean Absolute Error (MAE) for the FG-NET decreases when using the upper ROI approach.

Keywords Age estimation · Region of interest · Leave one picture out · Face recognition

H. A. Dahlan (✉) · S. Mashohor · S. M. S. A. A. Rahman · W. A. W. Adnan
Department of Computer and Communication Systems Engineering, University Putra Malaysia, 43400 Serdang, Selangor, Malaysia
e-mail: dysemail_1986@yahoo.com

S. Mashohor
e-mail: syamsiah@eng.upm.edu.my

S. M. S. A. A. Rahman
e-mail: s_mumtazah@eng.upm.edu.my

W. A. W. Adnan
e-mail: wawa@eng.upm.edu.my

Mohan S. and S. S. Kumar (eds.), *Proceedings of the Fourth International Conference on Signal and Image Processing 2012 (ICSIP 2012)*, Lecture Notes in Electrical Engineering 221, DOI: 10.1007/978-81-322-0997-3_56, © Springer India 2013

1 Introduction

Recently, there is a growing research in facial aging estimation. The research started when it was first discover as one of the limitation of the face recognition research area, and later open into another area of work. Currently, there is a needed for standard aging pattern recognition—an expert system on age estimation. The benefit of having a machine expert in age estimation is that it will reduce the need of human experts, who will have different interpretation since no same experts have the same opinion or experienced on detecting the age of a subject. Furthermore, this research will help greatly in the world application of biometric identification if it succeeds, such as in criminology; medical; and also age identification area.

Researchers have come up with different approaches in extracting age features from the facial images. Nevertheless, most researchers tend to concentrate on the overall region of the face in determining a person's age. Several have reported age estimation research work on a particular part of the face [1, 2]. The advantage of specific area extraction is that it reduces the total overall number features that need to be extracted, and this to some extend increases the speed of the system which is desirable particularly for real time response. In addition, it was reported by Lantis [1] and El Dib [2] that the upper facial region of the face (from the nose to the forehead) is the most prominent age features that affect the overall age classification/estimation performance. Thus, we hypothesize that the inclusion of features with low estimation capability, does not only consumes more computational resources, yet it can also degrade the accuracy of the system.

Previous researches that used the upper facial region mostly analyzed the entire upper facial region segment. However, none have chosen the part that really shows important age progression. Hence, an approach for age estimation by feature extraction at the specific part on the upper facial region is proposed. The proposed Gabor filter method is used for the feature extractions and the multi-Support Vector Machine (SVM) would be the classifier. Furthermore, this will be the first age estimation done on the Malaysian citizens' samples as to compare with the generally used Caucasian & European face database.

The organization of the rest of the paper is as follows. Section 2 will be the literature review of age estimation researches in term of medical and pattern recognition. Section 3 will be the methodology of the proposed method. Section 4 is the experimental setup done when making the experiment. And finally, the proposed method would be evaluated for its complexity and accuracy of the estimation will be discussed in detail in Sect. 5 and concluded in Sect. 6.

2 Background

Facial aging is actually an attribute of changes in the facial feature, shape and texture, bone structure, and other biological factors like weight loss or gain, and gray hair. Studies on aging were done by many computer scientist, psychologist [3, 4], and also in medical [5–7]. In the human aging progression, two biophysical change or growth happen on the face [3, 8], which are:

- The craniofacial bony aging—due to a relative change in bone expansion and loss, and;
- The change of texture and elasticity of the skin.

Figure 1 shows the wrinkle areas that may appear on an aging person. There were several methods used for estimating the human age. One of the techniques that have been proposed for estimating age was Kwon and Lobo [3, 9] who introduced the anthropometry of the face and the density of its wrinkles approach. In the research, they used the 'snakelet' which is a small snake segment designed to find a small individual curve segment in the image. Later on, the idea of using wrinkle has bring forth various methods of edge detection to be used by other researchers; like the Sobel filter in [10, 11] and Canny Edge operator [12]. There are other approaches that used different age feature extraction such as the Manifold analysis [13], patch-based hidden Markov model [14], and Local and global Discrete Cosinus Transformation (DCT) [15, 16]. They are also varies classification method such as mentioned in [17], and also the known Support Vector Machine (SVM) [18].

Choi et al. [19] proposed the hierarchical classifier in which the classification process is performed using several different extraction methods and classifiers

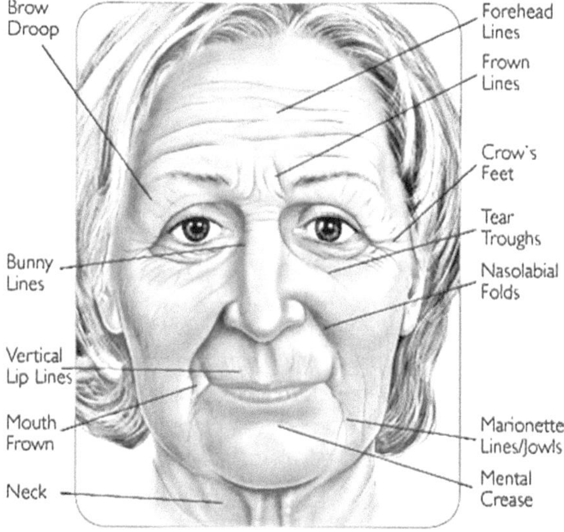

Fig. 1 Typical skin aging changes. Picture refer from [27]

instead of using one. The global features were extracted using the Active Appearance Model (AAM) method for the face shape while the local features by the Gabor filter for the wrinkle and Local Binary Patterns (LBP) for the skin texture. While the Support Vector Machine (SVM) and Support Vector Regression (SVR) was used as the classifier.

The Gabor filter was considered the most effective wrinkle extractor than any edge detectors. Its capability to extracted different orientation and scale edges; which is the wrinkle—were used by others such as Pirozmand et. al. [20] who proposed the Gabor PCA-LDA method which used 12 different Gabor wavelets for the wrinkle extraction that used less features. However, the method only categories age into three groups; baby, child, and adult; in which the range was to large if one wants to know specific age. Guo et. al. [21] used the Gabor by first investigated on the biologically inspired features (BIF) as to understand the human age estimation from faces, and then created the edge detector pyramid Gabor's filter. The BIF approach obtained high performance but, the method was complex and used large quantity of features. Meanwhile, Hironori et. al. [22] used the Gabor jet feature with the combination of Adapted Retinal Sampling Method (ARSM) for the use of age and gender classification. The errors obtained in [22] are small, and may robust to the variation facial pose. However, the combination of a skin feature extractor is needed creating more features uses for estimation.

There are researches that used different facial part for age estimation. As proposed by Lantis [1], they proposed a method of estimating age from feature taken from 4 different region of face (whole, internal, upper, and lower). Using the face model method on 0–35 years old samples, the results obtained was that the upper region gained the least mean error. Meanwhile, El Dib [2] also analyzed different facial parts for estimating age using the bio-inspired features (BIF). He concluded that the eye wrinkles contain the most important aging information, compared to internal and whole face based from the values of Mean Absolute Error (MAE). This remains the motivation of this research work which is to investigate the effectiveness of age estimation by concentrating on specific areas on the upper face.

3 Methodology

The wrinkle would be the aging feature in this work. The wrinkles are extracted from its Region Of Interest (ROI), base on their wrinkle orientation. In order to analyze the directional characteristics of the facial wrinkle, the Gabor filter with the direction and frequency parameters according to the selected facial area was used. Here, the Gabor filter with 6 orientations, and 4 scales was chosen based on [19, 23]. This was done after some analyzing with the method, and found out the method gave the best performance for different Gabor scales and orientations. For the classification process, the SVM was chosen due to it's designed for binary class problems which select the optimal linear decision hyper-plane. In the choice for

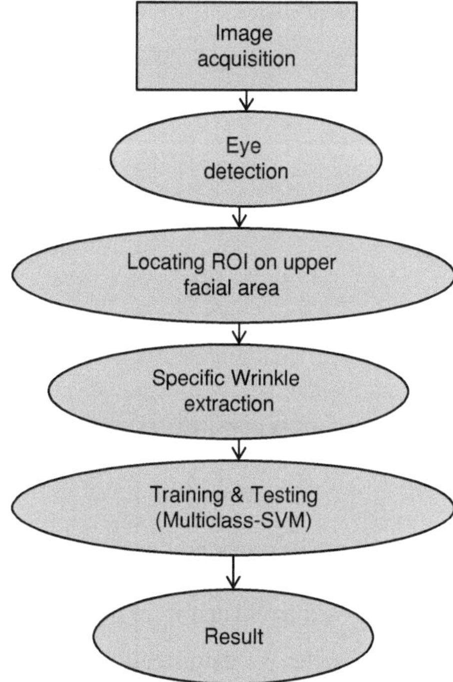

Fig. 2 Proposed methodology

the kernel function, Gaussian radial basis function kernel was chosen. The main aim of this research is to investigate the effectiveness of age estimation by focusing on ROI at the upper area region of the face using the specific Gabor filter orientation sets. Ages are grouped into 4 different sets which are 20-29; 30-39; 40-49; 50 and above. The proposed algorithm are trained and tested on actual facial databases of Malaysian and FG-NET [24].[1] The process of the methodology follows these steps shown in Fig. 2. The images are resized into 512 $\times$ 512 pixels with chosen face scale and image enhancement for better feature detection for the purpose of the methodology. The steps are as follows:

1. The eyes first located using the function Eigeneye courtesy by Adrian et. al. [25] which have been modified for experimental usage.
2. Next, calculation was made to locate the ROI at the upper area for wrinkle extraction as shown in Fig. 3.
3. After the ROI located, the wrinkle is extracted using the Gabor filter with specific orientation and scale.
4. With the data extracted, they are trained using the multi-SVM and afterward tested with sample set to determine which age group they belong to.

[1] In Malaysian database, a total of 170 image samples was used; for the FG-NET database, we only use sample image age 20 and above, giving a total of 172 image used.

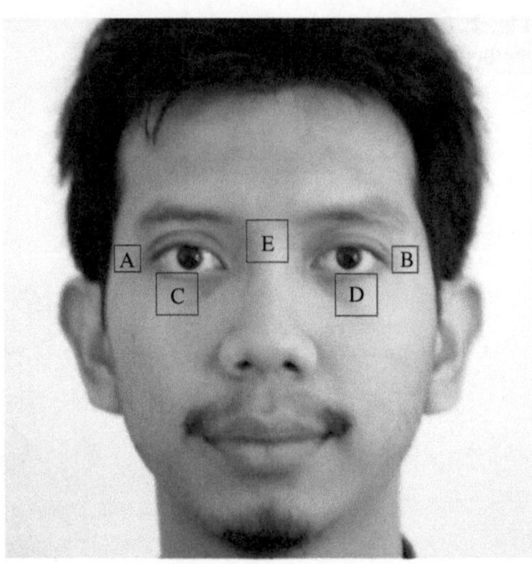

Fig. 3 The ROI chosen

5. Finally, the estimation Overall performance percentage (which analyze if the test sample is group into its true age group); and the Mean Absolute Error (MAE) are calculated from the result to see the overall performance.

The wrinkles are extracted based on their orientation. The Gabor wavelets are put according to the orientations of the wrinkles in the ROI chosen. In Fig. 3, on A and B ROI box, the orientations that are used are 0, 30, 60, 120, and 150 degree (5 orientation); on C, 30, and 60 (2 orientation); D, used 120 and 150 (2 orientation); and then E, 0 and 90 (2 orientation). All of the regions vary with 4 different scales. So a total of 64 features are collected from one image. These features are later organized into:

$$i_x = [f_1, f_2, \ldots, f_{64}];\ x = 1, 2\ldots n;\ age_g = \begin{bmatrix} i_1 \\ i_2 \\ \vdots \\ i_n \end{bmatrix};\ g = 20, 30, 40, 50 \quad (1)$$

where i_x is the image used for training and testing; $f_{1\text{-}64}$ is the wrinkle features; age_g are the set of images in certain age group used for training; and g is the age group category (e.g. 20 means 20-29 years range)

4 Experimental Setup

The training and testing method follow the Leave One Picture Out scheme (LOPO) [19]. It is done by taking one image out for test while the rest are use for training. The databases used for the experiment are our own collected database (Malaysian

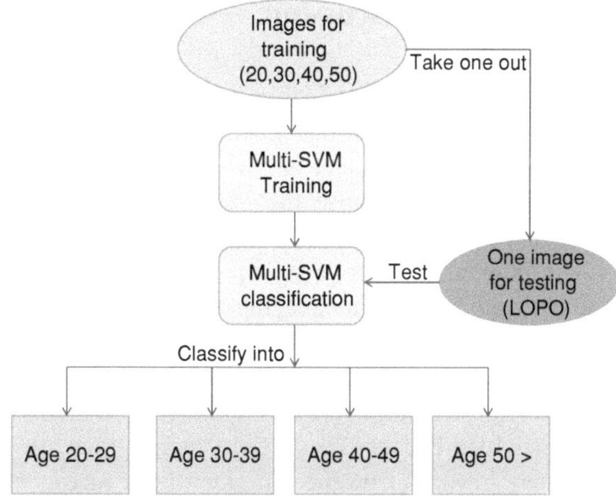

Fig. 4 Proposed multi-SVM test (age estimation) process

citizen) and the FG-NET database. The samples are tested according to their own gender and later both combine gender.

In the estimation process, the modified multi-SVM function developed by Mishra [26] was used. Figure 4 shows the training and testing process using the multi-SVM:

The performances of the estimation are taken based on the correct images estimated according to their age group; while the Mean Absolute Error (MAE) is calculated to gain its error in years. The MAE formula is:

$$MAE = \frac{1}{n}\sum_{i=1}^{n}|f_i - y_i| = \frac{1}{n}\sum_{i=1}^{n} e_i \tag{2}$$

The MAE is an average of the absolute errors $e_i = | f_i - y_i |$, where f_i is the prediction and y_i the true value.

The experiment then compares upper facial ROI (proposed) with full facial ROI to see the difference in performance; and also with the one-sided upper facial ROI for validation. The ROI are shown in Fig. 5 below:

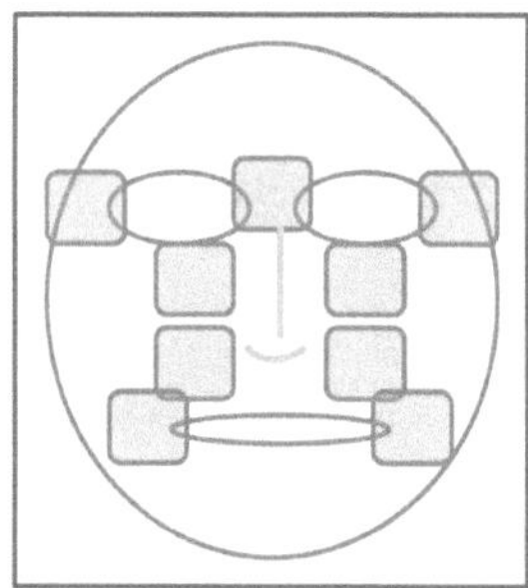
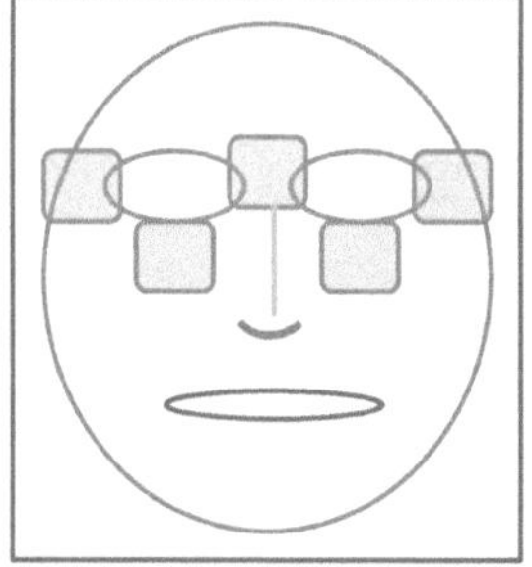
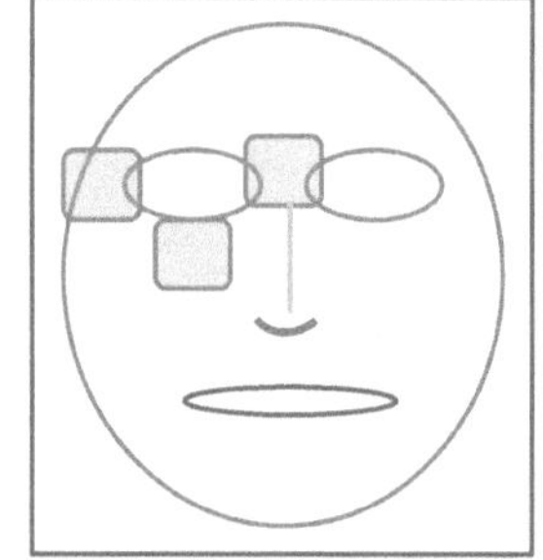

Fig. 5 Full; Upper (proposed); and One-sided Upper (on the left side only) ROI respectively

5 Result and Discussion

The results as presented in Fig. 6 shows that the proposed method performance outperformed full and one-sided. But for MAE, the results are comparable. Moreover, the performance and MAE of the Malaysian database are better than the FG-NET. This is because the FG-NET was based on scanned images that are dependant to its picture condition, while the Malaysian images are digital photos.

The comparison between upper and full facial have shown that taking less features with concentrated aging signature perform better and have comparable MAE than having more features that affect the average estimation performance. Meanwhile, the one-sided was performed to make validation for the upper part. However, in the purpose of estimating, both sided of the face are needed rather than one-sided because most of the human face are not entirely symmetrical (e.g. there may have scar on the left side of the face).

This approach can be beneficial in hybrid features extraction method since it can reduce the need of unwanted features and decrease time execution. However, the drawback is the image have to be frontal (without tilt) with no facial accessory

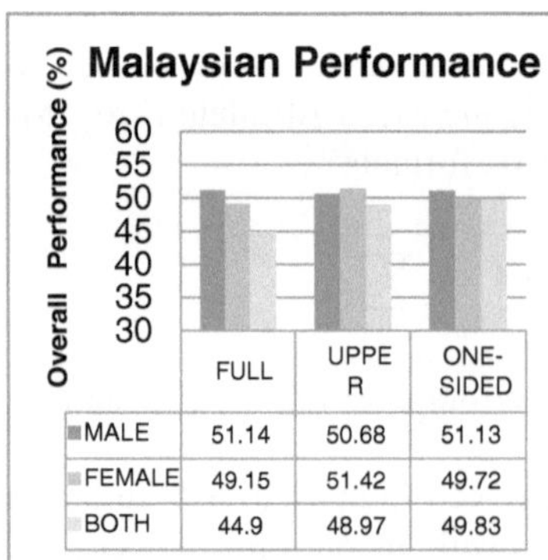

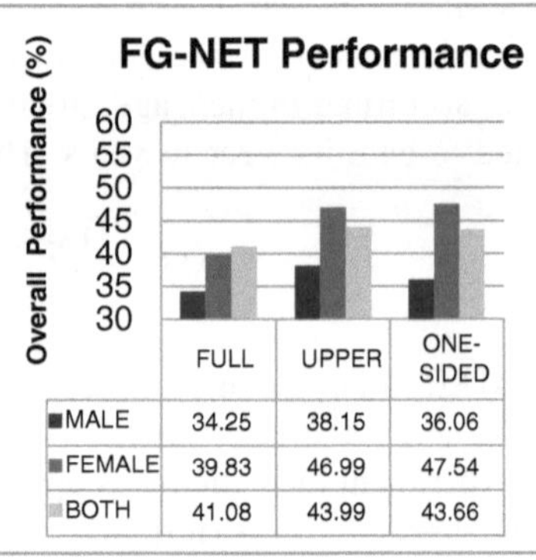

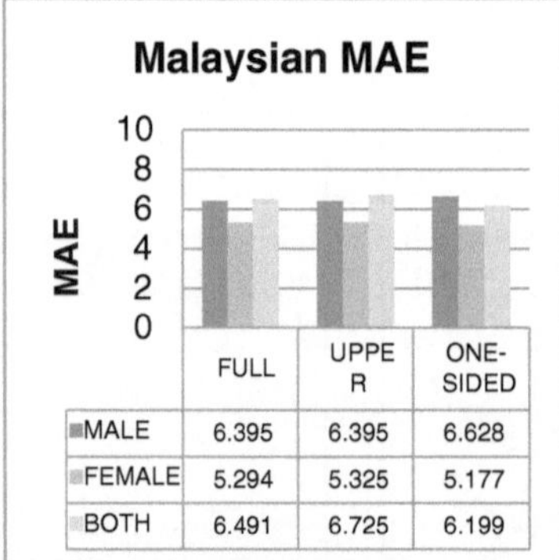

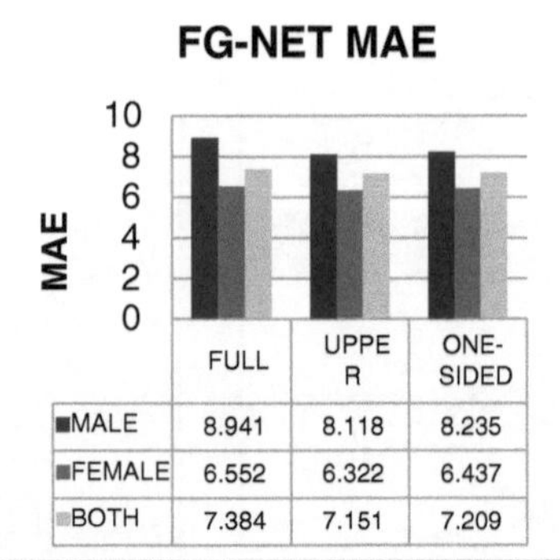

Fig. 6 The performance and the MAE of the proposed method

such as spectacle because these will results in low performances. Nevertheless, it is robust since it uses the function of an eye detection to calculate and locate the ROI.

6 Conclusion

Age estimation using wrinkle extraction on the Region of Interest (ROI), at the upper area of the face, with specific orientation of Gabor filter is proposed. The method is classified using the proposed multi-SVM. The database samples used are the Malaysian citizen, and the FG-NET database, and the Leave One Picture Out (LOPO) scheme is used for training and testing. The aim of obtaining better performance than the full facial ROI was achieved and the validation was obtained by testing with the one-sided ROI. Even though the image usages are limited to frontal position, fixed scale, and with no accessories (spectacle) on the face; a robust approach for age estimation was achieved. Moreover, the work can be applied on more complex approach of age estimation if one wants to create hybrid extractors. Future improvement of the work will be made to the SVM classification. This is to gain lower MAE and higher performance values than the current ones.

References

1. Lanitis A (2002) On the significance of different facial parts for automatic age estimation. In: 14th international conference on digital signal processing, vol 2, pp 1027–1030
2. El Dib MY, Onsi HM (2011) Human age estimation framework using different facial parts. Egypt Inform J 12(1):53–59
3. Ramanathan N, Chellapa R, Biswas S (2009) Computational methods for modeling facial aging: a survey. J Vis Lang Comput 20(3):131–144
4. Ramanathan N, Chellapa R, Biswas S (2009) Age progression in human faces: a survey. J Vis Lang Comput 15:3349–3361
5. Zimbler MS, Kokosa M, Thomas JR (2001) Anatomy and pathophysiology of facial aging. Facial Plast Surg Clin N Am 9:179–187
6. Chung JH (2003) Photoaging in Asians. Photodermatol Photoimmunol Photomed 19:109–121. doi:10.1034/j.1600-0781.2003.00027.x
7. Shirakabe Y, Suzuki Y, Lam SM (2003) A new paradigm for the aging Asian face. Aesthetic Plast Surg 27:397–402
8. Wu Y, Beylot P, Thalmann NM (1999) Skin aging estimation by facial simulation. In: Proceeding computer animation, pp 210
9. Kwon Y, Lobo N (1994) Age classification from facial images. In: Proceeding computer vision and pattern recognition, pp 762–767
10. Wen-Bing H, Cheng-Ping L, Chun-Wen C (2001) Classification of age groups based on facial features. Tamkang J Sci Eng 4(3):183–192
11. Txia J, Huang C (2009) Age estimation using AAM and local facial features. In: Fifth international conference on intelligent information hiding and multimedia signal processing, pp 885–888

12. Ramesha K et al (2010) Feature extraction based face recognition, gender and age classification. Int J Comput Sci Eng 02(1):14–23
13. Fu Y, Xu Y, Huang TS (2007) Estimating human age by manifold analysis of face pictures and regression on aging features. In: IEEE international conference on multimedia and expo, pp 1383–1386
14. Zhuang X, Zhou X, Hasegawa-Johnson M, Huang T (2008) Face age estimation using patch-based hidden markov model supervectors. In: 19th international conference on pattern recognition ICPR 2008, pp1–4
15. Mokadem A, Charbit M, Chollet G, Bailly K (2010) Age regression based on local image features. In: 2010 fourth pacific-rim symposium on image and video technology (PSIVT), pp 88–93
16. Guo J, Liou Y, Nguyen H (2011) Human face age estimation with adaptive hybrid features. In: 2011 international conference on system science and engineering (ICSSE), pp 55–58
17. Lanitis A, Draganova C, Christodoulou C (2004) Comparing different classifiers for automatic age estimation. IEEE Trans Syst Man Cybern B Cybern 34(1):621–628
18. Luu K, Ricanek K, Bui TD, Suen CY (2009) Age estimation using active appearance models and support vector machine regression. In: IEEE 3rd international conference on biometrics: theory, applications, and systems, pp 1–5
19. Choi SE, Lee YJ, Lee SJ, Park KR, Kim J (2011) Age estimation using a hierarchical classifier based on global and local facial features. Pattern Recogn 44(6):1262–1281
20. Pirozmand P, Amiri MF, Kashanchi F, Layne NY (2011) Age estimation, a gabor PCA-LDA approach. J Math Comput Sci (JMCS) 2(2):233–240
21. Guo G, Mu G, Fu Y, Huang TS (2009) Human age estimation using bio-inspired features. In: IEEE conference on computer vision and pattern recognition CVPR 2009, pp 112–119
22. Takimoto H, Mitsukura Y, Fukumi M, Akamatsu N (2008) Robust gender and age estimation under varying facial pose. Electron Commun Japan 91(7):32–40
23. Manjunathi BS, Ma WY (1996) Texture features for browsing, retrieval of image data. IEEE Trans Pattern Anal Mach Intell 18(8):837–842
24. FGNET (2012) The fg-net aging database (2002), http://www.fgnet.rsunit.com/. Accessed 20 April 2012
25. Aldrian P (2009) Matlab central : fast eye tracking. http://www.mathworks.com/matlabcentral/fileexchange/25056-fast-eyetracking. Accessed 2 Mei 2012
26. Mishra A (2011) Matlab central : multi class support vector machine. http://www.mathworks.com/matlabcentral/fileexchange/33170-multi-class-support-vector-machine. Accessed 10 June 2012
27. Pham TV (2012) Perth facial plastic and cosmetic surgery. http://www.perthcosmeticsurgery.com.au/wrinkle_management_procedure.do. Accessed 25 June 2012

Author Index

Mohan S. and S. S. Kumar (eds.), *Proceedings of the Fourth International Conference on Signal and Image Processing 2012 (ICSIP 2012)*, Lecture Notes in Electrical Engineering 221, DOI: 10.1007/978-81-322-0997-3,